Polymeric Liquids and Networks

Structure and Properties

Polymeric Liquids and Networks

Structure and Properties

William W. Graessley
Emeritus Professor, Princeton University
Adjunct Professor, Northwestern University

Garland Science
Taylor & Francis Group
NEW YORK AND LONDON

About the Author

William W. Graessley was born in Muskegon, Michigan in 1933. He holds degrees in both Chemistry and Chemical Engineering, as well as a Ph.D. from the University of Michigan. He has worked within academia and industry and has published extensively on radiation cross-linking of polymers, polymerization reactor engineering, molecular aspects of polymer rheology, rubber network elasticity, and the thermodynamics of polymer blends. He is currently a Professor Emeritus at Princeton University and an Adjunct Professor at Northwestern University. His honors include an NSF Predoctoral Fellowship, the Bingham Medal (Society of Rheology), the Whitby Lectureship (University of Akron), the High Polymer Physics Prize (American Physical Society), and membership in the National Academy of Engineering.

Denise T. Schanck, *Vice President*
Robert L. Rogers, *Senior Editor*
Liliana Segura, *Editorial Assistant*
Adam Sendroff, *Marketing Director*
Randy Harinandan, *Marketing Assistant*

Dennis P. Teston, *Production Director*
Anthony Mancini Jr., *Production Manager*
Brandy Mui, *STM Production Editor*
Mark Lemer, *Art Manager*

Published in 2004 by
Garland Science
A member of the Taylor & Francis Group
29 West 35th Street
New York, NY 10001

www.taylorandfrancis.com

Published in Great Britain by
Garland Science
A member of the Taylor & Francis Group
11 New Fetter Lane
London EC4P 4EE

www.taylorandfrancis.co.uk

Printed in the United States of America on acid-free paper.

10 9 8 7 6 5 4 3 2 1

Library of Congress Cataloging-in-Publication Data

Graessley, W.W. (William Walter)
Polymeric liquids and networks : structure and properties / by William W. Graessley.
p. cm.
Includes bibliographical references and index.
ISBN 0-8153-4169-5 (alk. paper)
1. Polymer solutions. 2. Polymer networks. I. Title.

QD381.9S65G73 2003
547'.70454–DC21 2003048324

To Helen

Contents

Preface

This book is the first of two volumes aimed at a unified view of flexible-chain polymer liquids and networks. The topics range from equilibrium properties, the subject of the present book, to dynamical response, finite deformation behavior, and non-Newtonian flow in Volume 2. Volume 2, subtitled "Dynamics and Rheology," will appear in about two years. These various aspects of the field were developed over the past seventy years by researchers from many academic disciplines. The infusion of fresh viewpoints continually invigorated and enriched the field, making polymeric liquids and networks a truly interdisciplinary subject. The lack of a common terminology and perspective, however, has led to compartmentalization, thus making it difficult for a newcomer, even one technically trained, to gain a broad appreciation of the field and to see the relationships among its various parts. I hope these two volumes, without diluting the substance, will go some way toward achieving a desirable unity.

The development of the topic emphasizes fundamental principles and the molecular viewpoint. The conceptual basis of the theories underlying each topical area is explained with the derivations sometimes outlined briefly and sometimes in detail; technical terminology is kept to the minimum necessary for a concise coherent presentation. The goal is informed understanding rather than detailed technical proficiency. Theory, experiment, and simulation are woven together as appropriate to achieve a balanced view. Both volumes are aimed to serve academic and industrial needs, consolidating the understanding of topics with both practical and fundamental significance, and written from a technical but nonspecialized perspective.

The books deal primarily with nonpolar and weakly polar species and with the results derived from experiments on structurally well-defined polymer systems. The object is not, of course, to ignore the more complex systems, which are pervasive in both nature and industry and important in their own right. Indeed, much space is devoted to structural distributions, their characterization, and their effect on properties. The object of this book is rather to provide a framework for the better understanding of all polymeric liquids by identifying, in the simplest possible circumstances, the universal attributes of a chainlike and flexible molecular structure.

Notable omissions from the books (aside from passing reference) are theories of the glass transition, properties of the glassy state, flow properties of multiphase liquids, crystallization phenomena, thermosetting resins, filled polymers, and highly polar polymers. Other topics that would seem natural for volumes like these—block copolymers, polyelectrolyte solutions, and elastically driven flow instabilities—are treated in rather cursory fashion. One reason for this is the author's inexperience in many of these areas; another is the newness and still rapidly evolving character of some topics. Still another is the lack of sufficient systematic experimental studies. The need to keep the size of the books within some reasonable bounds and still do justice to the subjects that are covered was another consideration. Even with the scope narrowed in this way, the amount of relevant material is enormous.

I have long felt that dynamics and flow behavior have been wrongfully neglected in general textbooks on polymers, being regarded as somehow too mathematical, too specialized, or perhaps simply less important in relation to other topics competing for the space. The structure–property relationships for dynamics and rheology abound in universal laws, especially those whose forms are independent of the polymeric species and are of comparatively recent discovery. Many of these are interrelated and can now be understood in quite simple terms. Others belong in introductory polymer textbooks, and I hope these volumes will assist the writers of these future polymer textbooks by giving them a place to find this information without the necessity of digging through a diverse, unfamiliar, and sometimes old literature to find examples and data.

I have also felt that the general subject of polymeric liquids and networks would benefit pedagogically by being developed from a background and language common with the molecular theory of liquids. Chapters 2 and 3 of this volume and the first chapter of Volume 2 begin by summarizing the relevant background for small-molecule substances in the dense liquid state. Many basic principles applied to polymeric liquids grew naturally from earlier considerations of monomeric liquids. It is unfortunate that these subjects are not part of the educational background of most people with interests in the field. Making such commonalities of the liquid state evident helps avoid the idea of polymeric liquids as things apart, somehow subject to different rules than other liquids. The freedom from disciplinary restriction also helps to make clear which features of behavior are unique to liquids and networks containing flexible chainlike molecules—rubberlike elasticity and easily observable viscoelastic response—and which are common to all liquids, such as the glass transition.

Some liberties have been taken, harmlessly I think, with the conventional subdivisions of topics, in order to proceed smoothly with the development while keeping related subjects of similar difficulty as close to one another as possible. Thus, although viscosity and diffusion in dilute polymer solutions are dynamic properties, they are considered along with the thermodynamic aspects of polymeric size in Chapters 5 and 6 of the present book. To do otherwise would have been artificial, the link between static and dynamic measures of size being otherwise so clear. Other dynamic properties in dilute solution are treated in Volume 2, on dynamics and rheology. Also, the chapters on flow behavior in simple shear histories, which includes nonlinear viscoelastic response, are grouped with those on linear viscoelasticity and diffusion, early in Volume 2. An alternative placement, with other aspects of nonlinear viscoelasticity, would have been formally more logical. However, when presented with some theoretical preliminaries, simple shear behavior, including normal stress effects, follows rather naturally after linear response. The alternative would have inserted some chapters of continuum mechanics between the two, thus spoiling the smooth progression.

I have assembled much of the material in both volumes while developing graduate courses on the various topics. The students came mainly from chemical engineering and materials science backgrounds but with a smattering of chemists and physicists and even some precocious undergraduates as well, first at Northwestern University and then at Princeton. A certain amount of background instruction was always necessary. Based on that experience, I feel the two volumes will quite nicely support a one-year graduate course. I have also written this first volume to stand alone, as a one-semester course, useful in its own right as an introduction to the nonrheologic aspects of the field.

Finally, I wish to thank the many people who have helped make this book possible: my students, associates and colleagues at Northwestern University and Princeton University and my coworkers during employment at Exxon. I am also grateful for the generous research support provided over the years by the National Science Foundation, the United States Department of Energy, and the Petroleum Research Fund. I particularly appreciate the advice of Buckley Crist, Jacques Roovers, Guy Berry, Robert Johnston, and Ralph Colby, who read and commented extensively on the draft of this book. I am also indebted to many others who generously contributed their data and time to discuss various technical matters, including Lew Fetters, Nikos Hadjichristidis, Gary Grest, Frank Bates, Chris Macosko, Scott Milner, Nitash Balsara, Ramanan Krishnamoorti, Tim Lodge, Alan Gent, Tony Habenschuss, Rick Register, David Lohse, Michael Rubinstein, Bruce Eichinger, Ben Chu, Ole Kramer, Claude Cohen, Thomas Sun, Pat Cotts, and Greg Dee.

William W. Graessley
Montague, Michigan
September, 2003

CHAPTER 1

Introduction

This chapter provides an introduction to the general subject of polymeric liquids and networks, referring to both equilibrium and dynamic properties for examples. It begins with the distinction between chemical microstructure and macrostructure in polymer molecules, then specializes the discussion to flexible chain macromolecules. The distinction between species dependence and architectural dependence is considered next, with examples chosen from among the dynamical properties. Polymeric heterogeneity ends the chapter—molecular weight and molecular size averages, the effects of long-chain branching, and crosslinking reactions on polydispersity and distribution functions, both theoretically based and empirical.

1.1 Molecular Nature of Polymers

Polymeric materials consist of *macromolecules*, made up of many more-or-less identical molecular subunits, the *mers* or *monomeric units* or *repeating units* that define the polymer species. Polymers are formed from *monomers*, substances whose molecules have the capacity to link chemically with at least two other molecules. The number of polymerizable substances is very large, as is the number of chemical reactions that have been used to form polymers. Molecular size increases with polymerization, and the material properties evolve away from those of the monomer, sometimes in unique and useful ways.

Polymers have been categorized on such attributes as shared properties, synthetic origin, or general areas of application, for example:

- Polymers of biological origin, such as proteins and polysaccharides.
- Covalent network solids, such as epoxy resins and phenolics.
- Flexible-chain polymers, such as acrylics and polyolefins.

Flexible-chain polymers are the substances of particular interest in this book. Some examples of flexible-chain species are listed in Table 1.1. Except at chain

TABLE 1.1 Monomer, monomeric unit, and common name for selected polymer species

Monomer	Monomeric Unit	Common Name	Acronym
$\begin{array}{c} H \quad\quad H \\ \diagdown\quad\diagup \\ C=C \\ \diagup\quad\diagdown \\ H \quad\quad H \end{array}$	$\begin{array}{c} H \quad H \\ \vert \quad \vert \\ -C-C- \\ \vert \quad \vert \\ H \quad H \end{array}$	polyethylene	PE
$\begin{array}{c} H \quad\quad CH_3 \\ \diagdown\quad\diagup \\ C=C \\ \diagup\quad\diagdown \\ H \quad\quad H \end{array}$	$\begin{array}{c} H \quad CH_3 \\ \vert \quad \vert \\ -C-C- \\ \vert \quad \vert \\ H \quad H \end{array}$	polypropylene	PP
$\begin{array}{c} H \quad\quad CH_2-CH_3 \\ \diagdown\quad\diagup \\ C=C \\ \diagup\quad\diagdown \\ H \quad\quad H \end{array}$	$\begin{array}{c} H \quad C_2H_5 \\ \vert \quad \vert \\ -C-C- \\ \vert \quad \vert \\ H \quad H \end{array}$	poly(1-butene)	PB
$\begin{array}{c} H \quad\quad CH_3 \\ \diagdown\quad\diagup \\ C=C \\ \diagup\quad\diagdown \\ H \quad\quad CH_3 \end{array}$	$\begin{array}{c} H \quad CH_3 \\ \vert \quad \vert \\ -C-C- \\ \vert \quad \vert \\ H \quad CH_3 \end{array}$	polyisobutylene	PIB
$\begin{array}{c} H \quad H \\ \vert \quad\diagup \\ C=C \\ \diagup\quad\diagdown \\ C=C \quad H \\ \diagdown \\ H \end{array}$	$\begin{array}{c} H \\ \vert \\ H \quad C=CH_2 \\ \vert \quad \vert \\ -C-C- \\ \vert \quad \vert \\ H \quad H \end{array}$	1,2 polybutadiene	PVE
$\begin{array}{c} H \quad H \\ \vert \quad\diagup \\ H \quad C=C \\ \diagdown\quad\quad\diagdown \\ C=C \quad\quad H \\ \diagup\quad\diagdown \\ H \quad\quad H \end{array}$	$\begin{array}{c} H \quad H \quad H \quad H \\ \vert \quad \vert \quad \vert \quad \vert \\ -C-C=C-C- \\ \vert \quad\quad\quad \vert \\ H \quad\quad\quad H \end{array}$	1,4 polybutadiene	PBD
$\begin{array}{c} H \quad\quad CH_3 \\ \diagdown\quad\diagup \\ C=C \\ \diagup\quad\quad\diagdown \\ H \quad\quad\quad H \\ \quad\quad C=C \\ \quad\diagup\quad\diagdown \\ \quad H \quad\quad H \end{array}$	$\begin{array}{c} H \quad CH_3 \;H \quad H \\ \vert \quad \vert \quad\; \vert \quad \vert \\ -C-C=C-C- \\ \vert \quad\quad\quad\quad \vert \\ H \quad\quad\quad\quad H \end{array}$	1,4 polyisoprene	PI
$\begin{array}{c} H \quad\quad H \\ \diagdown\quad\diagup \\ C\!-\!C \\ \diagup\; \diagdown\quad\diagdown \\ H \;\; O \;\; H \end{array}$	$\begin{array}{c} H \quad H \\ \vert \quad \vert \\ -C-C-O- \\ \vert \quad \vert \\ H \quad H \end{array}$	poly(ethylene oxide)	PEO
$\begin{array}{c} H \quad\quad H \\ \diagdown\quad\diagup \\ C=C \\ \diagup\quad\diagdown \\ H \quad\quad Cl \end{array}$	$\begin{array}{c} H \quad Cl \\ \vert \quad \vert \\ -C-C- \\ \vert \quad \vert \\ H \quad H \end{array}$	poly(vinyl chloride)	PVC

(Continued)

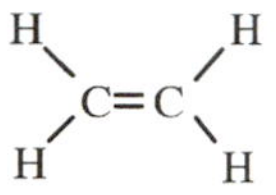

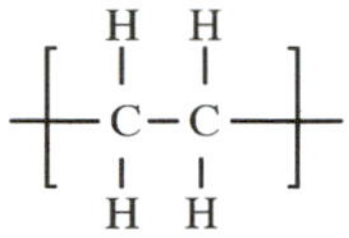

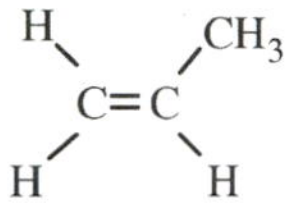

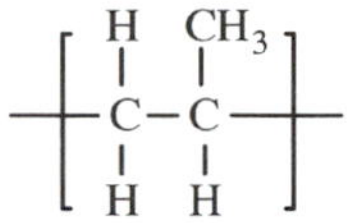

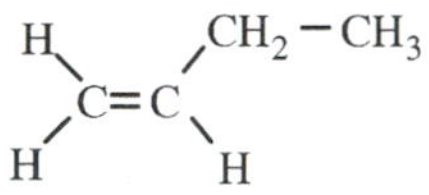

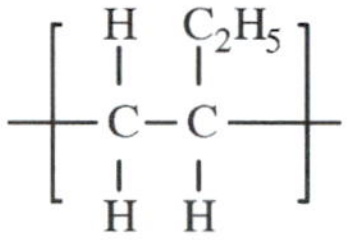

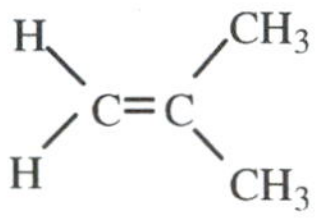

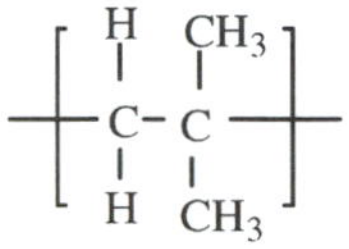

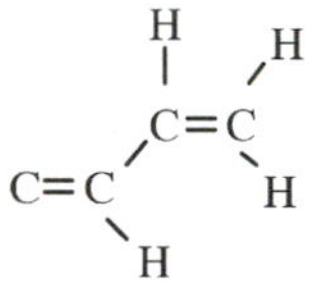

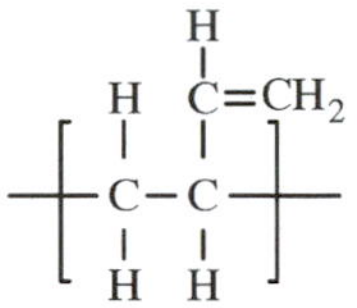

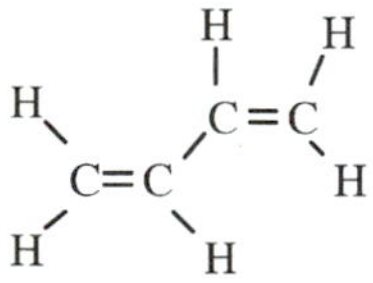

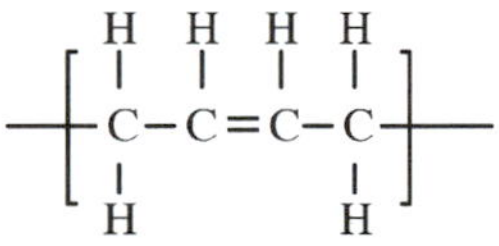

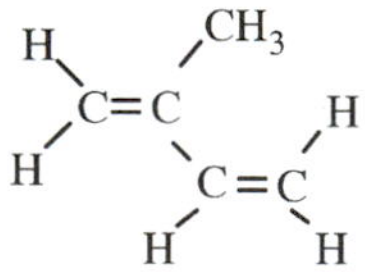

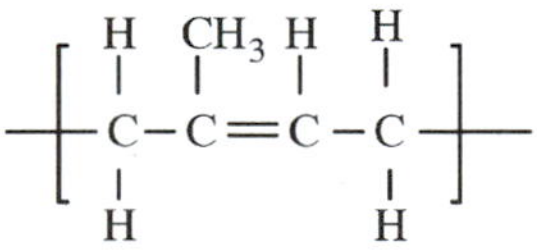

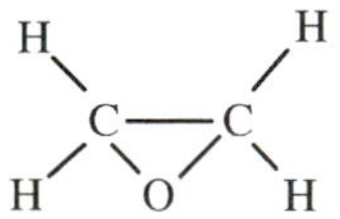

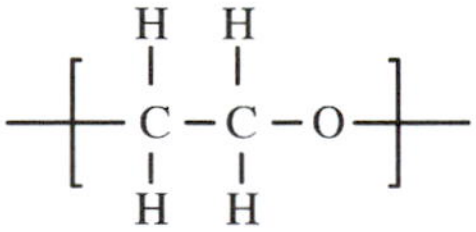

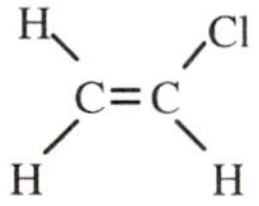

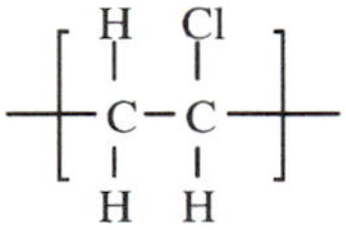

TABLE 1.1 *(Continued)*

Monomer	Monomeric Unit	Common Name	Acronym
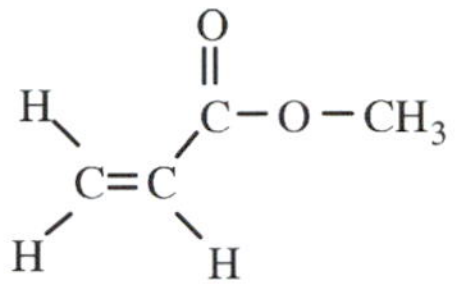	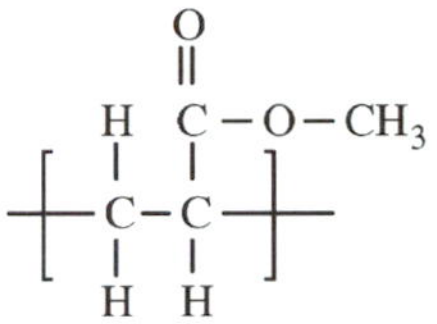	poly(methyl acrylate)	PMA
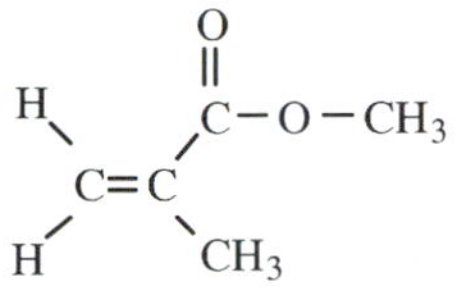	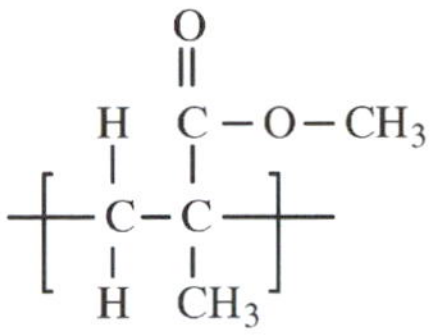	poly(methyl methacrylate)	PMMA
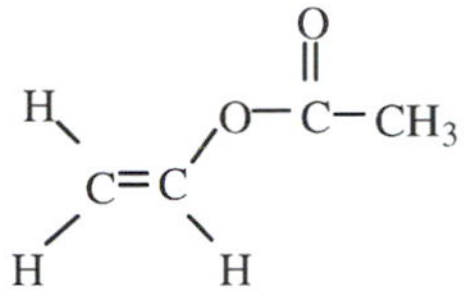	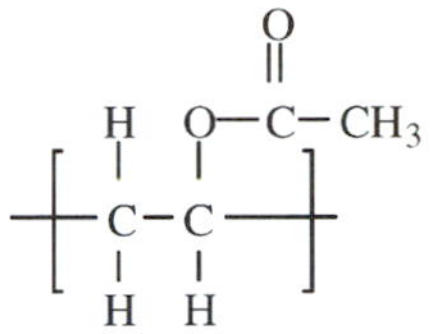	poly(vinyl acetate)	PVAC
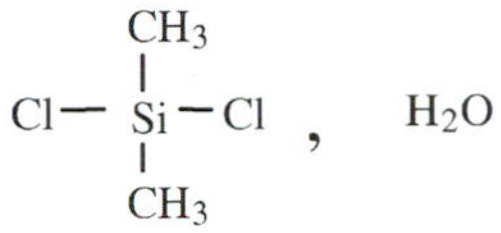	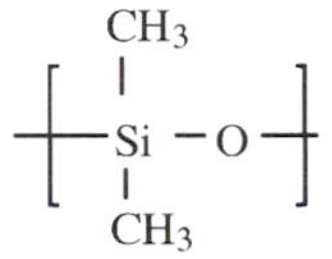	poly(dimethyl siloxane)	PDMS
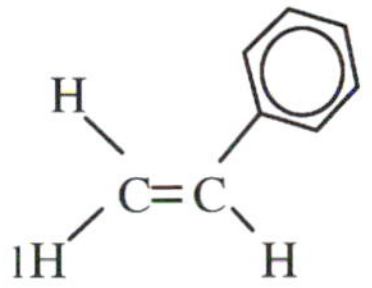	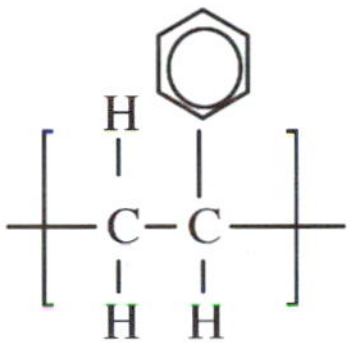	polystyrene	PS
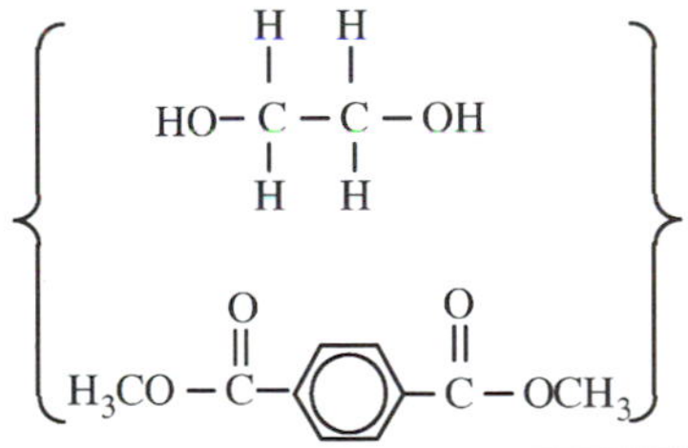	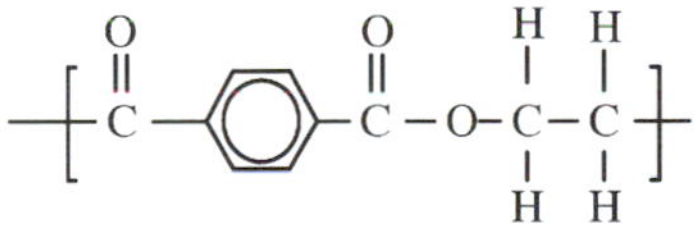	poly(ethylene terephthalate)	PET

ends, or at relatively sparse branch points, each monomeric unit is covalently linked with two others. The links define a molecular chain whose backbone, the skeletal structure of the macromolecule, consists of atoms joined by covalent bonds. Side groups complete the covalent bonding of the backbone atoms. These also serve to define the polymer species and to control its properties.

TABLE 1.2 Lengths of commonly encountered covalent bonds[1]

Structure	Name	Internuclear Distance (nm)
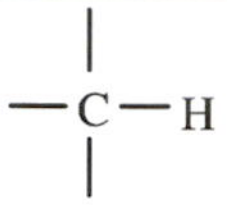	carbon-hydrogen	0.108 ± 1%
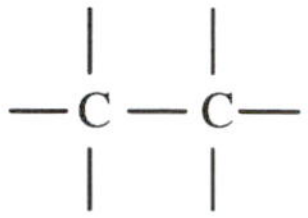	single carbon-carbon	0.153 ± 2%
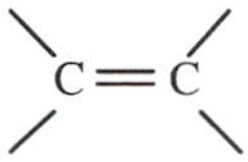	double carbon-carbon	0.133 ± 2%
	aromatic carbon-carbon	0.139 ± 1%
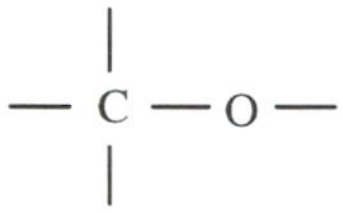	single carbon-oxygen	0.135 to 0.143
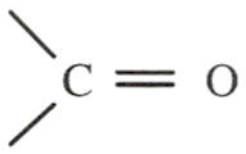	double carbon-oxygen	0.120 to 0.125
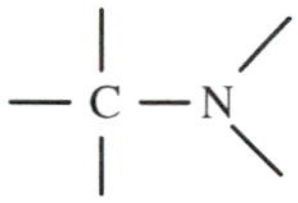	single carbon-nitrogen	about 0.15
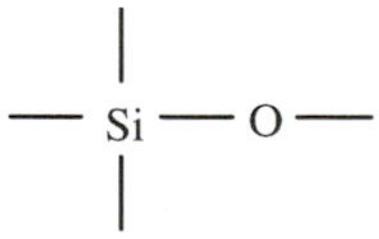	single silicon oxygen	about 0.17

Most flexible polymers are synthetic in origin, made by exploiting the covalent bonding capacity of carbon, hydrogen, and a relatively few other elements—mainly oxygen, nitrogen, chlorine, fluorine, and silicon. Much is known about the properties of covalent bonds—bond lengths, bond angles, and chemical reactivities—since they are of central importance in the broad and important subject of organic chemistry[1]. Polymerization reactions and their mechanisms vary widely even among those used to form this relatively limited group of materials[2]. Like all systems of competing chemical reactions, polymerization is statistical at the microscopic level, and this invariably leads to a distribution of molecular structures[3]. Only a few polymerization methods are capable of providing samples of sufficient uniformity to be useful for the fundamental studies of physical properties. Those methods, as well as separations of molecules according to size by fractionation from dilute solution, have supplied the *model polymers* that have been used to establish a scientific base for the field.

TABLE 1.3 Directional characteristics of covalent bonds[1]

Local Structure	Geometry	Bond Angle θ (degrees)
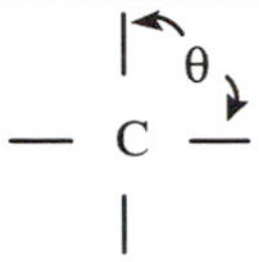	tetrahedral	109
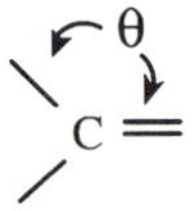	planar	120
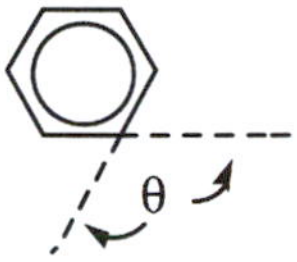	planar	120
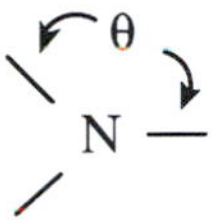	pyramidal	107
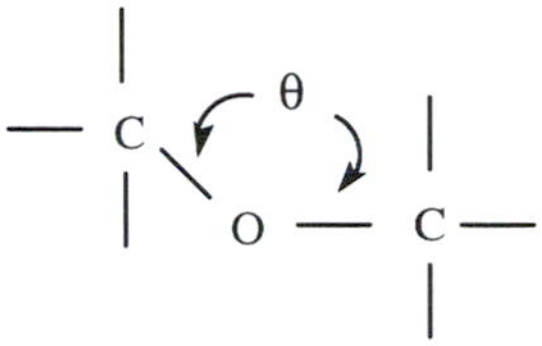	tetrahedral	109
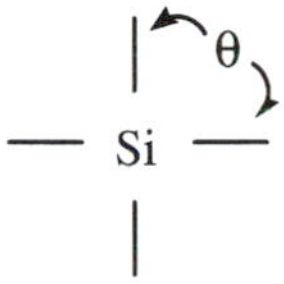	tetrahedral	109
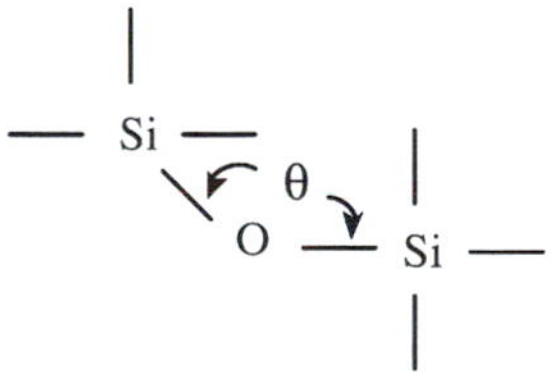	bent (soft)	130–160

Covalent bonds are stiff and directional: Both bond length (internuclear distance) and the angle between bonds on the same atom are fixed within very narrow limits by the rules of quantum mechanics. The lengths and angles for several common types of covalent bonds are given in Tables 1.2 and 1.3. Some bonds are weakly polar, so the molecules that contain them, although electrically neutral overall, may have regions of excess positive or negative charge distributed over their surfaces.

TABLE 1.4 Typical atomic radii in covalent molecules[1]

Element	Atomic Radius (nm)
C	0.077
H	0.037
O	0.074
N	0.075
Cl	0.098
F	0.072
Si	0.111

Carbon-hydrogen bonds and carbon-carbon single bonds are essentially nonpolar and only weakly polarizable. Carbon-carbon double bonds are also nonpolar but are more easily polarized by the electric fields created, for example, by adjacent polar bonds in the molecule. Carbon-oxygen, carbon-nitrogen, and hydrogen-oxygen bonds are permanent electric dipoles of various strengths, but all having much less polarity than fully ionic bonds.

For the covalent structures and conditions of common interest, rotation around single bonds, involving a single pair of shared electrons, is permitted. The electron clouds of the atoms also occupy space, as indicated by the typical radii in Table 1.4, so the steric (nonoverlap) restrictions imposed by more remotely connected parts of the molecule must also be considered. Multiple covalent bonds involve two or more electron pairs and do not have the rotational freedom of single bonds. Conjugated double bonds (sequences of alternating single and double bonds) are also rotationally stiff. When a double bond is formed, the angles between all bonds on the two atoms it connects are made permanent.

Freedom of rotation around single bonds confers molecular flexibility. Thus, the distance between a pair of atoms connected through a sequence of bonds having rotational freedom can be varied over some range without requiring distortion of covalent bond angles or lengths—that is, with little variation in the intramolecular energy. All flexible-chain polymer species have that characteristic, which is the origin of many useful properties.

1.2 Polymeric Structure

Flexible polymers in the liquid state are the central concern of this book. Liquid-state properties in general depend on the nature and strength of the dominant intermolecular forces. These properties are much less variable within groups of species selected on that basis—atomic liquids, ionic liquids, metallic liquids, quantum liquids, covalent liquids, nematic liquids, and the like. The category of interest here is *molecular liquids,* also variously called organic liquids, covalent liquids, and nonassociating liquids. Essentially all synthetic polymers, their monomeric precursors, and their solvents belong to this category. These molecules are held together by strong covalent bonds, they are uncharged but may be somewhat polar, and they have a variety of shapes. They interact with one another through the interplay of excluded volume repulsion and van der Waals attraction. (For simplicity, we omit detailed consideration of strongly polar or hydrogen-bonding substances such as water.)

The liquid state plays a prominent role in both the science and technology of synthetic polymers. Thus, for example, polymers are shaped into films, fibers, and a multitude of molded objects by liquid-state processing. The molecular characterization of polymers, whether to determine the local chemical structure or the large-scale architecture of the molecules, is generally conducted in dilute solution. Polymerization itself, and the chemical modification of already-formed polymers, such as those created by grafting and crosslinking, is conducted in the liquid state. Polymeric liquids and solutions are used directly as lubricants, water treatments, and oil field chemicals, and in various adhesive and coating applications. Many polymers cannot crystallize, owing to the geometric irregularity of their molecules. Their properties and applications thus depend on such universal properties of the liquid state as the glass transition. At least as important, the remarkable and highly useful property of rubber elasticity is the result of a flexible network superstructure combined with a liquidlike local mobility.

Some general understanding of molecular liquids is essential to most areas of polymer science. The equilibrium properties of the nonpolymeric variety are surveyed in Chapters 2 and 3. Their dynamics and mechanical properties are surveyed in Chapter 1 of Volume 2. The properties of polymeric liquids and networks depend on three rather distinct aspects of structure. One is the *chemical microstructure,* the molecular structure at the atomic scale of distances that defines the polymer species. Another is the *chemical macrostructure,* the pattern of chain connectivity over large scales of distance—the *macromolecular architecture,* or simply the large-scale architecture. The third aspect of liquid structure is the *physical structure,* the spatial arrangements of the molecules as governed by the competition of molecular forces and thermal energy. Scattering techniques are widely used to determine the physical structure; scattering principles are described and developed as needed throughout the book. Various features of chemical microstructure and macrostructure are introduced in the following sections of this chapter.

1.2.1 Chemical Microstructure

The chemical microstructure of a polymer depends on the monomeric species from which it is formed, but also on the precise particulars of local linking that were brought about by the repetitive chemical reactions involved in its polymerization. *Rotational isomers* are transient. They correspond to the various states of relative rotation for the single bonds in the backbone that make up the *molecular conformation. Enchainment isomers,* on the other hand, are effectively permanent and distinct in both structure

TABLE 1.5 Examples of polymeric structural isomers

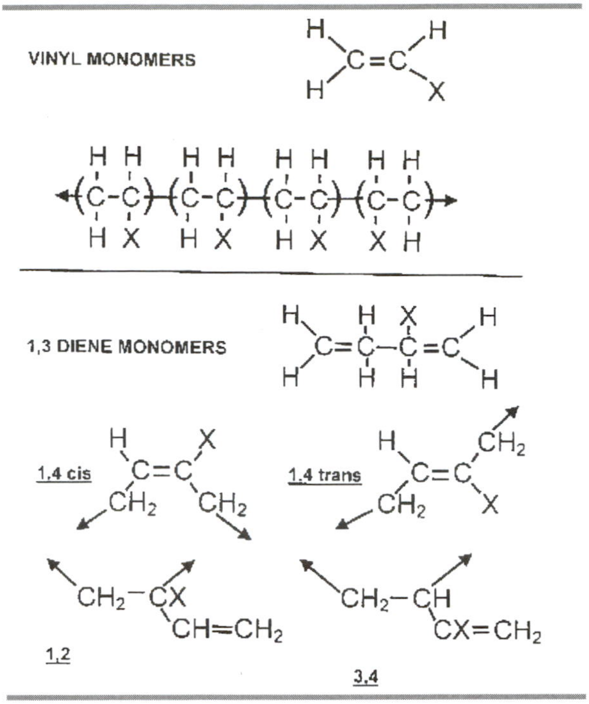

and the properties they confer. They specify the invariant spatial relationships within the chains, the *molecular configuration.*

Some of the more common examples of enchainment isomers in flexible-chain species are illustrated in Table 1.5. Vinyl monomers have the generic structure $CH_2 = CHX$, leading to monomeric units with the generic structure [—CH_2—CHX—], in which X stands for some side group other than H. Table 1.1 contains several examples of the vinyl family. The backbone of a vinyl polymer molecule is a sequence of carbon-carbon single bonds. The three monomeric units on the left in Table 1.5 are enchained *head-to-tail,* the third and fourth *head-to-head. Stereoisomers* are also a natural consequence of vinyl enchainment. The *Newman projection* helps to distinguish the various possibilities. All backbone bonds are rotated into the same plane, forming the *planar zigzag* conformation, and the result is then viewed parallel to the plane surface. A sequence of four chain units (eight backbone bonds) is shown in Figure 1.1. Successive side groups can be on the same side of the plane (*meso*) or on opposite sides (*racemic*). In the example, the first and second units are meso, whereas the second and third, and third and fourth, are racemic. A polymer is *isotactic* if all enchainments are meso, *syndiotactic* if all are racemic, and *atactic* if the meso–racemic sequencing is random. Polymerization of dienes can lead to two geometrically distinct

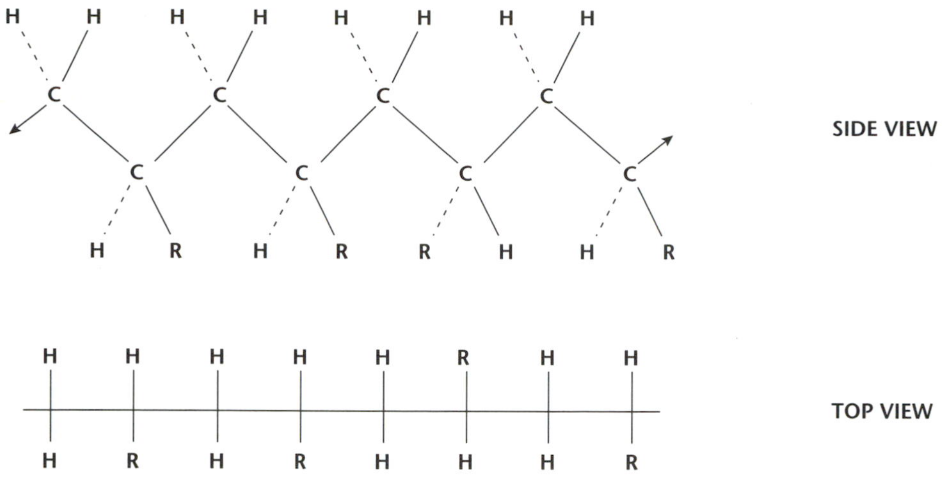

FIGURE 1.1 Two views of the planar zig zag conformation for a 4-mer sequence in a vinyl polymer chain. The top view is the Newman projection.

types of 1,4 enchainment, *cis* and *trans*. It can lead also to 1,2 or *vinyl* enchainment, with further possibilities in that case for meso and racemic sequencing. Table 1.5 illustrates the example of 1,3 butadiene. A geometrically distinct 3,4 enchainment is also possible for asymmetric dienes, such as isoprene. Head-to-head sequencing is relatively rare, but the meso–racemic, cis–trans or 1,4–1,2 ratios can sometimes be varied over wide ranges by the choice of polymerization catalyst and conditions, depending on the particular vinyl or diene species.

Copolymerization, the synthesis of macromolecules containing two or more species of monomeric units, adds other possibilities for varying the local structure through the frequency and sequencing of the monomeric units. Sequencing possibilities range from strict alternation, through statistical sequencing, which is the preferential incorporation of one of the monomer species, including the no-preference case of random sequencing, and finally to strict block copolymer formation. Other variants include compositional tapering along the chain and compositional differences among the chains. Not all these variations are possible for all species, but the variability is important enough to supply the need for special analytic techniques for sequence determination.

It is thus evident that two polymers made from the same monomer—or from the same pair of monomers in the same overall proportion—may differ significantly from one another in chemical microstructure. For that reason, they may also have very different physical properties. Polymerization chemistry dictates the range of choices. Spectroscopic techniques of various kinds are available for quantifying the chemical microstructure of polymers[4]. This aspect of the subject, important though it is, lies outside the scope of the book.

1.2.2 Chemical Macrostructure

The chemical macrostructure includes such polymeric features as the number of monomeric units in a molecule and the location and size of any long branches or macrocyclic rings it may contain. The distributions of monomeric units, branches, and rings among the various molecules in a polymer sample must also be considered. Even within the same species, a polymer with long branches may behave differently from a linear polymer with the same number of monomeric units per molecule. Likewise, a polymer containing a broad distribution of large-scale architectures may behave differently from one with a narrow distribution.

We will have many occasions to speak about the *chain length.* Doing this literally, such as by specifying the sum of backbone bond lengths, would obviously be awkward and is in fact seldom necessary. The length of a linear polymer molecule can be specified adequately by its *molecular weight,* M, a measure we will use frequently throughout the book, since it is commonly the directly measured quantity. The number of mers per chain—the *degree of polymerization* or *polymerization index P*—sometimes designated as n or r, serves the same purpose. The two are of course related,

$$P = \frac{\text{number of mers}}{\text{number of molecules}} = \frac{M}{m_0} \tag{1.1}$$

where m_0 is the molecular weight of the monomeric unit, or average molecular weight in the case of copolymers. Nonlinear architectures require a more detailed description, which will be developed as needed throughout the book.

1.2.3 Rotational States

Molecular architecture such as chain length, branching, and distribution provide only part of the information needed to characterize the large-scale features of polymer molecules. *Polymeric size,* which defines the space pervaded by the molecule by specifying either radius or volume, depends on chain conformation as well. The planar zig-zag depicted in Figure 1.1 is only one of many conformations available to chains that have some backbone bond rotational freedom while in the liquid state. Depending on species, rotation is always hindered to some extent by steric interference of the substituent groups. Even for the most flexible species, where the barriers are relatively low, differences in energy with rotational angle are still significant.

Consider a carbon-carbon single-bond backbone and the simplest example of a three-bond sequence. The first two bonds define a plane. Rotation around the third changes the distances separating the various backbone substituents. As shown in Figure 1.2, interaction energy varies with the angle ϕ. Three energy minima are typically present: the lowest corresponding to the zigzag conformation at $\phi = 0$ and, depending on the substituents, the other two differing in energy. The rotational

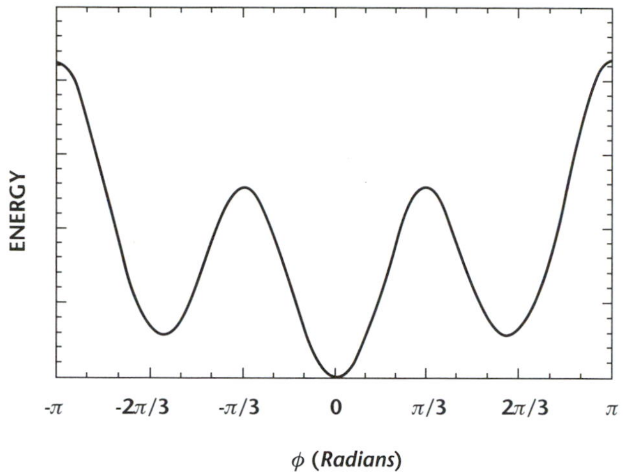

FIGURE 1.2 Intramolecular energy as a function of relative rotation angle around a carbon-carbon backbone bond.

angles along the chain are distributed according to their energies, as dictated by statistical thermodynamics; for flexible chains, the relative bond orientations shift rapidly among the various possibilities. Two distinct consequences arise from the multiplicity of rotational states and the rapid interchanges: Chain size on average is much smaller than the chain length, and the chain conformations vary spontaneously with time. The first of these is crucial in the equilibrium properties of polymeric liquids discussed in this volume, and the second is crucial to their dynamic properties, discussed in Volume 2.

At any instant, the various parts of a flexible-chain molecule are distributed in some manner around its center-of-mass. The average of their distances from that center is a measure of its *molecular size.* Unlike molecular weight, which depends only on the number and mass of the various parts, the molecular size depends on their spatial arrangement. In flexible molecules, the size can also vary with time and among molecules of identical chemical structure. The *radius of gyration* R_g, an experimentally accessible measure of molecular size, is an average over whatever variations arise owing to local rotational possibilities.

The average size of a flexible-chain polymer varies with conditions. In solution, R_g depends on concentration, temperature, and the choice of solvent. For long chains, the distance between remotely connected units can vary over a wide range. With increasing length, the backbone connecting the bonds can take up an increasingly large number of shapes, resulting in different relative bond directions and separation distances. Many properties of interest in the polymeric liquid state depend more directly on molecular size and shape than on chain length or molecular weight.

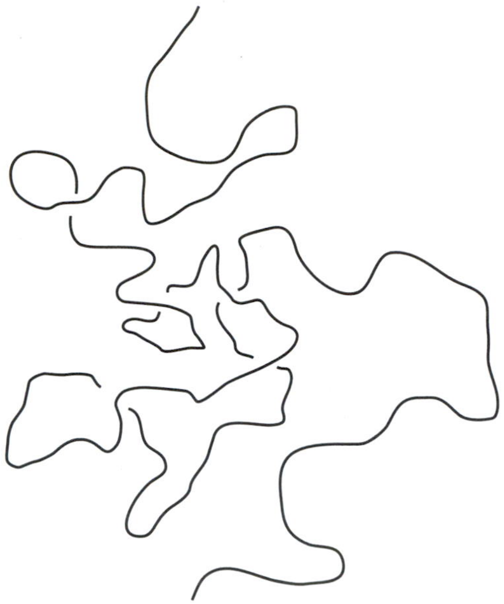

FIGURE 1.3 A representative backbone trajectory for random coil polymer molecules in the liquid state.

A representative trajectory for the backbone of a long flexible-chain molecule appears in Figure 1.3. For sufficiently long chains, and under conditions where local steric constraints alone are important, the various shapes distribute themselves according to the *random coil* model, a statistical description of large-scale conformations. The properties of the random coil are discussed in detail in Chapter 4 and applied extensively in later chapters.

1.3 Polymeric Properties

The distinction between species dependence and architectural dependence is an important one for the physical properties of flexible-chain polymers in the liquid state. The properties of a monomeric substance depend in a complex way on the details of molecular geometry and energetics. Semiempirical methods for property estimation are available, and computer simulation is becoming increasingly useful. Nevertheless, truly predictive microscopic theories are not available, and for precise work the species-dependent properties of molecular liquids must be measured. The situation for species dependence is no different for polymers than for monomeric substances. Fortunately, as discussed in Chapter 2, the properties of nonpolar or weakly polar molecular substances in the *dense liquid region*—the domain of interest to us here—do not change drastically from one species to another.

The architectural dependence of polymeric properties, on the other hand, frequently takes on the character of a universal law, pertaining to the behavior of long flexible strings and devoid of chemical details. What seems at first sight to be a grossly simplified picture of polymer molecules, long chains of flexibly connected "spherelets," turns out to be quite useful. The spherelets represent the monomeric units and absorb the local complexity associated with species. The species dependence can frequently be consolidated into a single spherelet parameter—a cohesive energy density or a Stokes friction coefficient—leaving the pattern of connectivity as that feature of structure whose effects are universal.

1.3.1 Species Dependence

Some properties of polymers have a long-chain limit that depends only on its chemical microstructure. Consider for example the melting temperature T_{mp} of the linear alkanes. All have structures of the form CH_3-$(CH_2)_r$-H ($r = 1, 2, 3 \cdots$), and T_{mp} increases monotonically with r, finally approaching a constant value beyond $r = 200$ or so. The members with odd values of r can be regarded as polymers of ethylene: ethane ($r = 1$), n-butane ($r = 3$), and the like, in which $(r + 1)/2$ is P, the number of monomeric units (–CH_2–CH_2–) per chain. How T_{mp} varies with N is shown by the linear-alkane data[5–7] in Figure 1.4. The values were obtained for ordinary rates of cooling and heating, and they give 136°C as the long-chain asymptote for melting temperature. Achieving equilibrium becomes difficult for long chains, however, and

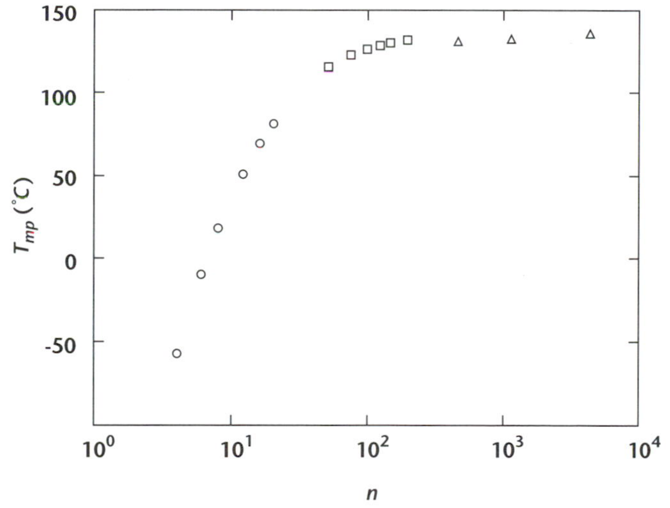

FIGURE 1.4 Melting temperature as a function of chain length for linear alkanes. Data taken from Ungar et al.[5] (□), Nicholson and Crist[6] (Δ), and Frenkel et al.[7] (O).

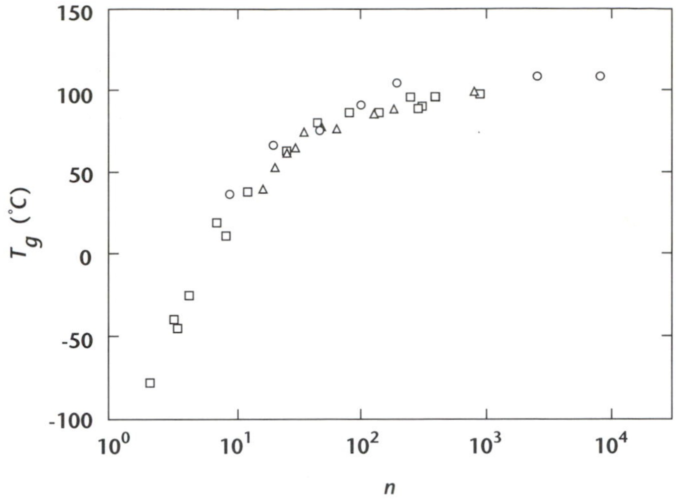

FIGURE 1.5 Glass transition temperature T_g versus chain length for polystyrene. Data taken from Ueberreiter and Kanig[9] (□), Fox and Flory[10] (Δ), and Cowie[11] (O).

special care is required to obtain the thermodynamic value for high molecular weight polyethylene[8], $(T_{mp})_\infty = 146°C$.

The *glass transition temperature* T_g is another example of limiting species dependence. As discussed in Volume 2, the glass transition is a universal property of supercooled or noncrystallizable liquids, an inherently rate-dependent phenomenon marked by changes in thermal and mechanical behavior. Data for polystyrene[9–11], a polymer species that, in its common atactic form does not crystallize, are shown as a function of the polymerization index in Figure 1.5. With ordinary rates of heating and cooling ($\sim 0.1°C\,s^{-1}$), an asymptote $(T_g)_\infty = 99°C$ is obtained.

1.3.2 Architectural Dependence

Species-dependent limiting values are observed for many properties besides T_{mp} and T_g. These depend on the chemical microstructure and must therefore be determined experimentally. Other properties, however, depend strongly on the large-scale architecture. The asymptotic result in those cases is a structure–property relationship rather than a species-dependent parameter.

Viscosity. The viscosity of polymers in the liquid state is a dynamic property that exemplifies such behavior. For a polymeric liquid, the viscosity η rises monotonically with molecular weight. In linear chains, the relationship settles into a power law, $\eta \propto M^{3.4\pm0.1}$, beyond a *characteristic molecular weight* M_c that depends on the polymer species. Data from Plazek and O'Rourke[12] for the viscosity of linear

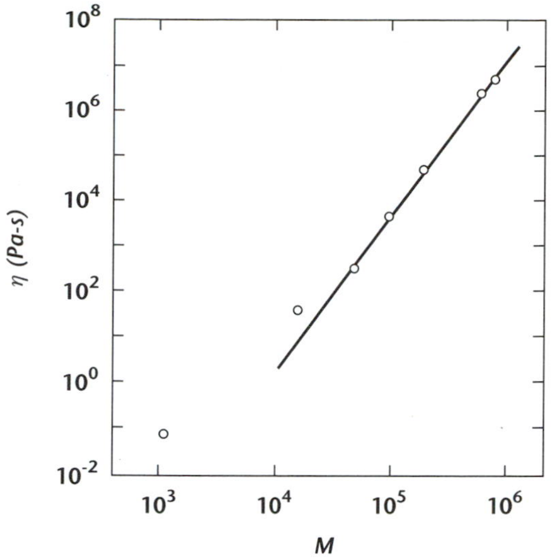

FIGURE 1.6 Viscosity as a function of molecular weight for undiluted polystyrene at 180°C. Data from Plazek and O'Rourke[12].

polystyrene at 180°C are shown as a function of molecular weight in Figure 1.6. The line in the figure indicates a least-squares fit to the data at high molecular weight:

$$\eta(Pa\,s) = 5.81 \times 10^{-14} M^{3.39} \tag{1.2}$$

Departures from Eq. 1.2 become significant below $M_c \sim 3.5 \times 10^4$. Even the form of the relationship varies with temperature in that region. Both M_c and the long-chain power law exponent are insensitive to temperature: Only the proportionality constant, $K = \eta/M^{3.39}$, depends significantly on the temperature. The power-law behavior and the exponent are universal properties for long, flexible-chain linear polymers in the undiluted liquid state. The temperature dependence of K and the value of M_c depend on the species. The viscosity of polymers with long branches obeys quite different limiting laws, which is discussed in some detail in Volume 2.

Other properties vary strongly with macromolecular architecture and obey universal laws in the long-chain limit. Examples include the size-dependent properties of polymers in dilute solutions (Chapters 5 and 6), phase equilibria for polymer solutions and polymeric mixtures (Chapters 7 and 8), and the elastic behavior of polymer networks (Chapters 9 and 10). Polymeric viscoelasticity, a collection of phenomena, introduced in the next section, but primarily dealt with in Volume 2, provides many additional examples of universal dependence on large-scale molecular architecture.

Viscoelastic behavior. Under the proper conditions—easily achieved in many cases and sometimes unavoidable experimentally—polymeric liquids behave in more complex and interesting ways than do monomeric liquids. Like monomeric liquids, polymeric liquids can assume any shape at equilibrium, and they deform and flow indefinitely in response to even the smallest of driving forces. However, they may also respond elastically, for example, by exhibiting a rubberlike recoil when the driving forces are suddenly removed.

The response to a mechanical disturbance by any substance, whether polymeric or not, depends on at least some of the time-dependent characteristics of the disturbance itself. For weak disturbances, the response depends on the relative magnitude of two time scales. One is the *experimental testing time* τ_E (or test rate τ_E^{-1}) characterizing the disturbance. The other is the *structural relaxation time* τ_S (or relaxation rate τ_S^{-1}), a material property that characterizes the persistence of molecular arrangements. The response is solidlike when τ_E is much smaller than τ_S, because the work expended in deforming the material is stored elastically and recoverable up to some time of order τ_S. The response is liquidlike when τ_E is much larger than τ_S, because the work is dissipated to heat beyond some time that is much shorter than τ_E. Ideal elastic solids can store mechanical energy for indefinitely long times, whereas ideal viscous liquids turn it into thermal energy instantaneously. The behavior of real substances, that is to say viscoelastic substances, can approach either extreme, and they can also behave in some intermediate way, depending on the ratio of time scales τ_S/τ_E.

The practical range of time scales for mechanical tests is about ten decades, $10^{-5}s \lesssim \tau_E \lesssim 10^5 s$. (Each limit could be extended by perhaps two more decades, with high frequency testing at one limit and patience at the other.) The structural relaxation time for a monomeric liquid near its melting temperature is very small, typically $\tau_S = 10^{-11}s$, so $\tau_S \ll \tau_E$, and its behavior is liquidlike. If the same substance is crystallized, the structural relaxation rate becomes extremely slow, so $\tau_S \gg \tau_E$, and the behavior is solidlike. Accordingly, monomeric substances at equilibrium pass in an effectively discontinuous way from one extreme of mechanical response to the other, thereby missing the viscoelastic intermediate, $\tau_S \sim \tau_E$.

The time for the full structural relaxation of a polymeric liquid is much longer than $10^{-11}s$. Under ambient conditions and using linear chains of about 10^4 mers, the times for full recovery range from $10^{-5}s$ in dilute solutions to $10^{-1}s$ or much longer in the undiluted liquid. The time of final relaxation for various solutions of linear 1,4 polybutadiene ($M = 925,000$) in phenyl octane at 25°C[13] is shown in Figure 1.7. The final relaxation time ranges from about 65s for the undiluted polymer (volume fraction $\phi = 1.0$) to about $10^{-3}s$ for a 1% solution ($\phi = 0.01$). Over this range, the relaxation time can be described by a power law:

$$\tau_o(s) = 65.2\,\phi^{2.38} \tag{1.3}$$

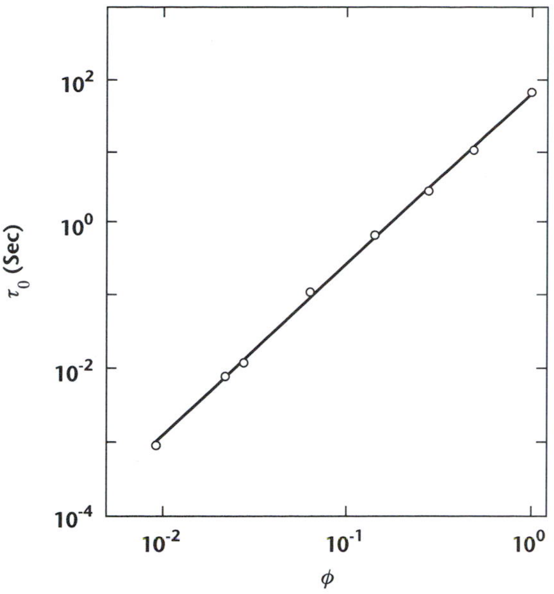

FIGURE 1.7 Relaxation time as a function of polymer concentration for solutions of 1,4 polybutadiene. ($M = 920,000$ in phenyl octane at 25°C, data from Colby et al.[13])

As discussed in Volume 2, the dependence of relaxation time on concentration is not universal. It depends on two not easily separated factors, one universal and the other varying with the species involved.

In contrast with the behavior of monomeric liquids, the relaxation of a polymeric liquid is spread over many decades of time. It ranges from the time required for the reorientation of individual mers (the fastest modes of relaxation, typically $10^{-11}s$) to the time required for the conformational rearrangement of the entire chain (the slowest mode, corresponding to τ_o above). The response of the liquid to any small disturbance depends on the distribution of its time-dependent characteristics. For any test rate τ_E^{-1}, a rough partitioning exists between a viscous response from fully relaxed faster modes (rate > τ_E^{-1}) and an elastic response from fully active slower modes (rate < τ_E^{-1}). Thus, the response is to some extent viscoelastic over the entire range of testing rates from $\tau_E^{-1} \sim 10^{11}$ sec $^{-1}$ to $\tau_E^{-1} \sim \tau_o^{-1}$. Viscoelastic response to both small and large disturbances is dealt with in Volume 2. The analogy between polymeric liquids and networks provides important insights in both cases.

1.4 Macromolecular Heterogeneity

As discussed in the previous section, the macrostructure of any synthetic polymer is heterogeneous. Any competition of chemical reactions is inherently statistical, as is the formation of polymer molecules. Thus, no matter how careful the synthesis, the polymer product is always a mixture of molecular structures, the distribution of which depends on the mechanism and conditions of polymerization. Commercial polymers vary widely in molecular weight distribution, and the breadth of the distribution in itself can have very important effects on physical properties. Thus, for example, many properties of polymers in the solid state are sensitive to the presence of low molecular weight components and hence to the characteristics of the low tail of the distribution. Liquid-state properties, especially those related to flow behavior, are sensitive to high molecular weight components and hence to the high tail characteristics. We introduce here the terminology for molecular weight distribution and averages that will be used throughout the book.

1.4.1 Averages

Let $M_i (i = 1, 2, \ldots)$ be the set of molecular weights represented in a sample, and W_i their respective weight fractions. The sum of weight fractions must be unity:

$$\sum W_i = 1 \tag{1.4}$$

In addition, the respective number of moles per unit sample mass is W_i/M_i. The ratio of total mass to total moles is $\sum W_i / \sum (W_i/M_i)$, the *number-average molecular weight* of the sample:

$$M_n = \frac{\text{mass of mers}}{\text{moles of polymer}} = \frac{1}{\sum (W_i/M_i)} \tag{1.5}$$

The number-average polymerization index is thus:

$$P_n = \frac{\text{moles of mers}}{\text{moles of polymer}} = \frac{M_n}{m_o} \tag{1.6}$$

Other averages of molecular weight are also used. The larger molecules are weighted more heavily in the *weight-average molecular weight:*

$$M_w = \sum W_i \, M_i \tag{1.7}$$

The large molecules are weighted more heavily still in the z and $z + 1$ *average molecular weights:*

$$M_z = \frac{\sum W_i \, M_i^2}{\sum W_i M_i} \tag{1.8}$$

$$M_{z+1} = \frac{\sum W_i M_i^3}{\sum W_i M_i^2} \tag{1.9}$$

As in Eq. 1.6, the corresponding average polymerization indices are simply the average molecular weights divided by the mer weight.

Because of these weightings, the averages for any distribution are always ordered as $M_n \leq M_w \leq M_z \leq M_{z+1}$. The averages all have the same value, M, if only one molecular weight is represented, that is, if the polymer is *monodisperse.* Averages of properties other than molecular weight are frequently of interest. For example, scattering experiments (Chapter 4) can provide the z-average radius of gyration:

$$(R_g^2)_z = \frac{\sum W_i M_i R_{g,i}^2}{\sum W_i M_i} \tag{1.10}$$

Other examples are the number-average and weight-average number of branches:

$$\begin{aligned} B_n &= \frac{\sum\sum j W_{i,j}/M_i}{\sum\sum W_{i,j}/M_i} \\ B_w &= \frac{\sum\sum j W_{i,j}}{\sum\sum W_{i,j}} \end{aligned} \tag{1.11}$$

where $W_{i,j}$ is the weight fraction of molecules with molecular weight M_i and j branch points.

The distribution breadth is commonly characterized by the *dispersion ratio* or *polydispersity* M_w/M_n. Sometimes the *heterogeneity index,* $M_w/M_n - 1$, is used. The *variance* of the weight distribution, $(M_z - M_w)/M_w$, is also used, as is the *rheological polydispersity,* $M_z M_{z+1}/M_w^2$. These various measures of distribution breadth vary in sensitivity to different parts of the distribution. Thus, M_w/M_n depends most strongly on the low molecular weight tail, whereas M_z/M_w depends most strongly on the high molecular weight tail. For historical reasons, M_w/M_n receives the most attention in all issues of distribution effects, despite its being almost useless, for example, as a gauge of melt flow behavior. All such averages are attempts to describe the shape of a distribution function by a number. They are convenient for discussion purposes (any information is probably better than no information at all), but they do present an overly simplified and sometimes highly misleading description. Averages are crude and should be used with caution.

The *weight distribution,* the M_i, W_i set, is commonly represented by a continuous function, $W(M)dM$, the fractional weight of a sample having molecular weights in the interval M, $M + dM$. The use of a continuous function to describe the distribution is harmless when the average number of units per molecule is large. It merely changes the sums over integer values to integrals over the range $0 < M < \infty$. It is not at all uncommon to have significant contributions to the distribution that extend over three or four decades of molecular weight. In such cases, it is much more convenient to express the weight distribution in terms of logarithmic increments in molecular weight.

Consider $W'(M)\,d\log M$ as the *logarithmic weight distribution*, the fractional weight of the sample having molecular weights in the interval $\log M$, $\log M + d\log M$. Then, with the identity $W(M)dM = MW(M)d\ln M$,

$$W'(M) = (\ln 10)MW(M) \tag{1.12}$$

The *number distribution*, the mole fraction F_i or $F(M)dM$, is an alternative way to describe distributions. The fraction of molecules F_i and the fractional weight W_i are easily converted one to the other:

$$\begin{aligned} F_i &= \frac{W_i/M_i}{\sum W_i/M_i} \\ W_i &= \frac{F_i M_i}{\sum F_i M_i} \end{aligned} \tag{1.13}$$

Distributions can, of course, be described just as well in terms of polymerization indices as molecular weights. The fraction of r-mer chains and the mass fraction of r-mers, $F(r)$ and $W(r)$ respectively, are in fact widely used in the derivation of distributions from polymerization mechanisms as in the examples below. The various sums or integrals whose ratios define the averages are *distribution moments.* The following are the moments of the mer number distribution in discrete and continuous form:

$$\begin{aligned} Q_j &= \sum_{i=1}^{\infty} (P_i)^j F_i \\ Q_j &= \int_0^{\infty} r^j F(r)dr \end{aligned} \tag{1.14}$$

Thus, in this terminology, the number-average and weigh-average polymerization indices are expressed as ratios of successive moments, Q_1/Q_o and Q_2/Q_1. Note that the power j in the definitions above need not be an integer, and indeed we will find need for noninteger moments at various places in this book. Note also that, because of cancellation, the choice of normalization for a distribution does not affect the molecular weight averages.

1.4.2 Distribution Functions

Molecular weight distribution can sometimes be predicted from the mechanism of polymerization[3]. Two methods are widely used: probability arguments[14] and population balances[15]. Both rely upon the *equal reactivity principle,* which states that, for a fixed set of external conditions, the chemical reactivity of a functional group is governed only by its locally bonded neighborhood. Accordingly, the propensity to react is the same, independent of chain length and distribution, for all polymeric functional groups with the same structure and nearby connected environment. The evidence supporting this now well-accepted principle is summarized by Flory[14].

Exponential distribution. As an example of the probability method, consider the distribution of the linear chains produced by a condensation polymerization of difunctional monomers. Call the monomers A_2 and B_2, having functional groups A and B that can react to form AB links, and start with the same number N of each monomer. After a fraction p of the A functions react, $2N(1-p)$ of the A functions and $2N(1-p)$ of the B functions remain unreacted. All unreacted functions must be chain ends, and each chain has two ends, so there must now be exactly $2N(1-p)$ molecules in the system. The system contains $2N$ mers in the system, which never changes (by convention, all monomers in condensation polymerizations at any conversion count as mers), so from Eq. 1.6,

$$P_n = 1/(1-p) \tag{1.15}$$

The distribution of mers among the molecules can be calculated using the equal reactivity principle. Select a chain end at random and calculate the probability that its molecule contains exactly i mers. The result will also be equal to the fraction of i-mer chains F_i, since all chains have the same number of ends. This i-mer chain outcome can happen in only one way. Beginning with the selected end, each of the first i-1 functional groups encountered along the chain must be reacted, thereby continuing the chain—but the ith one must be unreacted, thus providing the other end of the chain. With equal reactivity, the probabilities of reacted and unreacted for any functional group are simply p and $1-p$, so the probability of having selected an i-mer chain is the product $p^{i-1}(1-p)$, or

$$F_i = (1-\varepsilon)^{i-1}\varepsilon \tag{1.16}$$

where $\varepsilon = 1-p$.

The focus here is on long-chain systems, so from Eq. 1.15, we must have $\varepsilon \ll 1$, in which case $(1-\varepsilon)^i \approx \exp(-\varepsilon i)$ for large i. Thus, going to the continuous distribution function, using $\varepsilon = 1/P_n$ and $1-\varepsilon \approx 1$, and with r in place of i,

$$F(r)\,dr = \frac{1}{P_n}\exp\left(-\frac{r}{P_n}\right)dr \tag{1.17}$$

This is the *exponential distribution*, also known as the *Flory distribution*, and is encountered in many polymerization systems. It is often expressed as a weight distribution:

$$W(M)\,dM = \frac{M\exp(-M/M_n)}{(M_n)^2}\,dM \tag{1.18}$$

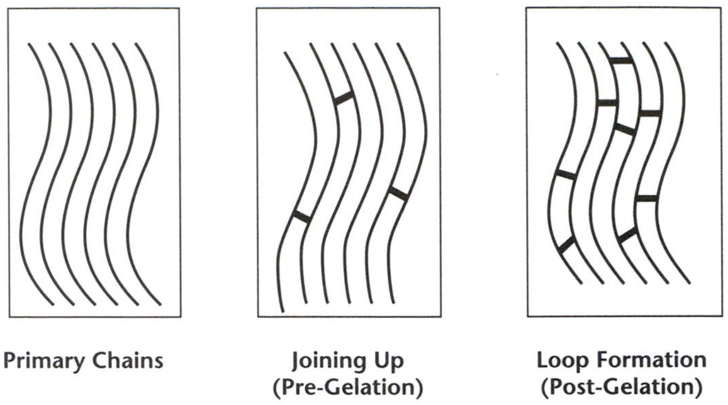

FIGURE 1.8 Gelation and network formation stages for crosslinking.

Compared with many distributions observed for polymers, it is relatively narrow: $M_w/M_n = 2$, $M_z/M_w = 3/2$, and $M_{z+1}/M_z = 4/3$.

Distributions from branching and scission. Crosslinking is a reaction that joins mers on different polymer chains (known as the *primary chains*) by a covalent bond, forming thereby in its early stages branched molecules of higher molecular weight. Crosslinking is one means of forming the polymer networks that are considered in detail in Chapters 9 and 10. Consider first, however, some qualitative features of a crosslinking process. Start with some collection of primary chains, each having a large number of mers that are able to form *crosslinks* with other mers. Two stages can be identified as crosslinks are added: pre-gelation and post-gelation. Figure 1.8 illustrates the evolution of structure using a system of six primary chains. In the pre-gel stage, there are many fewer crosslinks than primary chains, and each new crosslink reduces the number of molecules—the molecules simply grow bigger. In the post-gel stage, there are many more crosslinks than primary chains. The chains are already joined together to form a single giant molecule, so each new crosslink must of necessity be intramolecular, forming a ring by joining parts that already belong to the same molecule. These interconnections, the surplus beyond mere joining, turn the giant molecule into a *network* or *gel.* The *gel point* marks the onset of network behavior—corresponding to five crosslinks in Figure 1.8—and separates the pre-gel and post-gel stages.

All this is greatly simplified, of course. First, there are approximately 10^{18} primary chains per gram of starting material, not six. Second, the progression toward network formation, like any set of chemical reactions, is statistical rather than determinate at the molecular level. The gel point is attained with far fewer than one crosslink per primary chain, for example, and beyond the gel point, the system consists of both gel and unattached molecules or *sol.* The following analysis, an example of the population balance method, follows the evolution of molecular weight distribution from the primary chain distribution to the gel point as crosslinks are added between mers selected at random.

The interest here is long primary chains, $(P_n)_o \gg 1$ for primary chain distribution $F(r, 0)$, which is the distribution prior to crosslinking. The distribution changes with random crosslinking to $F(r, \alpha)$, where α is the *crosslink density*, the fraction of mers that participate in links between primary chains. Thus, α is the average number of crosslinked units per mer, and $\alpha/2$ is the average number of crosslinks per mer. It is convenient to treat all distributions as normalized on the per-mer basis, such that $\int_o^\infty rF(r, \alpha)dr = 1$ for all α.

With the assumption of equal mer reactivity crosslink formation a balance can be written expressing how the population of r-mer chains, $F(r, \alpha)$, changes with each small increment $\Delta\alpha$. For simplicity, let α be small ($\alpha \ll 1$), so that all primary chains retain the same reactivity, which is proportional to r, throughout the linking process. Then,

$$\Delta F(r, \alpha) = -\Delta\alpha r F(r, \alpha) + \frac{1}{2}\Delta\alpha \int_o^r sF(s, \alpha)(r - s)F(r - s, \alpha)\, ds \tag{1.19}$$

The term on the left is the change in r-mers. The first on the right is the loss of r-mers owing to their mers having been linked to other chains. The second term on the right is the gain in r-mers from the coupling of smaller molecules whose mer sum is just equal to r, the factor of 1/2 compensating for double counting. In the small $\Delta\alpha$ limit,

$$\frac{\partial F(r, \alpha)}{\partial \alpha} = -rF(r, \alpha) + \frac{1}{2}\int_o^r sF(s, \alpha)(r - s)F(r - s, \alpha)\, ds \tag{1.20}$$

with the conditions $\int_o^\infty rF(r, \alpha)dr = \int_o^\infty rF(r, 0)dr = 1$.

Solving Eq. 1.20 for $F(r, \alpha)$ is not easy, but it can be converted to a set of equations for the moments of $F(r, \alpha)$—Q_o, $Q_1, \ldots$ according to Eq. 1.14—and hence to expressions for the various average molecular weights. Multiply both sides of Eq. 1.20 by r^m and integrate from 0 to ∞. Then reverse the orders of integration, summation, and differentiation as needed to arrive at:

$$\frac{dQ_m}{d\alpha} = -Q_{m+1} + \frac{1}{2}\int_o^\infty \int_o^\infty uv(u + v)^m F(u)\, F(v)\, du\, dv \tag{1.21}$$

Now solve the equations obtained sequentially with $m = 0$, 1, and 2 and the condition $Q_1(\alpha) = 1$ from above to obtain expressions for Q_o, Q_1, and Q_2, and thus the average polymerization indices:

$$P_n = \frac{(P_n)_o}{1 - \alpha(P_n)_o/2}$$
$$P_w = \frac{(P_w)_o}{1 - \alpha(P_w)_o} \tag{1.22}$$

Note that P_w diverges to infinity at $\alpha(P_w)_o = 1$, a condition corresponding to the gel point or onset of network formation.

Crosslinking is frequently accompanied by backbone bond fracture, a process called *chain scission*. The effect of scission can be regarded as a change in the primary chain distribution. If all mers have the same probability of scission, the effect on the primary chains can be expressed in terms of the scission probability β, the fraction of fractured mers. If the primary chains are long enough, and the average number of fractures per chain is large enough, then regardless of the primary chain distribution, the fragments will approach and finally become indistinguishable from the exponential distribution:

$$p(s, \beta) = \beta \exp(-\beta s) \qquad (r \gg s \gg 1) \tag{1.23}$$

If the initial distribution of primary chains is exponential, that form is retained with scission, and only the length parameter changes:

$$F(r, \beta) = \frac{1}{P_n(\beta)} \exp\left(-\frac{r}{P_n(\beta)}\right)$$
$$\frac{1}{P_n(\beta)} = \frac{1}{(P_n)_o} + \beta \tag{1.24}$$

Both results are special cases that can be demonstrated using the general equation for random chain scission[16,17]:

$$F(r, \beta) = \exp(-\beta r)\left[F(r, 0) + \frac{\beta}{r}\int_r^{\infty} F(s, 0)[2 + \beta(s - r)]\, ds\right] \tag{1.25}$$

The effects of simultaneous crosslinking and scission are considered in Chapter 9.

The steady-state distribution for free-radical polymerization, in a continuous stirred reactor with branching by polymer transfer, can be predicted by either probability or the population balance method. The result is the *Beasley distribution*[18]:

$$W(r) = \frac{1}{\nu^2} \frac{(1 - b)r}{(1 + br/\nu)^{1+1/b}} \tag{1.26}$$

TABLE 1.6 Some properties of commonly used molecular weight distribution models

Name	Weight Distribution Formula	P_n	P_w/P_n	P_z/P_w	P_{z+1}/P_z
Zimm–Schulz	$W(r) = \frac{1}{\nu}\left(\frac{r}{\nu}\right)^2 \frac{\exp(-r/\nu)}{\Gamma(z+1)}$	νz	$\frac{z+1}{z}$	$\frac{z+2}{z+1}$	$\frac{z+3}{z+2}$
Wesslau	$W(r) = \frac{1}{B\pi^{1/2}r} \exp\left[-\frac{1}{B^2}\ln^2\frac{r}{\nu}\right]$	$\nu \exp\left(-\frac{B^2}{4}\right)$	$\exp\left(\frac{B^2}{2}\right)$	$\exp\left(\frac{B^2}{2}\right)$	$\exp\left(\frac{B^2}{2}\right)$
Beasley	$W(r) = \frac{1}{\nu^2}\frac{(1-b)r}{(1+br/\nu)^{1+1/b}}$	ν	$2\left[\frac{1-b}{1-2b}\right]$	$\frac{3}{2}\left[\frac{1-2b}{1-3b}\right]$	$\frac{4}{3}\left[\frac{1-3b}{1-4b}\right]$

where ν is P_n, a property that is independent of conversion in this case, and b is the branching parameter $\nu C_p x/(1-x)$, in which C_p is the polymer transfer constant for the system, and x is the steady-state fractional extent of reaction. Polymer transfer is one mechanism that leads to long-chain branching. The effect of long branches on molecular size is considered in Chapters 4, 5, and 6. The effect of long branches on dynamic properties is taken up in Volume 2.

Empirical distribution functions. Two empirical forms in common use are the *Zimm–Schulz distribution*[19] and the *log-normal* or *Wesslau distribution*[20]:

$$W(r) = \frac{1}{\nu}\left(\frac{r}{\nu}\right)^z \frac{\exp(-r/\nu)}{\Gamma(z+1)} \qquad \text{Zimm–Schulz} \qquad (1.27)$$

$$W(r) = \frac{1}{B\pi^{1/2}r} \exp\left[-\frac{1}{B^2}\ln^2\frac{r}{\nu}\right] \qquad \text{Wesslau} \qquad (1.28)$$

where ν is a chain length parameter, and z and B govern the distribution breadth. The Zimm–Schulz expression is most useful in representing narrow distributions between the exponential distribution ($z = 1$) and monodisperse ($z = \infty$). The Wesslau form is mostly used for broad distributions ($B > 2$), especially those with high molecular weight tails, such as polymers made with Ziegler–Natta catalysts. Some properties of the Beasley, Zimm–Schulz and Wesslau distributions are listed in Table 1.6.

To show the variety of distribution types, four calculated examples, all scaled to $M_w = 10^5$, are plotted logarithmically [Eq. 1.12] in Figure 1.9. Under the most favorable conditions, a carefully conducted anionic polymerization[2] can yield samples with a polydispersity smaller than 1.01. The effect of polydispersity on most properties is still rather small at $M_w/M_n = 1.05$, which is example A in Figure 1.9, modeled with the Zimm–Schulz form [Eq. 1.27]. Such materials are termed *nearly monodisperse.* The molecular weights of nearly monodisperse samples are denoted throughout the

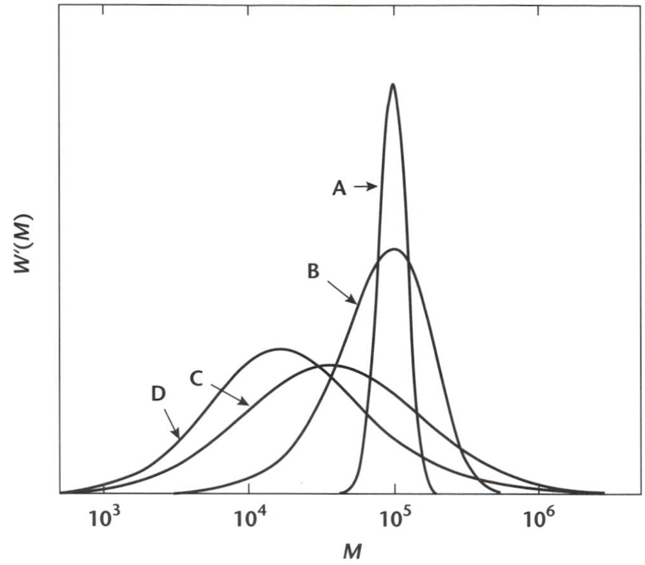

FIGURE 1.9 Weight distribution for polymers all with $M_w = 10^5$, but different molecular weight distributions. A and B are Zimm–Schulz distributions (Eq. 1.27) for $z = 20$ ($M_w/M_n = 1.05$) and $z = 1$ ($M_w/M_n = 2$), C is a Wesslau distribution (Eq. 1.28) with $B = 2$ ($M_w/M_n = 7.4$), and D is a Beasley distribution (Eq. 1.26) with $b = 5/11$($M_w/M_n = 12$).

book as if they were truly monodisperse by the use of a nonsubscripted M. (This polydispersity range can be achieved for some species and for many chain architectures by modern anionic methods. It can also be approached by the careful fractionation of more polydisperse linear samples [$M_w/M_n \sim 2 - 3$] in dilute solution.) Example B is the frequently encountered exponential distribution [Eq. 1.18]. Example C, represented in the Wesslau form [Eq. 1.28], corresponds to the products of heterogeneous polymerization catalysts. Example D, with $M_w/M_n > 10$ as modeled with the Beasley form [Eq. 1.26], is sometimes encountered in the products of polymerizations with long-chain branch formation.

Because distributions can vary considerably, even within the same polymer species, some understanding of how distribution affects properties is important. It is well established that both melting temperature and glass transition temperature for a given species depend primarily on the concentration of chain ends. Linear chains have two chain ends regardless of chain length, so otherwise identical linear samples with different distribution breadths should have the same T_{mp} or T_g if they have the same M_n. Although Figures 1.4 and 1.5 were constructed with data for nearly monodisperse linear samples, the correlations should also apply to polydisperse linear samples if M_n is used. The melt viscosity of linear polymers is found to depend primarily on M_w, at least over a modest range of polydispersities. There appears to be no theoretical explanation for this observation, but M_w is nevertheless commonly used as a correlating variable for the viscosity. The effect of polydispersity on relaxation time is complicated. Relaxation times also vary with polymer concentration, the choice of solvent, and the temperature. How to accommodate samples with different polydispersities in plots, such as that shown in Figure 1.7, is still an open question. The effect of distribution breadth on the properties of polymeric liquids is a recurring consideration in Volume 2.

1.5 Molecular Simulations

So far, this chapter has touched briefly upon polymeric structure and properties in the traditional context of laboratory experiments, data correlations, and molecular theories. Modern computational methods have made available another source of information—molecular simulations[21]. Simulation has grown in importance and sophistication as computer processing speed has increased. Now—and especially for the liquid state—simulation adds a crucial new element to traditional experiment and theory. The idea is first to create by computer program a collection of objects in memory that mimic molecules, manageably confined in computational space at some meaningful density and in contact with the computer equivalent of a constant temperature bath. All the motions and rearrangements that occur spontaneously in real liquids are permitted to occur, according to some set of programmed rules, and properties of interest are calculated from the collection by the methods of statistical mechanics. The beauty of simulation is that all microscopic information is available. The collection can be used to answer questions about the microscopic aspects of behavior that are not otherwise observable, but which may have been postulated in theoretical models. On the other hand, predicting properties and checking models in reliable fashion depends on many factors—molecular mimicking, programming rules, system size, computation time—that frequently conflict.

Two general methods are used for liquid-state simulations. One is *molecular dynamics,* the method just described, which uses programming rules involving the detailed enforcement of Newton's laws on the motions. *Monte Carlo simulation* is the second method. In the Monte Carlo simulation, no heat bath is represented, and molecules or parts of molecules are moved to adjacent positions with rules carefully chosen to represent the effects of random thermal agitation. Molecular dynamics simulation has the reassuring feel of microscopic reality. However, equilibration is usually much faster using Monte Carlo, thus providing a compromise to circumvent the time problem, but still quite capable of providing valid information. The practitioner's art in either method is like that of the classical theorist: choosing a molecular mimic as detailed as required—but not unnecessarily detailed—to capture the essence of the problem at hand.

Simulation practitioners also resemble experimentalists. Computers are their laboratory instruments, and the relevant issues, as always, are well-posed questions, proper data analysis and interpretation, and the broader applicability of the results obtained. Problems in collection size and equilibration time always exist, even with the fastest computer assemblies. Because the equivalence to a microsecond time scale in real liquids is extremely challenging in simulations for any collection that even resembles an observable droplet of liquid, the exploration of long time processes by simulation is difficult in general. The capacity of simulation to supply much more detailed microscopic information than laboratory experiments now plays an invaluable part in polymeric liquid research. The results obtained by simulation are prominently represented in several parts of this book.

REFERENCES

1. Morrison R.T. and Boyd R.N. 1992. *Organic Chemistry,* 6th ed. Englewood Cliffs NJ: Prentice Hall.
2. Hsieh H.L. and Quirk R.P. 1996. *Anionic Polymerization, Principles and Practical Applications,* New York: Marcel Dekker.
3. Dotson N.A., Galvan R., Laurence R.L., and Tirrell M. 1996. *Polymerization Process Modeling,* New York: VCH.
4. Koenig J.L. 1992. *Spectroscopy of Polymers,* Washington DC: American Chemical Society.
5. Ungar G., J. Stejny, A. Keller, I. Bidd, and M.C. Whiting. 1985. *Science* 229:386.
6. Nicholson J.C. and B. Crist. 1989. *Macromolecules* 22:1704.
7. Frenkel M., Q. Dong, R.C. Wilhoit, and K.R. Hall. 2001. *Int. J. Thermophys.* 22:215.
8. Rijke A.M. and L. Mandelkern. 1970. *J. Polym. Sci., Part A-2* 8:225.
9. Ueberreiter K. and G. Kanig. 1952. *J. Colloid Sci.* 7:569.
10. Fox T.G. and P.J. Flory. 1954. *J. Polym. Sci.* 14:315.
11. Cowie J.M.G. 1975. *Eur. Polym. J.* 11:297.
12. Plazek D.J. and V.M. O'Rourke. 1971. *J. Polym. Sci., Part A-2* 9:209.
13. Colby R.H., L.J. Fetters, W.G. Funk, and W.W. Graessley. 1991. *Macromolecules* 24:3873.
14. Flory P.J. 1953. *Principles of Polymer Chemistry,* Ithaca, NY: Cornell University Press.
15. Bamford C.H. and H. Tompa. 1954. *Trans. Faraday Soc.* 50:1097.
16. Mark H. and R. Simha. 1940. *Trans. Faraday Soc.* 36:611.
17. Montroll E.W. 1941. *J. Am. Chem. Soc.* 63:1215.
18. Beasley J.K. 1953. *J. Am. Chem. Soc.* 75:6123.
19. Zimm B.H. 1948. *J. Chem. Phys.* 16:1093.
20. Wesslau H. 1956. *Makromol. Chem.* 20:111.
21. Binder K. (ed.) 1995. *Monte Carlo and Molecular Dynamics Simulations in Polymer Science,* New York: Oxford University Press.

CHAPTER 2

Molecular Liquids

This chapter summarizes current ideas about the liquid state for molecular substances, with particular attention paid to the dense liquid region. The connection between intermolecular forces and liquid structure, the use of scattering to determine liquid structure, and various molecularly motivated models of the liquid state are introduced. Other topics include the simple liquid model and equation of state measurements, which relate pressure, volume, and temperature in the dense liquid state (PVT data). Reduced equations of state, cohesive energy, and internal pressure are still other aspects of the liquid state considered in this chapter.

Molecular liquids, polymeric or not, have many features in common[1]. This chapter focuses on the nonpolymeric variety, the subgroup we refer to as *monomeric liquids*. In the most elementary view, monomeric molecules are spheres and polymeric molecules are flexibly connected strings of spheres. The fundamental questions at this level of approximation are about the strength of the forces acting on the spheres, how the spheres are arranged, and how their spatial relationships evolve with time. The first two have to do with the energy and structure that govern the equilibrium (thermodynamic) properties of monomeric liquids and are the main concern of this chapter. The third also involves the random thermal motion that governs dynamical (transport) properties, and is the subject of Volume 2. The same questions apply to the spheres in polymeric strings, only now with the additional effect of permanent connectedness. A broad understanding of monomeric liquids, at even this most simplified level, carries over quite directly to polymeric liquids and permits the effects arising from chain connectivity per se to be more readily identified.

Even in the spherical approximation, however, two mathematically intractable complications dominate the molecular theory of liquids[2]. One is the nonanalytic behavior near vapor–liquid and liquid–liquid critical points, caused respectively by

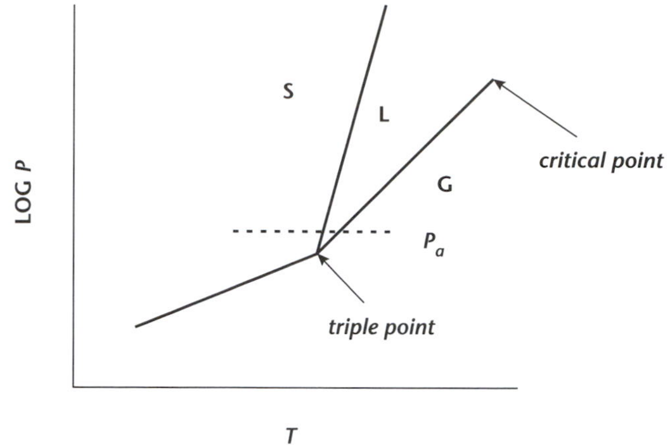

FIGURE 2.1 Typical phase diagram for a monomeric species.

density and composition fluctuations. As a result of fluctuation and phase separation competition, critical points and phase boundary shapes in the critical region can differ significantly from those predicted by classical thermodynamic methods. In some cases, the effects are small or irrelevant. Thus for example, a polymeric liquid is effectively nonvolatile, so its gas–liquid critical region is inaccessible and hence without practical interest*. For the liquid–liquid critical behavior of polymeric mixtures, the temperature range affected is relatively small owing to large molecular size[5]. However, gas–liquid and liquid–liquid critical phenomena are important in both monomeric mixtures and polymer solutions, and significant nonclassical behavior is both expected and found. This aspect of liquid-state complexity is considered briefly in Chapter 7.

The other universal complication in liquid-state theory is *volume exclusion*. The mutually excluded volume of the molecules governs local packing in liquids and is especially important in the *dense liquid* region, as noted in Chapter 1 and which is the domain of primary interest throughout the book. In the dense liquid region, the density is only slightly smaller than the crystalline solid density, but significantly larger than density at the gas–liquid critical point: $\rho_L \sim 0.9\rho_s$ and $\rho_L \sim 3\rho_c$ is typical for liquids in the dense region. The triple point and the critical point are characteristic liquid-state properties. As shown in Figure 2.1, these points anchor the ends of the gas–liquid coexistence curve. A liquid at the triple point is the quintessential dense

---

*The critical temperature T_c rises rapidly with chain length. In the linear alkane series, for example, T_c is already 775 K for dodecane ($C_{20}H_{42}$) and certainly well beyond the thermal decomposition range even for a C_{30} paraffin wax[3,4].

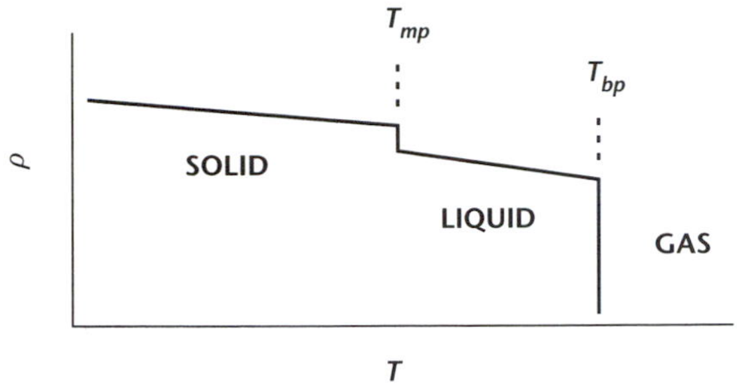

FIGURE 2.2 Density–temperature isobar near atmospheric pressure.

liquid. Critical pressure for molecular liquids is 1–10 MPa, so any liquid near or below atmospheric pressure ($\sim$0.1 MPa), monomeric or polymeric, is of necessity in the dense region. Excluded volume interactions play a major role in determining dense liquid *physical structure*—the molecular arrangements and distributions—and through the physical structure, their equilibrium properties.

In contrast to critical fluctuations, which can extend over large distances, excluded volume is a spatially localized interaction and susceptible to study by molecular simulation methods. This has permitted major advances to be made during the past forty years in understanding the behavior of dense liquids. Henceforth, unless otherwise stated, reference to liquids will be understood to indicate dense liquids.

The composition and some physical properties for a selection of monomeric species[3] are listed in Table 2.1. Their molecular sizes, shapes, and polarities are representative of the "typical solvent" category. All are dense liquids at atmospheric pressure, that is, over the range between normal melting and boiling points, $T_{mp} < T < T_{bp}$, for reasons already discussed. The relationship between density and temperature along the atmospheric pressure isobar, $P_a = 0.102\,MPa$, indicated by the dashed line in Figure 2.1, is shown in Figure 2.2. The variations in density with temperature and pressure near this isobar are governed by the thermal expansion coefficient, $\alpha = -(\partial \ln \rho/\partial T)_P$, and isothermal compressibility, $\beta = (\partial \ln \rho/\partial P)_T$. Values of α and β for some representative monomeric liquids[3,6] are listed in Table 2.2. They cluster in relatively narrow ranges, $\alpha \sim 10^{-3}K^{-1}$ and $\beta \sim 10^{-3}MPa^{-1}$, and vary only slowly with temperature and pressure.

The microscopic understanding of liquid-state properties, although vastly improved in the past forty years, remains more primitive than for the gaseous and crystalline states. A special difficulty for liquids is the absence of a thoroughly understood point of reference. No liquid-state equivalent of the perfect single crystal or the dilute gas limit exists. Thus, intermolecular interactions play a relatively minor role in the properties of dilute gases because the molecules are well separated on average. The

TABLE 2.1 Structure and properties at 20°C for selected molecular species[3]

Species	Acronym	Structure	Molecular weight	T_{mp}(°C)	T_{bp}(°C)	$\rho_L(g\,cm^{-3})$
n-hexane	HEX	CH_3-$(CH_2)_4$-CH_3	86.18	−95.	69.	0.660
n-octane	OCT	CH_3-$(CH_2)_6$-CH_3	114.23	−57.	125.7	0.702
n-decane	DEC	CH_3-$(CH_2)_8$-CH_3	142.29	−30.	174.	0.730
n-tetradecane	TET	CH_3-$(CH_2)_{12}$-CH_3	198.39	5.8	253.5	0.763
cyclohexane	CHN		84.16	6.6	80.7	0.778
benzene	BEN		78.12	5.5	80.1	0.879
toluene	TOL	 - CH_3	92.15	−95.	110.6	0.867
ethylbenzene	EB	 - CH_2 - CH_3	106.17	−95.	136.2	0.867
acetone	ACE	CH_3-C(=O)-CH_3	58.08	−95.4	56.2	0.790
2-butanone (methyl ethyl ketone)	MEK	CH_3-C(=O)-CH_2-CH_3	72.12	−86.	79.6	0.805
cyclohexanone	CHO	 = O	98.15	−45	155.7	0.948

TABLE 2.1 *Continued*

Species	Acronym	Structure	Molecular weight	$T_{mp}(^\circ C)$	$T_{bp}(^\circ C)$	$\rho_L (g\,cm^{-3})$
water	H_2O	H_2O	18.02	0.0	100.0	0.998
methanol	MeOH	CH_3OH	32.04	−97.5	64.5	0.791
ethanol	EtOH	$CH_3\text{-}CH_2\ OH$	46.07	−114.5	78.3	0.789
n-hexanol	HeOH	$CH_3\text{-}(CH_2)_5\text{-}OH$	102.18	−51.6	157.5	0.819
n-decanol	DoOH	$CH_3\text{-}(CH_2)_9\text{-}OH$	186.34	24	259	0.832
tetrahydrofuran	THF		72.12	−65.	65.	0.889
1,2,4 trichlorobenzene	TCB		181.45	17.	213.5.	1.459
chloroform	CHL	$HCCl_3$	119.38	−63.5	61.7	1.48
carbon tetrachloride	CCl_4	CCl_4	153.83	−23.	76.5	1.59
carbon disulfide	CS_2	CS_2	76.14	−111.	46.2	1.26
ethyl acetate	EtAc	$CH_3C(=O)\text{-}OCH_2\,CH_3$	88.12	−83.6	77.1	0.900
acetic acid	HAC	$CH_3\text{-}C(=O)\text{-}OH$	60.05	16.6	118.	1.049
methylene chloride	MEC	$ClCH_2Cl$	84.93	−95	40	1.327

TABLE 2.2 Thermal expansion coefficient and isothermal compressibility at 20°C for selected molecular liquids[3,6]

Liquid	ρ (g cm^{-3})	$\alpha(K^{-1}) \times 10^4$	β (MPa^{-1}) $\times 10^4$
water	0.9982	2.06	4.59
n-tetradecane	0.7628	8.7	9.1
n-decane	0.7299	10.2	10.9
ethylbenzene	0.8671	10.2	8.6
n-hexanol	0.8159	10.3	8.24
toluene	0.8669	10.5	8.96
acetic acid	1.0452	10.8	9.08
carbon disulfide	1.2632	11.2	9.38
benzene	0.8790	11.4	9.7
carbon tetrachloride	1.5940	11.4	10.5
cyclohexane	0.7784	11.5	11.3
chloroform	1.4890	12.1	9.96
2-butanone	0.8049	12.9	11.2
ethyl acetate	0.9006	13.5	11.3
ethanol	0.7894	14.0	11.2
n-hexane	0.6595	14.1	16.7
acetone	0.7906	14.6	12.6
methanol	0.7914	14.9	12.1
n-pentane	0.6262	16.4	21.8

forces that molecules exert upon one another in close approach are large, but their spatial range is limited, and the average effect can be made as small as desired by dilution. Accordingly, at the limit of infinite dilution any collection of molecules, no matter how complex their chemical structures, obeys the ideal gas law. The theory of gases begins at the low-density limit, with an ideal gas of featureless particles, then moves systematically to increasing density by considering the effect of molecular forces, first as pairwise interactions and then including higher order interactions as well[7,8].

Intermolecular forces play a dominant role at the much higher densities of the solid state. Each molecule in a crystalline solid interacts strongly with many others. However, the periodicity of molecular positions and small amplitude of thermal oscillations permits a fruitful zero-order approximation: Each molecule in a perfect crystal has the same environment and experiences the same field of force. Solid state theory begins with the perfectly ordered single crystal, then moves toward imperfection by adding, in a systematic way, such effects as lattice distortions, lattice site vacancies, thermally activated displacements and polycrystallinity[9].

Liquids occupy an awkward middle ground. The density is comparable to the solid-state density, and each molecule in the liquid feels strong forces from many neighbors. However, the simplification for crystals that comes from long-range order and the permanence of relative position is not appropriate in the liquid state. The molecular environment changes rapidly with time and position in liquids, and even nearest-neighbor separations are variable. Gases acquire some liquidlike character with increasing density, and solids do likewise with increasing disorder. Early theories of the liquid state were developed from both *lattice models* and *cell models*[10], which sought to capture the essentials of liquid behavior through various approximations about the physical structure of liquids. Refined versions are useful and still widely applied. The more recent *perturbation theories*[11], originating from the remarkable success of the hard-sphere models for atomic liquids[12], incorporate the effects of both molecular shape and a more realistic representation of the repulsive forces.

These various approaches to liquid theory are introduced in this chapter. The interest of this book lies mainly in the use of these approaches as a framework for understanding the properties of molecular liquids and mixtures.

2.1 Microscopic Origin of Liquid Properties

Several topics provide the basis for investigating and describing dense liquids microscopically. The fundamental ingredients, intermolecular interaction, statistical thermodynamics, and elastic scattering behavior are introduced in this section.

2.1.1 Intermolecular Forces

The forces that act between molecules, like the forces governing covalent bond lengths and angles, are quantum mechanical in origin[13,14]. The force varies strongly with separation distance r, and it depends on the constitution, conformations, and relative orientation of the interacting molecules. At large separation distances, and averaged over conformation and orientation, intermolecular forces act along the line of molecular centers and are always mutually attractive. The long-range interaction can accordingly be expressed as a *central force*, $\boldsymbol{f}(\boldsymbol{r}) = f(r)\boldsymbol{u}$, in which $\boldsymbol{u}$ is a unit vector in the direction of the separation vector $\boldsymbol{r}$. It can be expressed as the gradient of a potential energy, the *potential of mean intermolecular force* $\psi(r)$:

$$f(r) = -\frac{d\psi}{dr} \tag{2.1}$$

The minus sign is defined by the convention that attractive forces are negative.

In nonpolar molecules, the long distance attraction originates from the zero-point energy of the atoms, or, expressed classically, from a mutually induced electrostatic interaction of polarizable bodies. Called *London forces* or *dispersive interactions*[15,16], the potential is attractive with the universal form $-\psi(r) \propto 1/r^6$. Dispersive interactions also contribute significantly to the long-range attraction between molecules that contain permanent dipoles. The permanent charge displacements in *polar* molecules add an amount to the potential that depends on molecular orientation as well as distance. When averaged over all orientations, the polar contribution depends on temperature, but turns out to resemble dispersive interactions in both form and behavior—the potential at long range is attractive with an r^{-6} dependence. Except for highly polar species like water, the averaged polarity contribution is small relative to the dispersive contribution. The combined r^{-6} contribution at long range is the *van der Waals attraction:*

$$\psi(r) = -\frac{w}{r^6} \tag{2.2}$$

where w is some positive, species-specific coefficient.

The intermolecular potential becomes more complicated with decreasing intermolecular distance. The electron clouds of the molecules eventually begin to overlap; the mean force passes through zero and turns strongly repulsive. The potential energy, which first decreases from $\psi(\infty) = 0$ as separation distance decreases, then passes through a minimum near the overlap and finally rises very steeply with further decreases in distance. Even for monatomic substances like the inert gases, the rigorous theory of $\psi(r)$ becomes unmanageably complicated in the region near overlap. The geometric constraints on bond angle and length in multiatom species are enforced by such large energy penalties that molecular shape and local packing begin to play major roles. Polar molecules present special difficulty at short range, because orientational averaging becomes an increasingly unrealistic approximation. The only universal feature at short range is a steeply rising repulsive force. The repulsion changes much more rapidly with distance than the longer-range attraction and eventually dominates the interaction at short range.

Further progress requires that drastic approximations be made in expressing the interaction potential. A common practice is to represent each molecule as a single interaction site, with $\psi(r)$ written as the sum of a van der Waals attraction and a much steeper short-range repulsion. The *Lennard–Jones* 6-12 *potential* is a representative and computationally convenient example:

$$\psi(r) = 4\varepsilon\left[\left(\frac{\sigma}{r}\right)^{12} - \left(\frac{\sigma}{r}\right)^{6}\right] \tag{2.3}$$

That form, illustrated in Figure 2.3a, has a potential well of depth ε and a molecular separation σ, where the potential crosses through zero. Simpler representations of the intermolecular potential are sometimes used. *The hard sphere potential* is one example:

$$\begin{aligned} \psi(r) &= \infty \qquad r < \sigma \\ \psi(r) &= 0 \qquad r > \sigma \end{aligned} \tag{2.4}$$

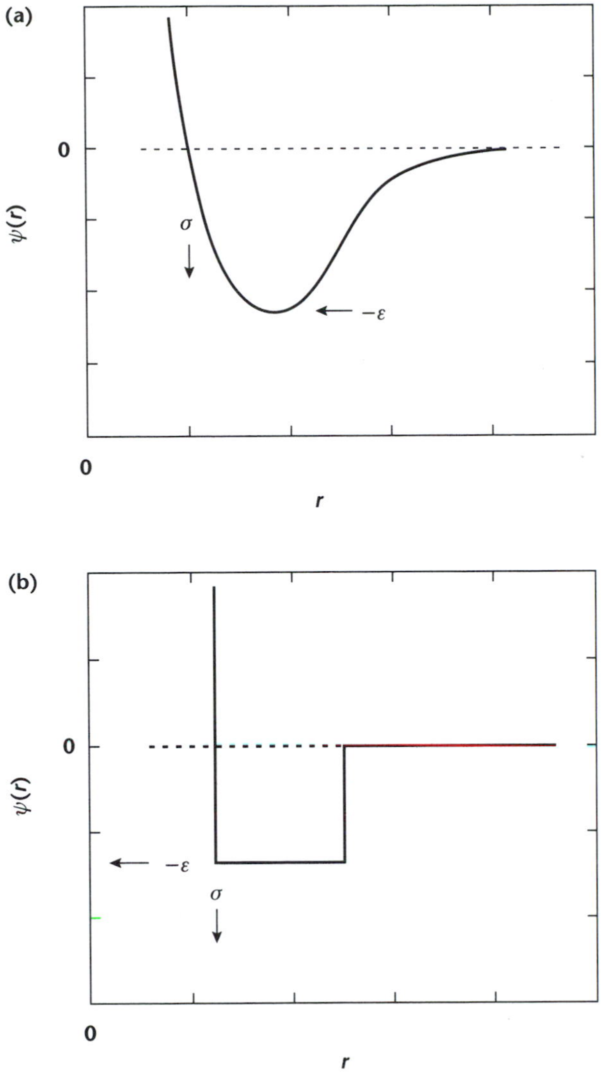

FIGURE 2.3 Typical forms for the single-site intermolecular potential. (a) The Lennard-Jones potential. (b) The square well potential.

The *hard sphere with square well attraction* is another:

$$\begin{aligned} \psi(r) &= \infty \qquad r < \sigma \\ \psi(r) &= -\varepsilon \qquad \sigma < r < \sigma \\ \psi(r) &= 0 \qquad r > \sigma \end{aligned} \tag{2.5}$$

The square well potential is illustrated in Figure 2.3b. Both of these simplified potentials are used in this volume.

TABLE 2.3 Lennard-Jones parameters for selected covalent molecules[17]

Substance	Formula	$\sigma(nm)$	$\varepsilon/k(K)$
chloroform	$CHCl_3$	0.54	340
carbon tetrachloride	CCl_4	0.595	323
water	H_2O	0.265	809
methanol	CH_4O	0.365	482
ethanol	C_2H_6 O	0.455	363
carbon disulfide	CS_2	0.45	467
n-hexane	C_6H_4	0.595	399
benzene	C_6H_6	0.535	412
cyclohexane	C_6H_{12}	0.62	297
acetone	C_3H_6O	0.46	560
ethyl acetate	$C_4H_8O_2$	0.52	521

Lennard-Jones parameters for several species[17] are listed in Table 2.3. These parameters were determined by fitting experimental data for dilute gas viscosities to the molecular theory prediction based on Eq. 2.3. Values have also been derived by similarly fitting data for gas-phase second virial coefficients. For a variety of reasons, the parameters from the two types of data do not always agree[13,18]. Other expressions for $\psi(r)$, some with additional adjustable parameters, are available. These have been used mainly to quantify the interactions in dilute gases, for which pairwise interactions dominate, the statistical mechanical theories are well established, and very accurate data of the necessary kinds are available. Even in those cases, however, inconsistencies persist between the model parameters obtained from fitting different properties. The point is that the interaction potential forms are themselves only approximations of the real interactions. One hopes to get from them, in some practical sense, relationships between properties based on some reasonable choice of $\psi(r)$ as intermediary.

Approaches of this sort are usefully employed in many liquid-state situations. The simpler forms of $\psi(r)$ suffice remarkably well for dense liquids, in which capturing essential features and establishing relationships among observables are common aims. A multiple interaction–site model is sometimes employed for molecules of many atoms, such as polymers. Small groups of atoms are the interaction sites in the united atom model[19], a Lennard–Jones based representation that is widely applied in numerical simulations[20].

2.1.2 Terminology

At various places throughout the book, beginning with Eq. 2.17, a shorthand notation is used in the discussions of such topics as intermolecular energy, scattering behavior, and the random coil model. Particles may represent molecules, parts of molecules, or scattering sites, depending on the context. These particles are numbered sequentially, $i = 1, 2, \ldots, n$, and vectors specify their individual positions in space. Vectors are represented in equations by bold symbols, and a convention is used to represent integrations involving the particles and their position vectors. Thus, a particle i at position $\boldsymbol{r}_i$ in the volume element $d\boldsymbol{r}_i$ means that the particle is at a point with Cartesian coordinates x_i, y_i, z_i that lie within the differential volume element $dx_i dy_i dz_i$. The symbol $\int_V$ stands for a triple integration $\int\int\int$, one for each of the three coordinates and extending over all possible x_i, y_i, z_i positions in the volume V. If no V is specified, the symbol means integration over all space. Spherical coordinates are also used to locate the particles, in which case particle i is located by a distance and two angles, r_i, θ_i, ϕ_i. The volume element in spherical coordinates is $r_i^2 \sin\theta_i d\phi_i d\theta_i dr_i$. The Cartesian and spherical coordinate systems are shown in Figures 2.4a and 2.4b.

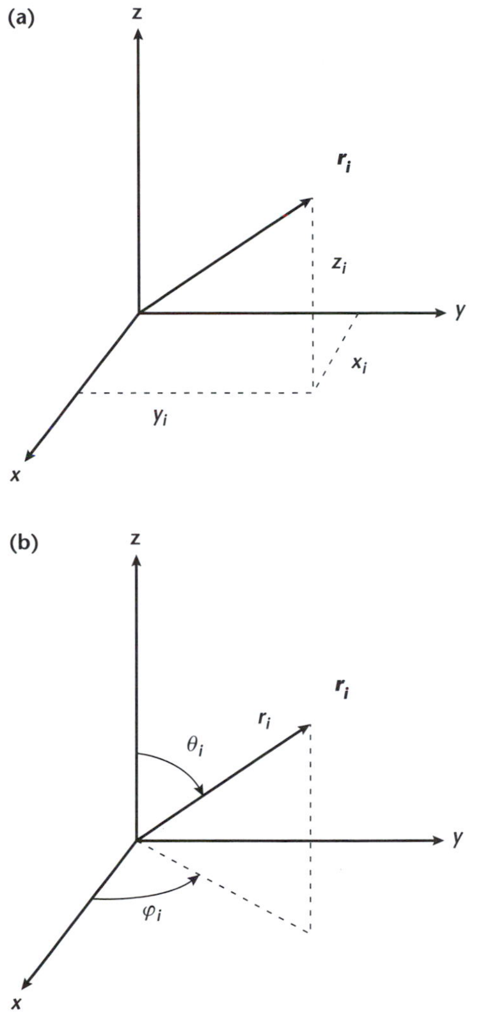

FIGURE 2.4 Coordinate systems. (a) The Cartesian coordinate system. (b) The spherical coordinate system.

The coordinate ranges for the two systems are:

$$\text{Cartesian} \quad -\infty < x_i < \infty, -\infty < y_i < \infty, -\infty < z_i < \infty \tag{2.6}$$

$$\text{Spherical} \quad 0 \le \varphi_i \le 2\pi, 0 \le \theta_i \le \pi, 0 \le r_i < \infty \tag{2.7}$$

As an example, the integral of a function of $\boldsymbol{r}$ for a one-particle system, expressed in the Cartesian and spherical coordinate systems, is:

$$\int f(\boldsymbol{r})d\boldsymbol{r} = \int_{-\infty}^{\infty}\int_{-\infty}^{\infty}\int_{-\infty}^{\infty} f(x, y, z)dxdydz = \int_{0}^{\infty}\int_{0}^{\pi}\int_{0}^{2\pi} f(r, \theta, \varphi)r^2 \sin\theta d\varphi d\theta dr \tag{2.8}$$

A function of the positions for a four-particle system is written simply as:

$$\int\int\int\int f(\boldsymbol{r}_1, \boldsymbol{r}_2, \boldsymbol{r}_3, \boldsymbol{r}_4)d\boldsymbol{r}_1 d\boldsymbol{r}_2 d\boldsymbol{r}_3 d\boldsymbol{r}_4 \tag{2.9}$$

or, for n particles,

$$\int \cdots \int f(\boldsymbol{r}_1, \ldots \boldsymbol{r}_n)d\boldsymbol{r}_1 \cdots d\boldsymbol{r}_n \tag{2.10}$$

Because many of the systems considered contain enormous numbers of particles, a compressed terminology is essential.

2.1.3 Statistical Thermodynamics

The physical properties of liquids depend strongly on intermolecular interactions, and statistical thermodynamics provides the formal connections. From classical thermodynamics[21], the pressure, entropy, and internal energy of a macroscopic system of n identical molecules in volume V at temperature T can be expressed as derivatives of the Helmholtz free energy, $A(n, V, T)$:

$$P = -\left(\frac{\partial A}{\partial V}\right)_{n,T} \tag{2.11}$$

$$S = -\left(\frac{\partial A}{\partial T}\right)_{n,V} \tag{2.12}$$

$$U = \left(\frac{\partial (A/T)}{\partial (1/T)}\right)_{n,V} \tag{2.13}$$

From statistical thermodynamics[7], the Helmholtz free energy is related to $Q(n, V, T)$, the *canonical partition function* of the system:

$$A = -kT \ln Q(n, V, T) \tag{2.14}$$

in which k is the Boltzmann constant. The partition function can be expressed as a product of intramolecular and intermolecular terms:

$$Q = Q_{\text{intra}}\, Q_{\text{inter}} \tag{2.15}$$

The first term in Eq. 2.15 is governed by interactions between the various parts of the same molecule, the second by the interactions between the various parts of different molecules. A good approximation for covalent molecules is that the interactions governing Q_{intra} are strong enough to be insensitive to the nonconnected surroundings, and hence Q_{intra} is independent of molecular concentration n/V. It is also a good approximation that the intermolecular interactions are weak enough to permit an analysis of their effects by classical mechanics. Accordingly, Q_{inter} can be expressed as:

$$Q_{\text{inter}} = \frac{Z(n,V,T)}{n!\; \Lambda^n} \tag{2.16}$$

The quantity Λ is a volume $(h^2/2\pi mkT)^{3/2}$, in which h is Planck's constant and m is the molecular mass. The quantity Z is the *classical configuration integral* for the system. Thus, using the conventions in Section 2.1.2:

$$Z(n,V,T) = \int_V \cdots \int_V \exp\left[-\frac{\Psi(\boldsymbol{r}_1,\ldots,\boldsymbol{r}_n)}{kT}\right] d\boldsymbol{r}_1 \cdots d\boldsymbol{r}_n \tag{2.17}$$

The total interaction energy of the n molecules—each treated as a single interaction site, labeled $i = 1, 2 \ldots, n$, and located at positions $\boldsymbol{r}_1, \ldots \boldsymbol{r}_n$ in volume elements $d\boldsymbol{r}_1, \ldots d\boldsymbol{r}_n$—is $\Psi(\boldsymbol{r}_1, \ldots, \boldsymbol{r}_n)$. The integrations provide the sum over all distinguishable configurations of the system, such that Q_{inter} is the sum of all distinguishable sets of positions for n indistinguishable molecules in the system volume V, each set weighted by its *Boltzmann factor*, $\exp(-\Psi/kT)$.

With Q_{intra} included, Eq. 2.14 through Eq. 2.17 lead to an expression for Helmholtz free energy:

$$A(n,V,T) = -kT\left[\ln\frac{Z(n,V,T)}{[\Lambda(n,T)]^n\, n!} + \ln[Q(n,T)]_{\text{intra}}\right] \tag{2.18}$$

Accordingly, from Eq. 2.11 through Eq. 2.13,

$$P(n,V,T) = kT\left(\frac{\partial \ln Z}{\partial V}\right)_{n,T} \tag{2.19}$$

$$S(n,V_2,T) - S(n,V_1,T) = k\frac{\partial}{\partial T}\left[T \ln \frac{Z(n,V_2,T)}{Z(n,V_1,T)}\right] \tag{2.20}$$

$$U(n,V_2,T) - U(n,V_1,T) = \langle \Psi(n,V_2,T)\rangle - \langle \Psi(n,V_1,T)\rangle \tag{2.21}$$

The pointed brackets indicate an average over all distinguishable configurations of the system. The average intermolecular potential energy is:

$$\langle \Psi \rangle = \frac{\int \cdots \int \Psi \exp(-\Psi/kT)\, d\mathbf{r}_1 \cdots d\mathbf{r}_n}{\int \cdots \int \exp(-\Psi/kT)\, d\mathbf{r}_1 \cdots d\mathbf{r}_n} \tag{2.22}$$

Thus, if Q_{intra} is independent of molecular concentration, the pressure for any set of n molecules depends only on $Z(n,V,T)$, through Eq. 2.19, and thus, on average intermolecular potential energy alone. For the same conditions, $Z(n,V,T)$ alone, and hence only the intermolecular potential energy, governs isothermal differences in both entropy and internal energy [Eqs. 2.20 and 2.21].

Finally, when all configurations of the system have the same energy Ψ_o, the configuration integral is simply $\Omega \exp(-\Psi_o/kT)$, in which Ω is the number of distinguishable configurations. Equation 2.20 for the entropy difference between two states of the system at the same temperature then reduces to the *Boltzmann entropy equation:*

$$S_2 - S_1 = k \ln \frac{\Omega_2}{\Omega_1} \tag{2.23}$$

This equation has many polymeric applications.

The excluded volume interaction is important for dense liquids. Each molecule occupies some volume, its *space-filling* volume, from which it excludes all others. More precisely, the intermolecular potentials grow rapidly to become strongly repulsive as the electron clouds begin to interpenetrate. Even slightly closer approaches have much higher energies relative to kT, so they are effectively prohibited. The real situation, a large but finite steepness in the repulsive part of the potential, is frequently approximated as an infinitely steep repulsion or *hard core,* thus turning a temperature-dependent energetic constraint into a temperature-independent geometric constraint.

The configuration integral for a system of fully excluding molecules, or hard core objects, can be expressed by assigning the energy $\Psi = 0$ to each of the configurations without overlap and $\Psi = \infty$ to each with one or more overlaps. Accordingly, the only available configurations are those without overlap, and all have zero energy. Equation 2.17 then becomes:

$$Z(n,V,T) = [Vf(n,V)]^n \tag{2.24}$$

in which $[f(n,V)]^n$ is the fraction of all distinguishable configurations that satisfy the no-overlap constraint, and $f(n,V) = 1$ for a system of noninteracting molecules. From Eq. 2.19 through Eq. 2.21,

$$P(n,V,T) = \frac{nkT}{V}\left[1 + \left(\frac{\partial \ln f}{\partial \ln V}\right)_n\right] \tag{2.25}$$

$$S(n,V_2,T) - S(n,V_1,T) = nk\left[\ln \frac{V_2}{V_1} + \ln \frac{f(n,V_2)}{f(n,V_1)}\right] \tag{2.26}$$

$$U(n,\ V_2,\ T) = U(n,\ V_1,\ T) \tag{2.27}$$

For a fixed number of molecules, the fraction of nonoverlapping configurations must certainly increase with increasing system volume: $(\partial \ln f/\partial V)_n > 0$. Thus, from Eq. 2.25 the mutual exclusion of volume by the molecules always increases the pressure relative to the pressure for the noninteracting system. Likewise, from Eq. 2.26, volume exclusion always reduces the entropy of expansion relative to that of the noninteracting system. Volume exclusion alone thus generates a repulsive, purely entropic interaction.

2.1.4 Pair Distributions and Intermolecular Energy

Two simplifications are used to represent the structure and calculate the intermolecular energy content of dense liquids from the molecular interactions. These simplifications are the *mean-field approximation* and *pairwise additivity*[7]. The distribution of all other molecules around an individual molecule is replaced by the average distribution. The energy is computed from that distribution by summing over the pair potentials as if they acted independently.

Consider a single-phase macroscopic system of volume V containing some large number n of identical molecules at equilibrium. The *average number density* of molecules is:

$$\nu = \frac{n}{V} \tag{2.28}$$

For gases and liquids, ν is the same at all points in the system. For purposes here, however, it is necessary that the average number density be expressed as a function of a distance that is measured from an origin on the center of a molecule selected at random. The location of molecule $i\,(1 \leq i \leq n)$ is specified by a position vector $\boldsymbol{r}_i$. The locations of the other molecules relative to molecule i—the environment of molecule i—is specified by the set of $n-1$ relative position vectors $\boldsymbol{r}_{ij} = \boldsymbol{r}_j - \boldsymbol{r}_i$. Both $\boldsymbol{r}_i$ and $\boldsymbol{r}_{ij}$ vary with time—all molecules move under the influence of the intermolecular forces and with the thermal distribution of velocities—so the environment of each molecule continually changes. The average, however, can be obtained as follows. At any instant, the $n-1$ neighbors of the selected molecule i are distributed through space in some manner. Group them according to $\boldsymbol{r} = \boldsymbol{r}_{ij}\,(j \neq 1)$, such that $[\nu_2(\boldsymbol{r})]_i$ is the number at position $\boldsymbol{r}$ in $d\boldsymbol{r}$. Now let each molecule in turn be the selected molecule and then calculate the average $\nu_2(\boldsymbol{r}) \equiv \sum_{i=1}^{n}[\nu_2(\boldsymbol{r})]_i/n$.

The average obtained is the *radial distribution of neighbors* $\nu_2(\boldsymbol{r})$, the average number of molecules that have centers located at position $\boldsymbol{r}$ in $d\boldsymbol{r}$ relative to some molecule at the origin. Each molecule has $n-1$ others that define its environment. Thus,

$$\nu_2(\boldsymbol{r}) = (n-1)p(\boldsymbol{r}) \tag{2.29}$$

where $p(\boldsymbol{r})\,d\boldsymbol{r}$ is the probability of finding a neighbor at $\boldsymbol{r}$ in the volume element $d\boldsymbol{r}$.

The *pair distribution function* $g(\boldsymbol{r})$ is a dimensionless quantity, defined in such a way as to display explicitly the structural aspects of the system:

$$g(\boldsymbol{r}) \equiv \frac{(n-1)p(\boldsymbol{r})}{\nu} \tag{2.30}$$

The *pair correlation function,*

$$h(\boldsymbol{r}) \equiv g(\boldsymbol{r}) - 1 \tag{2.31}$$

is sometimes used as well. If the system is without structure, the molecular locations are independent of one another. In this case, $g(\boldsymbol{r}) = 1$ and $h(\boldsymbol{r}) = 0$ for all $\boldsymbol{r}$. Likewise, if there is no correlation, the relative molecular locations have a *random distribution:*

$$p(\boldsymbol{r})\,d\boldsymbol{r} = \frac{d\boldsymbol{r}}{V} \qquad \text{(everywhere in } V) \tag{2.32}$$

Gases and liquids are isotropic, so the probability density $p(\boldsymbol{r})$ can be written $p(r)$ to indicate that it depends on distance but not on direction. The pair distribution for a dense liquid of hard spheres, obtained analytically with the Percus–Yevick approximation[22–24], is shown in Figure 2.5. The general features of $g(r)$ seen there typify the *liquidlike order* that characterizes all dense molecular liquids. Some of its properties are obvious: Short-range repulsions forbid significant molecular overlap, so $g(r)$ must go rapidly to zero at distances smaller than the molecular diameter d_m. At large enough distances, $r \gg d_m$, and in the absence of long-range order (crystallinity),

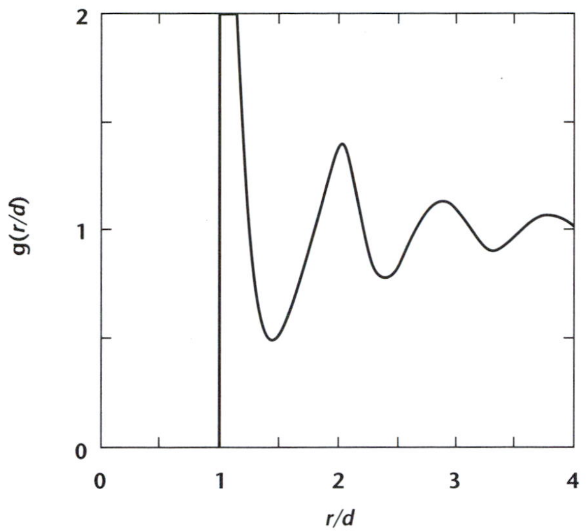

FIGURE 2.5 Pair distribution function for hard spheres, calculated for a packing fraction of 0.5 with the Percus–Yevick approximation[24,33].

$g(r)$ must approach unity: All correlation in the relative position of molecules must vanish in the large separation limit. The first peak corresponds to nearest neighbor positions. The abrupt cutoff for $r \leq d$ in Figure 2.5 reflects the infinitely steep repulsion of hard spheres. The onset of repulsion is less abrupt for real molecules. The successive peaks indicate next-near neighbor correlations, and so on. In dense liquids, including the hard-sphere liquid, the heights of the peaks decrease and their widths increase progressively as the correlation in molecular positions diminishes and finally vanishes with increasing distance.

With the pairwise-additivity approximation, the total intermolecular energy is simply the average potential energy per molecule multiplied by their number. The average intermolecular potential energy per molecule is obtained by integrating over the pair distribution:

$$\langle \psi \rangle = \frac{\nu}{2} \int_{0}^{\infty} \psi(r)\, 4\pi r^2 g(r)\, dr \tag{2.33}$$

in which the factor of ½ compensates for double counting. The total energy of the system, $\Psi = n\langle \psi \rangle$, can thus be written:

$$\Psi = \frac{V\nu^2}{2} \int_{0}^{\infty} \psi(r)\, 4\pi r^2 g(r)\, dr \tag{2.34}$$

The pair distribution function can be obtained by scattering experiments, as described in the next section.

2.1.5 Principles of Elastic Scattering

Elastic scattering is a widely used technique for obtaining structural information on materials of many kinds[25–29]. A schematic diagram for the scattering experiment is shown in Figure 2.6a. A collimated beam of elementary particles (photons, electrons, neutrons, etc.), all with the same energy, arrives as a plane wave of intensity I_o, propagating in a direction specified by the unit vector $\boldsymbol{u}_o$. The incident beam interacts with a scattering site located at some position $\boldsymbol{r}$ in an *illuminated volume* that contains many scatterers. The interaction, quantum-mechanical in origin, produces a spherical scattered wave propagating outwardly from $\boldsymbol{r}$. Similar interactions occur with many other scattering sites in the illuminated volume. The aperture of a detector at the position $\boldsymbol{r} + \boldsymbol{r}_D$ intercepts some fraction of the outgoing wave from these many

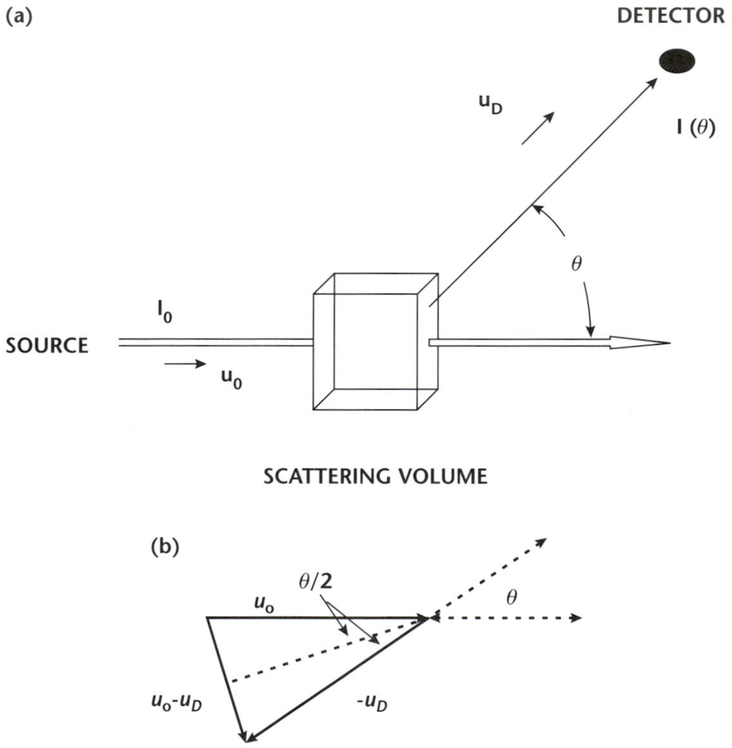

FIGURE 2.6 (a) Relationship between incident and scattered waves for an elastic interaction of elementary particles. (b) Definition of scattering vector q in terms of incoming and outgoing wave vectors.

scattering events. The unit vector $\boldsymbol{u}_D$ denotes the direction from the scattering site to the detector. The detector subtends some small solid angle $\Delta\Omega$ of the scattered wave and records $\Delta\Sigma$, the incident energy per unit time. The detector is mounted on a goniometer, or multiple detectors are mounted at various fixed angles. In either case, the object of the experiment is to determine scattering intensity at the detector I_D, the power (energy/time) delivered per unit solid angle $\Delta\Sigma/\Delta\Omega$(or $d\Sigma/d\Omega$, the differential scattering cross-section energy/time/solid angle in the small $\Delta\Omega$ limit) as a function of the scattering angle θ.

The incident and scattered waves, which are sinusoidal functions of time and position, are characterized by a phase angle and amplitude. In *elastic scattering*, the time dependence is the same for both waves. The scattering is *coherent* if there is a causal relationship between the phases of the incident and scattered waves. Only the coherent part of the scattering is of interest here. Other kinds of interaction can occur in scattering experiments (absorption, incoherent scattering, etc.), and corrections to the readings must be made accordingly. Finally, only the collective outcome of single scattering events is of interest. Multiple scattering, scattering of the scattered wave and the like, can be made negligible by operating with small enough path lengths through the scattering volume.

Complex algebra[30] is the natural language for discussing scattering phenomena. Thus, properties such as the additivity of waves from different scattering sites and the relationship between wave amplitude and intensity follow directly from the algebra of complex numbers. Thus, a wave can be written as a complex number:

$$\begin{aligned} u &= x + iy = z\,\exp(i\theta) \\ \exp(i\theta) &= \cos\theta + i\,\sin\theta \end{aligned} \tag{2.35}$$

in which $z = (x^2 + y^2)^{1/2}$ is the wave amplitude, θ is the phase angle, and $i = (-1)^{1/2}$. The conjugate of a complex number is obtained simply by replacing i by $-i$ everywhere it appears. Thus,

$$\begin{aligned} u^* &= x - iy \\ u^* &= z\,\exp(-i\theta) \end{aligned} \tag{2.36}$$

The intensity of a wave is the square of its amplitude and can be written as the product of the wave and its conjugate:

$$I = z^2 = u^*u = uu^* \tag{2.37}$$

Waves add according to:

$$\sum_j u_j = \sum_j x_j + i\sum_j y_j = \sum_j z_j\,\exp(i\theta_j) \tag{2.38}$$

The following relationships are useful:

$$\cos\theta = \frac{\exp(i\theta) + \exp(-i\theta)}{2}$$
$$\sin\theta = \frac{\exp(i\theta) - \exp(-i\theta)}{2i} \tag{2.39}$$

In complex notation, the incident or incoming wave for a scattering site at $\boldsymbol{r}$ is expressed as:

$$A_o(t, \boldsymbol{r}) = A_{in} \exp[-i(\omega t + \boldsymbol{k}_o \cdot \boldsymbol{r})] \tag{2.40}$$

in which A_{in} is the incident wave *amplitude*, $\omega t + \boldsymbol{k}_o \cdot \boldsymbol{r}$ is the *phase angle*, ω is the wave *frequency*, $\boldsymbol{k}_o = 2\pi \boldsymbol{u}_o/\lambda$ is the *wave vector*, and λ is the *wavelength*. When the object of interest is the intensity, the choice of a sign for the exponent is arbitrary. The coherent part of the scattered wave at the detector is:

$$A_D(t, \boldsymbol{r} + \boldsymbol{r}_D) = b A_o(t, \boldsymbol{r}) \frac{\exp(-i\boldsymbol{r}_D \cdot \boldsymbol{k}_D)}{r_D} \tag{2.41}$$

in which $\boldsymbol{k}_D = 2\pi \boldsymbol{u}_D/\lambda$ is the scattered or outgoing wave vector, $r_D = |\boldsymbol{r}_D|$ is the scatterer-to-detector distance, and b is the *scattering length*, a quantity that depends on the details of the interaction and must be supplied separately. Equations 2.40 and 2.41 thus give:

$$A_D(t, \boldsymbol{r} + \boldsymbol{r}_D) = b A_{in} \exp(-i\omega t) \exp[-i(\boldsymbol{k}_o - \boldsymbol{k}_D) \cdot \boldsymbol{r}] \frac{\exp(-i\boldsymbol{k}_D \cdot \boldsymbol{r}_D)}{r_D} \tag{2.42}$$

The incoming–outgoing difference, $\boldsymbol{k}_o - \boldsymbol{k}_D = 2\pi(\boldsymbol{u}_o - \boldsymbol{u}_D)/\lambda$, is the *scattering vector* $\boldsymbol{q}$. As indicated in Figure 2.6b, the magnitude of the scattering vector, $q = |\boldsymbol{q}| = 2\pi |\boldsymbol{u}_o - \boldsymbol{u}_D|/\lambda$, depends on both wavelength and scattering angle:

$$q = \frac{4\pi}{\lambda} \sin\frac{\theta}{2} \tag{2.43}$$

Accordingly, Eq. 2.42 can be expressed as:

$$A_D(t, \boldsymbol{r} + \boldsymbol{r}_D) = b A_{in} \frac{\exp(-i\omega t) \exp(-i\boldsymbol{q} \cdot \boldsymbol{r}) \exp(-i\boldsymbol{k}_D \cdot \boldsymbol{r}_D)}{r_D} \tag{2.44}$$

Thus, for scattering by a single particle, the incident intensity and the scattering intensity at the detector are:

$$I_o = A_o A_o^* = A_{in}^2$$
$$I_D = A_D A_D^* = \frac{b^2 A_{in}^2}{r_D^2} = \frac{b^2}{r_D^2} I_o \tag{2.45}$$

Note that $4\pi r_D^2 I_D$, independent of the detector distance r_D, is the power being extracted from the incident beam by scattering.

Consider now the scattering by two particles of identical properties, located at $\boldsymbol{r}_1$ and $\boldsymbol{r}_2$. The scattered waves are added, so from Eq. 2.44, and for large sample-to-detector distances ($|\boldsymbol{r}_1 - \boldsymbol{r}_2| \ll r_D$),

$$A_D = b A_{in} \frac{\exp(-i\omega t)\exp(-i\boldsymbol{k}_D \cdot \boldsymbol{r}_D)}{r_D}[\exp(-i\boldsymbol{q} \cdot \boldsymbol{r}_1) + \exp(-i\boldsymbol{q} \cdot \boldsymbol{r}_2)] \tag{2.46}$$

Finally, with Eq. 2.39,

$$I_D = \frac{b^2 I_o}{r_D^2}[2 + 2\cos[\boldsymbol{q} \cdot (\boldsymbol{r}_2 - \boldsymbol{r}_1)]] \tag{2.47}$$

Thus, the scattering intensity for two-particle systems varies with scattering angle and relative scatterer locations. The scattered waves from the two particles *interfere* and, depending on the value of $\boldsymbol{q} \cdot (\boldsymbol{r}_2 - \boldsymbol{r}_1)$, the intensity can range anywhere from 0 to $4b^2 I_o / r_D^2$.

The extension of a system of n scatterers is straightforward. The scatterer locations are $\boldsymbol{r}_1, \boldsymbol{r}_2, \ldots, \boldsymbol{r}_n$, and the amplitude at the detector is the sum of contributions:

$$A_D = \frac{b A_{in}}{r_D} \exp(-i\omega t)\exp(-i\boldsymbol{k}_D \cdot \boldsymbol{r}_D) \sum_{j=1}^{n} \exp(-i\boldsymbol{q} \cdot \boldsymbol{r}_j) \tag{2.48}$$

leading finally to:

$$I_D = \frac{b^2 I_o}{r_D^2} \sum_{j=1}^{n} \sum_{k=1}^{n} \exp[i\boldsymbol{q} \cdot (\boldsymbol{r}_j - \boldsymbol{r}_k)] \tag{2.49}$$

This expression is easily generalized to systems containing particles that have different scattering lengths:

$$\frac{I_D r_D^2}{I_o} = \sum_{j=1}^{n} \sum_{k=1}^{n} b_j b_k \exp[i\boldsymbol{q} \cdot (\boldsymbol{r}_j - \boldsymbol{r}_k)] \tag{2.50}$$

The n diagonal terms ($j = k$) contribute $\sum n_m b_m^2$ to the double sum, there being m scattering lengths represented, or nb^2 for the case of identical scatterers. This *self-scattering* for the system makes a q-independent contribution to the intensity. The $n(n-1)$ off-diagonal terms ($j \neq k$) depend in general on q and contain information on the structure of the system.

The interest here is always in scattering by systems of many scatterers, occupying some uniformly illuminated scattering volume V_{sc}. The scatterers move about continually, causing the detector intensity at each scattering angle to fluctuate around some average value. The flickering of intensity is rapid for liquids, and the excursions typically small. Under favorable circumstances, however, the time dependence can be analyzed and used to obtain useful information about the dynamic properties (see Section 5.2.1). Here we are interested in equilibrium properties, for which only the average intensity—the static intensity, the integrated intensity, or simply the intensity—is required:

$$I(\boldsymbol{q}) \equiv \langle I_D(\boldsymbol{q}, t)\rangle \tag{2.51}$$

In the language of light scattering, the intensity is expressed in a reduced form, as an intensive property called the *Rayleigh ratio* (see Section 5.1.2), the definition of which takes into account the scattering volume. In x-ray and neutron scattering, the main interest of this chapter, intensity is also expressed in reduced form, but as an extensive property, which is the differential scattering cross section of the illuminated volume:

$$\frac{d\Sigma}{d\Omega}(\boldsymbol{q}) = \frac{I(\boldsymbol{q})r_D^2}{I_o} \tag{2.52}$$

The *reduced intensity,* a quantity used throughout this book, is an intensive property:

$$I_r(\boldsymbol{q}) \equiv \frac{1}{V_{sc}}\frac{d\Sigma}{d\Omega}(\boldsymbol{q}) = \frac{I(\boldsymbol{q})r_D^2}{V_{sc}I_o} \tag{2.53}$$

The differential cross-section is directly proportional to the scattering volume, and its average value, based on Eq. 2.50, is expressed as:

$$\frac{d\Sigma}{d\Omega}(\boldsymbol{q}) = \sum_m n_m b_m^2 + \sum_{j \neq k}\sum b_j b_k \langle \exp[i\boldsymbol{q}\cdot(\boldsymbol{r}_j - \boldsymbol{r}_k)]\rangle \tag{2.54}$$

in which the diagonal contribution, being independent of $\boldsymbol{q}$, has been broken out as a separate term.

Each off-diagonal term in Eq. 2.54 can be expressed as the integral over a probability distribution:

$$\langle \exp[i\boldsymbol{q}\cdot(\boldsymbol{r}_j - \boldsymbol{r}_k)]\rangle = \int_{V_{sc}} p_{jk}(\boldsymbol{r}_{jk}) \exp(i\boldsymbol{q}\cdot\boldsymbol{r}_{jk})\, d\boldsymbol{r}_{jk} \tag{2.55}$$

where $p_{jk}(\boldsymbol{r}_{jk})d\boldsymbol{r}_{jk}$ is the probability of finding scatterers j and k with separation $\boldsymbol{r}_{jk} = \boldsymbol{r}_j - \boldsymbol{r}_k$ in $d\boldsymbol{r}_{jk}$, and the vector integral terminology introduced in Section 2.1.2 has been used. Thus, Eq. 2.54 can also be written:

$$\frac{d\Sigma}{d\Omega}(\boldsymbol{q}) = \sum_m n_m b_m^2 + \sum_{j \neq k}\sum b_j b_k \int_{V_{sc}} p_{jk}(\boldsymbol{r}_{jk}) \exp(i\boldsymbol{q}\cdot\boldsymbol{r}_{jk})\, d\boldsymbol{r}_{jk} \tag{2.56}$$

For one-component systems of molecules that act as single scatterers, the probability distribution must be the same function of separation for all j,k pairs. Hence,

$$\frac{d\Sigma}{d\Omega}(\boldsymbol{q}) = b^2 n + b^2 n(n-1) \int_{V_{sc}} p(\boldsymbol{r}) \exp(i\boldsymbol{q}\cdot\boldsymbol{r})\, d\boldsymbol{r} \tag{2.57}$$

in which $p(\boldsymbol{r})d\boldsymbol{r}$ is the pair probability distribution, introduced in Eq. 2.29. With the pair distribution replaced by the pair probability through Eq. 2.30, the relationship between scattering intensity and structure becomes:

$$I_r(\boldsymbol{q}) = b^2 \nu \left(1 + \nu \int_{V_{sc}} g(\boldsymbol{r}) \exp(-i\boldsymbol{q}\cdot\boldsymbol{r})\, d\boldsymbol{r}\right) \tag{2.58}$$

in which the $n - 1 \approx n$ approximation and the definition of scattering site number density $\nu = n/V_{sc}$ were used. Accordingly, the scattering behavior for a system of single-site molecules is related to the structure of the system through its pair distribution.

Equation 2.58 applies only to systems of one-particle molecules, each with the same scattering length. However, one-component systems of molecules with two or more identical scattering sites can be treated within the same framework. Consider, for example, the scattering by a one-component liquid of n molecules, each with z scattering sites of scattering length b. Scattering intensity can be expressed as the sum of two terms:

- The intramolecular contributions—the nz contributions from site self-scattering and the $nz(z-1)$ contributions from pairs of different sites on the same molecule, and
- The intermolecular contributions—the $n(n-1)z^2$ contributions from pairs on different molecules.

Thus, with Eq. 2.54 and $b_j = b_k = b$,

$$\frac{1}{b^2}\frac{d\Sigma}{d\Omega}(\boldsymbol{q}) = n\sum_{j=1}^{z}\sum_{k=1}^{z}\langle\exp[i\boldsymbol{q}\cdot(\boldsymbol{r}_j - \boldsymbol{r}_k)]\rangle_{\text{intra}} + n(n-1)\sum_{j=1}^{z}\sum_{k=1}^{z}\langle\exp[i\boldsymbol{q}\cdot(\boldsymbol{r}_j - \boldsymbol{r}_k)]\rangle_{\text{inter}} \quad (2.59)$$

or, proceeding as before, with $n - 1 \approx n$,

$$I_r(\boldsymbol{q}) = \nu b^2[P(\boldsymbol{q}) + nQ(\boldsymbol{q})] \quad (2.60)$$

in which:

$$P(\boldsymbol{q}) = \frac{1}{z^2}\sum_{j=1}^{z}\sum_{k=1}^{z}\langle\exp[i\boldsymbol{q}\cdot(\boldsymbol{r}_j - \boldsymbol{r}_k)]\rangle_{\text{intra}} \quad (2.61)$$

$$Q(\boldsymbol{q}) = \frac{1}{z^2}\sum_{j=1}^{z}\sum_{k=1}^{z}\langle\exp[i\boldsymbol{q}\cdot(\boldsymbol{r}_j - \boldsymbol{r}_k)]\rangle_{\text{inter}} \quad (2.62)$$

The intramolecular term $P(\boldsymbol{q})$ is the *form factor,* characterizing molecular size and conformations. The intermolecular term $Q(\boldsymbol{q})$ characterizes the relative molecular positions.

The intramolecular and intermolecular contributions to scattering are discussed in some detail in Chapters 4 through 8. For now, however, we note only that $P(\boldsymbol{q})$ and $Q(\boldsymbol{q})$ cannot be determined separately in one-component systems without the assistance of simulation, molecular modeling, or isotopic labeling. Such analysis is discussed briefly in Section 2.2.1.

2.1.6 The Structure Factor

From Eq. 2.58, all information that can be derived from elastic scattering about the equilibrium structure of a one-component system of single-site molecules is contained in $S(\boldsymbol{q})$, the *static structure factor:*

$$S(\boldsymbol{q}) \equiv 1 + \nu\int_{V_{sc}} g(\boldsymbol{r})\exp(-i\boldsymbol{q}\cdot\boldsymbol{r})\,d\boldsymbol{r} \quad (2.63)$$

As discussed in Section 2.1.4, whereas $g(\boldsymbol{r})$ is used for globally isotropic materials such as gases and liquids, $g(r)$ is a function of separation distance alone. The same is true for the dependence of the structure factor on the scattering vector: $S(\boldsymbol{q}) = S(q)$.

The structure factor can also be expressed in terms of the pair correlation function. Thus from Eq. 2.31, for liquids,

$$S(\boldsymbol{q}) = 1 + \nu \int h(r)\exp(-i\boldsymbol{q}\cdot\boldsymbol{r})\,d\boldsymbol{r} + \nu \int_{V_{sc}} \exp(-i\boldsymbol{q}\cdot\boldsymbol{r})\,d\boldsymbol{r} \tag{2.64}$$

in which the integral in the second term on the right side has been extended over all space, instead of the illuminated volume alone. Since $h(r)$ for liquids becomes effectively zero beyond a few molecular diameters, that integral, the *Fourier transform*[31,32] of the pair correlation function, must converge.

The infinite Fourier transform and its inverse are defined mathematically as follows:

$$\begin{aligned} \hat{f}(\boldsymbol{s}) &= \int f(\boldsymbol{r})\exp(-i\boldsymbol{s}\cdot\boldsymbol{r})\,d\boldsymbol{r} \\ f(r) &= \frac{1}{(2\pi)^3}\int \hat{f}(\boldsymbol{s})\exp(i\boldsymbol{r}\cdot\boldsymbol{s})\,d\boldsymbol{s} \end{aligned} \tag{2.65}$$

From a mathematical standpoint, the independent variables $\boldsymbol{r}$ and $\boldsymbol{s}$ are simply vectors with no particular physical interpretation attached. The "cap" indicates the transformed function, and the vector integral convention (Section 2.1.2) has been used. These formulas apply for any function $f(\boldsymbol{r})$ that satisfies the convergence requirements for integration. When $f(\boldsymbol{r}) = f(r)$, the integrations over the angular variables can be carried out directly. Thus, in spherical coordinates, Eq. 2.7, and with the r axis direction chosen to be aligned with $\boldsymbol{s}$,

$$\hat{f}(s) = \int_0^\infty r^2 f(r)\left[\int_0^{2\pi}\int_0^{\pi} \exp(-isr\cos\theta)\sin\theta\,d\theta\,d\phi\right]dr \tag{2.66}$$

After integrating over the angles,

$$\hat{f}(s) = 4\pi\int_0^\infty r^2 f(r)\frac{\sin sr}{sr}\,dr \tag{2.67}$$

Fourier transforms have useful properties in their own right, and they will be used elsewhere in this volume. Thus, for example, the *convolution* of two functions is:

$$c(\boldsymbol{s}) = \int f_1(\boldsymbol{r})f_2(\boldsymbol{s}-\boldsymbol{r})\,d\boldsymbol{r} \tag{2.68}$$

and the transform of a convolution is simply the product of the transforms of the two functions:

$$\hat{c}(\mathbf{s}) = f_1(\mathbf{s}) f_2(\mathbf{s}) \tag{2.69}$$

The *Dirac delta* function, or *infinite spike*, is sometimes useful:

$$\begin{aligned} \delta(\boldsymbol{r} - \boldsymbol{z}) &= \infty \quad \boldsymbol{r} = \boldsymbol{z} \\ \delta(\boldsymbol{r} - \boldsymbol{z}) &= 0 \quad \boldsymbol{r} \neq \boldsymbol{z} \\ \int \delta(\boldsymbol{r} - \boldsymbol{z}) d\boldsymbol{r} &= 1 \end{aligned} \tag{2.70}$$

where $\boldsymbol{z}$ is the position of the spike. Its transform is:

$$\hat{\delta}(\mathbf{s}) = \exp(-i\mathbf{s} \cdot \mathbf{z})$$

or

$$\hat{\delta}(s) = \frac{\sin sz}{sz} \tag{2.71}$$

for the isotropic case. The transforms of some other common functions are given in Table 2.4.

By this method, integration over the illuminated volume (a sphere of radius R for convenience) for the last term on the right in Eq. 2.64 leads to:

$$\nu \int_V \exp(-i\boldsymbol{q} \cdot \boldsymbol{r}) d\boldsymbol{r} = 3n \left[\frac{\sin qR - qR \cos qR}{(qR)^3} \right] \tag{2.72}$$

The result is a contribution proportional to the number of scatterers in the illuminated volume. It is multiplied by a quantity, which for typical values of interest

TABLE 2.4 Selected functions and their Fourier transforms

Name	$f(r)$	$\hat{f}(s)$
Exponential	$\exp\left(-\frac{r}{b}\right)$	$\frac{8\pi b^3}{(1 + b^2 s^2)^2}$
Power Law	$r^{-\alpha}$	$s^{\alpha-3}$
Screened Potential	$\frac{\exp(-r/b)}{r}$	$\frac{4\pi b^2}{1 + b^2 s^2}$
Gaussian	$\left(\frac{3}{2\pi b^2}\right)^{3/2} \exp\left(-\frac{3r^2}{2b^2}\right)$	$\exp\left(-\frac{b^2 s^2}{6}\right)$

$q \approx 1\,nm^{-1}$ and $R \approx 10^7\,nm$, is very small and moreover oscillates about zero so rapidly that the average over any real detector aperture would be essentially zero. Accordingly, that term can be omitted without harm. Thus from Eq. 2.64, the structure factor for single scattering-site liquids, after integrating over the angles, is related to the pair correlation as follows:

$$S(q) = 1 + \nu \int_0^\infty 4\pi r^2 h(r) \frac{\sin qr}{qr} dr \tag{2.73}$$

Moreover, the pair correlation function can be obtained from the inverse transform of the structure factor:

$$h(r) = \frac{1}{2\pi^2 \nu} \int_0^\infty q^2 [S(q) - 1] \frac{\sin rq}{rq} dq \tag{2.74}$$

2.2 Liquid Structure and Properties

The structure of dense liquids as determined by elastic scattering methods and as elucidated by application of the simplest models, hard sphere liquids, and the original van der Waals liquid are the subjects of this section.

2.2.1 Scattering by Pure Liquids

The structure factor for hard spheres at high density for the hard-sphere pair distribution in Figure 2.5[33] is shown in Figure 2.7. The structure factor for liquid methane near its triple point, determined by Habenschuss et al. with x-ray scattering[34] [wavelength $\lambda = 0.071\,nm$ for the range $3 < q(nm^{-1}) < 80$], is shown in Figure 2.8. The results with the scattering vector scaled by molecular diameter are essentially the same as those for the hard-sphere liquid (Figure 2.7). The pair distribution function for methane, obtained numerically by applying Eq. 2.74[35], is likewise similar to the hard-sphere result (Figure 2.5).

The scattering analysis developed in the previous section deals with systems of identical elementary scatterers, each acting as a unit with the same scattering power and having no accessible internal structure of its own. That situation applies directly for liquid argon data[36]. The scattering results for multiatom molecules with near-spherical symmetry, such as methane, can be interpreted by only a slight modification of the single scatterer analysis. Methane has internal structure, but its molecular shape is essentially fixed. Also, the atomic scattering power for x rays is proportional to the square of the atomic number, so the carbons, for example, dominate the scattering pattern for liquid methane. Of course, the carbon-carbon spatial correlations still

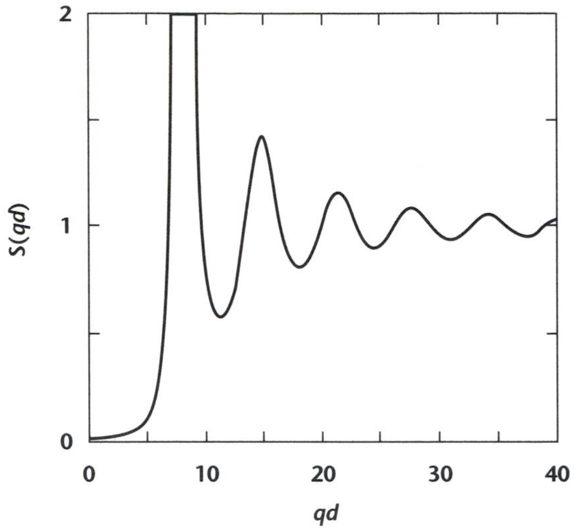

FIGURE 2.7 Static structure factor for hard spheres, calculated for a packing factor of 0.5 with the Percus–Yevick formula[33].

feel the presence of the hydrogens, so the pair distribution function, as deduced from the scattering, does indeed faithfully reflect the liquid methane structure. The similarities of Figures 2.7 and 2.8 exemplify a general result: The dense liquid-state structure for spherical molecules is well described by the hard-sphere model.

The interpretation of scattering data grows more difficult as the molecules increase in complexity. The x-ray scattering pattern for the C_7 linear alkane n-heptane, obtained by Habenschuss and Narten[37], is shown in Figure 2.9. Geometric asymmetry can lead to the local alignment of molecules and thus to a less direct connection between thermodynamic properties in the liquid state and the scalar correlation functions that scattering provides. The effect of molecular flexibility also increases with size. The interpretation must therefore take into account a distribution of molecular shapes and thus of relative scattering site locations.

Distinguishing between the intramolecular and intermolecular contributions to scattering for complex molecules presents a difficult problem, but one that must be overcome if connections between structure and thermodynamic properties are to be established. Thus, the pair distribution consists of at least two parts:

$$g(r) = [g(r)]_{\text{total}} = [g(r)]_{\text{intra}} + [g(r)]_{\text{inter}} \tag{2.75}$$

and even these parts may consist of terms that depend on which sites are involved. Scattering, unassisted by other information, provides at best only $[g(r)]_{\text{total}}$, but cohesive energies and mixing behavior depend primarily on $[g(r)]_{\text{inter}}$. An estimate of

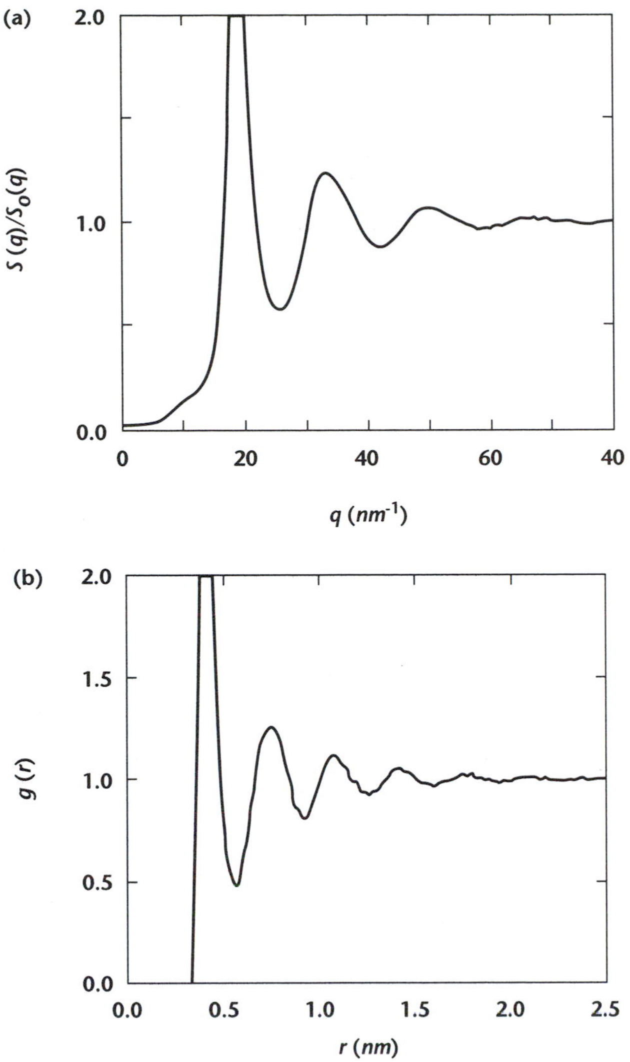

FIGURE 2.8 (a) Structure factor for liquid methane near the triple point density from x-ray scattering by Habenschuss et al.[34]. (b) Pair distribution function for liquid methane obtained from the structure factor data in (a)[34].

$[g(r)]_{inter}$ for n-heptane[37], based on the experimental $[g(r)]_{total}$ and $[g(r)]_{intra}$ as calculated from the various n-heptane rotational states[35], is compared with $[g(r)]_{total}$ in Figure 2.9. Figure 2.10 compares $[g(r)]_{total}$ for highmolecular weight linear polyethylene from x-ray scattering[35] and $[g(r)]_{inter}$, obtained by subtracting out $[g(r)]_{intra}$ from molecular dynamic simulations using the united atom model[38,39].

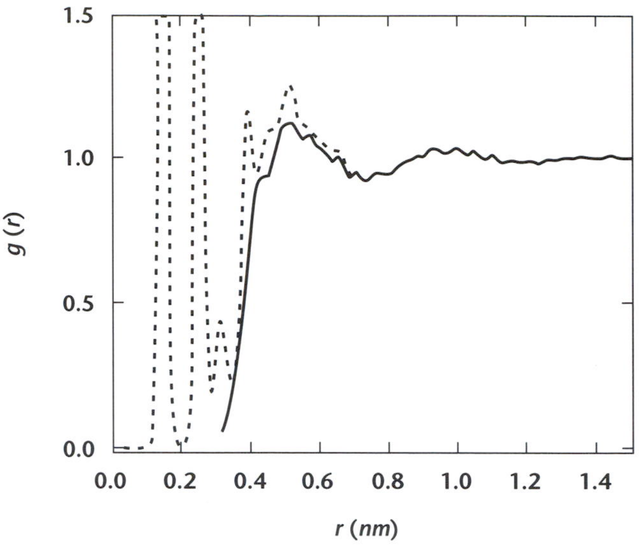

FIGURE 2.9 Total and intermolecular pair distribution functions for liquid n-heptane. From Habenschuss x-ray scattering data and calculations[35].

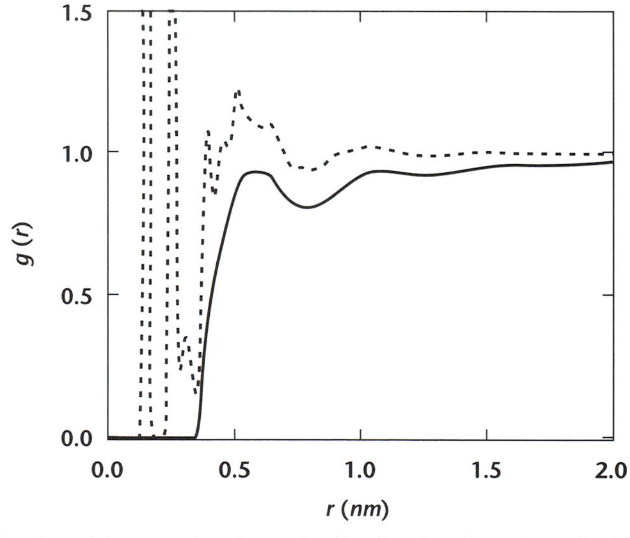

FIGURE 2.10 Total and intermolecular pair distribution functions for liquid polyethylene. From Curro calculations[38,39].

Methods based on neutron scattering and labeling techniques, which rely less on theory, molecular modeling, or simulation alone, offer some possibilities for obtaining $[g(r)]_{\text{inter}}$. These are being explored[40]. For now at least, the problems of sensitivity and uniqueness remain formidable, and less direct methods of establishing the properties that govern mixing must still be employed. This topic is reconsidered in Chapter 8, where the interactions that govern the phase behavior of polymer blends are considered in detail.

2.2.2 The van der Waals Liquid

The van der Waals equation of state[41] plays an important conceptual role in the theory of dense liquid structure and energy. Thus, it is appropriate to introduce the general subject with a simple derivation of this equation, then use the result to illustrate its relevance to the dense liquid region.

Consider a collection of n molecules in volume V at temperature T in which the molecules interact in two ways:

1. A mutual molecular attraction, producing an average potential energy density proportional to the square of the molecular number density:

$$\frac{\langle \Psi \rangle}{V} = -a\nu^2 \tag{2.76}$$

2. An excluded volume repulsion, owing to the finite size of the molecules. The volume available for occupation by each molecule is assumed to be the system volume less some excluded amount that is proportional to the number of molecules in the system:

$$V_a = V - bn \tag{2.77}$$

   The parameters a and b are assumed to be independent of temperature and the number density of molecules.

Now use Eq. 2.17 to obtain the configuration integral. Apply the average potential energy [Eq. 2.76] to all allowable molecular configurations so that the Boltzmann factor comes out of the integral. Then use Eq. 2.77 for each $d\boldsymbol{r}_i$ integral over the available volume, obtaining directly:

$$Z(n, V, T) = \exp\left(\frac{an^2}{VkT}\right)(V - bn)^n \tag{2.78}$$

Applying Eq. 2.19 then gives an expression for the pressure, the van der Waals equation:

$$P = \frac{\nu kT}{1 - b\nu} - a\nu^2 \tag{2.79}$$

The van der Waals equation predicts all the qualitative features of real gas and liquid behavior, including the existence of a gas–liquid transition and a gas–liquid critical point. As with any analytical equation of state, it does not correctly describe behavior in the critical region. It does predict, however, the properties of a dense liquid state at low temperatures and pressures in a qualitatively correct manner. Under those dense liquid conditions, a near-cancellation of the attractive and repulsive contributions to the pressure occurs, and in the zero-pressure limit, the van der Waals equation reduces to a relationship between number density and temperature:

$$kT = a\nu(1 - b\nu) \tag{2.80}$$

This relationship between liquid density and temperature is not quantitatively correct, due in part to the fundamentally incorrect handling of volume exclusion. However, its formulation principle—a near balance between attraction and repulsion in the dense liquid region—remains a central feature of the modern (post-1965) theory.

To understand the flaw in the treatment of excluded volume that leads to Eqs. 2.79 and 2.80, consider the example of a system of three hard spheres that interact only by mutual volume exclusion. The essential difficulty, evident in even this simple system, lies in handling the overlap of exclusion volumes, as made abundantly clear in a 1967 paper by Widom[2]. The space-filling volumes of the spheres themselves cannot overlap, of course, but the overlap of their exclusion volumes is permitted. Consider first the two-sphere system in Figure 2.11, in which each sphere has the same space-filling volume v, and then the change when a third identical sphere is introduced.

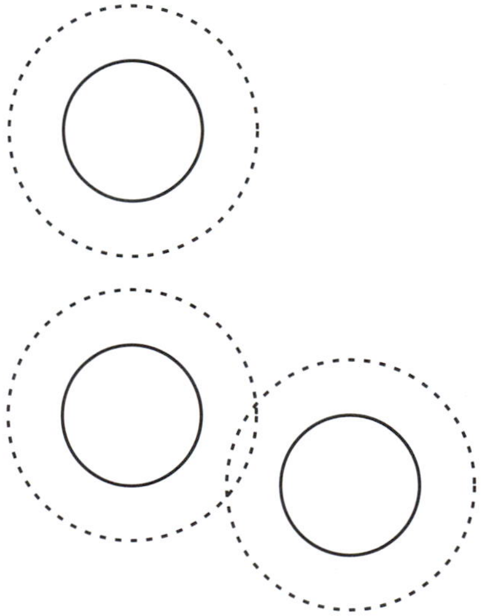

FIGURE 2.11 Excluded volume overlap for hard-sphere pairs. The inner circles indicate the three bare-sphere volumes, and the dashed circles indicate their exclusion volumes. The exclusion volumes of the lower pair overlap, so both are excluding the center of the third sphere from that overlap region. Either one would suffice, so the space excluded to the third sphere center is less than the simple sum of the space excluded separately by other two spheres.

In a two-sphere system, each sphere excludes the center of the other from a volume 8υ, so in Eq. 2.77, $V - V_a$ would be $2(8\upsilon) = 16\upsilon$. The volume excluded to any third sphere by the pair now depends on the overlap of their excluded volumes. Thus, the two spheres exclude a volume to the third of $8\upsilon + 8\upsilon = 16\upsilon$ if their own exclusion volumes do not overlap, but they exclude less than 16υ to the third if they do. Simple additivity would count the volume exclusion overlap twice, giving $V - V_a = 3(8\upsilon)$. Avoiding the double counting would certainly yield $V - V_a < 3(8\upsilon)$. The decrease continues as more spheres are added, and the consequences are clear: The assumption of an excluded volume per sphere that is independent of the density of spheres is incorrect. The parameter b in Eq. 2.77 decreases with increasing number density.

Even in the simplest case of monodisperse hard spheres, an exact relationship between excluded volume per molecule and molecular number density has never been derived analytically. However, the excluded volume problem for hard spheres has been studied with numerical simulations over a wide range of densities. As discussed in the next section, the results provide rather complete information about the structure, pressure, and phase behavior of hard-sphere liquids.

2.2.3 *Hard-Sphere Liquids*

Consider a system of n hard spheres confined in some volume V at temperature T. The spheres have diameter d and thus a space-filling volume $\upsilon = \pi d^3/6$. The concentration of spheres is expressed as the *packing fraction*, $\phi = n\upsilon/V$, and they interact only through mutual volume exclusion. Equation 2.24 applies, so the configuration integral for the system can be written:

$$Z(n,V,T) = [V f_{hs}(\phi)]^n \tag{2.81}$$

in which $f_{hs}(\phi)$ is the fraction of configurations with no overlap for monodisperse hard spheres. From Eq. 2.25, the pressure can be expressed:

$$P_{hs} = \nu kT\left(1 - \frac{d\ln f_{hs}}{d\ln\phi}\right) \tag{2.82}$$

Beginning with the molecular dynamics studies of Alder and Wainwright[42], and the Monte Carlo studies of Hoover and Ree[43], the relationship between pressure and packing fraction in hard-sphere systems, and the concomitant evolution of structure, has been investigated in considerable detail by numerical simulations. The pressure versus packing fraction relationship for monodisperse spheres is illustrated in Figure 2.12. The values discussed here are taken from a brief summary of hard-sphere data given by Debenedetti[44].

The equilibrium state is a single disordered phase, up to a packing fraction $\phi = 0.494 \pm 0.002$. A transition from this disordered phase to an ordered phase then occurs. The packing fraction of the ordered phase rises with increasing pressure from

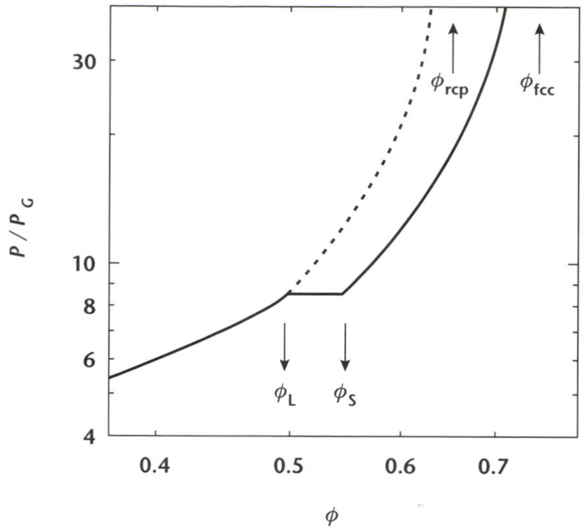

FIGURE 2.12 Pressure as a function of the packing fraction for monodisperse hard spheres. Pressure is normalized by ideal gas pressure at the temperature of interest. Solid and liquid packing fractions at equilibrium, and the limiting random close-packed and fully ordered packing factions are shown. The dashed curve indicates the estimated continuation of the disordered-state pressure through the metastable and unstable regions.

its initial value, $\phi = 0.545 \pm 0.002$, and finally approaches in the high pressure limit, the *face-centered cubic* close-packed value, $\phi_{fcc} = 0.740$. The disordered phase is metastable (stable to small fluctuations in local density) through the range $0.494 \leq \phi \leq 0.565$. Homogeneous nucleation takes place above $\phi \sim 0.565$, suggesting the appearance of local instability at that density, but in any case causing a transition to the ordered phase. The disordered phase can also be made to exhibit a glass transition beyond about $\phi = 0.562$ by imposing a rapid densification in the simulations. Various techniques for extrapolating the metastable branch beyond $\phi = 0.565$ suggest that the equilibrium pressure of the disordered phase diverges at a state attributed to *random close packing*, $\phi_{rcp} = 0.648$.

As it stands, the hard-sphere system is not a particularly useful liquid model since, for example, it shows no vapor–liquid transition. That would require the presence of an attractive interaction, and there is nothing here but the repulsion from volume exclusion. However, it does show a phase change at $\phi = 0.494$ that closely resembles the liquid–solid transition of real substances, even showing the typical L–S density change of about 10%. It also displays typical supercooling (the metastability range) and glass formation. As noted above, the pair distribution in the dense disordered region also resembles closely the pair distribution for dense liquids of spherelike molecules (Figures 2.5 and 2.8).

For purposes here, the hard-sphere simulations furnish two useful features. One is an accurate numerical description of the density dependence of excluded volume contribution where none existed before. Thus, pressure for hard-sphere systems in the stable disordered phase is described well by the *Carnahan–Starling equation*[45]:

$$P_{hs} = \nu kT\left[\frac{1+\phi+\phi^2-\phi^3}{(1-\phi)^3}\right] \qquad (0<\phi<0.494) \tag{2.83}$$

This equation is based on a combination of the simulation data and analytical expressions for the leading terms of the virial expansion for hard spheres. Pressure over the entire disordered range, including the metastable region, is described accurately by the algebraically more complex *Hall equation*[46]. As explained in Section 2.3.1, Longuet-Higgins and Widom[12] demonstrated the enormous value of hard-sphere data in understanding dense liquid thermodynamics.

The second useful feature of the hard-sphere studies is the concept of an upper density limit for the liquid phase—the random close-packed state—which, as explained in later sections, is used together with the density ρ_{rcp} as a dense liquid reference state.

2.3 Generalized van der Waals Theory

Modern versions of the van der Waals liquid are presented in this section. The Longuet-Higgins and Widom combination of van der Waals attraction and hard-sphere volume exclusion is introduced. The cell and lattice models are reformulated as reduced equations of state. The properties of cohesive energy density and internal pressure are developed, preparing the way for a discussion of liquid mixtures in Chapter 3.

2.3.1 The Longuet-Higgins and Widom Model

Longuet-Higgins and Widom[12] reformulated the van der Waals equation by substituting the observed hard-sphere result for the excluded volume term in Eq. 2.79:

$$P(\nu,T) = P_{hs}(\nu,T) - a\nu^2 \tag{2.84}$$

in which a is a species-dependent constant, and P_{hs} is obtained by the numerical simulations. They showed that the form of Eq. 2.84 alone—with no numerical value for a provided—was sufficient to predict with fair success a variety of properties for atomic liquids. The predictions for liquid argon at its triple point are compared with the experimental values[12] in Table 2.5. Agreement is within a few percent for the energy-related properties—vapor pressure, crystallization entropy, cohesive energy density. The differences are larger for volume-related properties, but considering

TABLE 2.5 Properties of liquid argon at the triple point[12] (83.81K, 0.0688 MPa)

Property	Dimensionless Form	Predicted	Observed
Vapor pressure p_o	$p_o V_L / RT$	2.7×10^{-3}	2.8×10^{-3}
Cohesive energy density Π_{CED}	$\Pi_{CED} V_L / RT$	8.6	8.53
Volume of crystallization	$-(\rho_L - \rho_C)/\rho_L$	0.19	0.114
Entropy of crystallization	$-(S_L - S_C)/R$	1.64	1.69
Thermal expansion coefficient	$\alpha_L T$	0.50	0.366
Isothermal compressibility	$\beta_L RT / V_L$	0.058	0.0495
Internal pressure Π_{IP}	Π_{IP}/Π_{CED}	1.00	0.867

V_L, S_L, S_S are molar quantities.

the range of properties predicted and the absence of adjustable parameters, the agreement is impressive.

The work of Longuet-Higgins and Widom gave strong support to the idea that the physical structure of dense liquids is governed primarily by the repulsive part of the intermolecular potential. Conversely, thermal properties such as the heat of vaporization, are determined primarily by the attractive part of the potential, acting upon the repulsion-determined structure. Accordingly, in the hard-sphere approximation, the pressure can be expressed as the sum of the hard-sphere pressure and a single-site attractive energy density.

$$P(v, T) = P_{hs} - \Pi_{CED} \tag{2.85}$$

The last term is Ψ/V, the *cohesive energy density* of the liquid, and from Eq. 2.34,

$$\Pi_{CED} = \frac{v^2}{2} \int_0^\infty [-\psi(r)] 4\pi r^2 g(r, \phi)\, dr \tag{2.86}$$

where $g(r, \phi)$ for hard spheres is $g_{hs}(r, \phi)$. Equation 2.85 is the mechanical equation of state for a hard-sphere model in the *generalized van der Waals* (GVDW) liquids category. All GVDW liquids have the same attractive contribution to the pressure, namely $-\Pi_{CED}$. Other GVDW models, described in subsequent sections, differ only in the excluded-volume contribution. A general definition of the GVDW liquid is given in Section 2.3.5.

The cohesive energy parameter, $a = \Pi_{CED}/v^2$, is a species-specific quantity. It is also insensitive to temperature throughout the dense liquid region. Thus, the attractive part of the intermolecular potential, which governs Π_{CED}, decreases rapidly with distance ($\psi \propto r^{-6}$). Accordingly, the value of Π_{CED} is dominated by contributions from the smallest available separation distances—distances only slightly beyond the sphere diameter d for hard spheres, but within some narrow, closest-approach range in any case. Although the average spacings certainly increase with thermal expansion, near closest-approach populations are unlikely to be changed greatly: The pair distribution is insensitive to density in the range of r that governs Π_{CED}. By this reasoning, a is a species-specific constant, and the cohesive energy density is approximately quadratic in ν (the molecular number density) and hence in the mass density:

$$\Pi_{CED} \propto \rho^2 \tag{2.87}$$

The consistency of this expectation with experimental data for nonpolar and weakly polar dense liquids has been thoroughly verified by Hildebrand and coworkers[47]. The corollary to Eq. 2.87—that energy per molecule in the dense liquid region varies approximately as ρ—was confirmed for the Lennard-Jones potential by Flory[48]. Equation 2.87 also implies that the cohesive energy density of a GVDW liquid can vary with both temperature and pressure, but only in accordance with the resulting variations in density. The quadratic density dependence of Π_{CED} is used extensively in this chapter and is examined for polymeric liquids in Chapter 8.

The perturbation theories of the liquid state begin from the dense liquid picture[11]. Thus, for example, the starting point for a *reference-interaction-site-model* (RISM) is the GVDW liquid, represented by a single interaction site[49]. Departures from spherical shape and from infinite steepness in the repulsive potential are treated as perturbations. Systematic procedures for overcoming the most difficult theoretical problems have been developed. Improved descriptions of monomeric liquid structure are achieved, although at a considerable sacrifice of simplicity. The *polymeric* extension of RISM is PRISM[50], in which the polymer is represented as a string of interaction sites. The polymeric problem of resolving the pair distribution function into its intermolecular and intramolecular components, a necessity for the calculation of polymeric cohesive energy for example, remains largely unsettled, but the combination of PRISM and scattering offers some reasonable hope for success. Perturbation models such as RISM and PRISM have contributed significantly to the fundamental understanding of liquids. However, they require specialized procedures and, although offering useful insights, are too difficult at present to apply meaningfully to liquid mixtures, the main interest here.

Two simpler classes of liquid models, whose species-dependent parameters can be obtained from *pressure-volume-temperature* (PVT) measurements, are widely used for mixing problems. *Cell models* treat the molecules as hard spheres, free to take up space independently but in a confined region defined by the average positions of their nearest neighbors. *Lattice models* also treat the molecules as hard spheres, but now located on a lattice with vacancies. Cell and lattice models, as described here, employ the *random close-packed structure* as a reference state; The n spheres in the system, each with a volume υ, occupy a total volume V_{rcp} at the reference state; the volume per sphere being $V_{rcp}/n = \upsilon/\phi_{rcp}$ and the cohesive energy density $(\Pi_{CED})_{rcp}$. Increasing the temperature expands the structure from its random close-packed limit and provides thereby some extra space, the *free volume*, $V - V_{rcp}$. The models differ mainly in how they distribute this free volume.

These models originated in the 1930s as efforts to describe the structure and properties of single component liquids. They were taken up later by others for application to the free volume theory of mixtures, as introduced in Chapter 3. The first of the early models appears to have been the Lennard-Jones and Devonshire cell model[51,52], which later became the basis for the Prigogine free volume theory [53]. The Eyring-Hirschfelder cell model[54] preceded the Flory-Orwoll-Vrij (FOV) version of free volume theory[55], and the Ceruschi-Eyring lattice model preceded the Sanchez-Lacombe (SL) lattice fluid version[56]. Pure component equation-of-state expressions for the latter two models are developed in the next sections.

2.3.2 Cell Models

In a cell model, the expansion of structure beyond the random close-packed state is uniform. The free volume is distributed equally among the cells, with the cell dimensions defined by the expanded nearest-neighbor locations, as shown in Figure 2.13. Each cell expands by the same amount, and the center of the sphere it contains can move in the cavity of extra space generated by the expansion. This cavity of free volume in each cell is represented by a sphere with a radius equal to the difference between the radii of the expanded and close-packed spheres. Thus,

$$\upsilon_f = \left[\upsilon^{1/3} - \upsilon_{rcp}^{1/3}\right]^3 \tag{2.88}$$

In the simplest versions, the potential energy of each sphere, resulting from its interactions with neighbors, is the same everywhere in the cavity. The potential energy of the system is the mean field value of the intermolecular energy, $\Psi = -V\Pi_{CED}$,

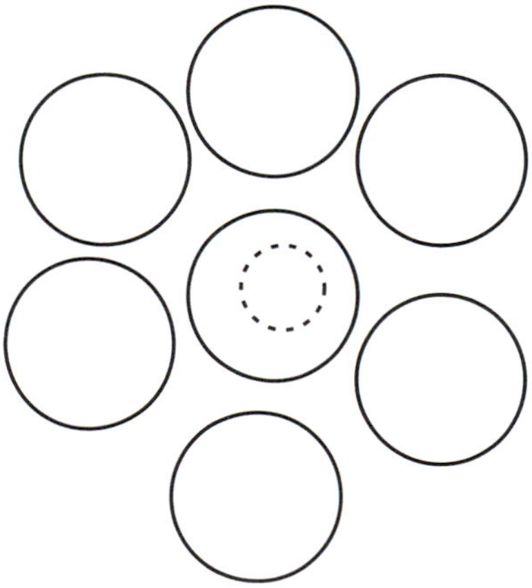

FIGURE 2.13 The cell model for liquids. Free volume is distributed uniformly. The center of the central sphere is confined to the dashed volume by the nearest neighbors.

with the inverse square dependence of cohesive energy density on volume, $\Psi = -(V^2\Pi_{CED})_{rcp}/V$. These substitutions lead to the following expression for the configuration integral for the system:

$$Z(n, V, T) = \left[v^{1/3} - v_{rcp}^{1/3}\right]^{3n} \exp\left[\frac{(V^2\Pi_{CED})_{rcp}}{VkT}\right] \tag{2.89}$$

The application of Eq. 2.19 then leads to the cell model expression for pressure:

$$P = \frac{\nu kT}{1 - (\phi/\phi_{rcp})^{1/3}} - (\Pi_{CED})_{rcp}\left(\frac{\phi}{\phi_{rcp}}\right)^2 \tag{2.90}$$

Cell models have been developed with other expressions for the cohesive energy density–volume relationship, namely the Lennard-Jones and Devonshire (LD) form[51,52] and the Dee-Walsh (DW) form[57].

2.3.3 Lattice Models

In a lattice model, the expansion from the random close-packed state takes place through the creation of lattice vacancies. Thus, as shown in Figure 2.14, the free volume is distributed nonuniformly, as vacant sites in the random close-packed lattice. The number of nearest neighbor sites occupied by spheres decreases with increasing

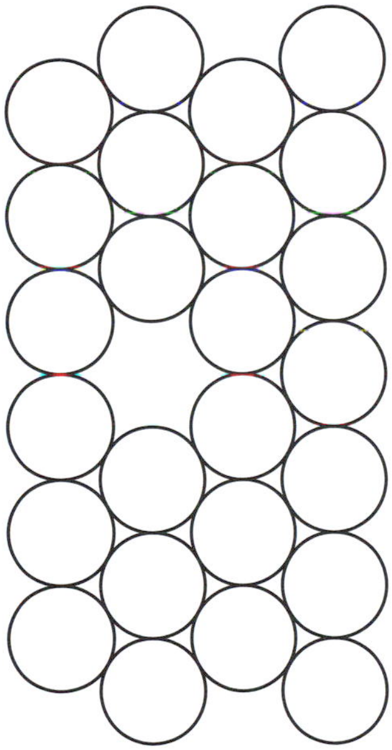

FIGURE 2.14 The lattice model for liquids. The free volume is distributed as randomly located sphere-sized holes in a close-packed site arrangement.

volume, but the lattice spacing does not change. The simplest versions assume the energy is independent of vacancy arrangement—that is, the vacancy distribution is random—and then use a mean field value for the lattice energy Ψ. The configuration integral then reduces to:

$$Z(n, V, T) = \Omega(n, n_o)\left(\frac{\upsilon}{\phi_{rcp}}\right)^{n+n_o} \exp\left[-\frac{\Psi}{kT}\right] \tag{2.91}$$

in which $\Omega(n, n_o)$ is the number of distinguishable ways to distribute n_o indistinguishable vacancies on $n + n_o$ lattice sites. The Boltzmann equation [Eq. 2.17] applies, and direct counting leads to:

$$\Omega = \frac{(n + n_o)!}{n!n_o!} \tag{2.92}$$

Both n_o and n are large for a macroscopic system, so the factorials in Eq. 2.92 can be expressed using *Stirling's approximation* for the factorials of large numbers:

$$\ln \Gamma! = \Gamma \ln \Gamma - \Gamma \tag{2.93}$$

The result is:

$$\ln \Omega = n \ln\left(\frac{n + n_o}{n}\right) + n_o \ln\left(\frac{n + n_o}{n_o}\right) \tag{2.94}$$

The number of vacancies is the ratio of the expansion volume to the volume per sphere at random close packing:

$$n_o = \frac{V - V_{rcp}}{\upsilon/\phi_{rcp}} \tag{2.95}$$

The lattice energy is the sum of contributions from nearest neighbor pairs only:

$$\Psi = \left(\frac{\varepsilon z n}{2}\right)\frac{n}{n + n_o} \tag{2.96}$$

in which z is the number of nearest neighbors per site, ε is the energy per nearest neighbor pair, and the quantity $\varepsilon zn/2$ is $-(V\Pi_{CED})_{rcp}$. Using these various substitutions in Eq. 2.91, $\Psi = -(V^2\Pi_{CED})_{rcp}/V$, and applying Eq. 2.19, leads to the expression for lattice model pressure:

$$P = \nu kT\left[\frac{\phi_{rcp}}{\phi}\ln\left(\frac{1}{1 - \phi/\phi_{rcp}}\right)\right] - (\Pi_{CED})_{rcp}\left(\frac{\phi}{\phi_{rcp}}\right)^2 \tag{2.97}$$

2.3.4 Reduced Equations of State

Four equation-of-state models for dense liquids have been developed so far—van der Waals [Eq. 2.79], Longuet-Higgins and Widom (LW) [Eq. 2.85], cell [Eq. 2.90], and lattice [Eq. 2.97]. All can be expressed in a reduced form. Pressure, volume, and temperature are written as reduced (dimensionless) variables:

$$\begin{aligned} \tilde{P} &= P/P^* \\ \tilde{v} &= v/v^* \\ \tilde{T} &= T/T^* \end{aligned} \qquad (2.98)$$

The asterisks indicate species-dependent parameters that describe liquid properties at some reference state, which we choose to be the state of random close packing. The following identifications then apply to the four models:

$$\begin{aligned} P^* &= (\Pi_{CED})_{rcp} \\ v^* &= V_{rcp}/n \\ T^* &= P^* v^*/k \end{aligned} \qquad (2.99)$$

where $a\nu^2 = \Pi_{CED}$ and $b = V_{rcp}/n$ have been used for Eq. 2.79. The following reduced equation-of-state expressions are obtained:

$$\tilde{P} = \frac{\tilde{T}}{\tilde{v} - 1} - \frac{1}{\tilde{v}^2} \qquad \text{(van der Waals)} \qquad (2.100)$$

$$\tilde{P} = \tilde{T}\left[\frac{1 + 0.648\tilde{v} + (0.648\tilde{v})^2 + (0.648\tilde{v})^3}{(1 - 0.648\tilde{v})^3}\right] - \frac{1}{\tilde{v}^2} \qquad \text{(LW)} \qquad (2.101)$$

$$\tilde{P} = \frac{\tilde{T}}{\tilde{v} - \tilde{v}^{2/3}} - \frac{1}{\tilde{v}^2} \qquad \text{(FOV cell)} \qquad (2.102)$$

$$\tilde{P} = \tilde{T}\ln\left[\frac{\tilde{v}}{\tilde{v} - 1}\right] - \frac{1}{\tilde{v}^2} \qquad \text{(SL lattice)} \qquad (2.103)$$

where the Carnahan-Starling equation [Eq. 2.83], using $\phi/\phi_{rcp} = 0.648\tilde{v}$ from the hard-sphere simulations, has been substituted for P_{hs} in the LW model.

Why is random close packing chosen as the reference state? At first sight, it seems a strange choice, since random close packing lies at an extreme of the dense liquid region, the state of maximum liquid density. It is also inaccessible experimentally. Either crystallization or a glass transition intervenes, and its location and properties can only be inferred from extrapolations, thus making it a kind of hypothetical state. On the other hand, it is appropriate for dense liquids because it is located in the dense liquid region, albeit at the extreme. Other seemingly natural choices for a reference state—the triple point, the critical point, and the normal melting point—depend on the properties of other phases of the species, and these are not necessarily relevant to dense liquid behavior. Random close packing is also a state that exists, at least in principle, for liquids of all species. Further, one might hope that dense liquid properties are linked in some universal way with how far the liquid is from its random close-packed state. A reference is necessary, and there being no better alternative, the random close-packed state is chosen.

The characteristic parameters P^*, T^*, and v^* can be obtained for a liquid by fitting its equation-of-state (P,V,T) data. The reduced forms of the various equations of state are different, so the characteristic parameters that are obtained depend on the choice of model as well as the species. Furthermore, in principle only two independently adjustable fitting parameters occur in the formulation, the parameters P^*, T^*, and v^* being interrelated through the last expression in Eq. 2.99. However, that relationship is seldom obeyed when the three parameters are determined independently. Some reduced forms fit data better than others, although none fits extended ranges of PVT data with great fidelity. As it happens, the FOV and SL models are among the poorest in fitting data, and the Dee-Walsh model is among the best. Interestingly, all the forms fail in about the same way. Many can accommodate the observed P-V isotherms in an acceptable manner, at least over modest ranges, but all depart systematically from the observed V-T isobars. All the forms overstate the rate of increase of thermal expansion coefficient with temperature.

The situation is not as desperate as this somewhat disjointed discussion would seem to suggest, however. The main interest here is in mixtures and the FOV and SL models have been extended to deal with mixing behavior. The aim is to derive the properties of mixtures from those of the individual components, and free volume concepts provide the vehicle. For such purposes, the inadequacy of the model in fitting pure component PVT data, unless extreme, may be less important—at least that is the hope. The free volume aspects of mixing behavior are introduced in Chapter 3 and applied extensively in Chapters 7 and 8.

2.3.5 Cohesive Energy and Internal Pressure

The various equations of state that have been considered so far in Section 2.3 are particular cases of the generalized van der Waals equation of state:

$$P = \frac{nkT}{V} f(\tilde{v}) - (\Pi_{CED})_{rcp} \frac{1}{\tilde{v}^2} \tag{2.104}$$

Only the density-dependent function $f(\tilde{v})$ varies among them. GVDW liquids automatically satisfy Eq. 2.87: Π_{CED} is quadratic in density and varies with T and P only through their effects on density. Another interesting and valuable relationship exists between the thermal and mechanical properties of any GVDW liquid, which is demonstrated as follows.

The internal energy of any substance $U(V, T)$ is the sum of its microscopic kinetic and potential energies. The internal energy of a dense liquid having volume V_L, which obeys Eq. 2.104 is thus given by:

$$U_L = \frac{3}{2}nkT + U_{intra} + \left(-\Psi_{rcp}\right)\frac{1}{\tilde{\upsilon}} \tag{2.105}$$

where the first term on the right is independent of $\tilde{\upsilon}$, the second is highly insensitive to $\tilde{\upsilon}$, and the third is the van der Waals attraction.

The vapor in equilibrium with a dense liquid is always dilute, so its internal energy U_G is simply $3nkT/2 + U_{intra}$. The vapor–liquid energy difference, $U_G - U_L$, is measurable quantity for volatile liquids, the *internal energy of vaporization.* It is the cohesive energy or the intermolecular potential energy that holds the liquid together. The density of this energy is the cohesive energy density of the liquid:

$$\Pi_{CED} = \frac{U_G - U_L}{V_L} \tag{2.106}$$

With Eq. 2.84, and $V_L/V_{rcp} = \tilde{\upsilon}$,

$$\Pi_{CED} = \left(\frac{-\Psi}{V_L}\right)_{rcp}\frac{1}{\tilde{\upsilon}^2} \tag{2.107}$$

The internal energy of any substance satisfies the general thermodynamic relationship[21]:

$$\left(\frac{\partial U}{\partial V}\right)_T = T\left(\frac{\partial P}{\partial T}\right)_V - P \tag{2.108}$$

The right side, a mechanical property, is called the *internal pressure:*

$$\Pi_{IP} \equiv T\left(\frac{\partial P}{\partial T}\right)_V - P \tag{2.109}$$

Using Eq. 2.107 and $V_L/V_{rcp} = \tilde{\upsilon}$,

$$\left(\frac{\partial U}{\partial V}\right)_T = \left(\frac{-\Psi}{V_L}\right)_{rcp}\frac{1}{\tilde{\upsilon}^2} \tag{2.110}$$

so from applying Eq. 2.108,

$$\Pi_{CED} = \Pi_{IP} \tag{2.111}$$

Accordingly, the cohesive energy density for any GVDW liquid is equal to the internal pressure.

From Eq. 2.109, the internal pressure is related directly to a measurable mechanical property of the liquid, the *thermal pressure coefficient* $(\partial P/\partial T)_V$. A thermodynamic identity,

$$(\partial P/\partial T)_V = \alpha(P,T)/\beta(P,T) \tag{2.112}$$

relates the pressure coefficient to the thermal expansion coefficient and the isothermal compressibility. Typical magnitudes of α and β for dense molecular liquids (Table 2.2) indicate $\Pi_{IP} \sim 300\ MPa$ at near-atmospheric pressures and modest temperatures, far in excess of $P_{atm} \sim 0.1\ MPa$. Accordingly, P in Eq. 2.109 is negligible, so the following is an excellent approximation:

$$\Pi_{IP} = T\frac{\alpha}{\beta} \tag{2.113}$$

Cohesive energy density plays a major role in the thermodynamics of liquid mixtures. Equations 2.111 through 2.113 are important because they offer a method for estimating the cohesive energy density of nonvolatile substances, such as polymeric liquids, for which it cannot be measured directly.

TABLE 2.6 Cohesive energy density and internal pressure at 20°C for selected molecular liquids[3,6]

Substance	$\Pi_{CED}\ (J\ cm^{-3})$	Π_{IP}/Π_{CED}	$\delta\ (J\ cm^{-3})^{1/2}$
n-hexane	222	1.11	14.9
n-octane	234	1.10	15.3
n-decane	243	1.13	15.6
cyclohexane	282	1.06	16.8
carbon tetrachloride	310	1.03	17.6
ethylbenzene	324	1.07	18.0
toluene	331	1.04	18.2
ethyl acetate	346	1.01	18.6
benzene	353	0.97	18.8
chloroform	361	0.99	19.0
methyl ethyl ketone	361	0.94	19.0
acetone	412	0.83	20.3
carbon disulfide	415	0.86	20.4
acetic acid	533	0.65	23.1
ethanol	703	0.52	26.5
methanol	882	0.41	29.7
water	2314	0.057	48.1

TABLE 2.7 Cohesive energy density and internal pressure at 20°C for various heptane isomers[3,58]

Liquid	ρ ($g\ cm^{-3}$)	Π_{CED} (MPa)		Π_{IP} (MPa)	
		expt	sim	expt	sim
n-heptane	0.6837	235	247	256	243
2-methylhexane	0.6789	221	230	249	—
3-methylhexane	0.6870	226	237	252	247
3-ethylpentane	0.6984	231	245	267	261
2,2-dimethylpentane	0.6737	203	208	234	—
2,3-dimethylpentane	0.6952	222	230	259	238
2,4-dimethylpentane	0.6745	206	212	236	238

The values of Π_{CED} and Π_{IP}/Π_{CED} for a variety of molecular liquids[3,6] are shown in Table 2.6. Internal pressure is significantly smaller than cohesive energy density for such highly polar species as water, acetic acid, and the alcohols. The ratio for the nonpolar and weakly polar species is near unity, with Π_{IP} being slightly larger on average in these cases. Similar trends are found in the results from molecular dynamics simulations[58]. Experimental[3] and simulation-derived values of Π_{IP} and Π_{CED} for various C_7 alkane isomers are compared in Table 2.7. Although not evident from the data in Tables 2.6 and 2.7, the values of Π_{IP}/Π_{CED} also grow slowly with increasing molecular size, a trend that is considered in more detail in Chapters 7 and 8.

For convenience of reference, the term *simple liquid* will be used to indicate the generalized van der Waals liquid, as defined by Eq. 2.104, throughout the remainder of this volume.

REFERENCES

1. Rowlinson J.S. and Swinton F.L. 1982. *Liquids in Liquid Mixtures*, 3rd ed. London: Butterworth.
2. Widom B. 1967. *Science* 157:375.
3. Lide D.R. (ed.). 1998. *Handbook of Chemistry and Physics*, 79th ed. Boca Raton, FL: CRC Press.
4. Frenkel M., Q. Dong, R.C. Wilhoit, and K.R. Hall 2001. *Int. J. Thermophys.* 22:215.
5. de Gennes P.-G. 1979. *Scaling Concepts in Polymer Physics*, Ithaca, NY: Cornell University Press.
6. Allen G., G. Gee, and G.J. Wilson. 1960. *Polymer* 1:456.
7. McQuarrie D.A. 1976. *Statistical Mechanics*, New York: HarperCollins Publishers.
8. Hill T. L. 1987. *Statistical Mechanics*, New York: Dover Publications.
9. Kittel C. 1986. *Introduction to Solid State Physics*, 6th ed., New York: Wiley & Sons.
10. Temperley H.N.V. and Trevena D. H. 1978. *Liquids and Their Properties*, Chichester: Wiley.

11. Hansen J.P. and McDonald I.R. 1986. *Theory of Simple Liquids,* 2nd ed. London: Academic Press.
12. Longuet-Higgins H.C. and B. Widom. 1964. *Mol. Phys.* 8:549.
13. Maitland G.C., Rigby M., Smith E.B., and Wakeham W.A. 1981. *Intermolecular Forces,* Oxford: Clarendon Press.
14. Israelachvili J. 1992. *Intermolecular and Surface Forces,* 2nd ed. San Diego, CA: Academic Press.
15. London F. 1930. *Z. Physik* 63:245.
16. London F. 1937. *Trans. Faraday Soc.* 33:8.
17. Reid R.C., Prausnitz J.M., and Sherwood T.K. 1987. *Properties of Gases and Liquids,* 4th ed. New York: McGraw-Hill.
18. Hirschfelder J.O., Curtiss C.F., and Bird R.B. 1954. *Molecular Theory of Liquids and Gases,* New York: John Wiley and Sons.
19. Jorgensen W.L., J.D. Madura, and C.J. Swenson. 1984. *J. Amer. Chem. Soc.* 106:6638.
20. Allen M. and Tildesley D. 1989. *Computer Simulation of Liquids,* 2nd ed. Oxford: Clarendon Press.
21. Prausnitz J.M., Lichtenthaler R.N., and Gomez de Azevedo E. 1986. *Molecular Thermodynamics of Fluid-Phase Equilibria,* 2nd ed. Englewood Cliffs, NJ: Prentice Hall.
22. Percus J.K. and G.J. Yevick. 1958. *Phys. Rev.* 110:1.
23. Newstein M.C., H. Wang, N.P. Balsara, A.A. Lefebvre, Y. Shnidman, H. Watanabe, K. Osaki, T. Shikata, H. Niwa, and Y. Morishima. 1999. *J. of Chem. Phys.* 111:4827.
24. Balsara N.P. 2002. Personal communication.
25. Chu B. 1991. *Laser Light Scattering,* 2nd ed. Boston: Academic Press.
26. Berry G.C. 1994. *Adv. Polym. Sci.* 114:233.
27. Guinier A. and Fournet G. 1955. *Small-Angle Scattering of X-Rays,* New York: Wiley.
28. Guinier A. 1994. *X-ray Diffraction in Crystals, Imperfect Crystals and Amorphous Bodies,* New York: Dover Publications.
29. Higgins J.S. and Benoit H.C. 1994. *Polymers and Neutron Scattering,* New York: Oxford University Press.
30. Kaplan W. 1953. *Advanced Calculus,* Cambridge, MA: Addison-Wesley.
31. Bracewell R. 1965. *The Fourier Transform and Its Applications,* New York: McGraw-Hill.
32. Sneddon I.N. 1995. *Fourier Transforms,* New York: Dover Publications.
33. Davis H.T. 1996. *Statistical Mechanics of Phase Interfaces and Thin Films,* New York: VCH Publishers.
34. Habenschuss A., E. Johnson, and A.H. Narten. 1981. *J. Chem. Phys.* 74:5234.
35. Habenschuss A. 2002. Private communication.
36. Schmidt P.W. and Tompson C.W. 1968. In *Simple Dense Liquids,* Frisch H.L. and Salsburg Z.W. (eds.). New York: Academic Press, pp. 31–110.
37. Habenschuss A. and A.H. Narten 1990. *J. Chem. Phys.* 92:5692.
38. Curro J.G., E.B. Webb, G.S. Grest, J.D. Weinhold, M. Putz, and J.D. McCoy. 1999. *J. Chem. Phys.* 111:9073.
39. Curro J.G. 2002. Private communication.
40. Londono J.D., B.K. Annis, A. Habenschuss, G.D. Smith, O. Borodin, C. Tso, E.T. Hsieh, and A.K. Soper. 1999. *J. Chem. Phys.* 110:8786.
41. van der Waals J.D. 1873. *On the Continuity of the Gaseous and Liquid States*. Doctoral dissertation. Leiden.
42. Alder B.J. and T.E. Wainwright. 1960. *J. Chem. Phys.* 33:1439.

43. Hoover W.G. and F.H. Ree. 1968. *J. Chem. Phys.* 49:3609.
44. Debenedetti P.G. 1996. *Metastable Liquids: Concepts and Principles,* Princeton, NJ: Princeton University Press.
45. Carnahan N.F. and K.E. Starling. 1969. *J. Chem. Phys.* 51:635.
46. Hall K.R. 1972. *J. Chem. Phys.* 57:2252.
47. Hildebrand J.H. and Scott R.L. 1964. *The Solubility of Nonelectrolytes,* 3rd ed. New York: Dover Publications.
48. Flory P.J. 1965. *J. Amer. Chem. Soc.* 87:1833.
49. Chandler D., J.D. Weeks, and H.C. Anderson. 1983. *Science* 220:787.
50. Schweizer K.S. and J.G. Curro. 1997. *Adv. Chem. Phys.* 98:1.
51. Lennard-Jones J.E. and A.F. Devonshire. 1937. *Proc. Roy. Soc.* A163:53.
52. Lennard-Jones J.E. and A.F. Devonshire. 1938. *Proc. Roy. Soc.* A165:1.
53. Prigogine I. 1957. *The Molecular Theory of Solutions,* Amsterdam: North-Holland.
54. Eyring H. and J.O. Hirschfelder. 1937. *J. Phys. Chem.* 41:249.
55. Flory P.J. 1970. *Discuss. Faraday Soc.* 49:7.
56. Lacombe R.H. and I.C. Sanchez. 1976. *J. Phys. Chem.* 80:2568.
57. Dee G.T. and D.J. Walsh. 1988. *Macromolecules* 21:811.
58. Maranas J.K., M. Mondello, G.S. Grest, S.K. Kumar, P.G. Debenedetti, and W.W. Graessley. 1998. *Macromolecules* 31:6991.

CHAPTER 3

Molecular Mixtures

This chapter begins with the classical thermodynamics of liquid mixtures and their phase behavior, emphasizing binary systems and specialized to the dense liquid state. It then introduces the simple mixture, a model encompassing chainlike molecules that is broad enough to include monomeric mixtures, polymer solutions and polymer blends, and containing a single, locally determined interaction parameter. The liquid–liquid phase behavior for simple mixtures is then developed, and some characteristics of the interaction parameter are considered. The particular case of regular mixtures, for which the net interactions arise from differences in component cohesive density, is explored. Another source of net interactions—differences in component free volumes—is introduced and related to PVT properties, preparatory to their consideration in later chapters.

3.1 Thermodynamics of Liquid Mixtures

Textbooks are used as needed to supply standard thermodynamic relationships in this chapter[1,2]. The thermodynamic properties of mixtures can be expressed concisely with the *Gibbs free energy of mixing*. Thus, for a single-phase liquid of C components,

$$\Delta G_m = G_{mix} - \sum_{i=1}^{C} G_i \tag{3.1}$$

in which G_{mix} and the G_i are the free energies of mixture and pure components, all at the same temperature, pressure, and component quantities, which here we specify by numbers of moles $N_i,\ i = 1,\ 2, \ldots C$. All properties associated with the formation of mixtures can be calculated if the free energy of mixing is known. Thus, for example, the changes on mixing of enthalpy, entropy, and volume are given by the following expressions:

$$\Delta H_m = \left(\frac{\partial(\Delta G_m/T)}{\partial(1/T)} \right)_{P,N_i} \tag{3.2}$$

$$\Delta S_m = -\left(\frac{\partial \Delta G_m}{\partial T} \right)_{P,N_i} \tag{3.3}$$

$$\Delta V_m = \left(\frac{\partial \Delta G_m}{\partial P} \right)_{T,N_i} \tag{3.4}$$

Similarly:

$$\mu_i - \mu_i^{o} = \left(\frac{\partial \Delta G_m}{\partial N_i} \right)_{P,T,N_{j \neq i}} \tag{3.5}$$

$$\Delta G_m = \sum_{i=1}^{C} (\mu_i - \mu_i^{o}) N_i \tag{3.6}$$

in which the μ_i and μ_i^{o} are the chemical potentials of component i in the mixture and pure liquid respectively. The *Gibbs-Duhem equation,*

$$\sum_{i=1}^{C} N_i \, d\mu_i = 0 \tag{3.7}$$

is a general relationship linking the composition dependence of the component chemical potentials.

The volume change on mixing can be written:

$$\Delta V_m = \sum_{i=1}^{C} (\bar{V}_i - V_i) N_i \tag{3.8}$$

in which:

$$\bar{V}_i = \left(\frac{\partial V}{\partial N_i} \right)_{P,T,N_{j \neq i}} \tag{3.9}$$

where $\bar{V}_i$ is the partial molar volume of component i in the solution, and V_i is the molar volume of pure component i. The fractional change in volume with mixing is:

$$\Delta \upsilon_e = \Delta V_m \Big/ \sum_{i=1}^{C} N_i V_i \tag{3.10}$$

If $\Delta \upsilon_e = 0$, then from Eq. 3.4 the mixing free energy is independent of pressure. If $\bar{V}_i = V_i$ for all components, then $\Delta \upsilon_e = 0$ is automatic, and the system behaves as if the mixture and all the pure components are *incompressible*. The fractional volume change for mixtures of nonpolar or weakly polar monomeric liquids is typically small: $\Delta \upsilon_e$ is seldom more than 1%. It is even smaller for polymeric mixtures, small enough that its effect on mixture properties can usually be ignored.

When $\Delta \upsilon_e$ is deemed small enough to neglect its effects, the Gibbs and Helmholtz free energies of mixing can be used interchangeably to obtain mixture properties. Thus, from their definitions, $G = U + PV - TS$ and $A = U - TS$, the Gibbs and Helmholtz free energies are related through:

$$A = G - PV \tag{3.11}$$

Accordingly, with Eq. 3.4,

$$\Delta A_m = \Delta G_m + P \Delta V_m \tag{3.12}$$

Thus, if $\Delta \upsilon_e$ is small enough, the two mixing free energies are essentially the same. The following is frequently assumed in this book:

$$\Delta G_m = \Delta A_m \tag{3.13}$$

Many theories of mixing begin with the incompressibility idealization and derive expressions for ΔA_m based on that stipulation. Equation 3.12 can then be applied if for some reason a formally correct expression for ΔG_m is required. Under the typical conditions of modest pressures, the effect of volume change, if considered at all, is treated as a small correction. For dense liquids and their mixtures and for moderate pressure variations, setting ΔG_m equal to ΔA_m is a safe approximation.

In this volume, the primary concern is two-component liquid mixtures—monomeric mixtures if both are monomeric, polymer solutions if one is a polymer, and polymer blends if both are polymers. In principle, the components can always be labeled simply 1 and 2 without further designation; this is the accepted system for monomeric mixtures and polymer blends. In the case of polymer solutions, however, it is customary to refer to the monomeric component as the *solvent* and the polymer as the *solute*, and by convention to label the solvent component 1 or s and the polymer component 2 or p. At high polymer concentrations, the monomeric component is sometimes termed the *diluent*.

Mixture composition is expressed in several ways. *Mole fraction* x is sometimes used. Mass density, the *weight concentration* c, is also commonly used, as is *mass fraction* w and *volume fraction* ϕ. Obtaining either w or x from weighed amounts of known components is unambiguous:

$$\begin{aligned} w_1 &= \frac{W_1}{W_1 + W_2}, \qquad w_1 + w_2 = 1 \\ x_1 &= \frac{W_1/M_1}{W_1/M_1 + W_2/M_2}, \qquad x_1 + x_2 = 1 \end{aligned} \tag{3.14}$$

where the W_i and M_i are component weights and molecular weights. However, c and ϕ require partial specific volumes as well and, in principle, $\bar{v}_1$ and $\bar{v}_2$ can vary with mixture composition in generally unknown ways. Fortunately, for the liquids and mixtures of interest here—in which compressibility effects are usually negligible and $\Delta v_e = 0$ is a safe approximation—the approximation that partial specific volume equals the reciprocal of the pure component liquid density $\bar{v}_i = 1/\rho_i$, is also safe. Thus, the following are used throughout the book:

$$\begin{aligned} c_i &= \frac{w_i}{w_1/\rho_1 + w_2/\rho_2} \\ \phi_i &= \frac{c_i}{\rho_i} = \frac{w_i/\rho_i}{w_1/\rho_1 + w_2/\rho_2}, \qquad \phi_1 + \phi_2 = 1 \end{aligned} \tag{3.15}$$

It is sometimes convenient to express the chemical potentials of the mixture components in terms of the *osmotic pressure*. As discussed in Section 5.1.1, the presence of a solute reduces the chemical potential of the solvent, but it can be restored to its pure component value by the application of mechanical pressure to the solution. The extra pressure required to achieve the balance, $\Delta P = \pi$, is the osmotic pressure of the solution. Using the convention that component 1 is the solvent and the thermodynamic relation $\partial\mu_1/\partial P = \bar{V}_1$, the chemical potential of solvent in the solution can be increased from its value at ambient pressure P_a by applying pressure:

$$\mu_1(P) = \mu_1(P_a) + \int_{P_a}^{P} \bar{V}_1 dP \tag{3.16}$$

2.3.4 Reduced Equations of State

Four equation-of-state models for dense liquids have been developed so far—van der Waals [Eq. 2.79], Longuet-Higgins and Widom (LW) [Eq. 2.85], cell [Eq. 2.90], and lattice [Eq. 2.97]. All can be expressed in a reduced form. Pressure, volume, and temperature are written as reduced (dimensionless) variables:

$$\begin{aligned} \tilde{P} &= P/P^* \\ \tilde{v} &= v/v^* \\ \tilde{T} &= T/T^* \end{aligned} \tag{2.98}$$

The asterisks indicate species-dependent parameters that describe liquid properties at some reference state, which we choose to be the state of random close packing. The following identifications then apply to the four models:

$$\begin{aligned} P^* &= (\Pi_{CED})_{rcp} \\ v^* &= V_{rcp}/n \\ T^* &= P^* v^*/k \end{aligned} \tag{2.99}$$

where $a\nu^2 = \Pi_{CED}$ and $b = V_{rcp}/n$ have been used for Eq. 2.79. The following reduced equation-of-state expressions are obtained:

$$\tilde{P} = \frac{\tilde{T}}{\tilde{v}-1} - \frac{1}{\tilde{v}^2} \qquad \text{(van der Waals)} \tag{2.100}$$

$$\tilde{P} = \tilde{T}\left[\frac{1+0.648\tilde{v}+(0.648\tilde{v})^2+(0.648\tilde{v})^3}{(1-0.648\tilde{v})^3}\right] - \frac{1}{\tilde{v}^2} \qquad \text{(LW)} \tag{2.101}$$

$$\tilde{P} = \frac{\tilde{T}}{\tilde{v}-\tilde{v}^{2/3}} - \frac{1}{\tilde{v}^2} \qquad \text{(FOV cell)} \tag{2.102}$$

$$\tilde{P} = \tilde{T}\ln\left[\frac{\tilde{v}}{\tilde{v}-1}\right] - \frac{1}{\tilde{v}^2} \qquad \text{(SL lattice)} \tag{2.103}$$

where the Carnahan-Starling equation [Eq. 2.83], using $\phi/\phi_{rcp} = 0.648\tilde{v}$ from the hard-sphere simulations, has been substituted for P_{hs} in the LW model.

Why is random close packing chosen as the reference state? At first sight, it seems a strange choice, since random close packing lies at an extreme of the dense liquid region, the state of maximum liquid density. It is also inaccessible experimentally. Either crystallization or a glass transition intervenes, and its location and properties can only be inferred from extrapolations, thus making it a kind of hypothetical state. On the other hand, it is appropriate for dense liquids because it is located in the

The following expressions for the interaction contribution to the mixing free energy and the volume change on mixing are finally obtained:

$$\frac{\tilde{v}\Delta G_{int}}{V} = \frac{3\phi_1 P_1^*}{T_1^*}\ln\frac{\tilde{v}^{1/3}-1}{\tilde{v}_1^{1/3}-1} + \frac{3\phi_2 P_2^*}{T_2^*}\ln\frac{\tilde{v}^{1/3}-1}{\tilde{v}_2^{1/3}-1} + \left(\phi_1\frac{P_1^*}{\tilde{v}_1} + \phi_2\frac{P_2^*}{\tilde{v}_2} - \frac{P^*}{\tilde{v}}\right) \tag{3.114}$$

$$\Delta v_e = \frac{\phi_1(\tilde{v}-\tilde{v}_1) + \phi_2(\tilde{v}-\tilde{v}_2)}{\phi_1\tilde{v}_1 + \phi_2\tilde{v}_2} \tag{3.115}$$

in which the reduction parameter definitions in Eqs. 3.110 have been applied in Eq. 3.114 as needed. The interaction contributions to the chemical potentials can be obtained from Eq. 3.114. Here, however, the effect of temperature on the component volume fractions in the close-packing energy balance is ignored. The exchange energy contribution is simply inserted at the temperature of interest. Thus, for example,

$$\mu_1 - \mu_1^{\circ} = V_1\left[(\delta_1-\delta_2)^2\phi_2^2 + \frac{P_1^*}{\tilde{v}_1}\left(3\tilde{T}_1\ln\frac{\tilde{v}_1^{1/3}-1}{\tilde{v}^{1/3}-1} + \frac{1}{\tilde{v}_1} - \frac{1}{\tilde{v}}\right)\right] \tag{3.116}$$

in which the molar volume V_1 is $N_a r_1 v^* \tilde{v}_1$.

The application of Eqs. 3.114 through 3.116 requires evaluation of $\tilde{v}(\phi)$, the reduced volume of the mixture as a function of mixture composition. This evaluation is made by solving the reduced equation of state [Eq. (2.102)] with $P^*(\phi)$ and $T^*(\phi)$ for the mixture, composition-dependent properties obtained with their pure component values, and mixing rules derived from Eq. 3.113 for energy balance and c parameter conservation. Thus, with $(\Pi_{CED})_{rcp} = P^*$ and $c \propto P^*/T^*$ for pure components and mixtures, Eq. 3.113 leads to:

$$\begin{aligned} P^*(\phi) &= \phi_1 P_1^* + \phi_2 P_2^* - \phi_1\phi_2(\delta_1-\delta_2)^2_{rcp} \\ T^*(\phi) &= \frac{P^*(\phi)}{\phi_1 P_1^*/T_1^* + \phi_2 P_2^*/T_2^*} \end{aligned} \tag{3.117}$$

in which $(\delta_1-\delta_2)^2_{rcp}$ refers to pure component values. The pure component values of P^*, T^*, and $\tilde{v}(T)$ can be obtained by fitting their PVT properties to Eq. (2.102). They can also be obtained from the values of isothermal compressibility and thermal expansion coefficient at ambient pressures. Thus, Eq. (2.102), and its partial derivatives

$(\partial v/\partial T)_P$ and $(\partial v/\partial P)_T$ in the low-pressure limit, lead to:

$$\tilde{v}^{1/3} = (1 + 4\alpha T/3)/(1 + \alpha T) \tag{3.118}$$

$$T^* = T\tilde{v}^{4/3}/\left(\tilde{v}^{1/3} - 1\right) \tag{3.119}$$

$$P^* = \alpha T\tilde{v}^2/\beta \tag{3.120}$$

Using experimental values of α and β, these equations can be solved sequentially to give P^*, T^*, and $\tilde{v}(T)$. If desired, the characteristic volume v^* can be calculated by using, for example, the known mass density, $v^* = [\rho(T)\tilde{v}(T)]^{-1}$. In fact, only $\tilde{v}$ itself appears in the expressions for observables, so the value of v^* itself is irrelevant.

3.6.2 The Free Volume Interaction Parameter

Even in the simplified form presented here, the FOV model is awkward for purposes of discussion—the free volume contribution is buried too deeply. However, a series expansion in powers of concentration for the chemical potential, given by Flory as Eq. 43 in a 1965 paper[60], turns out to be quite convenient for estimating the magnitude of free volume contributions. Simplified in the manner above, and specialized to components with the same cohesive energy density, $\delta_1 = \delta_2$ at all temperatures, the Flory expression becomes:

$$\frac{\left(\mu_1 - \mu_1^{\circ}\right)_{int}}{\phi_2^2} = \frac{\alpha_1 T\,(\Pi_{CED})_1\,V_1\left(1 - T_1^*/T_2^*\right)^2}{2}\left[1 + B_1(T)(1 - T_1^*/T_2^*)\phi_2 + \cdots\right] \tag{3.121}$$

in which:

$$B_1(T) = \frac{2}{9}\left(3 - 4\alpha_1 T - 4\alpha_1^2 T^2\right) \tag{3.122}$$

From Eq. 3.69,

$$\frac{\left(\mu_1 - \mu_1^{\circ}\right)_{int}}{\phi_2^2} = \frac{\mu_1 - (\mu_1)_{\chi=0}}{\phi_2^2} = \chi RTr_1 \tag{3.123}$$

Substituting χRTr_1 for $(\mu_1 - \mu_1^{\circ})_{int}/\phi_2^2$ into Eq. 3.121 and using $r_1 = V_1/V_{ref}$ leads to an expression for the free volume contribution in interaction parameter terminology:

$$\chi_{FV} = \frac{(\Pi_{CED})_1\left(1 - T_1^*/T_2^*\right)^2\alpha_1 V_{ref}}{2R}\left[1 + B_1(T)(1 - T_1^*/T_2^*)\phi_2 + \cdots\right] \tag{3.124}$$

For convenience in later use and discussion, this equation is expressed in a slightly different way:

$$\chi_{FV} = (\chi_{FV})_o + \left(\frac{\partial \chi_{FV}}{\partial \phi}\right)_o \phi_2 + \cdots \tag{3.125}$$

in which:

$$(\chi_{FV})_o = \frac{(\Pi_{CED})_1 \alpha_1 V_{ref}}{2R}\left(\frac{T_2^* - T_1^*}{T_2^*}\right)^2 \tag{3.126}$$

$$\left(\frac{\partial \chi_{FV}}{\partial \phi}\right)_o = (\chi_{FV})_o B_1(T)\left(\frac{T_2^* - T_1^*}{T_2^*}\right) \tag{3.127}$$

Patterson derived Eq. 3.126 in 1968[62]. Although his formula is expressed using property terms different from these, the final result is the same. Thus, from the Patterson equation, the free volume contribution to the interaction coefficient at low concentration of the minority component 2 is always positive (unfavorable to mixing), all the factors composing $(\chi_{FV})_o$ being positive quantities. The value of $(\chi_{FV})_o$ varies quadratically with the difference in the pure component characteristic temperatures, and it is also primarily entropic, being only indirectly dependent on temperature, through the partially compensating temperature dependence of cohesive energy density, molar volume, and thermal expansion coefficient. From Eq. 3.127, χ_{FV} depends on composition, but it may either increase or decrease with the increasing concentration of component 2, depending on whether T_2^* is larger or smaller than T_1^*.

As mentioned earlier, the contribution of cohesive energy mismatch to the interaction coefficient has been deliberately suppressed to isolate the contribution of free volume mismatch alone. The FOV model is a generalized van der Waals liquid (Section 2.3.5), so, from Eq. (2.111), cohesive energy density is equal to internal pressure. Thus, the solubility parameter is $(\Pi_{IP})^{1/2}$ from Eq. 3.94, and Eq. (2.113) provides a PVT-based value for the solubility parameter that is exact for the FOV model, but distinct from the *CED*-based value in general:

$$\delta_{\mathrm{PVT}} = \left(\frac{\alpha T}{\beta}\right)^{1/2} \tag{3.128}$$

Finally, using Eq. 3.93,

$$\chi_{\mathrm{PVT}} = \frac{V_{ref}}{RT}(\delta_1 - \delta_2)^2_{\mathrm{PVT}} \tag{3.129}$$

When necessary in later chapters, the interaction parameter with both free volume mismatch and cohesive energy mismatch effects included will be approximated by the sum:

$$\chi = \chi_{FV} + \chi_{CED} \tag{3.130}$$

with χ_{PVT} used in place of χ_{CED} if better information is not available.

3.6.3 Commentary

The unmodified FOV theory has been applied extensively to mixtures of monomeric liquids [48,66–68]. The outcome of these studies has generally favored the idea that free volume mismatch sometimes makes significant contributions to the thermodynamics of monomeric liquid mixtures. Experimental values for the enthalpies of mixing were used as input to the calculations in the Abe work[48,66]. For hydrocarbon–fluorocarbon mixtures, and a wide range of systems with roughly spherical molecules, the predicted values of both ΔS_m and Δv_e followed the experimental trends in an acceptable manner. Positive as well as negative values of Δv_e are observed and predicted, and even numerical agreement is found in some cases. The Orwoll study[67] demonstrated that FOV theory nicely predicts the trends of behavior with chain length in mixtures of linear alkanes. Höcker[68] found that FOV theory could explain the volume changes that occur when liquefied noble gases are mixed. These various studies also demonstrated, however, that the mixing behavior of even the simplest of monomeric liquids in the dense range can be maddeningly complex, and that free volume contributions are not the most important ones in many cases. They also demonstrated that unexpected departures from the calculations were found in some systems, thus making prediction somewhat uncertain.

Instruments for measuring the PVT properties of polymers are available[69], as are compilations of carefully evaluated PVT data for a multitude of species[70–73]. However, in estimating the interactions based on any free volume theory, original or simplified, two concerns must be kept in mind: systematic differences exist between experimental PVT data and the FOV equation of state, Eq. (2.102). The values of P^* and T^* vary with the ranges of P and T covered in the fitting. Relative to some other equation of state models, the FOV model fits PVT data rather poorly[74]. The second concern is obtaining PVT data of sufficiently high accuracy or at least high relative precision. The differences between component values ΔP^* and ΔT^*, or $\Delta\alpha$ and $\Delta\beta$, govern the interactions. Both thermal expansion coefficient and isothermal compressibility depend on pressure and temperature, and accurate evaluation of $\alpha(T)$ at ambient pressures is a difficult undertaking. Achieving sufficient accuracy in $\beta(T)$ is even more difficult[70,75]. Compressibility changes fairly rapidly with pressure in the moderate pressure range, and, as explained by Hayward[75] and

Zoller[70], even the best data require highly refined methods of data correlation and analysis.

As will become evident in Chapter 7, the importance and generality of free volume effects in polymer solution behavior are well established. Free volume mismatch seems also to account for some observations in polymer blends, as discussed in Chapter 8.

REFERENCES

1. Denbigh K. 1981. *The Principles of Chemical Equilibrium,* 4th ed. Cambridge, UK: Cambridge University Press.
2. Prausnitz J.M., Lichtenthaler R.N., and Gomez de Azevedo E. 1986. *Molecular Thermodynamics of Fluid-Phase Equilibria,* 2nd ed. Englewood Cliffs, NJ: P T R Prentice Hall.
3. Pozharskaya G.I., N.L. Kasapova, V.P. Skripov, and Y.D. Kolpakov. 1984. *J. Chem. Thermodynamics* 16:267.
4. Hildebrand J.H. 1929. *J. Am. Chem. Soc.* 51:69.
5. Scatchard G. 1931. *Chem. Rev.* 8:321.
6. Van Laar J.J. 1913. *Z. Physik Chem.* 72:723.
7. Van Laar J.J. and R. Lorenz. 1925. *Z. Anorg. Allgem. Chem.* 146:42.
8. Meyer K.H. 1939. *Z. Physik. Chem.* B44:383.
9. Meyer K.H. 1940. *Helv. Chim. Acta* 23:1063.
10. Chang T.S. 1939. *Proc. Roy. Soc.* A 169:512.
11. Staverman A.J. 1941. *Recl. Trav. Chim., Pays-Bas* 60:640.
12. Huggins M.L. 1941. *J. Chem. Phys.* 9:440.
13. Huggins M.L. 1942. *J. Phys. Chem.* 46:151.
14. Huggins M.L. 1942. *Ann. N.Y. Acad. Sci.* 43:1.
15. Flory P.J. 1941. *J. Chem. Phys.* 9:660.
16. Flory P.J. 1942. *J. Chem. Phys.* 10:51.
17. Guggenheim A.E. 1944. *Proc. Roy. Soc.* A 183:203.
18. Hildebrand J.H. 1947. *J. Chem. Phys.* 15:225.
19. Flory P.J. 1970. *Discuss. Faraday Soc.* 49:7.
20. Dickman R. and C.K. Hall. 1986. *J. Chem. Phys.* 85:4108.
21. Hildebrand J.H. and Scott R.L. 1964. *The Solubility of Nonelectrolytes,* 3rd ed. New York: Dover Publications.
22. Flory P.J. 1953. *Principles of Polymer Chemistry,* Ithaca, NY: Cornell University Press.
23. Scott R.L. 1949. *J. Chem. Phys.* 17:279.
24. Dobry A. and F. Boyer-Kawenoki. 1947. *J. Polym. Sci.* 2:91.
25. Guggenheim A.E. 1944. *Proc. Roy. Soc.* A 183:213.
26. Flory P.J. 1944. *J. Chem. Phys.* 12:425.
27. Scott R.L. and M. Magat. 1945. *J. Chem. Phys.* 13:172.
28. Rowlinson J.S. and Swinton F.L. 1982. *Liquids and Liquid Mixtures.* 3rd ed. London: Butterworth.
29. Paul D.R. and Bucknall C.B. (eds.). 2000. *Polymer Blends,* New York: Wiley-Interscience.

30. Booth C., G. Gee, M.N. Jones, and W.D. Taylor. 1964. *Polymer* 5:353.
31. Qian C., S.J. Mumby, and B.E. Eichinger. 1991. *Macromolecules.* 24:1655.
32. Hildebrand J.H. and D.R.F. Cochran. 1949. *J. Am. Chem. Soc.* 71:22.
33. Copp J.L. 1955. *Faraday Soc. Trans.* 51:1056.
34. Hudson C.S. 1904. *Z. Physik. Chem.* 47:113.
35. Binder K. 1995. In *Monte Carlo and Molecular Dynamics Simulations in Polymer Science,* K. Binder (ed.). New York: Oxford University Press, pp. 356–432.
36. Müller M., K. Binder, and L. Schäfer. 2000. *Macromolecules.* 33:4568.
37. Dudowicz J. and K.F. Freed. 1992. *J. Chem. Phys.* 96:9147.
38. Taylor J.K., P.G. Debenedetti, W.W. Graessley, and S.K. Kumar. 1996. *Macromolecules* 29:764.
39. Joanny J.F. and H. Benoit. 1997. *Macromolecules.* 30:3704.
40. Taylor-Maranas J.K., P.G. Debenedetti, W.W. Graessley, and S.K. Kumar. 1997. *Macromolecules* 30.
41. Kumar S.K., B.A. Veytsman, J.K. Maranas, and B. Crist. 1997. *Phys. Rev. Lett.* 79:2265.
42. Benoit H. and G. Jannink. 2000. *Eur. Phys. J. E* 3:283.
43. Melenkevitz J., Crist B., and Kumar S.K. 2000. *Macromolecules* 33:6869.
44. London F. 1930. *Z. Physik. Chem* B11:222.
45. London F. 1937. *Trans. Faraday Soc.* 33:8.
46. Maitland G.C., Rigby M., Smith E.B., and Wakeham W.A. 1981. *Intermolecular Forces,* Oxford: Clarendon Press.
47. Hildebrand J.H., Prausnitz J.M., and Scott R.L. 1970. *Regular and Related Solutions,* New York: Van Nostrand-Reinhold.
48. Abe A. and P.J. Flory. 1965. *J. Am. Chem. Soc.* 87:1838.
49. Scott R.L. 1958. *J. Phys. Chem.* 62:136.
50. Barton A.F.M. 1983. *Handbook of Solubility Parameters and other Adhesion Parameters,* Boca Raton, FL: CRC Press.
51. Coleman M.M., Graf J.F., and Painter P.C. 1991. *Specific Interactions and Miscibility of Polymer Blends,* Lancaster, PA: Technomic.
52. Krishnamoorti R. 1994. Doctoral dissertation, Princeton University.
53. Prigogine I., N. Trappeniers, and V. Mathot. 1953. *J. Chem. Phys.* 21:559.
54. Prigogine I., N. Trappeniers, and V. Mathot. 1953. *Discuss. Faraday Soc.* 15:93.
55. Prigogine I. 1957. *The Molecular Theory of Solutions,* Amsterdam: North-Holland.
56. Flory P.J., R.A. Orwoll, and A. Vrij. 1964. *J. Am. Chem. Soc.* 86:3507.
57. Flory P.J., R.A. Orwoll, and A. Vrij. 1964. *J. Am. Chem. Soc.* 86:3515.
58. Sanchez I.C. and R.H. Lacombe. 1976. *J. Phys. Chem.* 80:2352.
59. Lacombe R.H. and I.C. Sanchez. 1976. *J. Phys. Chem.* 80:2568.
60. Flory P.J. 1965. *J. Am. Chem. Soc.* 87:1833.
61. Eichinger B.E. and P.J. Flory. 1968. *Trans. Faraday Soc.* 64:2035.
62. Patterson D. 1968. *J. Polym. Sci.: Pt. C* 16:3379.
63. Patterson D. 1969. *Macromolecules* 2:672.
64. Sanchez I.C. 1978. In *Polymer Blends,* Vol. 1., D. R. Paul and S. Newman (eds.). New York: Academic Press, p. 115.
65. Boyd R.H. and Phillips P.J. 1993. *The Science of Polymer Molecules,* Cambridge: Cambridge University Press.
66. Abe A. and P.J. Flory. 1966. *J. Am. Chem. Soc.* 88:2887.
67. Orwoll R.A. and P.J. Flory. 1967. *J. Am. Chem. Soc.* 89:6822.
68. Höcker H. and P.J. Flory. 1968. *Trans. Faraday Soc.* 64:1189.

69. Zoller P., P. Bolli, V. Pahud, and H. Ackermann. 1976. *Review of Scientific Instruments* 47:948.
70. Zoller P. 1989. In *Polymer Handbook,* 3rd ed. J. Brandrup and H. Immergut (eds.). New York: John Wiley & Sons, pp. VI-475.
71. Dee G.T., T. Ougizawa, and D.J. Walsh. 1992. *Polymer* 33:3462.
72. Dee G.T., B.B. Sauer, and B.J. Haley. 1994. *Macromolecules* 27:6106.
73. Zoller P. and Walsh D.J. 1995. *Standard Pressure-Volume-Temperature Data for Polymers,* Lancaster, PA: Technomic Publishing Co.
74. Rodgers P.A. 1993 *J. Appl. Polym. Sci.* 48:1061.
75. Hayward A.T.J. 1967. *Brit. Appl. Phys.* 18:965.

CHAPTER 4

The Random Coil Model

This chapter deals with the properties of random coils, the primary model for flexible-chain polymers in the liquid state. The analogy with random walks is developed and used to express polymer size. Relationships between average size and molecular weight are then developed for linear and nonlinear chain architectures. The various definitions of step size, which includes matching chain molecules with local bond angle restrictions to the random walk formulation, are itemized and tabulated for many common polymer species. The distribution of conformations in the long chain approximation is derived, and the mechanical, thermodynamic, and scattering properties of random coil molecules are developed for later application. The chapter finishes with an introduction to the fluctuation theory of scattering and a formulation of the structure factor for simple mixtures of random coil molecules, an expression used extensively in later chapters.

The growth of molecular flexibility with increasing chain length which leads to the *random-coil* character of polymeric molecules, was discussed briefly in Chapter 1. The picture of chain molecules as physical embodiments of random walks has a long history. This concept emerged in the early 1930s, during the period when Staudinger's macromolecular hypothesis was beginning to gain wide acceptance. Already in 1932, Mayer, von Susich, and Valkó[1] were attributing the restoring force in rubber elasticity to an entropy decrease brought about by increasing the end-to-end distances of network chains. Shortly thereafter, Guth and Mark[2] and Kuhn[3–5] were explicitly associating polymer conformations with random walks, thus leading to predictions relating molecular dimensions to chain length. The random walk, applied much earlier in other contexts[6], evolved quickly from a tentative suggestion to the canonical model for flexible polymers. It was supported by light scattering measurements on dilute polymer solutions, beginning in the 1940s with Debye[7,8] and Zimm[9], and was confirmed in detail for melts and concentrated solutions by neutron scattering, beginning in the mid-1970s by groups in France[10], Germany[11], and England[12]. The random walk began as and remains the starting point for basic understanding of the polymeric liquid state.

The large-scale conformation of a covalent chain molecule is defined by the relative positions of its backbone atoms. Bond lengths and the angles between successive bonds are fixed within very narrow limits, but the rotational angles around the single bonds can vary over wide ranges. The rotational freedom of backbone bonds is the main source of molecular flexibility. For a backbone of carbon-carbon single bonds, the bond length l is 0.153 *nm,* and the angle β between successive bonds is near the tetrahedral angle, $\sim 109.5°(\cos\beta = -1/3)$, essentially independent of the polymer species. As discussed briefly in Section 1.2.3, the orientation of a bond relative to the plane of its two predecessors, its azimuthal angle ϕ, can vary over the full 360° range. The energy may depend on ϕ in a manner that reflects the *steric constraints*—the nonbonded interactions of side groups on the chain—and thus on the particular polymer species. Typically, three energy minima exist at intervals of approximately 120°, as shown in Figure 1.2. The energy range is commonly a few multiples of the thermal energy kT, thus permitting the possibility of a rapid interchange of rotational states and a rapid rearrangement of backbone conformation. Accordingly, long flexible chains in the liquid state occupy a wide range of backbone conformations that change continually.

To grasp the magnitudes, consider a set of chains with some large number n of backbone bonds, each with three possible rotational isomeric states. The number of distinct conformations can then be obtained by counting from one end. The first bond sets an initial direction in space, the second defines a plane, and, beginning with the third, each succeeding bond can take up three orientations with respect to the plane defined by the previous two. The number of conformations for an n-bond backbone is thus $\Omega = 3^{n-2}$. For a chain of even modest length, such as $n = 1{,}000$ (corresponding to a polyethylene molecular weight of 14,000), approximately 10^{476} conformations are possible. Even with an Avogadro's number of molecules, $\sim 6 \times 10^{23}$ (a mass of 14 kg for the polyethylene example), the probability that any two have exactly the same conformation is very small, only about $(6 \times 10^{23})^2 / 10^{476} \sim 10^{-429}$. Furthermore, for a rotational state lifetime of 10^{-11} sec (the structural lifetime for molecular liquids, discussed in Volume 2), each chain samples about $10^3/10^{-11} = 10^{14}$ conformations per second. An Avogadro's number of 1,000-bond molecules would require about $10^{476}/10^{14}$ sec or about 10^{454} years to sample a significant fraction of all available conformations!

The following sections deal with the properties of random walks; their connection with polymer chains; the effect of local correlations in bond orientations; nonlinear chain architectures; and some mechanical, thermodynamic, and scattering properties of random-coil molecules.

4.1 Random Walks

Consider the path through space described by any connected sequence of n displacements or steps, each of length l. Label the steps in order, from $i = 1$ at one end to $i = n$ at the other, and represent each by a vector $\boldsymbol{r}_i$ ($|\boldsymbol{r}_i| = l; \boldsymbol{r}_i \cdot \boldsymbol{r}_i = l^2$) as shown in Figure 4.1. The vector from the beginning of step i to the end of step j ($i < j$) is the sum of the intervening step vectors:

$$\mathbf{R}_{ij} = \boldsymbol{r}_i + \boldsymbol{r}_{i+1} + \cdots + \boldsymbol{r}_{j-1} + \boldsymbol{r}_j = \sum_{k=i}^{k=j} \boldsymbol{r}_k \tag{4.1}$$

Consider now the path through space of a *random walk*. Each step direction in a random walk is independent of the location and direction of all other steps in the walk, and all step directions are equally likely. Let $p(\theta,\phi)d\phi d\theta$ be the probability of a step with directions in the angular increments $d\phi$ and $d\theta$, where $0 \leq \phi \leq 2\pi$ and $0 \leq \theta \leq \pi$. For any probability distribution, the sum over all possible outcomes

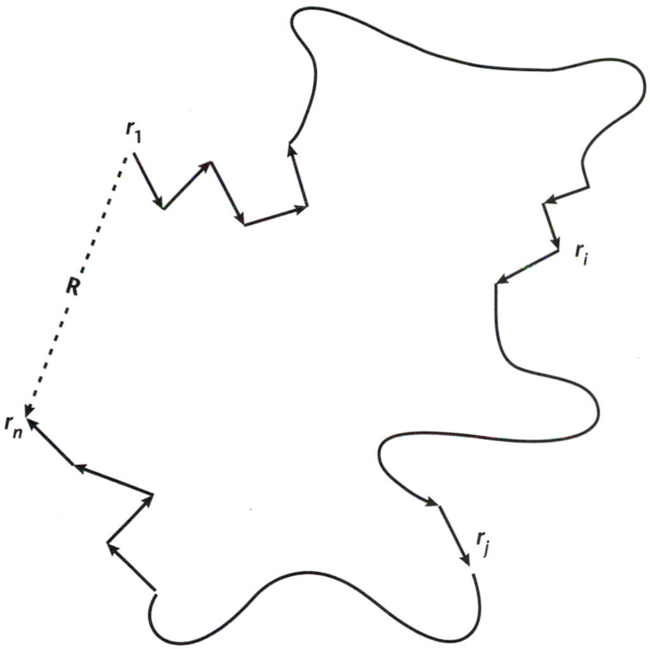

FIGURE 4.1 An *n*-step random walk and its end-to-end vector.

is unity:

$$\int_0^{\pi}\int_0^{2\pi} p(\theta,\phi)d\phi d\theta = 1 \tag{4.2}$$

For the random distribution of orientations,

$$p(\theta,\phi) = \frac{\sin\theta}{4\pi} \tag{4.3}$$

4.1.1 Average End-to-End Distance

Now consider some average sizes for a large collection of random walks, each with the same step length and number of steps. Some of the averages are known from symmetry alone. Thus, with pointed brackets indicating averages over the entire ensemble,

$$\begin{aligned}\langle \boldsymbol{r}_i \rangle &= 0\\ \langle \boldsymbol{R}_{ij} \rangle &= 0\end{aligned} \tag{4.4}$$

These are the direct consequences of randomness itself: All directions for a random step or sum of random steps have the same likelihood, so equal and opposite tend to cancel; for large collections, the average is zero. Consider now the scalar product of two steps in the same walk:

$$\boldsymbol{r}_i \cdot \boldsymbol{r}_j = l^2 \cos\alpha_{ij} \tag{4.5}$$

where α_{ij} is the angle between $\boldsymbol{r}_i$ and $\boldsymbol{r}_j$. For $i = j$, the directions are the same, so $\alpha_{ij} = 0$, $\cos\alpha_{ij} = 1$, and $\boldsymbol{r}_i \cdot \boldsymbol{r}_j = l^2$. For $i \neq j$, α_{ij} can be any angle between 0 and π, and its average value is $l^2 \int_0^{\pi}\int_0^{2\pi} \cos\alpha_{ij}\, p(\theta,\phi)d\phi d\theta$. Choose the axis to be parallel to one of the step vectors, making $\cos\alpha_{ij} = \cos\theta$. Thus, for the random distribution [Eq. 4.3],

$$\langle \boldsymbol{r}_i \cdot \boldsymbol{r}_j \rangle = \frac{l^2}{4\pi}\int_0^{\pi}\int_0^{2\pi} \cos\theta \sin\theta d\phi d\theta = 0 \tag{4.6}$$

Accordingly,

$$\begin{aligned}\langle \boldsymbol{r}_i \cdot \boldsymbol{r}_j \rangle &= l^2 \quad (i = j)\\ \langle \boldsymbol{r}_i \cdot \boldsymbol{r}_j \rangle &= 0 \quad (i \neq j)\end{aligned} \tag{4.7}$$

The scalar product of the end-to-end vector with itself—the square end-to-end distance—is always a positive quantity:

$$\boldsymbol{R}\cdot\boldsymbol{R} = \left(\sum_{i=1}^{n}\boldsymbol{r}_i\right)\cdot\left(\sum_{j=1}^{n}\boldsymbol{r}_j\right) = \sum_{i=1}^{n}\sum_{j=1}^{n}\boldsymbol{r}_i\cdot\boldsymbol{r}_j \tag{4.8}$$

where, for convenience here and henceforth, the end-to-end vector for the first i steps of an n-step walk $\boldsymbol{R}_{1i}$ is written $\boldsymbol{R}_i$, and that for the entire walk, $\boldsymbol{R}_{1n}$, is written $\boldsymbol{R}$. The *mean square end-to-end distance,* $\langle R^2\rangle = \langle\boldsymbol{R}\cdot\boldsymbol{R}\rangle$, is the average over many walks:

$$\langle R^2\rangle = \sum_{i=1}^{n}\sum_{j=1}^{n}\langle\mathbf{r}_i\cdot\mathbf{r}_j\rangle \tag{4.9}$$

Applying Eq. 4.7 to the terms, this expression then gives:

$$\langle R^2\rangle = nl^2 \tag{4.10}$$

The *contour length* of the walks, L, is nl, so

$$\langle R^2\rangle = Ll \tag{4.11}$$

The root-mean-square end-to-end distance, $\langle R^2\rangle^{1/2}$, is the geometric mean of contour length and step length. It is a measure of the extension in space of the walks.

Other averages can be obtained in like manner. Thus, for any connected subset of random walk steps,

$$\langle R_{ij}^2\rangle = |i-j|\,l^2 \tag{4.12}$$

Likewise, the average product of end-to-end vectors for different lengths of the same walk is the mean square end-to-end distance of their shared length. For the example of two walks that start at the same point and run for different distances along the same path,

$$\langle\mathbf{R}_i\cdot\mathbf{R}_j\rangle = \begin{matrix} il^2 & \text{if} & i\leq j \\ jl^2 & \text{if} & i\geq j \end{matrix} \tag{4.13}$$

4.1.2 Radius of Gyration

No general methods are available for directly measuring the end-to-end distance of polymer chains. When $\langle R^2\rangle$ is required for some purpose, it is inferred from data on other measures of molecular size. The root-mean-square distance of mass from

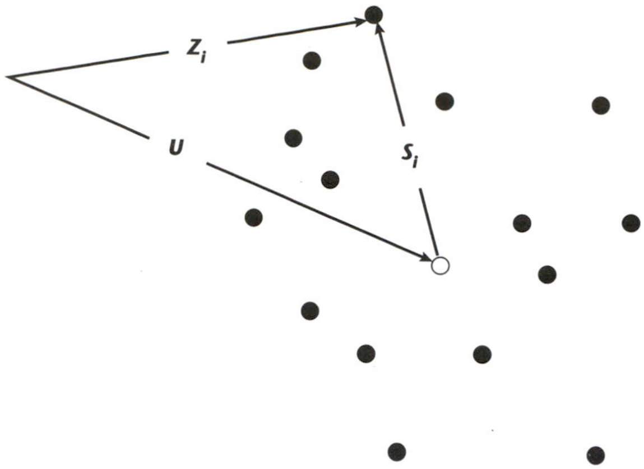

FIGURE 4.2 Points of equal mass representing some three-dimensional body showing their spatial locations relative to the center-of-mass.

the molecular center of mass or the *radius of gyration* R_g, can be determined by the scattering techniques described in Section 5.1.2. Thus, for future use we develop here an expression for R_g of random walks.

Consider first any array of n particles of equal mass ($i = 1, 2, \ldots n$), with locations in space specified by position vectors ($\mathbf{Z}_1, \mathbf{Z}_2, \ldots, \mathbf{Z}_n$), all relative to some common origin, as shown in Figure 4.2. The array has its center of mass at some location $\boldsymbol{U}$. The location of particle i, relative to the center of mass $\mathbf{S}_i$, is $\mathbf{Z}_i - \boldsymbol{U}$. Then, with $\sum_{i=1}^{n} \mathbf{S}_i = 0$ (the center of mass definition), $\sum_{i=1}^{n}(\mathbf{Z}_i - \boldsymbol{U}) = \sum_{i=1}^{n} \mathbf{Z}_i - n\boldsymbol{U} = 0$, so:

$$\boldsymbol{U} = \frac{1}{n}\sum_{i=1}^{n} \mathbf{Z}_i \tag{4.14}$$

The average square distance of particles from the center of mass of the array, $\mathbf{S}^2$, is the sum of the $\mathbf{S}_i \cdot \mathbf{S}_i$ values divided by their number: $S^2 = \sum_{i=1}^{n}(\mathbf{Z}_i - \boldsymbol{U}) \cdot (\mathbf{Z}_i - \boldsymbol{U})/n$. Thus, eliminating $\boldsymbol{U}$ by using Eq. 4.14,

$$S^2 = \frac{1}{n}\left[\sum_{i=1}^{n} \mathbf{Z}_i \cdot \mathbf{Z}_i - \frac{1}{n}\sum_{i=1}^{n}\sum_{j=1}^{n} \mathbf{Z}_i \cdot \mathbf{Z}_j\right] \tag{4.15}$$

With $\mathbf{Z}_i \cdot \mathbf{Z}_i = Z_i^2$ and $\mathbf{Z}_i \cdot \mathbf{Z}_j = Z_i Z_j \cos\theta_{ij}$, application of the law of cosines, $Z_{ij}^2 = Z_i^2 + Z_j^2 - 2Z_i Z_j \cos\theta_{ij}$, leads finally to:

$$S^2 = \frac{1}{2n^2}\sum_{i=1}^{n}\sum_{j=1}^{n} Z_{ij}^2 \tag{4.16}$$

The results so far are general, applying to any spatial distribution of n particles with equal mass. For flexibly connected sets of particles, like random walks or polymer chains, an ensemble average is required. The square root of this average, $\langle S^2 \rangle^{1/2}$, is the radius of gyration R_g. Thus,

$$R_g^2 = \frac{1}{2n^2} \sum_{i=1}^{n} \sum_{j=1}^{n} \langle Z_{ij}^2 \rangle \tag{4.17}$$

Now specialize this to random walks, with masses at the end of each step as measured from the walk origin. Thus, $\mathbf{Z}_i = \mathbf{R}_i$. So, using Eq. 4.12,

$$R_g^2 = \frac{l^2}{2n^2} \sum_{i=1}^{n} \sum_{j=1}^{n} |i - j| = \frac{l^2}{n^2} \sum_{i=1}^{n} \sum_{j=1}^{n} (i - j) \tag{4.18}$$

leading to:

$$R_g^2 = \frac{(n^2 - 1)l^2}{6n} \tag{4.19}$$

in which the sums $\sum_{j=1}^{n} (j - i) = (n - i)(n - i + 1)/2$ and $\sum_{i=1}^{n} (n - i)(n - i + 1) = n(n^2 - 1)/3$ were used[13].

The interior junctions and one chain end were weighted equally in these calculations, which is somewhat artificial and only an approximation for polymer chains. However, such detailed differences become unimportant when the chains are long. Thus,

$$R_g^2 = nl^2/6 \qquad (n \gg 1) \tag{4.20}$$

From Eq. 4.10, a general relationship exists between the two size measures for sufficiently long random walks and linear, uniform flexible-chain molecules:

$$\langle R^2 \rangle = 6R_g^2 \tag{4.21}$$

Other averages related to mass distribution can be calculated using the same methods, and the same simplifications can be used if the interest is long chain molecules. In such cases, sums can also be harmlessly replaced by integrals—for example, $R_g^2 = (l^2/n) \int_0^n \int_x^n (y - x)\,dy\,dx$ for linear chains [from Eq. 4.18, leading directly to Eq. 4.20]. Consider, for example, the average distance of chain unit i from the center of gravity:

$$\langle S_i^2 \rangle = \langle (\mathbf{R}_i - \mathbf{U}) \cdot (\mathbf{R}_i - \mathbf{U}) \rangle \tag{4.22}$$

Using Eq. 4.14 and some rearrangement,

$$\langle S_i^2 \rangle = \langle R_i^2 \rangle + \frac{1}{n^2} \sum_{j=1}^{n} \sum_{k=1}^{n} \langle \mathbf{R}_j \cdot \mathbf{R}_k \rangle - \frac{2}{n} \sum_{j=1}^{n} \langle \mathbf{R}_i \cdot \mathbf{R}_j \rangle \tag{4.23}$$

Using Eq. 4.13 and the replacement of sums by integrals,

$$\langle S_i^2 \rangle = \frac{\langle R^2 \rangle}{3} \left[1 - \frac{3i(n-i)}{n^2} \right] \tag{4.24}$$

Accordingly, $\langle S_i^2 \rangle$ is $2R_g^2$ for the chain ends and $R_g^2/2$ for the midpoint, indicating that chain units within the same molecule experience a modest range of average environments.

4.2 Polymer Chains

If the correspondence between chain conformations and random walks were exact, the size or mean extension in space of a polymer molecule—its coil dimensions—could be calculated using Eq. 4.10. Thus, for example, polyethylene has a molecular weight per backbone bond of 14, so for a polyethylene molecule with a molecular weight of 70,000, the number of backbone bonds is 70,000/14 = 5,000. The contour length of the chain, (5,000)(0.153) $\sim$ 770 *nm*, would be large enough to see with an optical microscope if its other dimensions were as big, but the cross section of the walk is atomic in size, <1 *nm*. From Eq. 4.10, the random walk size, $\langle R^2 \rangle^{1/2}$, $(0.153\ nm)(5{,}000)^{1/2}$, is about 11 *nm*, large compared with atomic dimensions but still submicroscopic. In fact, the coil size in solution for polyethylene molecules having $M = 70{,}000$ is larger, approximately 20 *nm*. The coil size also varies with the solvent species.

The estimate of size from the random walk expression has the right magnitude, but it is always too small. There are two distinctly different reasons for the discrepancy: long-range volume exclusion and local correlations in bond orientation, both arising from a breakdown in the analogy between random walks and polymer conformations.

First, a random walk is a mathematical path that occupies no volume and can therefore intersect itself, or come arbitrarily close to self-intersection, without consequence. A polymer chain is a physical object that takes up space. All its units have their space, and each unit repels the occupation of its space by any other unit, no matter how remotely the pair might be connected along the chain. This is the excluded volume interaction (Section 3.3), acting equally upon all pairs of units of the same chain. Being a repulsive intramolecular interaction, volume exclusion causes an expansion of coil dimensions in dilute solutions. For sufficiently long chains, it also causes a departure from the random walk relationship between length and coil size. Although still a power law relationship[14],

$$\langle R^2 \rangle = \Lambda l^2 n^\gamma \quad (4.25)$$

the exponent γ is larger than the random walk value of unity, and the coefficient Λ varies with the polymer–solvent system. At high polymer concentrations, the effect of self-repulsion on coil dimensions is cancelled by the mutual competition of many polymer molecules for the same space. The excluded volume effect can even be cancelled in dilute solutions by a judicious choice of solvent and temperature. The effect of self-avoidance on chain dimensions will be considered in some detail in Chapters 6 and 7.

The second difference between polymer chains and random walks—local correlations in the relative orientation of the backbone bonds—can also cause the coils to expand beyond the random walk size. The effect of correlation varies among polymer species, but it is relatively insensitive to environment and, unlike volume exclusion, it does not alter the form of the size–length relationship. The step directions in a random walk are independent of one another. The walk is *unrestricted,* and the path corresponds to that of a *freely jointed* or *random flight* chain. The backbone of a polymer chain, however, must obey bond angle restrictions and steric constraints that depend on chemical microstructure and hence on the polymer species. The steric constraints may also arise from mutual volume exclusion, for example, from the mutual denial of backbone rotational positions by side groups. That exclusion, however, unlike long-range exclusion, diminishes with increasing separation along the chain. As a result of these various constraints, the directions of successive bonds in polymer chains are correlated, so $\langle \boldsymbol{r}_i \cdot \boldsymbol{r}_{i+1} \rangle$ is not zero. The effect of this on the chain dimensions is considered in the following section.

4.2.1 Locally Restricted Conformations

A simple example of local correlation is the restriction to the tetrahedral bond angle ($\cos\beta = -1/3$) for successive carbon-carbon single bonds. For a freely rotating backbone of such bonds, the projection of $\boldsymbol{r}_{i+1}$ on $\boldsymbol{r}_i$ is always $l(-\cos\beta)$ except at the ends, so the scalar product $\boldsymbol{r}_i \cdot \boldsymbol{r}_{i+1}$ is always $l^2(-\cos\beta)$. Likewise, the projection of $\boldsymbol{r}_{i+2}$ on $\boldsymbol{r}_i$ is always $l(-\cos\beta)^2$, so $\boldsymbol{r}_i \cdot \boldsymbol{r}_{i+2}$ is always $l^2(-\cos\beta)^2$, and so on for the still more remotely connected bond pairs. The effect of a bond angle restriction on $\langle R^2 \rangle$ can thus be computed, term-by-term, for the double summation in Eq. 4.8. There are $n\langle \boldsymbol{r}_i \cdot \boldsymbol{r}_i \rangle$ terms, each contributing l^2, as in the unrestricted walk, but there are also $2(n-1)$ $\langle \boldsymbol{r}_i \cdot \boldsymbol{r}_{i+1} \rangle$ terms, each contributing $l^2(-\cos\beta)$; $2(n-2)\langle \boldsymbol{r}_i \cdot \boldsymbol{r}_{i+2} \rangle$

terms, each contributing $l^2(-\cos\beta)$; and finally $2\langle \boldsymbol{r}_i \cdot \boldsymbol{r}_{i+n/2}\rangle$ terms, each contributing $l^2(-\cos\beta)^{n-1}$. Using $x = -\cos\beta$ and some rearrangement,

$$\langle R^2\rangle = \left[1 + 2\sum_{i=1}^{i=n-1} x^i - \frac{2}{n}\sum_{i=1}^{i=n-1} i x^i\right] nl^2 \tag{4.26}$$

For large n, and with $\sum_{i=1}^{\infty} x^i = x/(1-x)$,

$$\langle R^2\rangle = \left[\frac{1-\cos\beta}{1+\cos\beta}\right] nl^2 \tag{4.27}$$

For the tetrahedral bond angle, $\cos\beta = -1/3$,

$$\langle R^2\rangle = 2nl^2 \tag{4.28}$$

Coil size is increased by the bond angle restriction, but its dependence on chain length, $\langle R^2\rangle^{1/2} \propto n^{1/2}$, is the same as for unrestricted walks [Eq. 4.10].

A more realistic chain, one having favored rotational positions as well, has also been considered. The energy depends on azimuthal angle ϕ for a rotationally hindered chain, so the angular average $\langle\cos\phi\rangle$ departs from its free-rotation value of zero. With ϕ measured from the planar zigzag position, the mean-square end-to-end distance for large enough n and small enough $\langle\cos\phi\rangle$, is[15],

$$\langle R^2\rangle = \left[\frac{1-\cos\beta}{1+\cos\beta}\right]\left[\frac{1+\langle\cos\phi\rangle}{1-\langle\cos\phi\rangle}\right] nl^2 \tag{4.29}$$

Again, the size–length relationship for random walks is preserved.

When unperturbed by excluded volume, the conformations of polymer chains are analogous to *Markoff walks*[16]. The distribution of orientations for step $i+1$ depends on the orientation of the preceding step i. The resulting correlation in step orientations extends over some number of steps along the walk, diminishing with separation towards independence at a rate that depends on the strength of the coupling between the successive orientations. In general, when long-range volume exclusion is cancelled, the form of the size–length relationship for long chains is universal. As expressed by Flory[17],

$$\langle R^2\rangle = C_\infty nl^2 \qquad (n \gg 1) \tag{4.30}$$

in which the *characteristic ratio* C_∞, referring to the asymptotic value ($n \to \infty$), depends on the detailed chemical microstructure of the chains.

TABLE 4.1 Effect of side group size and temperature on local length scale for selected polyolefins[18-20]

Species	Chain Unit	$C_\infty(140°C)$	$C_\infty(240°C)$
polyethylene	$-CH_2\text{-}CH_2-$	7.5	6.7
a-polypropylene	$\begin{array}{c} CH_3 \\ | \\ -CH_2\text{-}CH- \end{array}$	6.2	6.2
s-polypropylene	$\begin{array}{c} CH_3 \\ | \\ -CH_2\text{-}CH- \end{array}$	8.5	—
a-poly(1-butene)	$\begin{array}{c} CH_3 \\ | \\ CH_2 \\ | \\ -CH_2\text{-}CH- \end{array}$	7.3	7.6
a-poly(1-pentene)	$\begin{array}{c} CH_3 \\ | \\ CH_2 \\ | \\ CH_2 \\ | \\ -CH_2\text{-}CH- \end{array}$	8.1	8.3

With steric hindrance, the average, $\langle \cos\phi \rangle$, and thus chain dimensions become temperature dependent. In the absence of side groups and effects from more remotely connected units, the least excluded steric interference and thus the lowest energy for three bond sequences occurs for the planar zig-zag conformation, $\phi = 0$. The other two, less-deep minima may also center smaller, but still significant, populations that increase with increasing temperature. Those conformations also have smaller end-to-end lengths than the conformation, so C_∞ should decrease with increasing temperature. The steric repulsion of side groups can be more severe for the conformation, making its minimum relatively smaller and resulting in a reduction in C_∞. Increasing side group size should eventually reverse that trend, however, as the repulsions begin to force more extended conformations. As shown in Table 4.1, the C_∞ data for some members of the polyolefin family[18-20], measured at 140°C and estimated at 240°C from temperature coefficient information, generally support these expectations. Thus, C_∞ at 140°C is relatively large for polyethylene and even larger for s-polypropylene, the syndiotactic structure likely having furnished additional steric support for the lower energy planar zigzag conformation. Atactic polypropylene has much smaller C_∞, however, owing to meso side group interference with the planar zigzag conformation, but then C_∞ increases again with increasing side group size, as anticipated. At 240°C, C_∞ is significantly smaller for energetically dominated polyethylene but relatively unchanged for the steric dominated higher polyolefins.

Contributions from stereoregular placements and from interactions that are more remote along the chain than those considered above can also affect the chain dimensions. Attempts to relate size and contour length, based on detailed information about local molecular geometry and energetics for the species, have been successful in some species, for example polyethylene[17]. Such calculations, based on the *rotational isomeric state* (RIS) *model*[21], are described extensively elsewhere[17,21–23]. Methods based on PRISM or on computer simulations using a realistic representation of local structure and energetics are increasingly taking over the evaluation of rotational state probabilities, leaving to the RIS model the still very important task of generalization to long chains.

For the present at least, the unperturbed size–length relationship is best regarded as a polymer species property that must be determined experimentally. Values of C_∞ have been determined for many flexible-chain species by the scattering methods discussed in Chapters 5 and 7. Depending on the species, C_∞ can either increase or decrease with temperature. Usually, C_∞ is relatively insensitive to temperature and to the nonconnected environment. Variations with temperature are seldom more than 0.1% $°C^{-1}$ or more than 5% among different liquid media. Values of C_∞ and its temperature coefficient,

$$\kappa = \frac{d\ln C_\infty}{dT} \tag{4.31}$$

are given for some representative species in Table 4.2.

4.2.2 Step Length Definitions

The unperturbed size–length relationship is often expressed in terms other than the characteristic ratio. Several involve an *equivalent step length* for the species. For most purposes, the choice is more a matter of taste and convenience than substance. Some differences in usage among investigators have crept in over the years, however, so it is worthwhile to review briefly the various definitions.

In 1936, Kuhn[4] suggested an equivalent length $l' = \langle R^2 \rangle / L$ [Eq. 4.11] and a corresponding number of equivalent steps n', chosen to match the actual size and length of the chain. His choice to respect all backbone valence angles in expressing the chain length[24] turns out to be excessively intricate in some cases and was later simplified by Flory to $L = nl$, the sum of backbone bond lengths[15]. Equation 4.30 and the Flory modification are used here to define the *Kuhn step length:*

$$l_K \equiv C_\infty l \tag{4.32}$$

TABLE 4.2 Molecular properties related to chain dimensions for various polymer species[18,29]

Polymer	Acronym	$T(°C)$	$R_g/M^{1/2}(nm)$	$\rho\ (g\ cm^{-3})$	m_o	$l\ (nm)$	$l_\kappa(nm)$	$l_p(nm)$	C_∞	$10^3\kappa\ (K^{-1})$
polyisobutylene	PIB	140	0.0308	0.849	28	0.153	1.04	2.06	6.81	−0.27
1,4 polybutadiene	PBD	25	0.0382	0.895	13.5	0.148	0.80	1.27	5.40	
1,2 polybutadiene	1,2-PB	25	0.0346	0.889	27	0.153	1.27	1.56	8.28	
cis-polybutadiene	c-PBD	25	0.0355	0.900	13.5	0.148	0.69	1.46	4.66	0.4
1,4 polyisoprene	PI	25	0.0315	0.900	17	0.148	0.68	1.86	4.62	
cis-polyisoprene	c-PI	25	0.0332	0.910	17	0.148	0.76	1.65	5.13	0.4
polydimethyl siloxane	PDMS	25	0.0265	0.970	37	0.170	0.92	2.44	5.39	0.8
alt-poly(ethylene-propylene)	PEP	25	0.0392	0.856	17.5	0.153	1.05	1.26	6.89	−1.2
alt-poly(ethylene-butene)	PEB	25	0.0348	0.861	20.5	0.153	0.97	1.59	6.36	~0
poly(vinyl methyl ether)	PVME	25	0.0311	1.051	29	0.153	1.10	1.63	7.19	
polycarbonate	PC	25	0.0385	1.20	127	0.700	1.61	0.93	2.31	
poly(vinyl chloride)	PVC	30	0.0354	1.39	31.25	0.153	1.54	0.95	10.0	
poly(ethylene oxide)	PEO	80	0.0366	1.081	14.7	0.148	0.80	1.15	5.39	0.23
polystyrene	PS	140	0.0269	0.969	52	0.153	1.48	2.37	9.64	~0
poly(α-methyl styrene)	PαMS	25	0.0271	1.04	59	0.153	1.70	2.17	11.1	
poly(vinyl cyclohexane)	PVCH	140	0.0232	0.92	55	0.153	1.16	3.35	7.59	~0
poly(vinyl acetate)	PVAc	25	0.0286	1.08	43	0.153	1.38	1.88	9.02	
poly(methyl acrylate)	PMA	25	0.0270	1.11	43	0.153	1.23	2.05	8.03	0.1

(Continued)

TABLE 4.2 *(Continued)*

Polymer	Acronym	T(°C)	$R_g/M^{1/2}(nm)$	ρ (g cm^{-3})	m_o	l (nm)	$l_\kappa(nm)$	$l_p(nm)$	C_∞	$10^3\kappa$ (K^{-1})
poly(ethyl acrylate)	PEA	25	0.0278	1.13	50	0.153	1.52	1.90	9.90	
poly(methyl methacrylate)	PMMA	140	0.0266	1.13	50	0.153	1.39	2.08	9.07	0.1
polyethylene	PE	140	0.0456	0.785	14	0.153	1.14	1.02	7.46	−1.25
a-polypropylene	PP	140	0.0334	0.799	21	0.153	0.92	1.86	6.00	−0.1
i-polypropylene	iPP	190	0.0340	0.759	21	0.153	0.95	1.89	6.22	~0
s-polypropylene	sPP	180	0.0397	0.764	21	0.153	1.29	1.38	8.46	
head-to head polypropylene	hhPP	140	0.0339	0.810	21	0.153	0.95	1.78	6.19	~0
a-poly(1-butene)	PB	140	0.0289	0.813	28	0.153	0.92	2.44	5.99	0.45
i-poly(1-butene)	iPB	135	0.0320	0.813	28	0.153	1.12	1.99	7.35	
a-poly(1-pentene)	PPEN	85	0.0300	0.814	35	0.153	1.24	2.27	8.07	0.34
a-poly(1-hexene)	PH	61	0.0296	0.815	42	0.153	1.44	2.32	9.43	
a-poly(1-octene)	PO	50	0.0261	0.816	56	0.153	1.50	2.99	9.78	
a-poly(1-decene)	PDEC	30	0.0252	0.816	70	0.153	1.74	3.20	11.4	
a-poly(1-hexadecene)	PHDC	145	0.0188	0.796	112	0.153	1.55	5.90	10.2	
polytetrafluoroethylene	PTFE	280	0.0291	1.46	50	0.153	1.66	1.34	10.9	
poly(phenylene oxide)	PPO	220	0.0351	1.01	120	0.540	1.64	1.33	3.04	
poly(ethylene terephthalate)	PET	275	0.0386	0.99	32	0.268	1.07	1.13	3.98	

Other step length definitions have been used. The *statistical segment length* l_s is obtained by equating $\langle R^2 \rangle = C_\infty n l^2$ and $\langle R^2 \rangle = n l_s^2$:

$$l_s \equiv C_\infty^{1/2} l \tag{4.33}$$

The same information about the size–length relationships is conveyed by R_g^2/M and R_g^2/N, ratios of directly measurable quantities:

$$\begin{aligned} R_g^2/M &= l_s^2/6m_o \\ R_g^2/N &\equiv l_N^2/6 \end{aligned} \tag{4.34}$$

in which m_o is the polymeric molecular weight per backbone bond, N is the degree of polymerization, and l_N is the *mer-based statistical length.* The *persistence length,* or the average projection of bond vectors on the end-to-end vector, is still another definition of effective step length[25]. The persistence length definition is based on the representation of a polymer chain by an elastic rod, flexed all along its length by thermal agitation. The elastic energy of an arbitrarily flexed cylindrical rod with tensile modulus E, radius r, and length L is:

$$\Psi = \frac{\varepsilon}{2} \int_o^L \frac{dz}{\Gamma^2(z)} \tag{4.35}$$

in which $\Gamma(z)$ is the radius of curvature at location z along the rod, and $\varepsilon = \pi r^4 E/4$ is the flexural rigidity of the rod[26]. The projection of a vector of unit length tangent to the centerline on the end-to-end vector, averaged over the length of the rod, and then averaged over a large collection of identical rods at thermal equilibrium, yields the persistence length:

$$l_o = \frac{\varepsilon}{kT}\left[1 - \exp\left(-\frac{LkT}{\varepsilon}\right)\right] \tag{4.36}$$

The mean-square end-to-end distance and radius of gyration are:

$$\begin{aligned} \langle R^2 \rangle &= \frac{\varepsilon}{kT}(2L - l_o) \\ R_g^2 &= \frac{L\varepsilon}{6kT}\left[1 - \frac{3}{2}\left(\frac{\varepsilon}{LkT}\right) + \frac{3}{2}\left(\frac{\varepsilon}{LkT}\right)^2\left(\frac{L - l_o}{L}\right)\right] \end{aligned} \tag{4.37}$$

Rigid rods and flexible coil behavior are the limiting cases:

$$\begin{aligned} l_o &= L \\ \langle R^2\rangle &= L^2 \qquad\qquad L \ll \varepsilon/kT \\ R_g^2 &= L^2/4 \end{aligned} \tag{4.38}$$

$$\begin{aligned} l_p &= \varepsilon/kT \\ \langle R^2\rangle &= 2l_oL \qquad\qquad L \gg \varepsilon/kT \\ R_g^2 &= l_oL/3 \end{aligned} \tag{4.39}$$

The flexible rod forms the basis of the wormlike chain and the helical wormlike chain, models that can be used to deal with semiflexible macromolecules and also to permit a uniform description of the evolution of chain molecule properties from the very short to the very long[27]. Short-enough rods are stiff and long-enough rods become sufficiently flexible to have the statistical properties of random walks. For the flexible-coil limit, using Eqs. 4.30, 4.32, and 4.39,

$$l_o = l_K/2 \tag{4.40}$$

Some equivalent lengths depend on other properties in addition to step length. The *Helfand packing length*[28], a local size that frequently appears in physical property descriptions, is defined as:

$$l_p \equiv \frac{M}{N_a R_g^2 \rho} = \upsilon/l_s^2 \tag{4.41}$$

in which ρ is the mass density of the polymer liquid and υ is $M/nN_a\rho$, the molecular volume per backbone bond. The packing length is defined here with R_g^2, rather than the more recent usage of $\langle R^2\rangle$[18], thus reverting to Helfand's original definition. The preference in any case is always to use observables in the definitions wherever possible. The values of some local size measures for common polymer species[18,29] are given in Table 4.2.

4.3 Nonlinear Molecules

The polymeric architectures considered thus far are linear, and the random walk analogy has been exploited to discuss their size–length relationship. Some polymerization systems form branched chains, either through a conscious choice of reaction

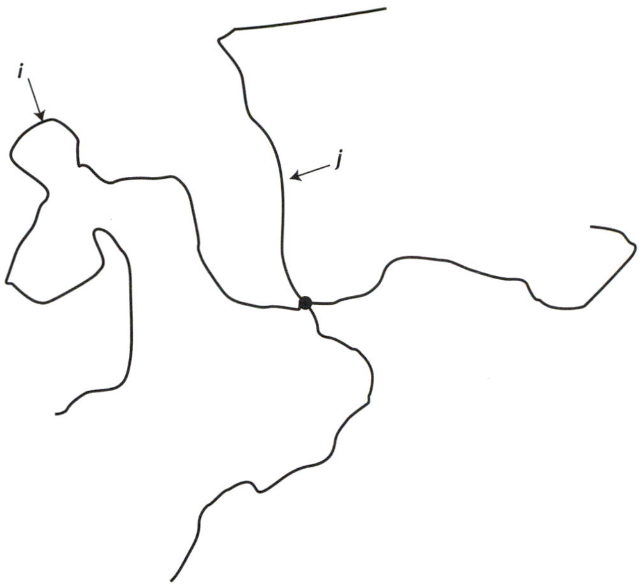

FIGURE 4.3 Random coil star with arms of equal length.

pathways or unbidden, through side reactions. In either case, the effect of long molecular branches on properties can be very large and sometimes even beneficial. This section considers the size relationships for several kinds of nonlinear structures, including macrocyclic rings.

4.3.1 Branched Chains

The methods for calculating chain dimensions described in Section 4.1.2 can be applied to chains with nonlinear architectures as well[30]. Consider, for example, the radius of gyration for f-arm *symmetric star* polymers, as illustrated in Figure 4.3. Each arm has a large number, n_a, units so that n, the total number in the molecule, is fn_a. Use Eq. 4.17 with the fact that each i, j pair is connected by a single random walk path, so that Eq. 4.12 is applicable. Replace the sums with integrals, and i and j by the continuous variables x and y, to obtain:

$$\left(R_g^2\right)_{ss} = \frac{l^2}{2n^2}\left[2f\int_0^{n_a}\int_x^{n_a}(y-x)dy\,dx + f(f-1)\int_0^{n_a}\int_0^{n_a}(n_a - x + y)dy\,dx\right] \tag{4.42}$$

In the first term, x and y are on the same arm, and in the second they are on different arms. Evaluation of the integrals leads to[30]:

$$\left(R_g^2\right)_{ss} = \frac{nl^2}{6}\frac{3f-2}{f^2} \tag{4.43}$$

Thus, $\left(R_g^2\right)_{ss}$ is the product of two terms, $nl^2/6 = R_g^2$, for a linear chain having the same number of chain units and a factor called the *size ratio*. The size ratio is characteristic of the large-scale chain architecture and is expressed for any nonlinear architecture using:

$$g \equiv \frac{\left(R_g^2\right)_{nonlinear}}{\left(R_g^2\right)_{linear}} \tag{4.44}$$

in which the sizes are compared at the same number of backbone bonds per molecule. For symmetric stars,

$$g_{ss} = \frac{3f-2}{f^2} \tag{4.45}$$

Note that $g = 1$ for both $f = 1$ and $f = 2$, which each correspond to linear chains, and that g decreases monotonically with f when the chain is branched; that is, $f > 2$. The size ratio g is less than one for all nonlinear architectures.

For random walk structures without closed loops, each location i on a structure of n units partitions it into two parts having n_i and $n - n_i$ units respectively. The sum of the products of these numbers is related to the radius of gyration according to the *Kramers equation*[30]:

$$\langle S^2 \rangle = \frac{l^2}{n^2}\sum_{i=1}^{n} n_i(n - n_i) \tag{4.46}$$

The Kramers equation applied to symmetric stars leads to:

$$R_g^2 = \frac{l^2 f}{n^2}\int_0^{n_a} x(n-x)\,dx \tag{4.47}$$

which integrates directly to give Eq. 4.43.

Higher moments of the mass distribution can be obtained[25,31]. The fourth moment is given for ring-free architectures by an extension of Eq. 4.46, the *Pearson-Raju equation*[32]:

$$\frac{\langle S^4 \rangle - \langle S^2 \rangle^2}{\langle S^2 \rangle^2} = \frac{24}{n^6}\sum_{i=1}^{n}\sum_{j=1}^{n} n_i^2 n_j^2 \tag{4.48}$$

from which terms of lower orders in n have been discarded. Unit i divides the structure into two parts, and n_i is the number of units in the part that does not contain unit j. Unit j also divides the structure into two parts, and n_j is the number in the part not containing unit i. For linear chains,

$$\frac{\langle S^4\rangle - \langle S^2\rangle^2}{24\langle S^2\rangle^2} = \frac{2}{n^6}\int_{0}^{n}\int_{x}^{n} x^2(n-y)^2 dy dx \tag{4.49}$$

which leads to:

$$\langle S^4\rangle = \frac{19}{15}\langle S^2\rangle^2 \tag{4.50}$$

in agreement with Yamakawa[25]. For symmetric stars,

$$\langle S^4\rangle_{ss} = \frac{135f^2 - 120f + 4}{15(3f-2)^2}\langle S^2\rangle_{ss}^2 \tag{4.51}$$

Size ratios have also been calculated for the two architectures shown in Figure 4.4. The following applies for *regular pom-poms*[33], which are two symmetric f-arm stars with centers linked by a single strand or "crossbar":

$$g_{rpp} = \lambda^3 + 3\lambda^2(1-\lambda) + \frac{3}{2}\left(\frac{f+1}{f}\right)\lambda(1-\lambda)^2 + \left(\frac{3f-1}{2f^2}\right)(1-\lambda)^3 \tag{4.52}$$

in which λ is the fraction of units in the crossbar and $f = 2$ corresponds to an H-shaped molecule.

Three types of regular combs can be distinguished. *Symmetric combs* have f arms equally spaced along a linear backbone, *random combs* have f arms randomly

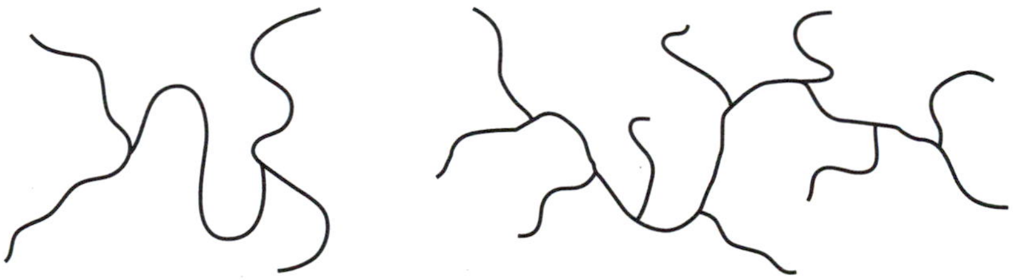

FIGURE 4.4 Random coil H-molecule and random coil comb.

distributed along a linear backbone, and *heterogeneous combs* have arms distributed randomly among backbones, such that $\bar{f}$ is the average number per backbone[34–36]:

$$g_{sc} = \lambda - \frac{1}{f+1}(1-\lambda)\lambda^2 + \frac{2}{f}(1-\lambda)^2\lambda + \frac{3f-2}{f^2}(1-\lambda)^3$$

$$g_{rc} = \lambda + \frac{2}{f}(1-\lambda)^2\lambda + \frac{3f-2}{f^2}(1-\lambda)^3 \tag{4.53}$$

$$g_{hc} = \lambda + \frac{3}{\bar{f}}(1-\lambda)^2 + \frac{1}{\bar{f}^2}(1-\lambda)^3$$

in which λ is the fraction of total monomeric units in the backbones. The difference among the three expressions is negligible for most cases of practical interest.

Branching to form treelike structures with broad distributions is the most commonly encountered class of nonlinear molecular architectures. In a remarkable paper, Zimm and Stockmayer in 1949 evaluated the distributed size-ratio characteristics for the products of random nonlinear polycondensations[37]. That analysis has served for more than 50 years as the basis of branch frequency estimations in polydisperse samples from dilute solution properties. The subject of long chain branching determination is touched on briefly in Chapter 6.

4.3.2 Macrocycles

Macrocyclic rings (see Figure 4.5) also have smaller dimensions than linear chains with the same number of units. Equation 4.17 still applies, but Eq. 4.12 and 4.46 do not, because some units are connected with one another along more than one chain path. Consider a ring of n units, in effect the subset of random walks of n steps that return exactly to their origin. For large n[30],

$$\langle R_{ij}^2 \rangle_{ring} = l^2[|i-j| - (i-j)^2/n] \tag{4.54}$$

Thus, using Eq. 4.17,

$$\left(R_g^2\right)_{ring} = \left(R_g^2\right)_{linear} - \frac{l^2}{2n^3}\int_0^n\int_0^n (y-x)^2 dy dx \tag{4.55}$$

leading immediately to $\left(R_g^2\right)_{ring} = nl^2/12$,

$$g_{ring} = \frac{1}{2} \tag{4.56}$$

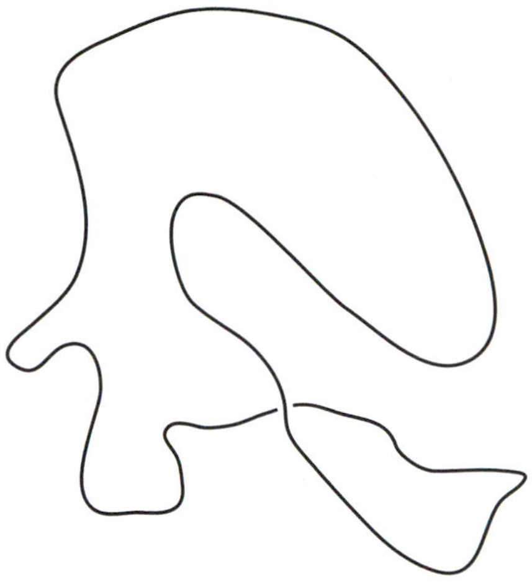

FIGURE 4.5 Random coil macrocycle.

Branched chains and rings invariably have smaller overall dimensions than linear chains with the same total number of chain units, but a significant distinction between the two remains. Thus, each pair of chain units in any acyclic structure is connected by only one path of chain units. Moreover, the mean-square separation distance for each pair is the same as that for linear chains with the same path length. More than one path connects pairs of units in rings, and their mean separation distance is smaller than the mean separation of linear chain ends having even the shortest of those paths.

The contraction effects for rings are expressed by a theorem that holds for arbitrary structures composed of random walk segments. Thus, let N_S be the number of linear chain segments that compose the structure and N_J the number of points in the structure whose locations are required to specify completely the end-to-end vectors of all N_S segments. Now let $\langle r_i^2 \rangle$ be the mean-square end-to-end distance for segment i in the structure and $\langle r_i^2 \rangle_o$ its free chain (ring-free) value. For any random walk architecture, $\langle r_i^2 \rangle$ and $\langle r_i^2 \rangle_o$ are related by the expression[38]:

$$\sum_{i=1}^{N_S} \frac{\langle r_i^2 \rangle}{\langle r_i^2 \rangle_o} = N_J - 1 \tag{4.57}$$

Thus, for example, all chain paths in any loop-free structure have the mean-square end-to-end distances of free chains, and the average contraction for loop-containing structures is simply the ratio of loops to segments. Applying Eq. 4.57 to an n-step ring represented by j linear segments, each with n/j steps, gives a segment contraction ratio $\langle r_j^2 \rangle / \langle r_j^2 \rangle_o$ equal to $(j - 1)/j$. Accordingly, the half-ring segments $(j = 2)$ are contracted by the factor 1/2, while the smaller ones are less contracted. The segments in rings approach their free chain dimensions in the many-segment limit.

4.4 Random Coil Asymmetry

The mean shape characteristics of random walks can be explored using these same methods. A simple measure of asymmetry can be obtained from the components of $\langle S^2 \rangle$ in directions parallel and normal to the end-to-end vector. Thus, for example, for linear chains and from symmetry considerations alone,

$$R_g^2 = \langle S^2 \rangle_P + 2\langle S^2 \rangle_N \tag{4.58}$$

A lengthy but straightforward calculation leads to $\langle S^2 \rangle_P = nl^2/12$, and with that result, Eq. 4.58 gives $\langle S^2 \rangle_N = nl^2/24$. The ratio $\langle S^2 \rangle_P / \langle S^2 \rangle_N = 2$ shows that, on average, random walks are not spherically symmetric objects even in the large n limit. Relative values of $\langle S^2 \rangle$ parallel to the principal axes of inertia indicate even greater asymmetries, and that linear chains are more asymmetric than nonlinear ones. The values for linear, star, and ring architectures in Table 4.3 were obtained by computer simulation for values of n between 60 and 100[39–42].

TABLE 4.3 Components of $\langle S^2 \rangle$ along the instantaneous principal axes [39–42]

Architecture	$\langle S_1^2 \rangle : \langle S_2^2 \rangle : \langle S_3^2 \rangle$
Linear Chains	11.8 : 2.5 : 1
Four-Arm Stars	6.2 : 2.6 : 1
Rings	6.2 : 2.3 : 1

4.5 Conformational Distributions

Until now, only some average properties of random walks have been considered. Here the properties of the distribution are treated, beginning with the distribution of end-to-end distances. The language of probability distributions and densities is used to discuss the subject. The purpose is to evaluate the fraction of all possible n-step walks in three dimensional space that begin at the origin and end within some specified differential volume. To understand the flow of calculations in what follows, consider first the distribution in one dimension as obtained by a simple probability formula.

Each walk takes place along the x-axis, starting at the origin, and has n steps. Each step has length l and an equal likelihood of choosing the $+x$ or $-x$ direction. The end-to-end coordinate of a walk end is its surplus of $+$ over $-$ step lengths, $x = ml$, where $m = n_+ - n_-$ and $n = n_+ + n_-$. The total number of different n-step walks is 2^n, the number with surplus m is the binomial coefficient $n!/m!(n-m)!$, as may be verified by direct enumeration. Thus, the probability of an n-step walk having surplus m is:

$$P(n,m) = \frac{n!}{2^n\left(\dfrac{n+m}{2}\right)!\left(\dfrac{n-m}{2}\right)!} \tag{4.59}$$

The formula of interest applies to walks in which n is very large, and where m is also large but still much smaller than n: $n \gg m \gg 1$. Large number factorials permit the use of Sterling's approximation, Eq. (2.93), with the result, after some algebra,

$$\ln P = -\frac{n}{2}\ln\left[1-\left(\frac{m}{n}\right)^2\right] - \frac{m}{2}\ln\left[\frac{1+(m/n)}{1-(m/n)}\right] \tag{4.60}$$

Because only the range $m/n \ll 1$ is considered, the logarithms can be expressed by first terms in power series, with the result:

$$\ln P = -\frac{m^2}{2n} \tag{4.61}$$

Exponentiating both sides, employing $x = ml$, and normalizing the function to satisfy $\int_{-\infty}^{+\infty} P(x)dx = 1$ leads to the 1-dimensional Gaussian distribution:

$$P(x) = \left(\frac{1}{2\pi nl^2}\right)^{1/2} \exp\left(\frac{-x^2}{2nl^2}\right) \tag{4.62}$$

4.5.1 The End-to-End Vector Distribution

Consider the end-to end distribution of a random walk in three dimensions[6,25]. Let $V_n(\boldsymbol{R})d\boldsymbol{R}$ be the probability of an n-step walk having end-to-end vector $\boldsymbol{R}$ in the differential volume $d\boldsymbol{R}$, and let $p(\boldsymbol{r})d\boldsymbol{r}$ be the probability of individual steps having step vector $\boldsymbol{r}$ in $d\boldsymbol{r}$. Both represent fractions and must satisfy the normalization condition:

$$\begin{aligned}\int V_n(\boldsymbol{R})d\boldsymbol{R} &= 1\\ \int p(\boldsymbol{r})d\boldsymbol{r} &= 1\end{aligned} \tag{4.63}$$

The object is to express the probability density $V_n(\boldsymbol{R})$ in terms of the step probability density $p(\boldsymbol{r})$ and the number of steps in the walk. The main interest is in steps of random orientations with some fixed length l, but it is no more difficult to carry through the calculation for randomly oriented steps of variable length. Thus, in spherical coordinates (see Section 2.1.2),

$$p(\boldsymbol{r})d\boldsymbol{r} = r^2 p(r)\sin\theta\, d\theta\, d\phi\, dr \tag{4.64}$$

with the normalization $\int_0^\infty \int_0^{2\pi} \int_0^\pi r^2 p(r)\sin\theta\, d\theta\, d\phi\, dr = 1$.

The walk and step probability densities, $V_n(\boldsymbol{R})$ and $p(\boldsymbol{r})$, are related as indicated in Figure 4.6. A displacement $\boldsymbol{R}$ occurs after n steps whenever any displacement $\boldsymbol{R} - \boldsymbol{R}'$ after $n-1$ steps is followed by the step displacement $\boldsymbol{R}'$:

$$V_n(\boldsymbol{R}) = \int V_{n-1}(\boldsymbol{R} - \boldsymbol{R}')p(\boldsymbol{R}')d\boldsymbol{R}' \tag{4.65}$$

This is a convolution equation [see Eq. (2.68)] and can be solved by the use of Fourier transforms [see Eq. (2.65)]. Thus, transforming both sides of Eq. 4.65 to Fourier space and using Eq. (2.68),

$$\hat{V}_n(\boldsymbol{s}) = \hat{V}_{n-1}(\boldsymbol{s})\hat{p}(\boldsymbol{s}) \tag{4.66}$$

Applying Eq. 4.66 successively for $n = 2, 3, \ldots$ then leads to:

$$\hat{V}_n(\boldsymbol{s}) = [\hat{p}(\boldsymbol{s})]^n \tag{4.67}$$

Transforming Eq. 4.67 to real space [note Eq. (2.65)] then gives a formal solution to the problem:

$$V_n(\boldsymbol{R}) = \frac{1}{(2\pi)^3} \int [\hat{p}(\boldsymbol{s})]^n \exp(-i\boldsymbol{R} \cdot \boldsymbol{s})d\boldsymbol{s} \tag{4.68}$$

The step probability is isotropic, $p(\boldsymbol{r}) = p(r)$, so $\hat{p}(\boldsymbol{s}) = \hat{p}(s)$ as well, and integration

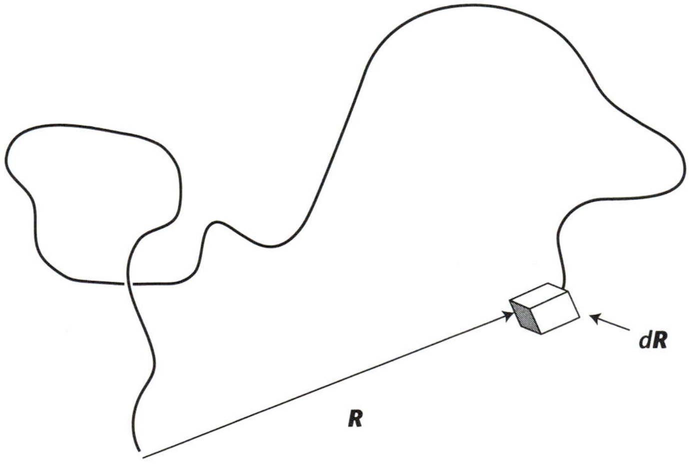

FIGURE 4.6 Trajectory of an *n*-step random walk, beginning at an origin and ending in an elemental volume.

over the angles [see Eq. (2.66)] yields:

$$\hat{p}(s) = \int_{0}^{\infty} 4\pi u^2 p(u) \frac{\sin s\,u}{s\,u} du \tag{4.69}$$

Applying that same procedure to the inverse transform, Eq. 4.68 then gives:

$$V_n(\boldsymbol{R}) = \frac{1}{2\pi^2|\boldsymbol{R}|} \int_{0}^{\infty} \left[\int_{0}^{\infty} 4\pi u^2 p(u) \frac{\sin s\,u}{s\,u} du \right]^n s \sin|\boldsymbol{R}|s\, ds \tag{4.70}$$

in which $|\boldsymbol{R}| = (\boldsymbol{R} \cdot \boldsymbol{R})^{1/2}$. For the case of fixed step length, $p(u)$ is the *Dirac delta function,* so using Eq. (2.70), $\hat{p}(s) = \sin ls/ls$, and Eq. 4.70 becomes:

$$V_n(\boldsymbol{R}) = \frac{1}{2\pi^2|\boldsymbol{R}|} \int_{0}^{\infty} \left[\frac{\sin ls}{ls} \right]^n s \sin|\boldsymbol{R}|s\, ds \tag{4.71}$$

Equation 4.71 is an exact expression for the end-to-end vector probability density for random walks having steps of fixed length, applicable for any n and $\boldsymbol{R}$.

Exact but awkward to use, Eq. 4.71 can be approximated with negligible error by a simpler expression in nearly all situations of polymeric interest. The main concern is walks of many steps, for which case the overwhelming majority have end-to-end distances that are much smaller than the contour length. A form that accurately describes the distribution in that regime, $n \gg 1$ and $|\boldsymbol{R}| \ll nl$, is sufficient for most purposes.

The approximation can be obtained as follows. Note that the magnitude of $\sin ls/ls$ for $ls > 0$ is less than 1, so the term $[\sin ls/ls]^n$ in Eq. 4.71 becomes vanishingly small for $n \gg 1$, except when ls very near ϕ. Accordingly, the integral is dominated by the behavior of $\sin ls/ls$ in the immediate $ls = 0$ neighborhood; thus, only the first terms of its series expansion around ϕ are needed:

$$\sin ls/ls = 1 - (ls)^2/3! + \cdots \tag{4.72}$$

This permits the approximation:

$$(\sin ls/ls)^n \Rightarrow (1 - (ls)^2/6)^n \Rightarrow \exp(-nl^2s^2/6) \tag{4.73}$$

Carrying out the integration in Eq. 4.71 using this approximation gives, for both

$n \gg 1$ and $|\boldsymbol{R}| \ll nl$,

$$V_n(\boldsymbol{R}) = \left(\frac{3}{2\pi\langle \boldsymbol{R}^2\rangle}\right)^{3/2} \exp\left(-\frac{3\boldsymbol{R}\cdot\boldsymbol{R}}{2\langle \boldsymbol{R}^2\rangle}\right) \tag{4.74}$$

in which $\langle|\boldsymbol{R}|\rangle^2 = \langle \boldsymbol{R}\cdot\boldsymbol{R}\rangle = \langle R^2\rangle = nl^2$.

Equation 4.74 is the *Gaussian probability density*, which in the polymeric context is sometimes called the *random coil distribution* or simply the *Gaussian distribution*. Expressed in spherical coordinates,

$$V_n(\boldsymbol{R})d\boldsymbol{R} = \left(\frac{3}{2\pi\langle R^2\rangle}\right)^{3/2} \exp\left[-\frac{3r^2}{2\langle R^2\rangle}\right] r^2 \sin\theta\, d\theta\, d\phi\, dr, \tag{4.75}$$

and in Cartesian coordinates,

$$V_n(\boldsymbol{R})\,d\boldsymbol{R} = \left(\frac{3}{2\pi\langle R^2\rangle}\right)^{3/2} \exp\left[-\frac{3(x^2+y^2+z^2)}{2\langle R^2\rangle}\right] dx\,dy\,dz \tag{4.76}$$

with the ranges $-\infty < x, y, z < +\infty$ in the latter. Notice that Eq. 4.76 is factorable—the distribution is the product of independent functions for each of the three coordinate directions. Thus,

$$V_n(\boldsymbol{R}) = V_n(x)V_n(y)V_n(z) \tag{4.77}$$

in which:

$$V_n(u) = \left(\frac{1}{2\pi\langle u^2\rangle}\right)^{1/2} \exp\left(-\frac{u^2}{2\langle u^2\rangle}\right) \tag{4.78}$$

This is the formula for a one-dimensional random walk, Eq. 4.62. The following symmetry property of the Gaussian is easily demonstrated:

$$\langle x^2\rangle = \langle y^2\rangle = \langle z^2\rangle = \langle R^2\rangle/3 \tag{4.79}$$

The Gaussian distribution in one dimension is sketched in Figure 4.7.

The factorability of the distribution of end-to-end distances is not obvious from the physical situation and deserves comment. Factorability requires that the distribution of each component be independent of component values in the other two directions. This cannot be correct as a general proposition. Walks with constant step lengths have finite contour lengths, so the sufficiently high extension of a walk in one direction is certain to affect its extension in other directions. Factorability is indeed correct, however, when the number of steps is large and the overwhelming majority

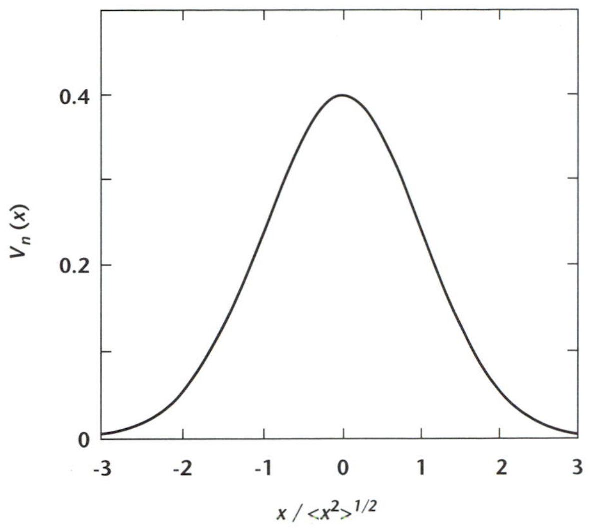

FIGURE 4.7 The Gaussian function in one dimension [Eq. 4.78].

of walks satisfy $|\boldsymbol{R}| \ll nl$, exactly the situation described by the Gaussian form. The errors that result from using the Gaussian approximation for flexible polymer chains are negligible as long as highly extended conformations are sufficiently rare. Considerations of the errors and better approximations for describing the situations in which they may be significant are described elsewhere[43].

4.5.2 The Gaussian Form

The Gaussian approximation is also not restricted to random walks, or even to fixed step lengths. It is a consequence of the *central limit theorem*[16] and applies to any collection of sufficiently long Markoff walks having the same structure. It also applies to any set of sufficiently long polymer chains when the large-scale conformations are not perturbed by excluded volume. The restriction of end-to-end distances much smaller than contour length is always in force, and the many-steps requirement is merely replaced by, for example, a requirement of many Kuhn steps.

Does the central limit theorem—a mathematical relationship that leads inexorably to the Gaussian approximation in the limit of many steps and small end displacements—still apply to polymer chains if the Kuhn length depends on temperature? As explained in Section 4.2.1, any measure of molecular size for a species depends on temperature if its various local conformers have lengths with different energies. Increasing end-to-end distances requires increasing the fractions of the more extended local conformers: Large-scale conformations have different energies if they have different end-to-end distances. One might then argue that the end-to-end distances require a Boltzmann weighting and thus that their distribution departs from the Gaussian form. On the other hand, in the long chain limit, arbitrary sequences of bonds can be grouped to form the step lengths, and if the sequences are taken long enough, each such step will contain a conformational average at the temperature of interest. When the chains are long enough to contain many such redefined steps, no additional end-to-end Boltzmann weighting is required, the central limit theorem surely applies, and the distribution of large-scale conformations must be Gaussian.

The second conclusion is correct—the distribution is Gaussian even if $\langle R^2 \rangle$ varies with temperature due to the various conformers having different lengths and energies. A physically satisfying demonstration of this, using a rotational isomeric model is given in Volkenstein's book[21]. Treating flexible-chain molecules as individual thermodynamic systems, he shows that three terms dominate the free energy change with end-to-end extension:

- a configurational entropy decrease that always accompanies end-to-end extension (the source of the "spring force" discussed in Section 4.6.1 below);
- a change in potential energy with extension, either an increase or a decrease, depending on whether the longer of the isomers has higher or lower energy; and
- a mixing-like entropy, arising from the change with extension in the distribution of isomers along the chain.

Volkenstein shows that, for long chains and not-too-large extension, the second and third terms cancel, leaving only the configurational entropy and hence the Gaussian form. This conclusion has important consequences for many topics considered in this volume. It means, for example, that the effect of local correlations in backbone bond direction, including their temperature dependence, is absorbed entirely into the parameter $\langle R^2 \rangle$ of the Gaussian expression, Eq. 4.74.

4.5.3 Mass Distribution for Gaussian Coils

The average distribution of chain units about the center of mass, $\nu(\mathbf{s})$, is another property of interest for polymeric systems. Even in the limit of long flexible chains, however, no simple expression analogous to the Gaussian for the end-to-end vector, emerges. The exact expression for long chains is known[25], and resembles the shape of a Gaussian distribution. It is frequently approximated by a Gaussian:

$$\nu(s) = n\left[\frac{3}{2\pi R_g^2}\right]^{3/2} \exp\left[-\frac{3s^2}{2R_g^2}\right] \tag{4.80}$$

in which $s^2 = \mathbf{s} \cdot \mathbf{s}$, and n is the number of units per chain.

A comparison of the Gaussian approximation with the exact distribution can be made by examining the relative values of distribution moments. The averages of successive powers of s^2 provide a useful signature of the Gaussian distribution. Thus, for the Gaussian,

$$\langle s^4 \rangle = \frac{5}{3} \langle s^2 \rangle^2 \tag{4.81}$$

whereas the exact distribution for linear chains gives:

$$\langle s^4 \rangle = 2 \langle s^2 \rangle^2 \tag{4.82}$$

By this criterion, the mass distribution for linear chains is slightly broader than Gaussian.

4.6 Properties of Random Coil Ensembles

Until now, the discussion has focussed on the architectural aspects of flexible chain polymers emphasizing mainly molecular size and size distribution of random coil systems. This section introduces the physical properties conferred by random coils, the microscopic forces they can exert and some of the special thermodynamic features they generate. It presents the basis for discussions of volume exclusion and network elasticity in later chapters and the molecular aspects of dynamics and rheology in Volume 2.

4.6.1 Mechanical Properties

Consider a large collection of identical molecules, each consisting of two particles that interact with one another classically through a force acting along their line of centers with a magnitude that depends only on their separation distance. The force $f(r)$ and the associated potential energy $\psi(r)$ are related by Eq. (2.1). If no other forces are involved, the fraction of the molecules having separation distance r in the range dr is related to the potential energy through the Boltzmann factor[44]:

$$p(r)dr = \frac{r^2 \exp[-\psi(r)/kT]\, dr}{\int_0^\infty r^2 \exp[-\psi(r)/kT]\, dr} \tag{4.83}$$

Direct substitution of the quadratic attraction,

$$\psi(r) = \frac{3kT}{2\langle R^2 \rangle} r^2 \tag{4.84}$$

leads to a Gaussian distribution. The result is in fact identical to the end-to-end distance distribution for random coil molecules, [Eq. 4.75] after angular averaging. Accordingly, the distribution of end-to-end distances for random coils is the same as would be obtained if the entire chain connecting the ends were replaced by an attractive quadratic potential acting on the ends alone. From Eq. 4.84, the force associated with the potential is

$$\boldsymbol{F}(\boldsymbol{r}) = -\frac{3kT}{\langle R^2 \rangle}\boldsymbol{r} \tag{4.85}$$

Equation 4.85 describes the restoring force in a linear spring having zero rest length and spring constant $K = 3kT/\langle R^2 \rangle$. Microscopically, the *spring force* $\boldsymbol{F}(\boldsymbol{r})$ is the mean of the fluctuating force that the Brownian motion of the chain applies to its ends if a fixed separation $\boldsymbol{r}$ of the ends is maintained. The potential of this mean force is $\psi(r)$ in Eq. 4.84. As discussed in Section 4.5.2, its origin is purely configurational entropy for long enough chains and short enough displacements: the same requirements for chains obeying the Gaussian approximation. The spring force is also configurational in origin and given by Eq. 4.85, again providing the displacements are not too large. The spring constant K is directly proportional to absolute temperature, but it also contains a temperature dependence through $\kappa = d\ln\langle R^2 \rangle/dT$, the temperature coefficient of chain dimensions for free chains of the species (see Table 4.2).

The spring force $K\boldsymbol{r}$ arises from the statistical mechanical properties of flexible chains, and it has important macroscopic consequences. It enters into the elastic properties of polymer networks in many ways and plays a central role in network thermoelasticity (Chapter 10). The spring force is also central to the theory of intramolecular excluded volume (Chapter 6). It is also a crucial element in the molecular theory of polymer dynamics and a variety of recoil phenomena in polymeric liquids, discussed in Volume 2.

4.6.2 Thermodynamic Properties

The distribution of end-to-end vectors $V_n(\boldsymbol{R})d\boldsymbol{R}$ was obtained for random walks by evaluating the probability of an n-step walk starting at the origin and ending at position $\boldsymbol{R}$ in $d\boldsymbol{R}$. Because each walk corresponds to a chain conformation, $V_n(\boldsymbol{R})d\boldsymbol{R}$ is also the fraction of all conformations that have end-to-end vector $\boldsymbol{R}$ in $d\boldsymbol{R}$. Let $\Omega(\boldsymbol{R})d\boldsymbol{R}$ be the number of chain conformations that produce an end-separation vector $\boldsymbol{R}$ in $d\boldsymbol{R}$. If these various conformations have the same energy, then all contribute equally to the average, and

$$V_n(\boldsymbol{R})d\boldsymbol{R} = \frac{\Omega(\boldsymbol{R})d\boldsymbol{R}}{\int \Omega(\boldsymbol{R})d\boldsymbol{R}} \tag{4.86}$$

The ratio of numbers of conformations for different end-to-end vectors, $\boldsymbol{R}$ in $d\boldsymbol{R}$ and $\boldsymbol{R}'$ in $d\boldsymbol{R}'$, is then the ratio of their probabilities:

$$\frac{\Omega(\boldsymbol{R}')d\boldsymbol{R}'}{\Omega(\boldsymbol{R})d\boldsymbol{R}} = \frac{V_n(\boldsymbol{R}')d\boldsymbol{R}'}{V_n(\boldsymbol{R})d\boldsymbol{R}} \tag{4.87}$$

When the Gaussian approximation applies, such relationships become more broadly useful. Consider an ensemble of noninteracting polymer chains with identical structure as the system. Eq. (2.14) applies, such that:

$$A(V, T) = -kT \ln Q(V, T) \tag{4.88}$$

where $Q(V, T)$ is the partition function of the ensemble. Consider now that fraction of chains with end-to-end coordinates in the range $\boldsymbol{R}, \boldsymbol{R} + d\boldsymbol{R}$. If the Gaussian approximation applies, their contribution to the partition function is proportional to $V_n(\boldsymbol{R})d\boldsymbol{R}$. Accordingly, from Eq. 4.76 and for the full ensemble,

$$A_n = A_o + kT \iiint_{-\infty}^{\infty} V_n(x, y, z)\left[3\left(x^2 + y^2 + z^2\right)/2\langle R^2\rangle_n - \ln(dxdydz)\right]dxdydz \tag{4.89}$$

in which $A_n(V, T)$ is the free energy of the ensemble per n-mer chain and $A_o(V, T)$ is a quantity that is independent of the end-to-end coordinates.

Now consider the system transformed to a new set of end-to-end coordinates, related to the initial coordinates as follows:

$$x' = \lambda_x x \quad ; \quad y' = \lambda_y y \quad ; \quad z' = \lambda_z z \tag{4.90}$$

If the differential volumes transform in the same manner as the coordinates, the difference in free energy per chain for the two states, $A' - A$, is then given by:

$$\Delta A_n = kT \iiint_{-\infty}^{\infty} V_n(x, y, z)\Phi_n dxdydz \tag{4.91}$$

where:

$$\Phi_n = 3\left[x^2\left(\lambda_x^2 - 1\right) + y^2\left(\lambda_y^2 - 1\right) + z^2\left(\lambda_z^2 - 1\right)\right]/2\langle R^2\rangle_n - \ln(\lambda_x\lambda_y\lambda_z) \tag{4.92}$$

After integrating over the distribution [Eq. 4.76] and applying Eq. 4.79,

$$\Delta A = \frac{kT}{2}\left(\lambda_x^2 + \lambda_y^2 + \lambda_z^2 - 3\right) - kT\ln\left(\lambda_x\lambda_y\lambda_z\right) \tag{4.93}$$

For the special case of an isotropic uniform expansion or contraction, $\lambda_x = \lambda_y = \lambda_z = \lambda$,

$$\Delta A = kT\left[\frac{3}{2}\left(\lambda^2 - 1\right) - 3\ln\lambda\right] \tag{4.94}$$

Equation 4.94 is used in Section 6.1.1 where the intramolecular excluded volume effect on chain dimensions in dilute solution is considered. When chain ends cannot take up relative locations independently, as in the case of the strands in a network swollen by solvent, the free energy expression differs from Eq. 4.93 only in the logarithmic term. According to one theory of network elasticity, the requirement to satisfy network connectivity reduces its magnitude relative to its independent chain value. According to another theory, the logarithmic term is identically zero. The logarithmic term, a contentious issue for many years, is considered in Chapters 9 and 10.

Consider finally a related question: how R_g for a collection of random coils at equilibrium changes if their end-to-end distances are all changed by the same factor. Thus, if $\boldsymbol{R}\cdot\boldsymbol{R} \to \lambda^2(\boldsymbol{R}\cdot\boldsymbol{R})_o$, so that $\langle R^2\rangle = \lambda^2\langle R^2\rangle_o$, how does R_g depend on λ? The answer is:

$$R_g^2 = \frac{\left(R_g^2\right)_o}{2}\left(1+\lambda^2\right) \tag{4.95}$$

Deriving this equation is not a simple matter. It was given in 1976 by Benoit and Levy[45], but without a derivation. It is directly obtainable from the affine model result in the Pearson 1977 paper on the scattering properties of stretched Gaussian networks[46]. It supplies useful estimates for the effects of swelling and supercoiling effects on strand dimensions in networks (Section 10.4.1).

4.7 Scattering Properties

The equations describing elastic scattering by single site scatterers were developed in Section 2.1.5. Here the analysis is extended to the form factors for multisite scatterers, beginning with two-particle molecules and finishing with scattering by random coils, and to simple mixtures in which the components interact by volume exclusion alone.

4.7.1 Form Factors

We consider here a special case: scattering by molecules that do not interact with one another. Accordingly, the molecules move independently of one another, they are distributed randomly in space, they contribute in an additive manner to the scattering intensity, and only their intramolecular interferences need be taken into account. Independence always applies to scattering by solutes in sufficiently dilute solutions, but it arises in other contexts as well. Molecular models are frequently used to interpret the data, and the calculations in this section demonstrate some examples. The model results are commonly expressed as a molecular scattering function or form factor $P(q)$, defined as the scattering intensity for a collection of independent molecules as a function of scattering angle, normalized by the intensity extrapolated to the forward direction:

$$P(q) = \left[\frac{I(q)}{\lim\limits_{q \to 0} I(q)} \right]_{ind} \tag{4.96}$$

The detector intensity for scattering by a pair of identical scatterers, located at $\boldsymbol{r}_1$ and $\boldsymbol{r}_2$, is given by Eq. (2.47). Suppose the two scatterers constitute a molecule, meaning that some force law binds them to one another. Because of that interaction, the site locations are correlated, and the nature of the correlation affects the scattering intensity for an ensemble of such molecules. Consider now the scattering by an ensemble of n noninteracting molecules, each consisting of two sites of identical scattering power, whose separation $\boldsymbol{r}$ is described by some pair probability distribution $p(\boldsymbol{r})d\boldsymbol{r}$. Owing to the absence of intermolecular interactions, only the $2n$ terms for sites on the same molecule contribute to the angular dependence of scattering. Thus, applying Eq. (2.55) for $b_j = b_k = b$, integrating over the angles, omitting the self-scattering term, and using $v = 2n/V_{sc}$, (the scattering site density in the illuminated volume), leads to an expression for the reduced intensity Eq. (2.53):

$$I_r(q) = b^2 v \int_0^\infty 4\pi r^2 p(r) \frac{\sin(qr)}{qr} dr \tag{4.97}$$

This formula is applied below to the following force-law examples—the *rigid dumbbell*, the *tethered dumbbell*, and the *Gaussian dumbbell:*

$$p(r) = \frac{\delta(l - r)}{4\pi l^2} \qquad \text{(rigid)} \tag{4.98}$$

$$\begin{aligned} p(r) &= \frac{3}{4\pi r_m^3} \qquad r \le r_m \\ &= 0 \qquad\qquad r \ge r_m \end{aligned} \qquad \text{(tethered)} \tag{4.99}$$

$$p(r) = \left[\frac{3}{2\pi \langle r^2 \rangle} \right]^{3/2} \exp\left(-\frac{3r^2}{2\langle r^2 \rangle} \right) \qquad \text{(Gaussian)} \tag{4.100}$$

All three models are purely attractive; the volume exclusion interaction, always present in real molecules for small enough separations, has been omitted for simplicity. The following form factors are obtained by applying Eq. 4.97 and normalizing to unity at $q = 0$:

$$P(q) = \frac{\sin ql}{ql} \qquad \text{(rigid)} \qquad (4.101)$$

$$P(q) = \frac{3(\sin qr_m - qr_m \cos qr_m)}{(qr_m)^3} \qquad \text{(tethered)} \qquad (4.102)$$

$$P(q) = \exp\left(-\langle r^2\rangle q^2/6\right) \qquad \text{(Gaussian)} \qquad (4.103)$$

In all cases, $P(q)$ remains near unity as long as qd is much less than 1, the value d being some measure of average molecular size. It begins to decrease rapidly near $qd = 1$, and goes finally to 0 for $qd \gg 1$. The shape of $P(q)$ depends on the force law—on the distribution of scattering site separation distances—but the qualitative features described above are the same for all. In general, the form factor is a property of molecules that depends on their environment, as will be seen in Chapters 6 through 8.

Random coil molecules. Generalizing Eq. 4.97 to obtain form factors for more complex molecules is straightforward. Thus, using Eq. (2.50) and for n independent molecules with m identical scatterers per molecule,

$$\frac{d\Sigma}{d\Omega}(\boldsymbol{q}) = b^2 \sum_{j \neq k}^{mn} \sum^{mn} \langle \exp[i\boldsymbol{q} \cdot (\boldsymbol{r}_j - \boldsymbol{r}_k)] \rangle \qquad (4.104)$$

Picking out the $2n(m-1)$ terms for $j \neq k$ pairs on the same molecule, then averaging over the angles, leads to:

$$P(q) = \frac{1}{(m-1)^2} \sum_{j \neq k}^{m} \sum^{m} \int_0^{\infty} 4\pi r_{jk}^2 \frac{\sin qr_{jk}}{qr_{jk}} p(r_{jk}) dr_{jk} \qquad (4.105)$$

Equation 4.105 is easily applied to ring-free, random coil molecules. Only one path connects any pair of points on such molecules, and only the number of steps along that path influences $p_{ij}(r_{ij})$, and [from Eq. 4.12], $\langle r_{ij}^2 \rangle = |i - j| l^2$. Thus, using Eq. 4.105 and the Gaussian approximation for $p_{ij}(r_{ij})$, the following expression is obtained for the form factor of random coil molecules with acyclic architectures and

m scattering sites:

$$P(q) = \frac{1}{(m-1)^2} \sum_{i=1}^{m} \sum_{j=1}^{m} \exp\left(-|i-j| l^2 q^2/6\right) \tag{4.106}$$

where $|i-j|$ is the number of steps connecting scatterers i and j. For linear chains, large m and summation replaced by integration,

$$P(q) = \frac{2}{m^2} \int_{0}^{m} \int_{z}^{m} \exp\left[-(ql)^2 (y-z)/6\right] dy dz \tag{4.107}$$

leading finally to:

$$P(q) = \frac{2}{u^2} \left[\exp(-u) - 1 + u\right] \tag{4.108}$$

in which $u = (qR_g)^2$. This is the *Debye equation*, the molecular scattering factor for linear, monodisperse random coil molecules[47]. Molecular scattering factors have been calculated for many chain architectures. For symmetric f-arm star polymers[48],

$$P(q) = \frac{2}{u^2} \left[u - f\left[1 - \exp(-u/f)\right] + \frac{f(f-1)}{2} \left[1 - \exp(-u/f)\right]^2\right] \tag{4.109}$$

in which $u = q^2 R_g^2 f^2/(3f-2)$. A formula has also been obtained for macrocyclic rings[49]:

$$P(q) = \int_{0}^{1} \exp\left[-\lambda u(1-u)\right] du \tag{4.110}$$

in which $\lambda = 2q^2 R_g^2$. Form factors for various types of comb molecules are also available[36].

Pair distributions. For scattering unit i on a random coil molecule as reference, consider the probability density $p_i(\boldsymbol{r})$ of finding some other unit of the same molecule at a relative location $\boldsymbol{r}$. The probability that unit j is there is given by Eqs. 4.12 and 4.75, so the sum over all $j \neq i$ is the total probability density:

$$p_i(r) = \sum_{j \neq i}^{m} \left(\frac{3}{2\pi l^2 |i-j|}\right)^{3/2} \exp\left[-\frac{3r^2}{2l^2 |i-j|}\right] \tag{4.111}$$

With the substitution $u = 1/|i - j|$ and summation replaced by integration,

$$p_i(r) = \left(\frac{\alpha}{\pi r^2}\right)^{3/2} \left[\int_{1/i}^{1} \frac{\exp(-\alpha u)}{u^{1/2}} du + \int_{1/(m-i)}^{1} \frac{\exp(-\alpha u)}{u^{1/2}} du \right] \tag{4.112}$$

For $m \gg 1$ and over distances for which the Gaussian approximation is valid, pair distribution examples, such as for chain-end and mid-chain references, can be evaluated:

$$\begin{aligned} p_1(r) &= \frac{3}{2\pi l^2 r} \\ p_{\frac{m}{2}}(r) &= \frac{3}{\pi l^2 r} \end{aligned} \tag{4.113}$$

Note that Eq. 4.113 applies for random coil polymers only over some range of intermediate distances. It is invalid for an r smaller than approximately the step length of the species and for an r larger than approximately the coil radius. The result, $p(\boldsymbol{r}) \propto |\boldsymbol{r}|^{-1}$ is a general characteristic of random coils at intermediate distances. It appears in scattering behavior as $I_r(q) \propto q^{-2}$, easily demonstrated using Eq. 4.97, and will prove useful in the later chapters of this volume.

General properties. Form factors share some general characteristics. Thus, for example, the initial slope of the form factor, $(dp(q)/dq^2)_{q=0}$, is related to molecular size in a simple and very useful way. Consider the behavior of $P(q)$ in the low q, region, as expressed by Eq. 4.105. Near $q = 0$, $\sin qr_{jk}/qr_{jk} = 1 - q^2 r_{jk}^2/6 + \cdots$. Term-by-term integration of Eq. 4.105 using this expression then leads to:

$$P(q) = \frac{1}{(m-1)^2} \sum_{j \neq k}^{m} \sum^{m} \left[1 - q^2 \frac{\langle \boldsymbol{r}_{jk}^2 \rangle}{6} + \cdots \right] \tag{4.114}$$

For identical, noninteracting molecules, Eq. 4.17 leads to:

$$P(q) = 1 - \frac{q^2}{6} \frac{1}{m^2} \sum_{i=1}^{m} \sum_{j=1}^{m} \langle \boldsymbol{r}_{ij}^2 \rangle + \cdots = 1 - \frac{q^2 R_g^2}{3} + \cdots \tag{4.115}$$

For a polydisperse system of noninteracting molecules, a weight fraction W_m, consisting of molecules with m scattering sites of identical scattering power, Eq. 4.115 leads to:

$$P(q) = 1 - \frac{q^2 \langle R_g^2 \rangle_z}{3} + \cdots$$

or

$$\frac{1}{P(q)} = 1 + \frac{q^2 \langle R_g^2 \rangle_z}{3} + \cdots \tag{4.116}$$

where $\langle R_g^2 \rangle_z$ is the z-average square radius of gyration Eq. (1.10). These are equivalent expressions of a relationship first propounded by Guinier[50]. The angular dependence of scattering intensity in the low q limit for any collection of noninteracting objects whose scattering power is proportional to mass is a precisely defined measure of their size:

$$\langle R_g^2 \rangle_z = \frac{3}{I(0)} \left(\frac{d[I(q)]^{-1}}{dq^2} \right)_{q=0} \tag{4.117}$$

The Gaussian approximation has been used throughout the calculations for random coil molecules, not just where the chain distance $|i-j|$ is large. In effect, the connectors between successive scattering sites are treated as linear springs [Eq. 4.85]. This is, of course, incorrect for locally connected monomeric units, because the connectors in this case are one or a relative few covalent bonds and certainly not springlike. Accurately representing the local connector mechanics, however, would make the calculation prohibitively difficult. Choosing longer chain segments, such as groups of several Kuhn steps, to be the scatterers would make the Gaussian representation more appropriate, but for most purposes it does not matter. Thus, a rigorous representation of local structure is not required if, as is usually the case, the information of interest is in a range of q such that the product ql_K is much smaller than unity. Local structure influences scattering behavior in the high-q range, typically $1 < q(nm^{-1}) < 10$, as it does for molecular liquids, but at smaller q the specifics of species structure are absorbed into such global parameters as the radius of gyration R_g. In that range Eq. 4.108, for example, is satisfactory. The scattering behavior for long chains parallels the distribution of conformations itself—universal in form at large-enough distances or small-enough q, but species-specific for small-enough distances or large-enough q. Three regimes are typically distinguished for scattering by macromolecules, Guinier, intermediate, and local:

$$P(q) = 1 - \frac{(qR_g)^2}{3} \qquad qR_g \lesssim 1 \qquad \text{(Guinier)} \tag{4.118}$$

$$P(q) = \frac{2}{(qR_g)^2} \qquad qR_g \gtrsim 1 \gtrsim ql_K \qquad \text{(intermediate)} \tag{4.119}$$

$$P(q) = P_i(q) \qquad ql_K \gtrsim 1 \qquad \text{(local)} \tag{4.120}$$

Expressions for the various random-coil architectures go over to the correct expressions for R_g in the Guinier range and the architecture-independent expression in the intermediate range. The equations contain no useful information about the local range.

4.7.2 Structure Factors

In Chapter 2, the coherent scattering intensity for a collection of scattering sites was expressed as an equilibrium average of contributions from pairs at specific sets of locations. It was shown that scattering depends on the pair-wise spatial correlation of site locations, such that if the site positions were independent of one another, no significant (q-dependent) scattering would occur. The *discrete scatterer description* incorporates detailed information on correlation extending down to the atomic level and is mainly useful for structural determinations in that size range, for example, crystals. In a *continuum scatterer description*, scattering is related to fluctuations, or excursions from the average, in the numbers and types of scattering sites in the volume elements of the system, (for example, liquids). For single component systems, these variations are the result of fluctuations in the local density. For multicomponent systems, they result from fluctuations in local concentrations as well as density.

Consider, for example, an illuminated volume V_{sc} of a binary liquid having some macroscopic mass density ρ and mass concentration of solute c. Imagine this volume to be divided into a large number of equal volume elements, each with some small volume δV. Each volume element is an open system, freely exchanging mass and energy with the other elements. Accordingly, the mass density and concentration of each, ρ_i and c_i will fluctuate about the macroscopic (mean) values, $\rho = \langle \rho_i \rangle$ and $c = \langle c_i \rangle$, but their excesses, $\Delta\rho_i = \rho_i - \rho$ and $\Delta c_i = c_i - c$ must average to zero:

$$\begin{aligned} \langle \Delta\rho_i \rangle &= \langle (\rho_i - \rho) \rangle = 0 \\ \langle \Delta c_i \rangle &= \langle (c_i - c) \rangle = 0 \end{aligned} \tag{4.121}$$

The squares of the fluctuations must always be positive, however, so that their averages, $\langle (\rho_i - \rho)^2 \rangle$ and $\langle (c_i - c)^2 \rangle$, are not zero. They are related to macroscopic properties and the size of the volume elements by the *Einstein formulas*[51–53]:

$$\left\langle (\Delta\rho)^2 \right\rangle = \frac{\rho^2 kT}{\delta V}\beta \tag{4.122}$$

$$\left\langle (\Delta c)^2 \right\rangle = \frac{c^2 kT}{\delta V}\gamma \tag{4.123}$$

$$\left\langle (\Delta\rho)(\Delta c) \right\rangle = 0 \tag{4.124}$$

in which β is $(\partial \ln \rho/\partial P)_T$,the isothermal compressibility of the liquid, and γ is $(\partial \ln c/\partial \pi)_T$, sometimes called the *osmotic compressibility* or *osmotic susceptibility*. These expressions are general and applicable to binary systems at all levels of concentration. Although all three are important, Eq. 4.124 is especially noteworthy in establishing the independence of density and concentration fluctuations, a restriction always to be respected in theories of scattering behavior. Note that the average magnitude of the fluctuations increases with decreasing size of the elemental volumes. As a result, a liquid might appear to be clear and hence optically uniform to the eye, but it always grows nonuniform when viewed at increasingly higher magnifications.

In the continuum formulation, no scattering occurs without such fluctuations. It is a coarse-grained description and does not lend itself to structural determinations at the atomic level. Its great virtue is the generality of the relationships it provides, through the Einstein formulas [Eqs. 4.122 through 4.124] between scattering behavior and thermodynamic properties.

Multicomponent systems are frequently considered in Chapters 5 through 8 in the context of one-component dispersions of nearly independent objects (macromolecules in dilute solution, screening volumes in the semidilute region, correlation volumes near critical points). Here we develop a formula for multicomponent scattering where two or more types of molecules at finite concentration interact significantly and the independence approximation cannot be applied. The only interaction considered here is mutual volume exclusion. Extensions to include other types, such as exchange interactions, are added as required in later chapters. Nonetheless, the development here completes the basic background in static neutron and x-ray scattering needed in this volume and will be drawn upon extensively. The fundamentals particular to light scattering are developed in Chapter 5.

When information on thermodynamic properties and large-scale structure is sought, it is natural to use the continuum description. The focus is dense liquids, with the simple mixture model (Section 3.4) used as needed. The development broadly follows Higgins and Benoit[54] for scattering first by one-component liquids and then by two-component liquids.

One-component systems. The expression for scattering intensity in one-component systems, as the averaged sum of particle-pair contributions in the discrete pairs method [see Eqs. (2.60) through (2.62)], translates to the averaged sum of fluctuation-pair contributions in the continuum method. The following is the continuum description of Eq. (2.60), for a system of n molecules each with volume υ and having z scattering sites of scattering length b:

$$I_r(q) = b^2 \int_V \int_V \langle \delta\nu(\boldsymbol{r})\delta\nu(\boldsymbol{r}')\rangle \exp[i\boldsymbol{q}\cdot(\boldsymbol{r}-\boldsymbol{r}')]d\boldsymbol{r}'d\boldsymbol{r} \qquad (4.125)$$

where ν is z/υ, the average number density of scattering sites in the system, $\delta\nu(\boldsymbol{r}) = \nu(\boldsymbol{r}) - \nu$ is the fluctuation from that average in volume element $d\boldsymbol{r}$, the integrations extend over the illuminated volume V_{sc}, and the intramolecular-intermolecular distinction has been omitted. As always, the q-independent contributions are ignored. The structure factor for the one-component liquid is therefore:

$$S(q) = \int_V \int_V \langle \delta\nu(\boldsymbol{r})\delta\nu(\boldsymbol{r}')\rangle \exp[i\boldsymbol{q}\cdot(\boldsymbol{r}-\boldsymbol{r}')]d\boldsymbol{r}'d\boldsymbol{r} \tag{4.126}$$

Notice that such molecular parameters as υ and z appear nowhere in these expressions. Indeed, for one-component systems in the continuum description, the only possible source of scattering is fluctuations in density. Thus, in the incompressible limit, which is used extensively below, a one-component single-phase liquid system does not scatter at all.

Two-component systems. Consider now a two-component liquid: n_1 and n_2 molecules each occupying volumes υ_1 and υ_2 and each having z_1 and z_2 sites, each with scattering lengths b_1 and b_2. It is convenient to express the intensity as the sum of contributions from the 1,1, 1,2 and 2,2 pairs of sites. Thus,

$$\begin{aligned}\frac{I(q)r_D^2}{I_o} &= b_1^2\sum_{j=1}^{n_1z_1}\sum_{k=1}^{n_1z_1}\langle\exp[i\boldsymbol{q}\cdot(\boldsymbol{r}_j-\boldsymbol{r}_k)]\rangle_{1,1} + 2b_1b_2\sum_{j=1}^{n_1z_1}\sum_{k=1}^{n_2z_2}\langle\exp[i\boldsymbol{q}\cdot(\boldsymbol{r}_j-\boldsymbol{r}_k)]\rangle_{1,2}\\ &+ b_2^2\sum_{j=1}^{n_2z_2}\sum_{k=1}^{n_2z_2}\langle\exp[i\boldsymbol{q}\cdot(\boldsymbol{r}_j-\boldsymbol{r}_k)]\rangle_{2,2}\end{aligned} \tag{4.127}$$

or

$$\frac{I(q)r_D^2}{I_o} = b_1^2 S_{11}(q) + 2b_1b_2S_{12}(q) + b_2^2S_{22}(q) \tag{4.128}$$

in which the $S_{jk}(q)$ are called *partial structure factors.* In the continuum representation,

$$S_{jk}(q) = \int_{V_{sc}}\int_{V_{sc}} \langle \delta\nu_j(\boldsymbol{r})\delta\nu_k(\boldsymbol{r}')\rangle \exp[i\boldsymbol{q}\cdot(\boldsymbol{r}-\boldsymbol{r}')]d\boldsymbol{r}'d\boldsymbol{r} \tag{4.129}$$

where the j, k stand for 1,1, 1,2 and 2,2, and where, as introduced for one-component systems in Eq. 4.125, the ν_j and ν_k are average scattering site densities z_j/υ_j and z_k/υ_k, and the $\delta\nu_j$ and $\delta\nu_k$ are the local excursions.

Incompressible mixtures. Although scattering in one-component systems arises only from density fluctuations, and hence volumetric compressibility, scattering in multicomponent systems can also be produced by composition fluctuations. Thus scattering in these systems depends on the osmotic susceptibility as well [see Eqs. 4.122 through 4.124]. Both types of contribution can be important, depending on the system and q range of interest. Information about the thermodynamics of mixing for polymeric systems in the dense liquid regime resides in the scattering behavior at low q ($q \ll 1\,nm^{-1}$). In this range, and for neutron scattering at least, the contribution of density fluctuations is negligible[55,56]: the assumption of *volumetric incompressibility* in interpreting the low-q scattering data for dense liquids is harmless.

With the stipulation of incompressibility, fluctuations in the scatterer concentrations in any volume element of a multicomponent system must be coupled in such a way as to conserve the volume. To impose that constraint, imagine the molecules distributed on a lattice having some reference volume υ_o per lattice site. Accordingly, $z_j \upsilon_o/\upsilon_j$ scattering sites exist per lattice site for component j. Thus, an excess number of scattering sites per unit volume, $\delta\nu_j$, corresponds an extra volume $\delta\nu_j/(z_j\upsilon_o/\upsilon_j)$ from component j. Incompressibility requires the sum of such extra volumes to be zero, so:

$$\frac{\upsilon_1}{z_1}\delta\nu_1(\boldsymbol{r}) + \frac{\upsilon_2}{z_2}\delta\nu_2(\boldsymbol{r}) = 0 \tag{4.130}$$

Multiplying this volume constraint by $\delta\nu_j(\boldsymbol{r})\exp\left[i\boldsymbol{q}\cdot(\boldsymbol{r}-\boldsymbol{r}')\right]$, then averaging and integrating leads to:

$$\begin{aligned}&\frac{\upsilon_1}{z_1}\int_V\int_V \langle\delta\nu_j(\boldsymbol{r})\delta\nu_1(\boldsymbol{r}')\rangle \exp[i\boldsymbol{q}\cdot(\boldsymbol{r}-\boldsymbol{r}')]d\boldsymbol{r}'d\boldsymbol{r}\\ &+\frac{\upsilon_2}{z_2}\int_V\int_V \langle\delta\nu_j(\boldsymbol{r})\delta\nu_2(\boldsymbol{r}')\rangle \exp[i\boldsymbol{q}\cdot(\boldsymbol{r}-\boldsymbol{r}')]d\boldsymbol{r}'d\boldsymbol{r} = 0\end{aligned} \tag{4.131}$$

so, first for $j = 1$ and then for $j = 2$, and using Eq. 4.129,

$$\begin{aligned}\frac{\upsilon_1}{z_1}S_{11}(q) + \frac{\upsilon_2}{z_2}S_{12}(q) = 0\\ \frac{\upsilon_1}{z_1}S_{12}(q) + \frac{\upsilon_2}{z_2}S_{22}(q) = 0\end{aligned} \tag{4.132}$$

With these relationships and Eq. 4.128, the scattering intensity for a two-component system can be expressed in terms of any one of the three partial structure factors alone. Thus,

$$\begin{aligned}\frac{I(q)r_D^2}{I_o} &= \left(\frac{b_1z_1}{\upsilon_1}-\frac{b_2z_2}{\upsilon_2}\right)^2\left(\frac{\upsilon_1}{z_1}\right)^2 S_{11}(q)\\ \frac{I(q)r_D^2}{I_o} &= -\left(\frac{b_1z_1}{\upsilon_1}-\frac{b_2z_2}{\upsilon_2}\right)^2\left(\frac{\upsilon_1}{z_1}\right)\left(\frac{\upsilon_2}{z_2}\right) S_{12}(q)\\ \frac{I(q)r_D^2}{I_o} &= \left(\frac{b_1z_1}{\upsilon_1}-\frac{b_2z_2}{\upsilon_2}\right)^2\left(\frac{\upsilon_2}{z_2}\right)^2 S_{22}(q)\end{aligned} \tag{4.133}$$

Return now to the discrete-sites representation to complete the derivation of an expression for $S(q)$. The procedure is first to resolve each $S_{jk}(q)$ into its intramolecular and intermolecular parts [see Eqs. (2.61) and (2.62)], then to obtain independent expressions for the intermolecular parts, and finally to use the results in Eq. 4.133.

There are $(n_1 z_1 + n_2 z_2)^2$ scattering interactions in all, $n_1 z_1^2 + n_2 z_2^2$ intramolecular and $n_1(n_1 - 1)z_1^2 + n_2(n_2 - 1)z_2^2$ intermolecular. Thus, with the factors of $n_j - 1$ harmlessly replaced by n_j,

$$S_{11}(q) = n_1 \sum_{j=1}^{z_1} \sum_{k=1}^{z_1} \langle \exp[i\boldsymbol{q} \cdot (\boldsymbol{r}_j - \boldsymbol{r}_k)] \rangle_{\text{intra},1} + n_1^2 \sum_{j=1}^{z_1} \sum_{k=1}^{z_1} \langle \exp[i\boldsymbol{q} \cdot (\boldsymbol{r}_j - \boldsymbol{r}_k)] \rangle_{\text{inter},11}$$

$$S_{22}(q) = n_2 \sum_{j=1}^{z_2} \sum_{k=1}^{z_2} \langle \exp[i\boldsymbol{q} \cdot (\boldsymbol{r}_j - \boldsymbol{r}_k)] \rangle_{\text{intra},2} + n_2^2 \sum_{j=1}^{z_2} \sum_{k=1}^{z_2} \langle \exp[i\boldsymbol{q} \cdot (\boldsymbol{r}_j - \boldsymbol{r}_k)] \rangle_{\text{inter},22}$$

$$S_{12}(q) = n_1 n_2 \sum_{j=1}^{z_1} \sum_{k=1}^{z_2} \langle \exp[i\boldsymbol{q} \cdot (\boldsymbol{r}_j - \boldsymbol{r}_k)] \rangle_{\text{inter},12} \tag{4.134}$$

As in Eqs. (2.61) and (2.62), the intramolecular and intermolecular sums can be expressed as component form factors and intermolecular factors:

$$P_j(q) = \frac{1}{z_j^2} \sum_{j=1}^{z_j} \sum_{k=1}^{z_j} \langle \exp[i\boldsymbol{q} \cdot (\boldsymbol{r}_j - \boldsymbol{r}_k)] \rangle_{\text{intra}}$$
$$Q_{jk}(q) = \frac{1}{z_j z_k} \sum_{j=1}^{z_j} \sum_{k=1}^{z_k} \langle \exp[i\boldsymbol{q} \cdot (\boldsymbol{r}_j - \boldsymbol{r}_k)] \rangle_{\text{inter}} \tag{4.135}$$

The following relationships are then obtained:

$$S_{11}(q) = n_1 z_1^2 P_1(q) + n_1^2 z_1^2 Q_{11}(q)$$
$$S_{22}(q) = n_2 z_2^2 P_2(q) + n_2^2 z_2^2 Q_{22}(q) \tag{4.136}$$
$$S_{12}(q) = n_1 z_1 n_2 z_2 Q_{12}(q)$$

Thus, using Eq. 4.132,

$$v_1 P_1(q) + v_1 n_1 Q_{11}(q) + v_2 n_2 Q_{12}(q) = 0$$
$$v_2 P_2(q) + v_2 n_2 Q_{22}(q) + v_1 n_1 Q_{12}(q) = 0 \tag{4.137}$$

A third equation relating the various $Q_{jk}(q)$ would permit the elimination of all such quantities from the expression for scattering intensity [Eq. 4.133]. The simple mixture representation supplies that need, filling all space with scattering sites while mutual volume exclusion enforces incompressibility. Formal expressions can be developed for the intermolecular factors as follows:

$$Q_{jk} = n_j \left\langle \int_V \exp(i\boldsymbol{s} \cdot \boldsymbol{r}) f_j(\boldsymbol{r}) p_k(\boldsymbol{r}) d\boldsymbol{r} \right\rangle \tag{4.138}$$

in which $f_j(\boldsymbol{r})$ is the fraction of sites at position $\boldsymbol{r}$ from a randomly selected j-scatterer that are intermolecular, and $p_k(\boldsymbol{r})$ is the fraction of the intermolecular sites at $\boldsymbol{r}$ that are occupied by k-scatterers. The brackets indicate an average over all conformations of the j-scatterer chain and all j-scatterer locations along the chain. As illustrated in Figure 4.8, the number of intramolecular sites diminishes with increasing $\boldsymbol{r}$, finally becoming negligible for $|\boldsymbol{r}| \gtrsim R_g$. For the large distance range, one would expect $p_k(\boldsymbol{r}) \approx \phi_k$ in general, where $\phi_k = n_k v_k/(n_1 v_1 = n_2 v_2)$ (the volume fraction of sites with k-scatterers in the system). Suppose, however, that $p_k(\boldsymbol{r}) \approx \phi_k$ is assumed to apply at all distances, specifying that, regardless of the j-scatterer proximity, the intermolecular site has the same likelihood of occupation by a k-scatterer as the system-average. The approximation is equivalent to the average vacancy approximation used by Flory in deriving an expression for the combinatorial entropy of simple

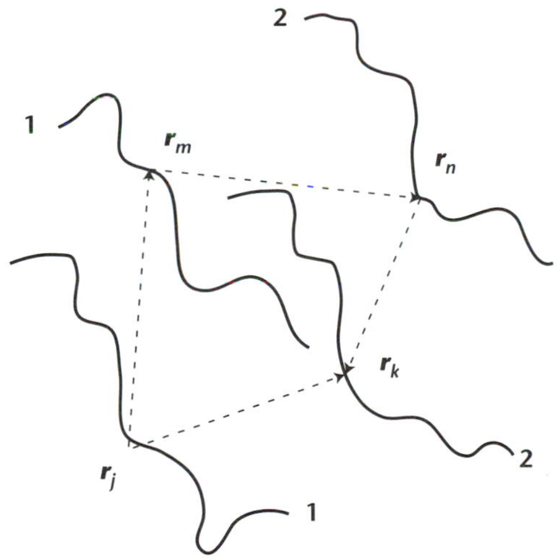

FIGURE 4.8 Schematic representation of a relative position vector in the calculation of intramolecular and intermolecular contributions to the structure factor.

mixtures (Section 3.3). When that approximation is made, $p_k(\boldsymbol{r})$factors out of the integral and the averaging to give the following expressions for the $Q_{jk}(q)$:

$$\begin{aligned}
Q_{11} &= n_1\phi_1\left\langle \int_V \exp(i\boldsymbol{s}\cdot\boldsymbol{r}) f_1(\boldsymbol{r})d\boldsymbol{r} \right\rangle \\
Q_{12} &= n_1\phi_2\left\langle \int_V \exp(i\boldsymbol{s}\cdot\boldsymbol{r}) f_1(\boldsymbol{r})d\boldsymbol{r} \right\rangle \\
Q_{22} &= n_2\phi_2\left\langle \int_V \exp(i\boldsymbol{s}\cdot\boldsymbol{r}) f_2(\boldsymbol{r})d\boldsymbol{r} \right\rangle \\
Q_{21} &= n_2\phi_1\left\langle \int_V \exp(i\boldsymbol{s}\cdot\boldsymbol{r}) f_2(\boldsymbol{r})d\boldsymbol{r} \right\rangle
\end{aligned} \tag{4.139}$$

Because $Q_{12} = Q_{21}$, it is evident from Eq. 4.139 that the intermolecular factors satisfy

$$Q_{12}^2(q) = Q_{11}(q)Q_{22}(q) \tag{4.140}$$

The combination of Eq. 4.140 with Eq. 4.137 permits the $Q_{jk}(q)$, and thus the $S_{jk}(q)$, to be expressed in terms of the z_i, n_i, v_i, and $P_i(q)$ properties alone. After some algebraic manipulations,

$$S_{11}(q) = \left(\frac{z_1}{v_1}\right)^2 \left[\frac{1}{n_1 v_1^2 P_1(q)} + \frac{1}{n_2 v_2^2 P_2(q)}\right]^{-1} \tag{4.141}$$

and similarly for $S_{12}(q)$ and $S_{22}(q)$. Using Eq. 4.133 and illuminated volume $V_{sc} = n_1 v_1 + n_2 v_2$, component molar volumes $V_i = N_a v_i$, and volume fractions $\phi_i = n_i v_i/(n_1 v_1 + n_2 v_2)$:

$$\frac{I(q)r_D^2}{I_o} = \frac{V_{sc}}{N_a}\left(\frac{b_1 z_1}{v_1} - \frac{b_2 z_2}{v_2}\right)^2 \left[\frac{1}{\phi_1 V_1 P_1(q)} + \frac{1}{\phi_2 V_2 P_2(q)}\right]^{-1} \tag{4.142}$$

or

$$I_r(q) = C_{sc} S(q) \tag{4.143}$$

where $I_r(q)$ is the reduced intensity, defined by Eq. (2.53),

$$C_{sc} = \frac{1}{N_a}\left(\frac{b_1 z_1}{v_1} - \frac{b_2 z_2}{v_2}\right)^2 \tag{4.144}$$

is the *contrast factor,* and

$$S(q) = \left[\frac{1}{\phi_1 V_1 P_1(q)} + \frac{1}{\phi_2 V_2 P_2(q)}\right]^{-1} \tag{4.145}$$

is the structure factor for a two-component simple mixture with $\chi = 0$. The contrast factor is written here as C_{sc}, indicating any kind of coherent scattering process. Sometimes this is written as C_{LS}, C_{SANS}, or C_{SAXS} to indicate the particular kind of scattering process. Equation 4.145 and a version using $\chi \neq 0$ are discussed and used extensively in Chapters 7 and 8.

REFERENCES

1. Meyer K.H., G. von Susich, and E. Valkó. 1932. *Kolloid Z.* 59:208.
2. Guth E. and H. Mark. 1934. *Monatsh.* 65:83.
3. Kuhn W. 1934. *Kolloid Z.* 68:2.
4. Kuhn W. 1936. *Kolloid Z.* 76:258.
5. Kuhn W. 1939. *Kolloid Z.* 87:3.
6. Chandrasekhar S. 1943. *Rev. Mod. Phys.* 15:1.
7. Debye P. 1944. *J. Appl. Phys.* 15:338.
8. Debye P. 1946. *J. Chem. Phys.* 14.
9. Zimm B. 1948. *J. Chem. Phys.* 16:1099.
10. Cotton J.P., G. Farnoux, G. Jannink, J. Mons, and C. Picot. 1972. *Compt. Rend., Ser. C* 275:175.
11. Kirste R.G., W.A. Kruse, and J. Shelten. 1972. *Makromol. Chem.* 162:299.
12. Ballard D.G.H., G.D. Wignall, and J. Shelten. 1973. *Eur. Polym. J.* 9:965.
13. Jolley L.B.W. 1961. *Summation of Series,* 2nd ed. New York: Dover Publications.
14. Li B., N. Madras, and A.D. Sokal. 1995. *J. Stat. Phys.* 80:661.
15. Flory P.J. 1953. *Principles of Polymer Chemistry,* Ithaca, NY: Cornell University Press.
16. Feller W. 1971. *An Introduction to Probability Theory and Its Applications,* 2nd ed. New York: J. Wiley & Sons.
17. Flory P.J. 1969. *Statistical Mechanics of Chain Molecules,* New York: Wiley-Interscience.
18. Fetters L.J., D.J. Lohse, D. Richter, T.A. Witten, and A. Zirkel. 1994. *Macromolecules* 27:4639.
19. Graessley W.W. and L.J. Fetters. 2001. *Macromolecules* 34:7147.
20. Jones T.D., K.A. Chaffin, F.S. Bates, B.K. Annis, E.W. Hagaman, M.-H. Kim, G.D. Wignall, W. Fan, and R. Waymouth. 2002. *Macromolecules* 35:5061.
21. Volkenstein M.V. 1963. *Configurational Statistics of Polymer Chains,* New York: J. Wiley & Sons.
22. Mattice W.L. and Suter, U.W. 1994. *Conformational Theory of Large Chain Molecules: The Rotational Isomer State Model in Macromolecular Systems,* New York: J. Wiley & Sons.
23. Smith G.D. and D. Yoon. 1995. *J. Amer. Chem. Soc.* 117:530.
24. Kuhn W. 1946. *Helv. Chim. Acta* 29:1095.
25. Yamakawa H. 1971. *Modern Theory of Polymer Solutions,* New York: Harper & Row.
26. Den Hartog J.P. 1961. *Strength of Materials,* New York: Dover Publications. pp. 40–80.

27. Yamakawa H. 1997. *Helical Wormlike Chains in Polymer Solutions,* New York: Springer-Verlag.
28. Helfand E. and A.M. Sapse. 1975. *J. Chem. Phys.* 62:1327.
29. Fetters L.J., D.J. Lohse, and W.W. Graessley. 1999. *J. Polym. Sci.: Pt. B: Polym. Phys.* 37:1023.
30. Kramers H.A. 1946. *J. Chem. Phys.* 14:415.
31. Eichinger B.E. 1980. *Macromolecules* 13:1.
32. Pearson D.S. and V.R. Raju. 1982. *Macromolecules* 15:294.
33. Hadjichristidis N. et al. 2000. *Macromolecules* 33:2424.
34. Orofino T.A. 1961. *Polymer* 2:305.
35. Berry G.C. 1962. Doctoral dissertation, University of Michigan.
36. Casassa E.F. and G.C. Berry. 1966. *J. Polym. Sci. Pt. A-2* 4:881.
37. Zimm B. and W.H. Stockmayer. 1949. *J. Chem. Phys.* 17:1301.
38. Graessley W.W. 1975. *Macromolecules* 8:865.
39. Solc K. 1971. *J. Chem. Phys.* 55:335.
40. Solc K. 1973. *Macromolecules* 6:378.
41. Solc K. 1977. *Polymer News* 4:67.
42. Bishop M. and J.P.J. Michaels. 1986. *J. Chem. Phys.* 84:444.
43. Treloar L.R.G. 1975. *The Physics of Rubber Elasticity,* 3rd ed. Oxford: Oxford University Press.
44. McQuarrie D.A. 1976. *Statistical Mechanics,* New York: HarperCollins Publishers.
45. Benoit H., D. Decker, R. Duplessix, C. Picot, P. Rempp, J.C. Cotton, B. Farnoux, G. Jannink, and R. Ober. 1976. *J. Polym. Sci.: Polym. Phys. Ed.* 14:2119.
46. Pearson D.S. 1977. *Macromolecules* 10:696.
47. Debye P. 1947. *J. Phys. Colloid Chem.* 51:18.
48. Benoit H. 1953. *J. Polym. Sci.* 11:507.
49. Casassa E.F. 1965. *J. Polymer Sci., Pt. A* 3:605.
50. Guinier A. and Fournet, G. 1955. *Small-Angle Scattering of X-Rays,* New York: John Wiley & Sons.
51. Einstein A. 1910. *Ann. Physik* 33:1275.
52. Einstein A. 1993. In *The Collected Papers of Albert Einstein (English Translation Supplement),* Vol. 3. M. J. Klein et al. (eds.), Princeton, NJ: Princeton University Press, p. 231.
53. Hill T.L. 1987. *Statistical Mechanics,* New York: Dover Publications.
54. Higgins J.S. and Benoit, H.C. 1994. *Polymers and Neutron Scattering,* New York: Oxford University Press.
55. Taylor J.K., P.G. Debenedetti, W.W. Graessley, and S.K. Kumar. 1996. *Macromolecules* 29:764.
56. Joanny J.F. and H. Benoit. 1997. *Macromolecules* 30:3704.

CHAPTER 5

Dilute Solution Characterization

This chapter describes the principal methods of polymer characterization in the dilute regime, from infinite dilution to the coil overlap concentration. The fundamentals and application of equilibrium, dynamic, and chromatographic techniques for evaluating molecular weight, molecular weight distribution, chain dimensions, diffusion coefficient, intrinsic viscosity, and second virial coefficient are explained. The theme of scattering for structure determination is continued, with a discussion of light scattering and its relationship to neutron and x-ray small angle scattering, the other major techniques of interest.

The molecular weight and size of polymers is determined by measurements on dilute solutions. The major methods—static, dynamic, and chromatographic—are introduced and described briefly in Sections 5.1 through 5.3. Identifying the molecular overlap concentration, which is the upper limit of the dilute regime, is also considered briefly. The self-concentration of random coils and the effect of excluded volume and hydrodynamic interactions, including the determination of the theta condition, evaluation of unperturbed dimensions, and introduction of the impenetrable coil approximation, are treated in Chapter 6. The relationships between dilute solution properties and molecular structure for linear and branched polymers, and parallels with microgels and colloidal dispersions, are also dealt with in Chapter 6.

5.1 Thermodynamic Characterization Methods

All methods for measuring molecular mass, size and interactions in dilute solution rely on assumptions of various kinds. The most reliable model-independent techniques, capable of providing absolute information, are based on thermodynamic principles. The two most commonly used thermodynamic methods, osmometry and static light scattering, are described in this section.

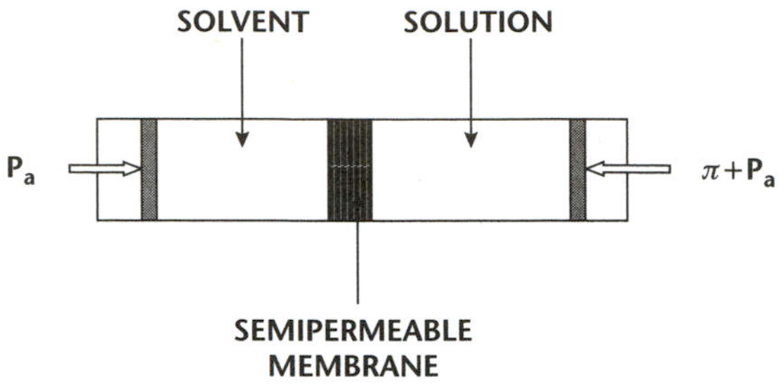

FIGURE 5.1 The principle of osmotic pressure measurement. The ambient pressure is P_a. The extra pressure on the solution side that is needed to achieve the no-flow equilibrium condition is the osmotic pressure π.

5.1.1 Osmotic Pressure

The measurement of osmotic pressure was the first technique used to estimate the absolute molecular weights of macromolecules[1]. As illustrated in Figure 5.1, the determination of osmotic pressure involves placing pure solvent in contact with the solution through a semipermeable membrane[2], one that permits the free passage of solvent but not solute. The chemical potential of solvent μ_1 is reduced by the presence of a solute. If the same pressure were maintained on both sides of the membrane, solvent would flow through the membrane and into the solution indefinitely. The extra pressure that must be applied to the solution side to achieve the no-flow condition at equilibrium is the *osmotic pressure* of the solution π.

The general relationship between osmotic pressure and chemical potential is given by Eq. (3.18). For the modest pressures and the negligible volume changes on mixing of interest here, the partial molar volume of solvent $\overline{V_1}$, is equal to its molar volume V_1, so:

$$\pi = -\frac{\mu_1 - \mu_1^{\circ}}{V_1} \tag{5.1}$$

Under these conditions, Eq. (3.21) is an exact relationship between solution composition and osmotic pressure in the dilute limit. Expressed macroscopically,

$$\pi = \frac{cRT}{M} \qquad (c \to 0) \tag{5.2}$$

in which c is the mass concentration of solute and M is its molecular weight. Accordingly, the absolute molecular weight of any solute can be determined from osmotic pressure data on its solutions in the dilute limit.

Osmotic pressure is a strictly additive property in the dilute limit, so that even if the solute is not a single component, a well-defined average molecular weight can still be obtained. Thus, for a polydisperse solute,

$$\pi = \sum \pi_i = RT \sum \frac{c_i}{M_i} \tag{5.3}$$

where the sum is over the i components that comprise the solute. Because $\sum c_i$ is c; and c_i/c is W_i, the mass fraction of solute species i, [Eq. 5.3] can be written:

$$\pi = cRT \sum \frac{W_i}{M_i} = \frac{cRT}{M_n} \tag{5.4}$$

in which M_n is the number-average molecular weight of the solute [see Eq. (1.5)].

For practical reasons, it is often not possible to measure osmotic pressure at low enough concentrations to apply Eq. 5.4 directly. Solute–solute interactions contribute to π as well, and some extrapolation procedure, providing π/c at the limit of zero concentration, is required to eliminate their effects. As we shall see, extrapolation to obtain interaction-free solute contributions is a routine procedure in all dilute solution investigations of polymers. In the dilute range, the concentration dependence of osmotic pressure is commonly expressed as a power series in concentration, called the *virial expansion:*

$$\frac{\pi}{cRT} = \frac{1}{M_n} + A_2 c + A_3 c^2 + \cdots \tag{5.5}$$

in which $1/M_n$ is the *first virial coefficient,* A_2 the *second virial coefficient,* and so on. A linear extrapolation, π/c versus c, is sometimes used, but such plots usually have some residual curvature, indicating that contributions from the A_3 and higher terms cannot be ignored. A procedure based on an approximation for the third virial coefficient[3], $A_3 = A_2^2 M/4$, is commonly used in such cases. With that approximation and terms beyond A_3 omitted, Eq. 5.5 can be expressed as:

$$\left(\frac{\pi}{cRT}\right)^{1/2} = \left(\frac{1}{M_n}\right)^{1/2}\left(1 + \frac{A_2 M_n}{2}c\right) \tag{5.6}$$

Accordingly, a plot of $(\pi/c)^{1/2}$ versus c is commonly used for the extrapolation, providing an intercept and slope that can be used to obtain values of M_n and A_2. The second and higher virial coefficients also depend on polydispersity[4].

The osmotic pressures for cyclohexane solutions of a 1,4 polybutadiene sample[5] are given in Table 5.1. Linear and square-root plots of these data are shown in Figure 5.2. Straight lines fit both plots fairly well, but on closer examination, a slight positive curvature in the linear plot is discernible. That difference is systematic, not random error. Moreover, the two methods of analysis can lead to different values of M_n and A_2, and the differences become significant for high molecular weights, $M_n \gtrsim 10^5$. The molecular weight listed in Table 5.1 was obtained, using data expressed in consistent units, by the square-root method. The linear method leads to a value of M_n that is 16% larger in this case.

These two methods of data analysis would have in fact agreed had the range of concentrations been low enough. Experimental uncertainty in measuring small

TABLE 5.1 Osmotic pressures at 27°C for a 1,4-polybutadiene sample in cyclohexane[5]

Properties	c, $g\ cm^{-3} \times 10^3$	π, cm Solvent	π, (Pa)
	1.99	0.69	52
$M_n = 124,000$	2.88	1.16	88
$A_2 = 11.6 \times 10^{-4} cm^3 g^{-1}$	3.89	1.67	126
$c^* \sim 7 \times 10^{-3}\ g\,cm^{-3}$	4.93	2.40	181
	5.92	3.20	242
	7.81	5.02	379
	9.79	7.50	566

osmotic pressures forced the use of concentrations high enough for π to depend significantly on the solute–solute interactions. Solute behavior at the dilute limit reflects the contribution of solute molecules that, although interacting plentifully with solvent, do not interact with one another. The solute molecules are far enough apart for the contributions from solute–solute interactions to become negligible. With increasing concentration, the separation distances decrease, and departures from the limiting behavior eventually appear as solute–solute interactions grow in importance. For systems dominated by short-range forces, the crossover from negligible to important occurs in the concentration region where solute size and center-to-center separation become comparable. That region can be specified by a characteristic concentration, obtained, for example, by setting the average separation distance of solute molecules equal to their size using, for example, their radius of gyration. This

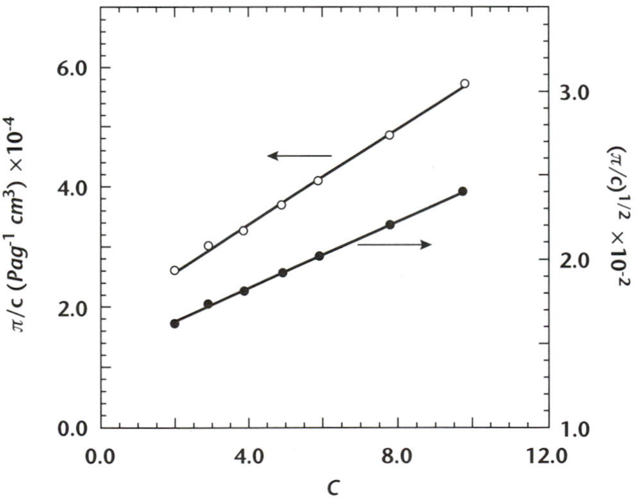

FIGURE 5.2 Linear and square-root plots of osmotic pressure-concentration data for a sample of 1,4-polybutadiene in cyclohexane at 25°C. Data from Hadjichristidis[5] (see Table 5.1).

distance comparability defines the *overlap concentration* c^*, which locates the region where the solution has just become filled with molecular coils.

Viewed from the standpoint of experimental observables, c^* corresponds to the equality of the relationships describing limiting behavior and the departures from limiting behavior caused by solute–solute interactions. For the case at hand, cRT/M_n is the limiting relationship for π and $RT(A_2c^2 + A_3c^3 + \cdots)$ is the departure. For simplicity, setting aside the ternary and higher terms, equality corresponds to $cA_2M_n = 1$, or:

$$c^* = \frac{1}{A_2 M_n} \tag{5.7}$$

The value of c^* listed in Table 5.1, obtained using this equation, shows that some of the data were obtained in the overlap region.

This method for defining overlap concentration as an intersection of limiting behavior and interaction contribution will be applied to each new dilute solution technique as introduced throughout the chapter. It usually works satisfactorily for locating distance comparability—the defined meaning of c^*—because molecular characterization is usually conducted in *good solvent* systems (see Chapter 6). However, near the *theta condition*, where A_2 goes to zero although molecular size does not, it may no longer apply. Equation 5.7 is a case in point; in general, any definition that depends significantly on A_2 will fail near the theta condition.

Regenerated cellulose is the membrane material typically used for solutions of organic polymers[2]. The range of molecular weights measurable by membrane osmometry is typically $20,000 < M_n < 500,000$, with the lower limit set by solute effusion through the membrane and the upper limit by instrument sensitivity. Modern osmometers can achieve the equilibration of individual osmotic pressure measurements in as little as a few seconds.

An alternative method for determining M_n, *vapor pressure osmometry*[2], is applicable to still lower molecular weights. Unlike membrane osmometry, vapor pressure osmometry is not strictly a thermodynamic method. The readings depend on the difference in vapor pressure between solvent and solution, but that difference is deduced from a differential rate of evaporation. Vapor pressure osmometry works best when calibrated with chemically similar samples of known molecular weight. With care, accurate values in the range $500 \lesssim M_n \lesssim 20{,}000$ can be obtained.

5.1.2 Light Scattering

Light scattering provided the first direct measurements of the size of polymer molecules[6]. It is still the most widely used method for obtaining the absolute values of both coil size and molecular weight[7–10]. The elastic scattering of radiation by ensembles of scatterers, and especially the effects of their spatial correlations on the angular

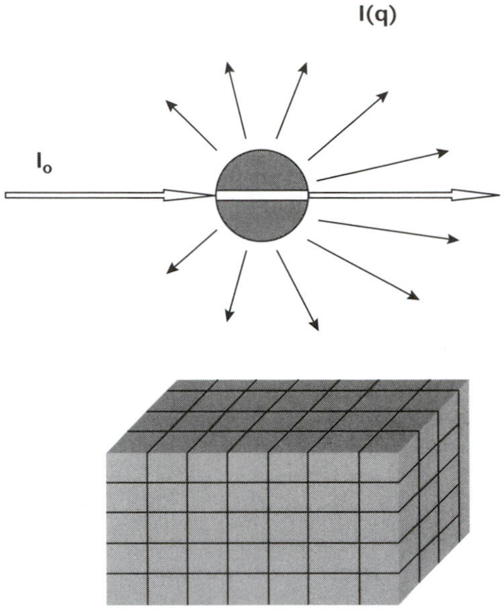

FIGURE 5.3 The light scattering experiment. The elemental volumes show schematically the basis for analyzing the data in terms of spatial fluctuations in density and concentration.

dependence of scattering intensity, was considered in Chapter 2. Those results were used in Chapter 4 to develop general expressions for the form factor of multisite molecules, including random coils. The theory of thermodynamic fluctuations was also introduced there and applied to the scattering properties of multicomponent liquids with only excluded volume interactions. In this chapter, fluctuation theory is applied to the special case of dilute solutions, and interactions in addition to excluded volume are considered.

Fundamentals. Consider a collimated beam of monochromatic light passing through a cell that contains a solution, in an arrangement like that in Figure 2.6a. A small fraction of the incident light is scattered elastically by the solution, leaving the illuminated volume with some angular distribution of scattered light intensity. Light scattering arises from the interaction between the electric field of the light and spatial variations in the electrical polarizability of the solution. In dilute solutions, the scattering resulting from such local inhomogeneities can be used to determine the molecular weight of the solute M, its size R_g, and the second virial coefficient of the solution A_2.

As discussed in Section 4.7, the fluctuations in local density and concentration of a liquid are related to its thermodynamic properties through the Einstein formulas, Eqs. (4.122) through (4.124). The illuminated volume V_{sc} in a scattering experiment, partitioned into small elements of volume δV, is depicted in Figure 5.3. The elemental volumes are chosen to be much bigger than the largest of the molecules in the solution and the scattering angles small enough to make intramolecular interference effects negligible:

$$(\delta V)^{1/3} \gtrsim R_g \tag{5.8}$$

$$q^{-1} \gtrsim R_g \tag{5.9}$$

The second of these restrictions will be relaxed later. [The size of δV drops out of the equations; it is arbitrary as long as the inequalities in Eq. 5.8 and Eq. 5.9 are satisfied.]

Fluctuations in density and concentration result in fluctuations of the electrical polarizability, the property that governs light scattering behavior. From the laws of electrostatics, the excess polarizability of an elementary volume, $\Delta\alpha$, depends on its excess dielectric constant $\Delta\varepsilon$: $\Delta\alpha = (\delta V/4\pi)\Delta\varepsilon$. Well away from absorption peaks, the dielectric constant is related to the *refractive index:* $\varepsilon = n^2$, or $\Delta\varepsilon = 2n\Delta n$ for small enough fluctuations. Thus,

$$\Delta\alpha = \left(\frac{n\,\delta V}{2\pi}\right)\Delta n \tag{5.10}$$

in which all properties are to be evaluated at the light frequency. The refractive index depends on both density and concentration; so for small fluctuations,

$$\Delta n = \left(\frac{\partial n}{\partial \rho}\right)_T \Delta\rho + \left(\frac{\partial n}{\partial c}\right)_T \Delta c \tag{5.11}$$

Squaring, averaging, then applying Eq. (4.122) through Eq. (4.124) and Eq. 5.10 leads finally to an expression for the mean-square excess polarizability:

$$\langle(\Delta\alpha)^2\rangle = \frac{n^2 kT}{4\pi^2}\left[\left(\frac{\partial n}{\partial \rho}\right)_T^2 \rho^2\beta + \left(\frac{\partial n}{\partial c}\right)_T^2 \frac{c}{(\partial\pi/\partial c)_T}\right]\delta V \tag{5.12}$$

in which β is the isothermal compressibility of the solution and π is its osmotic pressure.

Consider now the light scattering experiment in Figure 5.3. A collimated beam of monochromatic light ($400 < \lambda_o(nm) < 800$ with frequency ω and intensity I_o), usually from a laser source, illuminates some scattering volume V_{sc}. In a typical arrangement, the incident beam propagates in the x-direction. A photodetector viewing the scattering volume is mounted on a stage that can rotate in the xy (horizontal) plane, normal to the z (vertical) axis, at some fixed sample-to-detector distance r_D. The beam is usually vertically plane-polarized. Scattering intensity is measured as a function of angle θ from the incident beam direction.

The incident beam is a plane wave with electric field and intensity given by:

$$E_z(t) = A_z \exp(i\omega t) \tag{5.13}$$

$$I_o = E_z E_z^* \tag{5.14}$$

The scatterers are the $V_{sc}/\delta V$ volume elements, and intensity at the detector depends on their excess polarizabilities. The variations of $\Delta\alpha$ from one volume element to another are uncorrelated, so the intensity at the detector is the sum of the elemental scattering intensities:

$$I_D(\theta) = \sum_{i=1}^{V_{SC}/\delta V} (I_D(\theta))_i = \frac{V_{sc}}{\delta V} \langle \delta I_D(\theta) \rangle \tag{5.15}$$

Elastic light scattering is a quantum mechanical event that, in classical terms, originates from the acceleration of electric charge by the electric field of the incident beam acting upon the atomic electrons[6]. The charge displacement polarizes the molecule and, for isotropic scatterers, the polarization is the product of excess molecular polarizability and the incident field:

$$\mathbf{p}(t) = \Delta\alpha \mathbf{E}(t) \tag{5.16}$$

According to the laws of electromagnetism, a spherical scattered wave is generated by the charge acceleration, producing a field at the detector with amplitude:

$$[E_z(t)]_D = \frac{\sin\varphi_z}{\mathrm{v}_c^2 r_D} \frac{d^2 p}{dt^2} \tag{5.17}$$

in which φ_z is the angle between the incident electric field direction and sample-to-detector direction, and v_c is the velocity of light in vacuum. For a vertically polarized incident beam, $\varphi_z = \pi/2$ or $\sin\varphi_z = 1$. Thus, using Eqs. 5.13 and 5.16, and for vertical polarization,

$$[E_z(t)]_D = -\frac{A_o}{r_D}\left(\frac{2\pi}{\lambda_o}\right)^2 \Delta\alpha \exp(i\omega t) \tag{5.18}$$

so the detector intensity $(E_z E_z^*)_D$ is:

$$I_D(\theta) = \frac{I_o V_{sc}}{r_D^2}\left(\frac{2\pi}{\lambda_o}\right)^4 \frac{\langle(\Delta\alpha)^2\rangle}{\delta V} \tag{5.19}$$

Accordingly, I_D is independent of the scattering angle, corresponding to the $q = 0$ case $I_D(0)$, as long as conditions Eqs. 5.8 and 5.9 are fulfilled. Finally, using Eq. 5.12, and with Eq. (2.53) for the reduced intensity,

$$I_r(0) = \frac{4\pi^2 n^2 kT}{\lambda_o^4}\left[\left(\frac{\partial n}{\partial \rho}\right)^2 \rho^2 \beta + \left(\frac{\partial n}{\partial c}\right)^2 \frac{c}{\partial \pi/\partial c}\right] \tag{5.20}$$

The first term in Eq. 5.20 is the contribution from density fluctuations:

$$[I_r(0)]_{\delta\rho} = \frac{4\pi^2 kT n^2 \rho^2 (\partial n/\partial \rho)_T^2 \beta}{\lambda_o^4} \tag{5.21}$$

The second term is the contribution from concentration fluctuations:

$$[I_r(0)]_{\delta c} = \frac{4\pi^2 n^2 (dn/dc)^2}{N_a \lambda_o^4} \frac{cRT}{(\partial \pi/\partial c)_T} \tag{5.22}$$

The concentration fluctuations term contains information on the thermodynamics of mixing. Equation (4.143) and the contrast factor for light scattering defined as

$$C_{LS} = \frac{4\pi^2 n^2 (dn/dc)^2}{N_a \lambda_o^4} \tag{5.23}$$

gives an expression for the structure factor at $q = 0$:

$$S(0) = \frac{cRT}{(\partial \pi/\partial c)_T} \tag{5.24}$$

With mass concentration replaced by volume fraction, $\phi_2 = c\bar{v}_2$ and the solutions convention $\phi = \phi_2$, the following are useful equivalent formulas for the static structure factor at $q = 0$:

$$S(0) = \frac{1}{\bar{v}_2^2} \frac{\phi_2 RT}{(\partial \pi/\partial \phi_2)_T} \tag{5.25}$$

$$S(0) = \frac{1}{\bar{v}_2^2} \frac{V_1 \phi RT}{(-\partial \mu_1/\partial \phi)_T} \tag{5.26}$$

$$S(0) = \frac{1}{\bar{v}_2^2} \frac{RT}{\left(\partial^2 g_m/\partial \phi^2\right)_T} \tag{5.27}$$

The Gibbs-Duhem relation [Eq. (3.7)] makes the particular component choices in these expressions irrelevant. The factor $1/\bar{v}_2^2$ is commonly absorbed into the contrast factor.

Dilute solutions. The light scattering analysis to this point applies to binary systems of all concentrations. In light scattering terminology, the reduced intensity as defined by Eq. (2.53) is the *Rayleigh ratio,* a name that carries over to reduced intensities

TABLE 5.2 Some optical properties at $\lambda_o = 633$ *nm* and 23°C for various polymer–solvent systems

Polymer–Solvent System	n_s	$(R_\theta)_s$, $cm^{-1} \times 10^6$	dn/dc, cm^3g^{-1}
1,4 polyisoprene-cyclohexane	1.425	5.1	0.106
1,4 polybutadiene-cyclohexane			0.107
1,2 polybutadiene-cyclohexane			0.087
polyisobutylene-cyclohexane			0.095
1,4 polyisoprene-tetrahydrofuran	1.405	4.4	0.124
a polypropylene-tetrahydrofuran			0.079
a polypropylene-cyclohexane	1.425	5.1	0.059
a poly(1-butene)-cyclohexane			0.067
polyethylene-trichlorobenzene (135°C)	1.502	35.7	−0.109[(a)]
poly(dimethyl siloxane)-toluene	1.492	14.2	−0.104[(b)]
polystyrene-toluene			0.105
polystyrene-cyclohexane	1.425	5.1	0.168
poly(ethylene oxide)-methanol	1.327	2.9	0.135
poly(methyl methacrylate)-acetone	1.357	4.3	0.138[(b)]

[(a)]Obtained at 488 *nm*.
[(b)]obtained at 436 *nm*.

that vary with scattering angle.

$$R_\theta \equiv I_r(q) \tag{5.28}$$

Equation 5.20 expresses the intensity for optically isotropic scatterers and a small-enough product of scattering angle and scatterer size to satisfy Eq. 5.9. Accordingly, the Rayleigh ratio for pure solvent ($c = 0$) is:

$$(R_\theta)_s = \frac{4\pi^2 kTn^2\rho^2 (\partial n/\partial \rho)_T^2 \beta}{\lambda_o^4} \tag{5.29}$$

The experimental values for molecular liquids are in rather good agreement with values predicted by Eq. 5.29[11]. Differences occur, but these can be understood from the optical anisotropy of the solvent molecules. Experimental Rayleigh ratios obtained with a vertically polarized incident beam are given for some representative polymer-solvent systems in Table 5.2.

Scattering intensity is typically measured at several angles for several concentrations and pure solvent. The pure solvent scattering is then subtracted to give the *excess Rayleigh ratio,* the contribution from concentration fluctuations alone[7,10]:

$$(\Delta R_\theta)_c = (R_\theta)_c - (R_\theta)_s \tag{5.30}$$

The contribution of density fluctuations to scattering intensity is insensitive to polymer concentration in the dilute range. (The scattering contribution from random coils is not appreciably affected by the optical anisotropy of the mers. Many mer orientations are represented within each chain if they are long enough, and the contributions of local anisotropy tend to cancel, such that the molecular polarizability is essentially isotropic.) From Eqs. 5.20 and 5.29,

$$(\Delta R_\theta)_c = \frac{C_{LS}\, c RT}{(\partial \pi / \partial c)_T} \qquad (q R_g \ll 1) \tag{5.31}$$

in which the contrast factor C_{LS}, defined in Eq. 5.23, is commonly called the *optical constant* K in the light scattering literature. The *refractive index increment* dn/dc, required to evaluate C_{LS}, is a property of the polymer–solvent pair and is obtainable with a differential refractometer. Except at low molecular weights, C_{LS} varies only weakly with temperature, concentration, and molecular weight[12]. Some representative values of dn/dc are given in Table 5.2.

The quantity $c/(\partial \pi / \partial c)_T$ in Eq. 5.31 contains the thermodynamic information of interest. In the dilute range, and for $q R_g \ll 1$, the scattering equation can be expressed as a virial series using Eq. 5.5,

$$\frac{C_{LS}\, c}{(\Delta R_\theta)_c} = \frac{1}{M} + 2A_2 c + 3A_3 c^2 + \cdots \tag{5.32}$$

This formulation is easily generalized to polydisperse solutes by a procedure similar to that used for osmotic pressure. Thus, in the dilute limit, the various solute species contribute independently to the scattering intensity:

$$\Delta R_\theta = \sum_i (\Delta R_\theta)_i \tag{5.33}$$

From Eq. 5.32 in the dilute limit, and assuming C_{LS} to be independent of i,

$$(\Delta R_\theta)_i = C_{LS}\, c_i M_i \tag{5.34}$$

Combining Eqs. 5.33 and 5.34 gives:

$$(\Delta R_\theta)_c = C_{LS}\, c \left[\sum_i \frac{c_i}{c} M_i \right] = C_{LS}\, c M_w \tag{5.35}$$

Thus, independent of solute architecture and distribution, but subject to the species independence of the contrast factor C_{LS} and to $q R_g \ll 1$, the weight-average molecular weight is obtained:

$$\frac{C_{LS}\,c}{(\Delta R_\theta)_c} = \frac{1}{M_w} + 2A_2c + 3A_3c^2 + \cdots \tag{5.36}$$

Polydispersity also affects the second and higher virial coefficients, and in a manner from that of different the osmometry-derived and scattering-derived values[4].

The first term on the right in Eq. 5.36 is the independent solute contribution. The remainder reflects the influence of spatial correlations among solute molecules, induced by thermodynamic interactions and giving rise to intermolecular interference. The two contributions can be separated by a plot of $c/(\Delta R_\theta)_c$ versus c, which becomes a straight line in the dilute range, with M_w and A_2 obtained from its slope and intercept. Berry[13] demonstrated the value of a square-root plot, analogous to that used with osmotic pressure. That method is now used commonly in the analysis of light scattering data in the dilute range. An analysis paralleling that for the osmometry data [Eq. 5.6] but with a slightly different approximation, $A_3 = A_2^2 M_w/3$, leads to:

$$\left[\frac{C_{LS}\,c}{(\Delta R_\theta)_c}\right]^{1/2} = \left(\frac{1}{M_w}\right)^{1/2}(1 + A_2\,M_w\,c) \tag{5.37}$$

Rayleigh ratios for cyclohexane solutions of 1,4 polyisoprene[5] are given in Table 5.3, as is a value for the overlap concentration for this system, obtained in this case as:

$$c^* = \frac{1}{2A_2M_w} \tag{5.38}$$

Both linear and Berry plots of the data are shown in Figure 5.4. Similarly to Figure 5.2, the linear plot suggests a slight positive curvature, whereas a straight line fits the Berry plot rather well, with only a hint of an opposite curvature. The value of M_w in Table 5.3 was obtained with the Berry plot. In this case, M_w from the linear plot is about 47% larger. This rather unsettling comparison points to the

TABLE 5.3 Light scattering intensity at 25°C for a 1,4-polyisoprene sample in cyclohexane[5]

Properties	c, $g\,cm^3 \times 10^3$	$(\Delta R_\theta)_c$, $cm^{-1} \times 10^6$
$dn/dc = 0.106\,cm^3g^{-1}$	1.64	8.4
$\lambda_o = 633\,nm$	3.19	11.7
$M = 156{,}000$	4.84	13.6
$A_2 = 8.0 \times 10^{-4}\,cm^3g^{-1}$	6.44	14.4
$c^* \sim 4 \times 10^{-3}\,g\,cm^{-3}$	8.11	14.6

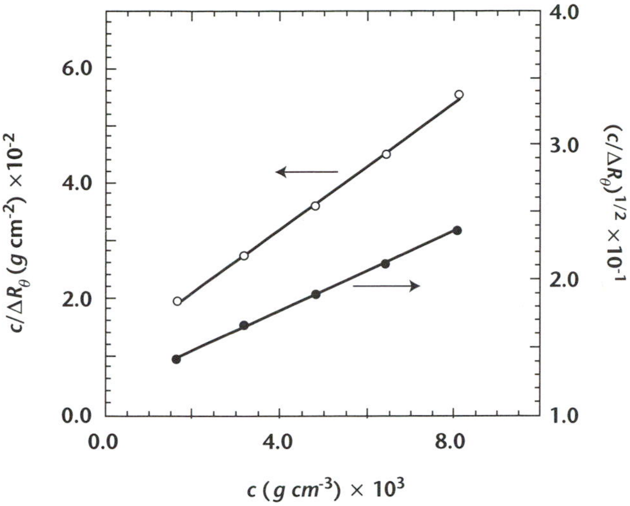

FIGURE 5.4 Linear and square root plots of light scattering intensity-concentration data for a sample of 1,4-polyisoprene in cyclohexane at 25°C. Data from Hadjichristidis[5] (see Table 5.3).

desirability of obtaining data at lower concentrations. Here only the lowest two concentrations are below c^*, as defined here by Eq. 5.38.

Until now, only conditions for which *intramolecular interference* is negligible ($qR_g \ll 1$) have been considered, but the analysis can be extended to include quite generally the effects of finite size. In the dilute limit, the solute molecules are far apart and their relative locations and orientations are random. Thus, they behave as independent scatterers, so that the only interference is intramolecular and can be described by a solute form factor $P(q)$. Using Eq. (4.96) and Eq. 5.36, the excess Rayleigh factor in the dilute limit becomes:

$$(\Delta R_\theta)_c = C_{LS}\, c M_w\, P(q) \tag{5.39}$$

The first corrections for intramolecular interference depend only on the radius of gyration of the solute. Thus, from Eq. (4.116):

$$\frac{cC_{LS}}{(\Delta R_\theta)_c} = \frac{1}{M_w} + \frac{\left(R_g^2\right)_z}{3M_w} q^2 \qquad (q(R_g)_z \lesssim 1) \tag{5.40}$$

When both intramolecular and intermolecular interference contribute, the scattering behavior for solutes of arbitrary structure and polydispersity can be written,

$$\frac{cC_{LS}}{(\Delta R_\theta)_c} = \frac{1}{P(q)RT}\left(\frac{\partial \pi}{\partial c}\right)_{P,T} \tag{5.41}$$

TABLE 5.4 Scattering data as a function of angle and concentration for solutions at 25°C for a polystyrene sample in toluene[5]

	$Kc/(\Delta R_\theta)_c \times 10^6$				
$c\ (g\,cm^{-3} \times 10^4)$	1.45	2.12	2.81	4.30	5.66
θ					
45°	1.268	1.296	1.342	1.439	1.527
60°	1.321	1.358	1.382	1.491	1.592
75°	1.384	1.410	1.442	1.546	1.620
90°	1.441	1.479	1.505	1.607	1.689
105°	1.504	1.538	1.577	1.664	1.766
120°	1.579	1.612	1.637	1.733	1.814
135°	1.641	1.670	1.699	1.793	1.886

$\lambda_o = 623 nm$; $n_s = 1.492$; $dn/dc = 0.105\ cm^3 g^{-1}$
$M_w = 920{,}000$; $(R_g)_z = 40.8 nm$; $A_2 = 3.2 \times 10^{-4} cm^3 g^{-1}$; $c^* = 17 \times 10^{-4}\ g\,cm^{-3}$

and, to lowest order in scattering angle and concentration,

$$\frac{cC_{LS}}{(\Delta R_\theta)_c} = \frac{1}{M_w} + \frac{16\pi^2 n_s^2}{3\lambda_o^2}\left(\frac{\left(R_g^2\right)_z}{M_w}\right)\sin^2\theta/2 + 2A_2 c + \cdots \tag{5.42}$$

Accordingly, scattering intensity in the dilute range—and for sufficiently small scattering vectors—can provide model-free values of three fundamental properties: the mass and size of the solute and a thermodynamic parameter governed by solute–solute interactions.

A *Zimm plot* [14] can be used to display visually the effects of intermolecular and intramolecular interference when both are significant. Values of $C_{LS}\,c/(\Delta R_\theta)_c$ for the various scattering angles and concentrations are plotted as a function of a composite variable, $\sin^2\theta/2 + c/c_o$, where c_o is an arbitrary parameter used for convenience to adjust the relative angle and concentration scales on the plot. Two graphic extrapolations, to obtain the angular dependence at $c = 0$ and the concentration dependence at $\theta = 0$, then yield the three fundamental properties from their initial slopes and common intercept.

Such results can also be obtained in other ways. Scattering data for solutions of a polystyrene sample in toluene[5] are given in Table 5.4. The values of $C_{LS}\,c/(\Delta R_\theta)_c$ are plotted against angle for the various concentrations in Figure 5.5, and their least-squares intercepts and slopes were recorded. A Berry plot of those intercepts was then used to obtain $M_w = 912{,}000$ and $A_2 = 3.02 \times 10^{-4}\,cm^3 g^{-1}$. Slopes from the Figure 5.5 plots, which vary slightly with concentration, were extrapolated to zero concentration, which leads to $(R_g)_z = 40.7\ nm$. The slightly different values given with the data in Table 5.4 were obtained using the fitting software provided with the scattering instrument. The values of R_g have been determined for dilute solutions of

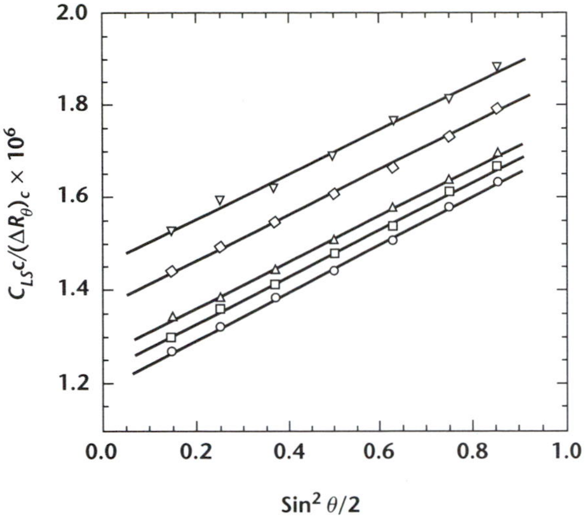

FIGURE 5.5 The angular dependence of light scattering intensity at several concentrations for a sample of polystyrene in toluene at 25°C. Data from Hadjichristidis[5] (see Table 5.4). Concentrations in units of gcm^{-3}, $10^4c = 1.45(\bigcirc)$, $2.12(\square)$, $2.81(\triangle)$, $4.30(\diamond)$, and $5.66(\nabla)$.

polystyrene in toluene over an extended molecular weight range [see Figure 6.1 and Eq. (6.6) in Chapter 6.] Note that c^* from Eq. 5.38 is significantly larger than even the highest concentration used, a more favorable situation than for the scattering data in Table 5.3.

Light scattering methods are sensitive to the presence of dust and adventitious particulates that are everywhere and compete with the solute scattering if not removed. Filters for clarifying solvents and solutions have improved significantly over the years, mainly through better control of the pore size and distribution. The scrupulous cleansing of scattering cells and other equipment contacting the solutions is also essential. Continuous flow systems with in-line filtration have been particularly effective in flushing and maintaining relatively dust-free environments.

The design of light scattering instruments has also evolved over time, and several types are now available[15–17]. Some are designed to provide data at very small scattering angles ($1.5° \lesssim \theta \lesssim 6°$), in which case the q-dependent term in Eq. 5.42 is negligible unless the solute is unusually large. Of course, $(R_g)_z$ is unavailable to such *small-angle light scattering* (SALS) instruments, but rather precise values of M_w and A_2 can be obtained over wide ranges. The Rayleigh ratios in Table 5.3 were obtained using a SALS instrument. Other instruments provide the angular dependence of intensity, in one case permitting both static and dynamic scattering measurements, in another by using an array of fixed detectors in place of the single detector on a rotating arm. With the development of physically stable (non-"shedding") chromatographic gels, light

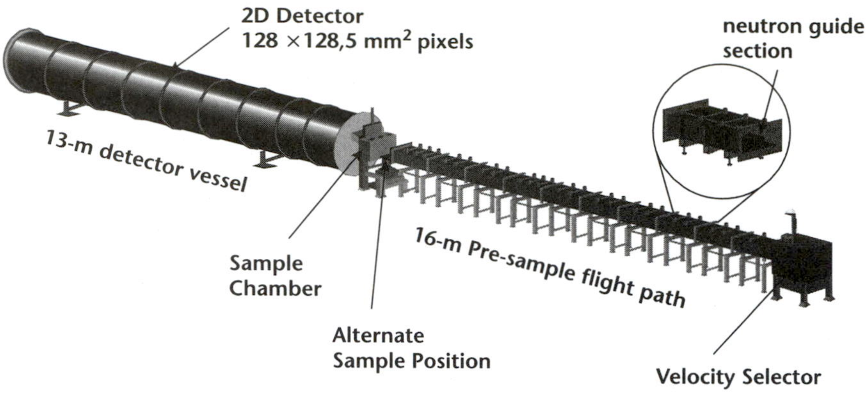

FIGURE 5.6 One of the beam lines for small-angle neutron scattering at the National Institutes of Standards and Technology in Gaithersburg, Maryland.

scattering instruments are finding increasing use for the in-line analysis of streams emerging from size exclusion chromatographs (See Section 5.3.2).

Instruments are available for evaluating the refractive index increment. Values of dn/dc are known for many polymer–solvent pairs[18], although not always at the wavelength of interest. In most practical cases, they lie in the range $-0.10 < dn/dc\,(cm^3 g^{-1}) < +0.20$. The solute scattering intensity is proportional to $(dn/dc)^2$, so large magnitudes are desirable for sensitivity reasons; that is, to obtain large values of $(\Delta R_\theta)_c$ at low concentrations.

Scattering method comparisons. The application of *small-angle neutron scattering* (SANS) to obtain of M_w, $(R_g)_z$, and A_2 in dilute solutions proceeds along lines rather similar to those used for light scattering intensity measurements. The practical considerations, however, are very different[19,20]. Neutron scattering requires a plentiful supply of neutrons, such as provided by a nuclear reactor, and the experiments themselves require a large physical space. In addition, extensive supporting equipment and personnel are required to produce collimated high-flux beams and to acquire and process the scattering data. Such experiments can only be conducted at a relatively few international facilities dedicated to neutron scattering.

One configuration for performing SANS experiments is shown in Figure 5.6. The neutrons are cooled (thermalized near absolute zero), collimated, and velocity-selected to supply an incident beam (diameter $\sim 2\,cm$) with an average wavelength in the range $0.5 \lesssim \lambda\,(nm) \lesssim 1.5$ and a wavelength spread in the range $0.5 \lesssim \Delta\lambda/\lambda \lesssim 1.5$. The scattering data are acquired by a two-dimensional neutron detector array (spatial resolution $\sim 1\,cm$, overall size $\sim 1\,m$). The sample-to-detector distance is adjustable and typically in the range $6 \lesssim r_D\,(m) \lesssim 30$. Scattering data can be collected over about one decade of q in a single SANS experiment. The range $0.05 \lesssim q(nm^{-1}) \lesssim 0.5$ is commonly employed, and $0.01 \lesssim q\ (nm^{-1}) \lesssim 0.1$ is achievable, although at some sacrifice in data acquisition rate.

The q range accessible by light scattering is smaller. Thus, $\lambda_o/n_s \sim 400\ nm$ and $15^\circ \lesssim \theta \lesssim 150^\circ$ for light scattering gives the span $0.003 \lesssim q\ (nm^{-1}) \lesssim 0.03$, overlapping only slightly (at best) with SANS. Light scattering and SANS are both useful for obtaining R_g, but over somewhat different ranges of solute size. Even the best of modern light scattering instrumentation cannot provide reliable values of R_g smaller than about 10 nm; the q-dependent term in Eq. 5.42 contributes less than ~5% of the total excess scattering in that range:

$$(R_g)_{min} \sim \frac{\lambda_o}{4\pi n_s}\left[\frac{(0.1)\,(3)}{\sin^2(\theta/2)_{max} - \sin^2(\theta/2)_{min}}\right]^{1/2} \tag{5.43}$$

Light scattering is limited to the Guinier range except for polymers of unusually large size. The lower limit on R_g for SANS, even in the commonly employed q range, is about 3 nm. Thus, SANS can routinely provide data in the Guinier range, $q R_g \lesssim 1$, only for polymers of low and moderate size. Without special efforts, data in the intermediate range, $1 \gtrsim q R_g \gtrsim q l_\kappa$, at best only approaches the Guinier range for the more typical polymeric sizes.

Another significant difference between light scattering and SANS is the source of scattering contrast. A difference in refractive index provides the contrast for light scattering; a difference in scattering length provides the contrast for SANS. The scattering lengths for hydrogen and deuterium nuclei are quite different, yet physical properties change only slightly when hydrogen is replaced by deuterium. The possibility of isotopic labeling by deuterium substitution confers an enormous advantage to SANS relative to other scattering techniques. Deuterium labeling permits the creation of contrast between molecules, or even between different parts of the same molecule, without profoundly disturbing the other physical properties of the system. Some uses and effects of deuterium substitution for melts and nondilute solutions are described in Chapters 7 and 8.

Small-angle x-ray scattering (SAXS) is an old but still valuable and widely practiced scattering technique[20, 21], only recently finding extensive use in dilute solution characterization[22, 23]. The interaction of x-ray photons ($\lambda \sim 10^{-1} nm$) and atomic electrons forms the scattering origin; contrast between species is produced by differences in electron density. The available q-range is essentially the same as for SANS. Like light scattering, and unlike SANS, SAXS experiments can be conducted with relatively inexpensive and small-scale equipment. Nevertheless, the importance of central SAXS facilities and synchrotron sources is growing rapidly as faster data acquisition rates and lower q-ranges are sought.

Both light scattering and SANS are absolute methods, whereas SAXS intensities are subject to corrections that can vary from sample to sample. Parallels and differences among the three methods, when applied in the dilute range, can be expressed in terms of their governing equations:

Light scattering:

$$R_{q,c} = R_{q,0} + C_{LS}\frac{cRT}{(\partial\pi/\partial c)_T}P(q) \tag{5.44}$$

Small-angle neutron scattering:

$$I_r(q,c) = I_r(q,0) + \frac{1}{V_{sc}}\left(\frac{d\Sigma}{d\Omega}\right)_{incoh} + C_{SANS}\frac{cRT}{(\partial\pi/\partial c)_T}P(q) \tag{5.45}$$

Small-angle x-ray scattering:

$$I_r(q,c) = I_r(q,0) + \mu C_{SAXS}\frac{cRT}{(\partial\pi/\partial c)_T}P(q) \tag{5.46}$$

On the left is the light-scattering Rayleigh factor for the solution and its equivalent for SANS and SAXS. The first term on the right is the contribution from solvent density fluctuations. This value is important in light scattering and SAXS but usually insignificant in SANS. The second term is the incoherent SANS scattering contribution[19], a q-independent intensity with no counterpart in either light scattering or SAXS. It is proportional to the number density of protons in the scattering volume and behaves operationally like a density-fluctuations contribution; that is, as an angle-independent additive term.

The last term on the right side of Eqs. 5.44 through 5.46 carries the size and thermodynamic information about the sample. It is scaled by the contrast factor C_{sc}, in principle determinable independently and without adjustable constants for all three types of scattering. The parameter μ for SAXS is a sample-dependent intensity correction. Even for SAXS, however, the quantities $(R_g^2)_z$and A_2M can be obtained independently of intensity scale, from slope-to-intercept ratios in plots of $c/[(d\Sigma/d\Omega)_{c,q} - (d\Sigma/d\Omega)_{o,q}]$ versus q^2 and c respectively, in analogy with Eq. 5.42. The SAXS data can also be placed on an absolute scale if M for the sample is independently known[22].

5.2 Dynamic Characterization Methods

In this section, some dynamic methods of dilute solution characterization are introduced. Like the thermodynamic methods, these provide information about molecular size. They are included here so that all the size measures can be discussed together. These same techniques—dynamic light scattering and viscometry—are also used to

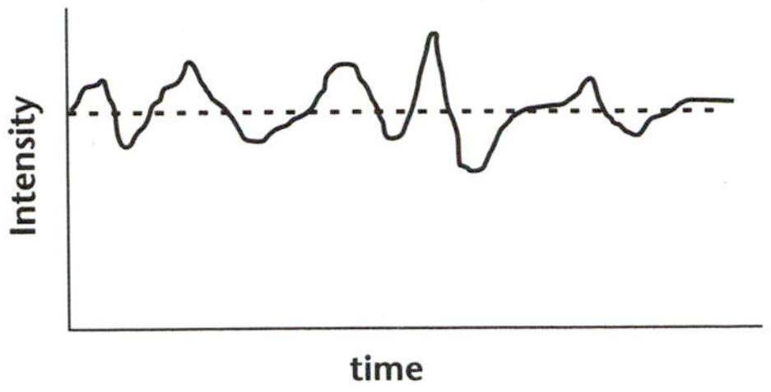

FIGURE 5.7 Schematic of light scattering intensity fluctuations.

investigate the internal dynamics of polymers and their response to flow, as explained in Volume 2.

5.2.1 Dynamic Light Scattering

Dynamic light scattering, also called *quasi-elastic light scattering* (QELS) or *photon correlation spectroscopy,* is the newest general technique for characterizing polymer molecules in dilute solution[10,17,24]. The statistical properties of the temporal fluctuations in scattering intensity are determined and analyzed to obtain the solute diffusion coefficient.

A typical photodetector output for the "homodyne" experiment, now used almost exclusively, is illustrated in Figure 5.7. The ensemble-average scattering intensity $\langle I(q)\rangle$ and the average over long times, $\overline{I(q)}$, are the same for any system at equilibrium[25]. Accordingly, the long-time average of the departure from the average intensity, $\overline{I(q,t)-\overline{I(q)}}$, must be ϕ. The average magnitude of the departures depends on the size of the scattering volume; for the cases of interest, these departures are much smaller than the long-time average intensity: $\overline{|I(q,t)-\overline{I(q)}|} \ll \overline{I(q)}$.

Variations in intensity at the detector reflect variations in scattering field amplitude, which result from time-dependent fluctuations of the excess polarizability $\Delta\alpha$ in the scattering volume. The time-dependent decay of fluctuations can be characterized statistically by a *field correlation function:*

$$g_1(q,t) = \overline{E_D(q,t')\,E_D^*(q,t'+t)}/\overline{I(q)} \tag{5.47}$$

The field correlation function decreases monotonically with the "delay time" t, going from $g_1(q,0)=1$ to $g_1(q,\infty)=0$. Its properties characterize the lifetime of fluctuations in the local polarizability, and hence the fluctuations of concentration, in the illuminated volume. Intensity fluctuations also reflect concentration fluctuations in the scattering volume and can be characterized by an *intensity correlation function:*

$$g_2(q,t) = \overline{I(q,t')\,I(q,t'+t)}/\overline{I(q)}^2 \tag{5.48}$$

Intensity fluctuations also decrease with increasing delay time, going in this case from $g_2(q,0) = \overline{I^2(q,t')}/\overline{I(q)}^2$, which is a value greater than unity, down to $g_2(q,\infty) = 1$.

The two correlation functions are related. For sufficiently dilute solutions of monodisperse solutes, and for solute size and scattering vector such that $qR_g \ll 1$, the field correlation function is a single exponential:

$$g_1(q,t) = \exp(-q^2 D_o t) \tag{5.49}$$

in which D_o is the solute diffusion coefficient in the dilute limit. The intensity correlation function can be expressed as:

$$g_2(q,t) = 1 + B g_1^2(q,t) \tag{5.50}$$

in which B is an instrumental parameter, $\overline{I(q)^2}/\overline{I(q)}^2 - 1$, that may depend weakly on q. The instruments and procedures for determining $g_2(t)$ are described elsewhere[10,24].

As when using other dilute solution methods, it is usually necessary to allow for possible concentration effects and then extrapolate the data to zero. The decay of concentration fluctuations is governed by the *mutual diffusion coefficient* $D(c)$:

$$D(c) = \frac{m}{Z}\frac{\partial \pi}{\partial c} \tag{5.51}$$

in which m is the diffusant molecular mass, M/N_a, and Z is the *molecular friction coefficient*. The mutual diffusion coefficient is affected by concentration in two distinct ways. Thus, for example, a concentration fluctuation is driven towards zero by the osmotic pressure gradient $\partial\pi/\partial c$ but resisted by the friction coefficient Z, both of which depend on solute concentration. For dilute solutions,

$$Z(c) = Z_o[1 + k_f c + \cdots] \tag{5.52}$$

in which k_f is the concentration coefficient, and Z_o is the *Stokes coefficient,* or the molecular friction coefficient at zero concentration. The diffusion coefficient at zero concentration is related to Z_o by the Einstein equation[26]:

$$D_o = kT/Z_o \tag{5.53}$$

From Eq. 5.5,

$$\frac{\partial \pi}{\partial c} = RT\left(\frac{1}{M} + 2A_2 c + \cdots\right) \tag{5.54}$$

From Eq. 5.51, and using Eqs. 5.52 through 5.54, the mutual diffusion coefficient in the dilute range becomes:

$$D(c) = D_o[1 + k_D\, c + \cdots] \tag{5.55}$$

TABLE 5.5 Dynamic scattering data at 25°C for a 1,4-polyisoprene sample in cyclohexane[5]

Properties	$c\ (g cm^{-3}) \times 10^3$	$D(c)(cm^2 s^{-1}) \times 10^7$
$\lambda_o = 632.8\ nm$	1.53	2.05
$M = 156{,}000$	2.95	2.29
$D_o = 1.85 \times 10^{-7}\ cm^2 s^{-1}$	4.46	2.48
$k_D = 76.6\ cm^3\ g^{-1}$	5.91	2.69
$c^* \sim 13 \times 10^{-3}\ g cm^{-3}$	7.42	2.90

in which k_D, the *mutual diffusion concentration coefficient,* is

$$k_D = 2A_2M - k_f \tag{5.56}$$

The first effect of finite scatterer size depends on R_g and introduces a q-dependence into the correlation function. With the leading contributions of both size and concentration included,

$$g_2(q,t) = 1 + B\exp[-2D_oq^2t(1 + k_Dc + \alpha q^2R_g^2 + \cdots)] \tag{5.57}$$

in which the coefficient α depends on both molecular structure and the internal dynamics of the solute species[24]. The data for each concentration and scattering angle (typically in the range, $20° \lesssim \theta \lesssim 120°$) are fitted to a three-term cumulant expansion in the delay time:

$$\ln(g_2 - 1) = \mu_o - \mu_1 t + \frac{\mu_2}{2!}t^2 + \cdots \tag{5.58}$$

in which μ_o is $\ln B$ and of no special interest here, $\mu_1 t$ is the exponent in Eq. 5.49, and the parameter μ_2/μ_1^2 characterizes the deviation of $g_1(q,t)$ from a single exponential form. Effects of solute size and dispersity are negligible if μ_1/q^2 is independent of scattering angle and μ_2/μ_1^2 is small relative to unity. A dependence of μ_1/q^2on q indicates a solute size effect, in which case an extrapolation to $q = 0$ is necessary: $D(c) = \lim_{q\to 0}[\mu_1(q)/2q^2]$. The effect of polydispersity in the dilute limit is complicated but understood[24].

Mutual diffusion coefficients for several cyclohexane solutions of a nearly monodisperse polyisoprene sample[5] are listed in Table 5.5. A plot for obtaining the values of D_o and k_D is shown in Figure 5.8. The data are fitted well by a straight line, which yields the values of D_o and k_D listed in Table 5.5. The overlap concentration given there was calculated using:

$$c^* = \frac{1}{k_D} \tag{5.59}$$

which was obtained in the manner of previous estimates from Eq. 5.55. The value indicates that all data were well within the dilute range. Note that this definition of

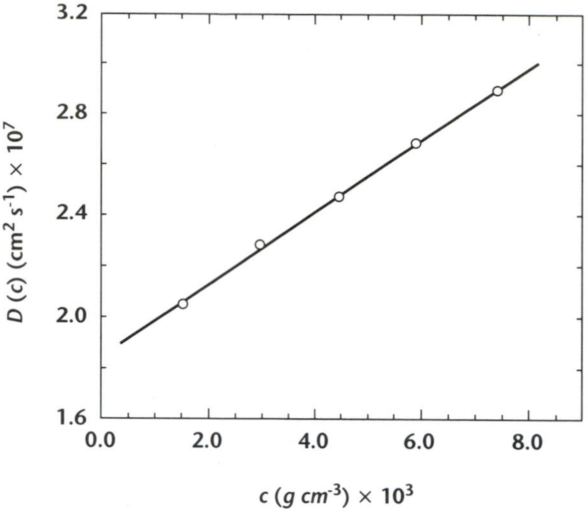

FIGURE 5.8 The mutual diffusion coefficient as a function of concentration for a sample of 1,4-polyisoprene in cyclohexane at 25°C. Data from Hadjichristidis[5]. (see Table 5.5).

c^* depends on A_2 through Eq. 5.56, and hence only applies in good solvent systems, as in the case here.

For linear polymers, D_o is a monotonically decreasing function of chain length, and the relationship for long chains is a power law:

$$D_o = K_d M^{-b} \tag{5.60}$$

The exponent b ranges between 0.55 and 0.59 for most polymer–solvent systems except at or near the theta-condition, an important special case that will be considered in Chapter 6. The result for linear polyisoprene in cyclohexane[27] is shown in Figure 5.9. A power law fits the data:

$$D_o(cm^2 s^{-1}) = 1.98 \times 10^{-4} M^{-0.584} \qquad (1.5 \times 10^4 < M < 3.4 \times 10^6) \tag{5.61}$$

Values of D_o can be obtained by other experiments. Thus, measurements of the solute *sedimentation velocity* in the intense field of force produced by an ultracentrifuge have been used to evaluate the molecular friction coefficient[28,29]. The net centrifugal force F acting on the solute molecules causes them to move relative to solvent with velocity U. According to Stokes' law (Chapter 6), the ratio of imposed force to relative velocity is equal to the molecular friction coefficient, so in the dilute limit,

$$Z_o = \lim_{c \to 0} \frac{F}{U} \tag{5.62}$$

and D_o can then be obtained using Eq. 5.53 if desired.

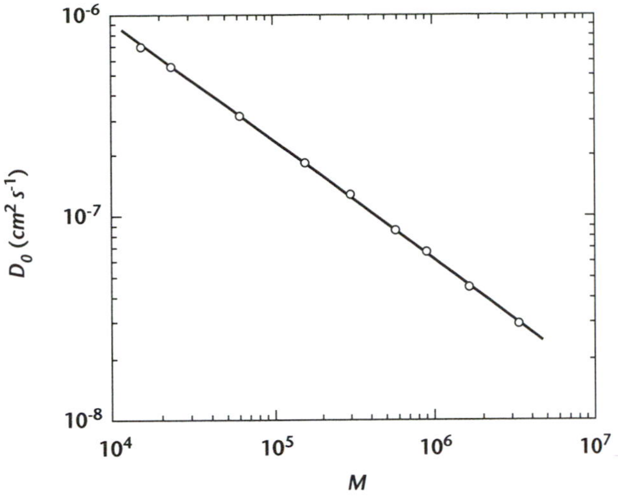

FIGURE 5.9 Relationship between diffusion coefficient and molecular weight for 1,4-polyisoprene in cyclohexane at 25°C.

5.2.2 Viscometry

Solution viscosity measurement was the first method used to estimate the relative molecular weights of polymers. Dissolving even small amounts of polymer in a solvent can observably increase its viscosity. Relative viscosity $\eta(c)/\eta_s$ can be accurately determined using quite simple equipment, such as the Ubbelohde capillary viscometer shown in Figure 5.10. It was known already in 1930[30] that, for a fixed mass concentration of polymer c, the increment in viscosity relative to pure solvent, $(\eta(c) - \eta_s)/\eta_s$, increases with increasing molecular weight. In the dilute range, and with viscosity represented by a power series in concentration[31],

$$\eta(c) = \eta_s(1 + ac + bc^2 + \cdots) \tag{5.63}$$

Both coefficients a and b increase monotonically with the polymer molecular weight, b more rapidly than a, and they have closely related values for a given polymer–solvent system. The parameter a is the *intrinsic viscosity* $[\eta]$, and the parameter b is now expressed as $k_H[\eta]^2$, where k_H, the *Huggins coefficient,* is substantially independent of molecular weight for many polymer–solvent systems.

Values of $[\eta]$ are obtained from the *specific viscosity,* $\eta_{sp} = (\eta(c) - \eta_s)/\eta_s$, at several concentrations in the dilute range, usually by measuring electronically the flow times, t and t_s for some fixed volume of solution in a viscometer. Then, a plot of $\eta_{sp}/c = (t - t_s)/t_s c$ versus c is constructed and extrapolated to zero concentration:

$$\left(\frac{\eta - \eta_s}{\eta_s c}\right) = [\eta] + k_H[\eta]^2 c + \cdots \tag{5.64}$$

to give $[\eta]$ and k_H from the intercept and slope.

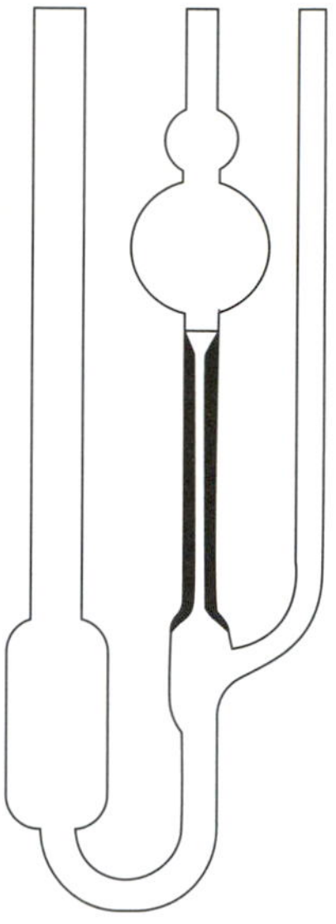

FIGURE 5.10 The Ubbelohde capillary viscometer.

Values of η_{sp} for a sample of linear polystyrene in toluene[5] are listed in Table 5.6. These values are plotted in Figure 5.11 and fitted with a straight line to obtain the values of $[\eta]$ and k_H given in the table. The overlap concentration listed in the table was obtained using:

$$c^* = \frac{1}{k_H[\eta]} \tag{5.65}$$

as estimated in the usual manner using Eq. 5.64. All concentrations lie well below c^*, a situation made possible by the high precision of the relative viscosity measurement, and also accounting for the excellence of the straight line fit. As it happens, this c^* definition does not depend on A_2, so it is valid for good solvents and theta systems alike.

TABLE 5.6 Specific viscosity data at 25°C for a polystyrene sample in toluene[5]

Properties	$c\,(g\,cm^{-3}) \times 10^3$	η_{sp}
$[\eta] = 72.6\ cm^3 g^{-1}$	0.789	0.0587
$k_H = 0.39$	1.098	0.0822
$c^* \sim 14 \times 10^{-3}\ g\,cm^{-3}$	1.78	0.135
	2.49	0.193
	3.20	0.253
	3.94	0.318
	4.70	0.387

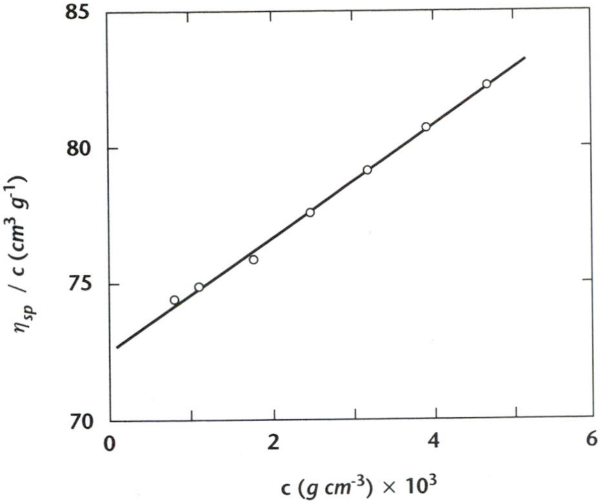

FIGURE 5.11 The ratio of specific viscosity to concentration as a function of concentration for a sample of polystyrene in toluene at 25°C. Data from Hadjichristidis[5] (see Table 5.6).

Neither k_H nor $[\eta]$ have the units of viscosity, k_H being dimensionless and $[\eta]$ having units of reciprocal mass concentration. Also, in contrast with η_s and $\eta(c)$, which like any liquid viscosity are sensitive functions of temperature, the values of $[\eta]$ and k_H are practically independent of temperature for most polymer–solvent systems. The fundamental meaning and significance of intrinsic viscosity will be considered in Chapter 6. Its practical utility lies in the ease of measuring accurate and reproducible values combined with its simple relationship to polymer molecular weight.

A power law relationship between intrinsic viscosity and molecular weight was discovered independently by several early workers[32–36]:

$$[\eta] = KM^a \tag{5.66}$$

For unknown historical reasons, this expression came to be known as the *Mark-Houwink* equation, and that name is retained here. The Mark-Houwink coefficients, K and a, depend on the polymer–solvent system. These coefficients must be obtained by experiments with samples whose molecular weight are known from some absolute determination. For linear polymers beyond some minimum molecular weight, the power law description is remarkably robust, applying over two or more decades in molecular weight and including the range of most practical interest. For flexible polymers, the range of exponents is narrow, $0.60 < a < 0.75$ for most systems. As in Section 5.2.1, however, solutions near the liquid–liquid critical point are exceptional. The values of $[\eta]$ obtained for samples of polystyrene in toluene[37] are fit well by a

Mark-Houwink formula for the range $4.8 \times 10^4 < M < 2.4 \times 10^7$:

$$[\eta](cm^3 g^{-1}) = 9.27 \times 10^{-3} M^{0.734} \tag{5.67}$$

A different range and data selection yields a slightly different result [See Figure 6.10 and Eq. (6.50).]. A compilation of K and a for the linear chains of many species in various solvents is available[38].

Intrinsic viscosity is affected by macromolecular architecture as well as molecular weight, a feature discussed in Chapter 6. The intrinsic viscosity of linear polymers obeying the Mark-Houwink equation is affected by polydispersity in a simple way. In the dilute limit, the increments of viscosity are additive, so $\eta - \eta_s = \Sigma(\eta - \eta_s)_i$. With $(\eta - \eta_s)_i = \eta_s[\eta]_i c_i$ in the dilute limit from Eq. 5.64, that becomes $\eta - \eta_s = \eta_s \Sigma[\eta]_i c_i$, leading to $[\eta] = \Sigma[\eta]_i W_i$. Accordingly, using Eq. 5.66,

$$[\eta] = K(M_v)^a \tag{5.68}$$

in which:

$$M_v = \left[\sum M_i^a W_i\right]^{1/a} \tag{5.69}$$

is the *viscosity average molecular weight*. Although smaller than M_w, M_v does not differ greatly from it unless the distribution is very broad.

5.3 Chromatographic Methods

Size exclusion chromatography (SEC), originally called *gel permeation chromatography* (GPC) was invented by John Moore. After he published the method in 1964[39], SEC rapidly became the method of choice for determining the molecular weight distribution of polymers. The discovery of the universal calibration principle, the creation of ever more efficient separation media, and the introduction of various on-line detectors have greatly enhanced its usefulness. It is now the primary method for the characterization of distributions for various molecular heterogeneities in polymers. Only a brief account of SEC analysis is given here and in Section 6.3.2. More detailed descriptions are available elsewhere[40–43].

5.3.1 Size Exclusion Chromatography

Like all chromatographic techniques, size exclusion chromatography depends on a separation principle. The SEC principle is the size-dependent partitioning of macromolecules between a dilute solution phase and a gel (swollen network) phase. The meshwork of network strands in the gel preferentially excludes large molecules from the gel phase. The partition coefficient, the ratio of polymer concentration in the solution and gel phase, $P = c_{gel}/c_{solution}$, depends on the size of the polymer molecules relative to the average "pore size" of the network. The partition coefficient varies monotonically with the size ratio, $\Lambda = R_{polymer}/R_{gel}$, between two limits:

$$\begin{aligned} P(\Gamma) &\to 1 \qquad (\Gamma \ll 1) \\ P(\Gamma) &\to 0 \qquad (\Gamma \gg 1) \end{aligned} \tag{5.70}$$

Chromatographic separations according to molecular size are achieved through this partitioning in a flow system. Small particles of gel packed in a column and swollen by solvent constitute the *stationary phase*. A stream of the solvent, flowing around the gel particles as it proceeds through the column at some macroscopically steady volumetric flow rate $\dot{q}$, constitutes the *mobile phase*. Upstream of the column, a small aliquot of a dilute polymer solution, $c_o \lesssim c^*$, is injected into the flowing solvent stream at some time t_o. As they pass through the column, the polymer molecules in the aliquot are then sorted according to size by a continual partitioning between the mobile and stationary phases. The largest molecules emerge first from the column, having spent the least time on average in the stationary phase; the other molecules emerge later, in order of decreasing size. The spreading out of the injected solution "plug," caused by the size-based separation in the column, reduces the local concentration significantly: $c < c_o/10$ is not uncommon, concentrations deeply in the dilute range are typical.

A downstream in-line detector, typically a differential refractometer, provides a signal that is proportional to $c(t)$, the mass density of polymer in the flowing stream as a function of time t. With *differential refractive index* (DRI) detection, the signal $S(t)$ is proportional to $\Delta n(t)$, the difference in refractive index between the stream and pure solvent, and hence $S(t) \propto (dn/dc)c(t)$. (For long chain homopolymers and statistically uniform copolymers, dn/dc is insensitive to molecular weight.) The time lapse since injection, $t - t_o$, can be expressed as an *elution volume* $v = (t - t_o)\dot{q}$. Accordingly, $S(t)$ becomes $S(v)$ and the chromatograph, as shown by the chromatogram for a high-density polyethylene sample in Figure 5.12, expresses the weight fraction of injected polymer in each of many elution volume "slices." Accordingly, in the continuous limit,

$$W(v) = \frac{S(v)}{\int_0^\infty S(v)dv} \tag{5.71}$$

where $W(v)dv$ is the weight fraction of polymer in the elution volume element dv.

For quantitative purposes, a column or column set is calibrated with a series of narrow-distribution standards of some species, typically polystyrene, having molecular weights that have been determined by an absolute method, typically light scattering. Molecular size varies monotonically with molecular weight for linear chains of

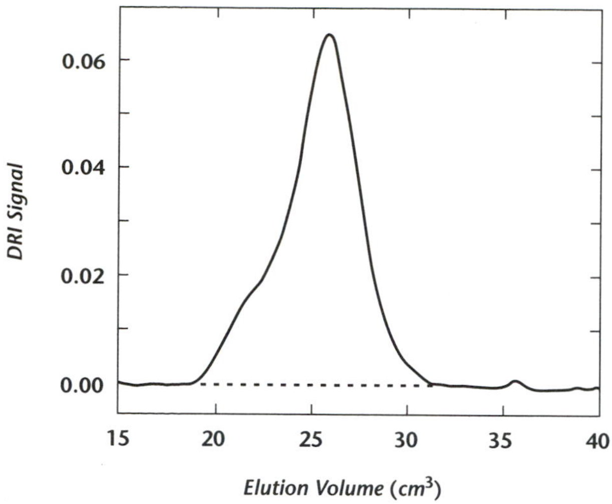

FIGURE 5.12 The size exclusion chromatography output of a differential refractometer reading as a function of elution volume for a sample of linear polyethylene in 1,2,4-trichlorobenzene at 135°C. Data from Cotts[51].

the same species, so the elution peak locations of the standards provide a calibration curve:

$$M = f(v) \tag{5.72}$$

A typical calibration curve, obtained with polystyrene standards in the elution solvent 1,2,4-trichlorobenzene at 135°C[44], is shown in Figure 5.13. As is true of this example, elution volume typically varies linearly or nearly linearly with the logarithm of molecular weight. For a linear variation,

$$\log M = -\alpha v + \beta \tag{5.73}$$

Molecular weight resolution thus varies inversely with the slope parameter α. Departures at high and low v define the limits of resolution for the chosen column set, but with the currently available column packings, the span for good molecular weight resolution can be very wide. The calibration curve in Figure 5.13 indicates good molecular weight resolution for polystyrene throughout the range $10^3 \lesssim M \lesssim 10^7$.

For a linear sample of the same species as the standard, the calibration curve permits the conversion of the chromatograph to the weight distribution function $W(M)$. The principle of *universal calibration*, proposed by the Strasbourg group[45,46] relaxes the requirement of calibration standards for each polymer species. It states that, for a given solvent and column set, the SEC elution volume for the solute molecules depends only on their dilute solution size in the elution solvent. As discussed in

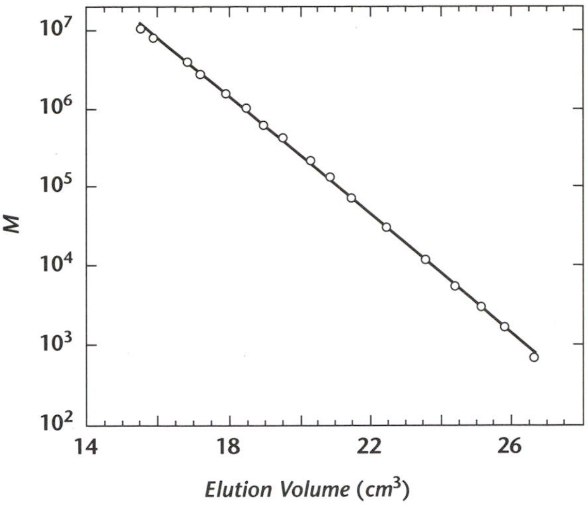

FIGURE 5.13 SEC elution volume for nearly monodisperse polystyrenes in 1,2,4-trichlorobenzene at 135°C. Data from Sun[44]. The line is a fit of Eq. 5.73 to the data.

Chapter 6, measures of molecular size in dilute solution are readily available, two based on equilibrium properties, R_g and A_2, and two based on dynamic properties, $[\eta]$ and D_o. The product $[\eta]M$ is proportional to the viscometric volume in the elution solvent and requires only properties that can be easily and accurately measured over wide ranges. The proposed universality was indeed demonstrated: Data covering a wide range of molecular weights for several polymer species and for various branched architectures fell along a single curve when expressed as $[\eta]M$ versus υ. Accordingly, universal calibration can be expressed by an extension of Eq. 5.73, whereby:

$$\log[\eta]M = -\alpha'\upsilon + \beta' \tag{5.74}$$

in which the parameters α' and β' are independent of polymer species, reflecting now only instrumental, packing, and elution solvent variables.

The universal calibration principle makes it possible to calibrate a column set using the polymer standards of some species, obtain $[\eta]M$ versus υ, then convert to any other species that has a known Mark-Houwink equation in the elution solvent. Molecular weight distributions for linear polymers can thus be obtained with the converted M versus υ relation. Universal calibration also permits, using only SEC and $[\eta]$ data, the estimation of molecular weight for polymers containing long chain branches. With additional assumptions, information can be gathered on how the branches are distributed with molecular weight in polydisperse samples. The universal calibration principle has been widely used for all these purposes. Section 6.3.2 describes the branched polymer applications in more detail.

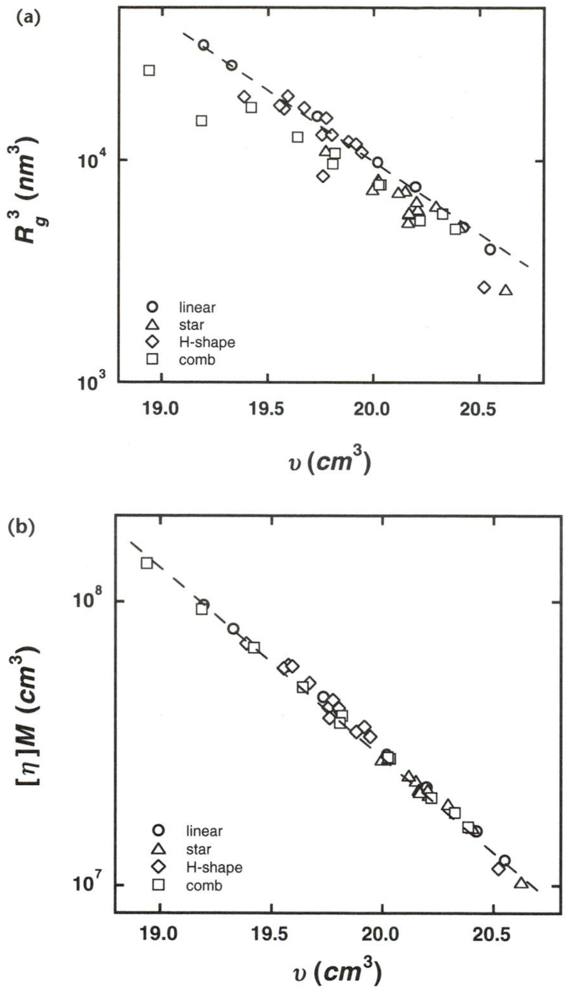

FIGURE 5.14 Correlation of elution volume for model polyethylenes of various molecular weights and architectures. Values of M, R_g, and $[\eta]$ from Hadjichristidis et al.[52] and Sun[44]. Linear chains, symmetric stars, H-shaped polymers, and combs are represented. (a) Correlation with equilibrium size; (b) correlation with dynamic size. Dashed lines are fits to the linear chain data.

Theory suggests that SEC separation depends fundamentally on an equilibrium size such as R_g[47–49], not on a dynamic size such as $[\eta]M$. For linear polymers, the differences caused by using a dynamic size rather than an equilibrium size to convert from standards to species of interest are probably small because the two size measures themselves are closely related (see Section 6.2.1). Exceptions are in fact found in the low molecular weight range[12], where the various size measures begin to diverge. However, as noted by the Benoit group[46], differences are more important in polymers with long branches because the relationship between dynamic size and equilibrium size varies with large-scale molecular architecture. Accordingly, they concluded that, given the success of $[\eta]M$ versus υ, a plot R_g versus υ would not produce a single curve. Recent studies by Sun[44] demonstrate that the differences are indeed significant. Thus, the calibration based on $[\eta]M$ in Figure 5.14a is more nearly universal for branched and linear systems than that based on R_g in Figure 5.14b. An informal survey of a few SEC users indicates this to be their experience as well. These various experimental affirmations of $[\eta]M$-rather than R_g-based universality, though supported by theory for stars[48], remain to be fully understood.

5.3.2 Multidetector Methods

Achieving a size-based separation and making the result "viewable" downstream opens possibilities for addressing questions of polymeric structure beyond just molecular weight distribution and long-chain branch frequency. A downstream detector presents data as slices of the SEC-resolved composition. The polymer in these slices is far into the dilute range, as noted earlier, so finite concentration corrections to the detector outputs are typically unnecessary. The uniformity of composition in statistical copolymers over the molecular weight distribution can be investigated by in-line infrared spectrometry combined with the differential refractometer output. The quantification of departures from compositional uniformity is difficult, but qualitative questions, such as deciding which of two copolymer samples is compositionally more uniform, can be addressed.

The in-line measurements of light scattering intensity at small angles (SEC-SALS) can be used to evaluate molecular weight distribution without relying on universal calibration and usually without finite concentration corrections. Light scattering has the advantage of being an absolute method not requiring calibration standards to obtain molecular weight. It has the additional advantage of being more sensitive to components in the high tail of the distribution than does SEC alone. Light scattering intensity at small angles $I(t)$ is proportional to $cM(dn/dc)^2$ [see Eqs. 5.23 and 5.34], whereas the refractometer signal $S(t)$ is proportional to $c(dn/dc)$. Accordingly, $I(t) \propto M(dn/dc)S(t)$, so scattering intensity can remain strong in the high tail even though the refractometer signal is decreasing toward its resolution limit.

In-line measurement of multiangle light scattering (SEC-MALS) can provide both size and molecular weight data throughout the high tail of the distribution. Its slices thus offer a way to detect and characterize long-chain branching. In-line viscometry also has an enhanced sensitivity to the high molecular weight tail. The combination of in-line light scattering and viscometry (triple detection, SEC-SALS-VISC) can also provide information about long-chain branching in the high molecular weight tail of the distribution, as discussed in Section 6.3.2. All three SEC-SALS-VISC outputs for the linear polyethylene example in Figure 5.12 are shown in Figure 5.15.

As shown in Figure 5.16 for SEC-MALS and Figure 5.17 for SEC-SALS-VISC, when applied to linear homopolymers or statistically uniform copolymers, the outputs provide respectively the R_g versus M and $[\eta]$ versus M relationships for the species,

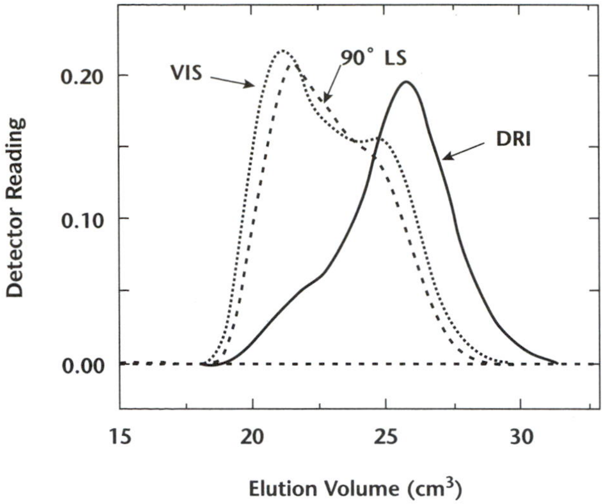

FIGURE 5.15 Detector outputs from an SEC-SALS-VISC study of a linear polyethylene sample. Data from Cotts[51].

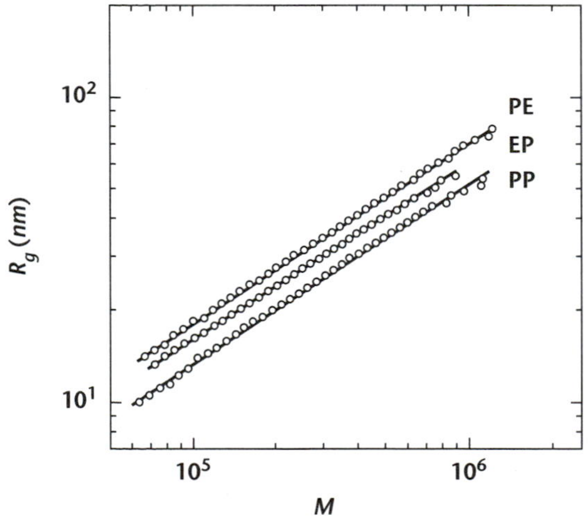

FIGURE 5.16 Radius of gyration as a function of molecular weight from SEC-MALS analysis for three samples of linear polymer, polyethylene (O), an ~50 wt % ethylene–propylene copolymer (Δ), and isotactic polypropylene (∇). The power-law exponents obtained from the data are 0.592, 0.575, and 0.588. From Sun et al.[50].

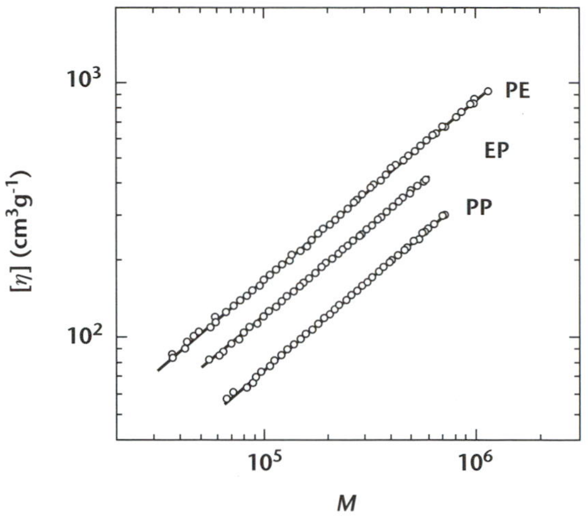

FIGURE 5.17 Intrinsic viscosity is shown as a function of molecular weight from SEC-SALS-VISC analysis for three samples of linear polymer, polyethylene (O), an ~50 wt % ethylene–propylene copolymer (Δ), and isotactic polypropylene (∇). The power-law exponents obtained from the data are 0.704, 0.691, and 0.703. From Sun et al.[50].

without requiring the use of model materials. The following results are obtained for solutions in trichlorobenzene at 135°C[50]:

$$\begin{aligned} R_g(nm) &= 0.0230M^{0.58} && \text{(Linear polyethylene)} \\ R_g(nm) &= 0.0202M^{0.58} && \text{(EP copolymer [53 wt \% P])} \\ R_g(nm) &= 0.0173M^{0.58} && \text{(Isotactic polypropylene)} \end{aligned} \tag{5.75}$$

$$\begin{aligned} [\eta](cm^3g^{-1}) &= 0.0579M^{0.695} && \text{(Linear polyethylene)} \\ [\eta](cm^3g^{-1}) &= 0.0420M^{0.695} && \text{(EP copolymer [53 wt \% P])} \\ [\eta](cm^3g^{-1}) &= 0.0262M^{0.695} && \text{(Isotactic polypropylene)} \end{aligned} \tag{5.76}$$

Average exponents were used in both cases because their uncertainties over all samples studied are larger than the differences.

REFERENCES

1. Schulz G.V. 1936. *Z. Physik. Chem.* A176:317.
2. Mays J.W. and Hadjichristidis N. 1991. In *Modern Methods of Polymer Characterization.* H.G. Barth and J.W. Mays (eds.), New York: John Wiley & Sons, p. 201.
3. Yamakawa H. 1971. *Modern Theory of Polymer Solutions,* New York: Harper & Row.

4. Casassa E.F. and Berry G.C. 1989. In *Comprehensive Polymer Science.* Vol. 2. G. Allen and J.C. Bevington (eds.), Oxford, UK: Pergamon Press, p. 71.
5. Hadjichristidis N. 1989. Private communication.
6. Debye P. 1947. *J. Phys. Colloid Chem.* 51:18.
7. Kratochvil P. 1981. *Classical Light Scattering from Polymer Solutions,* Amsterdam: Elsevier.
8. Burchard W. 1983. *Adv.Polym. Sci.* 48:1.
9. Berry G.C. 1987. *Encyclopedia of Polymer Science and Engineering,* 2nd ed., edited by H. Mark, N.M. Bikales, C.S. Overberger, and G. Menges, New York: John Wiley & Sons, p. 721.
10. Berry G.C. and Cotts P.M. 1999. *Experimental Methods in Polymer Characterization,* edited by R.A. Pethrick and R.S. Stein, Sussex, U.K.: John Wiley & Sons.
11. Kaye W. and J.B. McDaniel. 1974. *Appl. Optics.* 13:1934.
12. Chance R.R., S.P. Baniukiewicz, D. Mintz, and G. Ver Strate. 1995. *Int. J. Polym. Anal. Charact.* 1:3.
13. Berry G.C. 1966. *J. Chem. Phys.* 44:4550.
14. Zimm B.H. 1948. *J. Chem. Phys.* 16:1093.
15. Lindner J.S. and Huang S.S. 1991. In *Modern Methods of Polymer Characterization.* H.G. Barth and J.W. Mays (eds.), New York: J. Wiley & Sons, p. 313.
16. Kim S.H. and P.M. Cotts. 1991. *J. Appl. Polym. Sci.* 42:217.
17. Burchard W. 1999. *Adv. Polym. Sci.* 143:113.
18. Huglin M.B. 1989. *Polymer Handbook,* 3rd ed., edited by J. Brandrup and E.H. Immergut, New York: Wiley-Interscience, p. VII–409.
19. Higgins J.S. and Benoit H.C. 1994. *Polymers and Neutron Scattering,* New York: Oxford University Press.
20. Roe R.-J. 2000. *Methods of X-Ray and Neutron Scattering in Polymer Science,* New York: Oxford University Press.
21. Guinier A. 1994. *X-ray Diffraction in Crystals, Imperfect Crystals and Amorphous Bodies,* New York: Dover Publications.
22. Konishi T., T. Yoshizaki, T. Saito, Y. Einaga, and H. Yamakawa. 1990. *Macromolecules.* 23:290.
23. Tamai Y., T. Konishi, Y. Einaga, M. Fujii, and H. Yamakawa. 1990. *Macromolecules.* 23:4067.
24. Chu B. 1991. *Laser Light Scattering,* 2nd ed. Boston: Academic Press.
25. Hill T.L. 1987. *Statistical Mechanics,* New York: Dover Publications.
26. Einstein A. 1906. *Investigations on the Theory of the Brownian Movement,* New York: Dover Publications (1956 publication of 1906 doctoral dissertation, translated and edited by R. Furth , 1926).
27. Davidson N.S., L.J. Fetters, W.G. Funk, N. Hadjichristidis, and W.W. Graessley. 1987. *Macromolecules.* 20:2614.
28. Cowie J.M.G. and S. Bywater. 1965. *Polymer.* 6:197.
29. Roovers J. and P.M. Toporowski. 1980. *J. Polym. Sci.: Polym. Phys. Ed.* 18:1907.
30. Staudinger H. and R. Nodzu. 1930. *Berichte.* 63B:721.
31. J.W. Mays and Hadjichristidis N. 1991. In *Modern Methods of Polymer Characterization,* H.G. Barth and J.W. Mays (eds.), New York: J. Wiley & Sons, p. 227.
32. Haller W. 1931. *Kolloid Z.* 56:257.
33. Kuhn W. 1934. *Kolloid Z.* 68:2.
34. Houwink R. 1940. *J. Prakt. Chem.* 157:15.

35. Sakurada I. 1940. *Kasen Koenshu.* 5:33.
36. Sakurada I. 1941. *Kasen Koenshu.* 6:177.
37. Fetters L.J., N. Hadjichristidis, J.S. Lindner, and J.W. Mays. 1994. *J. Phys. Chem. Ref. Data.* 23:619.
38. Kurata M. and Tsunashima Y. 1989. In *Polymer Handbook,* 3rd ed., J. Brandrup and E.H. Immergut (eds.), New York: Wiley-Interscience.
39. Moore J.C. 1964. *J. Polym. Sci.* A2:835 p. VII–1.
40. Yau W.W., Bly D.D., and Kirkland J.J. 1979. *Modern Size-Exclusion Chromatography: Practice of Gel Permeation and Gel Filtration Chromatography.* New York: J. Wiley & Sons.
41. Wu G.-S. (ed.). 1995. *Column Handbook of Size Exclusion Chromatography,* New York: Marcel Dekkér.
42. Mori S. and Barth H.G. 1999. *Size Exclusion Chromatography.* New York: Springer-Verlag.
43. Pasch H. 2000. *Adv. Polym. Sci.* 150:1.
44. Sun T. 2003. Private communication.
45. Benoit H., Z. Grubisic, P. Rempp, D. Decker, and J.G. Zilliox. 1966. *J. Chim. Phys.* 63:1507.
46. Grubisic Z., P. Rempp, and H. Benoit. 1967. *J. Polym. Sci., Polym. Lett. Ed.* 5:753.
47. Casassa E.F. 1967. *J. Polym. Sci. Pt. B* 5:773.
48. Casassa E.F. 1971. *Separation Sci.* 6:305.
49. Boyd R.H., R.R. Chance, and G. Ver Strate. 1996. *Macromolecules.* 29:1182.
50. Sun T., P. Brant, R.R. Chance, and W.W. Graessley. 2001. *Macromolecules.* 34:6812.
51. Cotts P.M. Private communication, 2001.
52. Hadjichristidis N. et al. 2000. *Macromolecules.* 33:2424.

CHAPTER 6

Dilute Solution Properties

This chapter introduces two intramolecular contributions to the dilute solution properties of polymers, volume exclusion and hydrodynamic interaction. Coil expansion due to self-exclusion and the theta condition, in which pairwise exclusion is cancelled thermodynamically, are formulated in the Flory manner. The dilute solution properties introduced in the previous chapter are recast, through considerations of coil self-concentration and pervaded volume, as molecular sizes, and their relationships and chain length dependence are examined at the theta condition and in the commonly encountered good solvent system. The systematics of linear chain behavior, then macrocycles and branched chains of various sorts, and finally colloidlike macromolecules with large self-concentrations are discussed. The chapter ends by defining those regions of concentration beyond coil overlap—semidilute solutions and concentrated solutions including the undiluted liquid.

6.1 Intramolecular Interactions

Several sorts of interaction between units on the same chain are important in dilute solutions. Chapter 4 dealt with species-dependent restrictions on the local backbone trajectory—those interactions transmitted along the chain and characterized by a persistence length that specifies the interaction range (Section 4.2.1). It also introduced the spring interaction (Section 4.6.1), an average force that acts between chain units and depends on both spatial separation and distance as measured along the chain. Two other intrachain interactions are considered here, both of them dependent on spatial separation rather than distance along the chain and thus, like the spring interaction, offering a possibility for interaction between even the most remotely connected units of the chain. One is the *excluded volume interaction,* a mutual repulsion of chain units that leads to a net expansion of equilibrium size. Its characteristics and a simple theory of its effects are presented in Section 6.1.1. The other is *hydrodynamic interaction,* which causes a mutually induced reduction of velocity differences within the pervaded volume of the chain, lending an "impenetrable-sphere" character to its dynamic response. The interpretation of dynamic experiments is introduced in Section 6.1.2, mainly the relationship between coil size and dynamic properties such as D_o and $[\eta]$. (The dynamic response of chain molecules is discussed in detail in Volume 2.)

6.1.1 Excluded Volume

The size of polymer chains without excluded volume can be expressed in terms of Eqs. 4.21, 4.30, and 4.33:

$$(R_g)_o = l_s \left(\frac{n}{6}\right)^{1/2} \quad (n \gg 1) \tag{6.1}$$

in which $(R_g)_o$ is the *unperturbed radius,* l_s is the statistical segment length of the species, and n is the number of backbone bonds. Volume exclusion causes the average size to increase and changes the form of the size–length relationship. An excluded volume interaction converts an intricate but manageable problem of locally restricted walks into the analytically intractable problem of self-avoiding walks. Fortunately, like the intermolecular excluded volume problem for dense molecular liquids (Section 2.2.3), self-avoidance is relatively easy to investigate by computer simulation. The limiting form for self-avoiding walks, like that for locally restricted walks, is a power law, but with a larger exponent. From the most comprehensive Monte Carlo studies to date[1],

$$R_g \propto n^{0.5877 \pm 0.0006} \quad (5 \times 10^3 \lesssim n \lesssim 80 \times 10^3) \tag{6.2}$$

The excluded volume exponent obtained by *renormalization group* theory, $\gamma = 0.5880 \pm 0.0015$[2], is in excellent agreement with the simulations-based value. The limiting relationship between the end-to-end distance and radius of gyration is also changed, albeit only slightly, from $\langle R^2 \rangle = 6R_g^2$ for locally restricted walks to $\langle R^2 \rangle \sim 6.25R_g^2$ for self-avoiding ones[1].

Power laws such as Eqs. 6.1 and 6.2 are frequently encountered in the size–length relations of polymer molecules, often with exponents other than these two. It is often convenient as well to work out the general consequences of power-law behavior without specifying the exponent, in which case we use:

$$R = l_\gamma n^\gamma \tag{6.3}$$

where R and n may be any size and length measures.

The effects of self-exclusion are widely observed in dilute polymer solutions. Consider the relationship between size and molecular weight for linear polyisobutylene (PIB) in cyclohexane (CHN) and n-heptane (HEP) at 25°C and isoamylvisovalerate

(IAV) at $22.1°C$ (original sources given in a recent review[3]):

$$
\begin{aligned}
&R_g(nm) = 0.0137\, M^{0.595} \quad (0.18 < 10^{-6}M < 14) \quad (\text{CHN})\\
&R_g(nm) = 0.0138\, M^{0.582} \quad (0.14 < 10^{-6}M < 1.8) \quad (\text{HEP})\\
&R_g(nm) = 0.0262\, M^{0.511} \quad (0.4 < 10^{-6}M < 4.7) \quad (\text{IAV})
\end{aligned}
\tag{6.4}
$$

Both the sizes and molecular weights were determined by light scattering. The values of R_g obtained for one of the PIB samples having a molecular weight of 815,000, are listed below, together with R_g for random walks of the backbone bonds (Eq. 4.20; $n = 29{,}000$, $l_o = 0.153\ nm$):

$$
\begin{aligned}
&R_g = 43.0\, nm \quad (\text{CHN})\\
&R_g = 37.6\, nm \quad (\text{HEP})\\
&R_g = 27.4\, nm \quad (\text{IAV})\\
&R_g = 10.7\, nm \quad (\text{random walk})
\end{aligned}
\tag{6.5}
$$

The exponents for CHN and HEP solutions are close to the value for self-avoiding walks, Eq. 6.2; the one for IAV solutions is near the random walk exponent. The sizes themselves are different in the three solvents, and are also much larger than the size of backbone random walks. Both local correlations and self-exclusion contribute in varying degrees to the coil dimensions of long chains.

Figure 6.1 shows the relationship over an extended range of R_g and M for polystyrene in two solvents[3]. Scattering measurements were used to obtain the values of both R_g and M, SANS and SAXS for R_g below $\sim 20\ nm$ and light scattering above those values. In toluene at $30°C$, volume exclusion has a significant effect on size in the long-chain region. Thus,

$$
R_g(nm) = 0.0119\, M^{0.596} \quad 0.4 \leq 10^{-6}M \leq 20 \tag{6.6}
$$

In cyclohexane at $35°C$ however, volume exclusion is insignificant, and R_g for long chains varies nearly as $M^{1/2}$. The solvent-related size difference diminishes with decreasing chain length, disappearing finally into the noise for $M \lesssim 10{,}000$.

It perhaps seems odd at first sight that the effect of volume exclusion, an essentially geometrical restriction on polymer chain conformation, should vary so widely from one solvent to another. Flory[4], however, showed that these differences can be understood in terms of polymer–solvent thermodynamics, a discovery of major significance. He went on to develop a thermodynamic theory of self-exclusion, the main features of which are outlined below.

Pervaded volume and self-concentration. Consider a polymer molecule in dilute solution, as depicted in Figure 6.2. From Section 4.5.3, the average distribution of

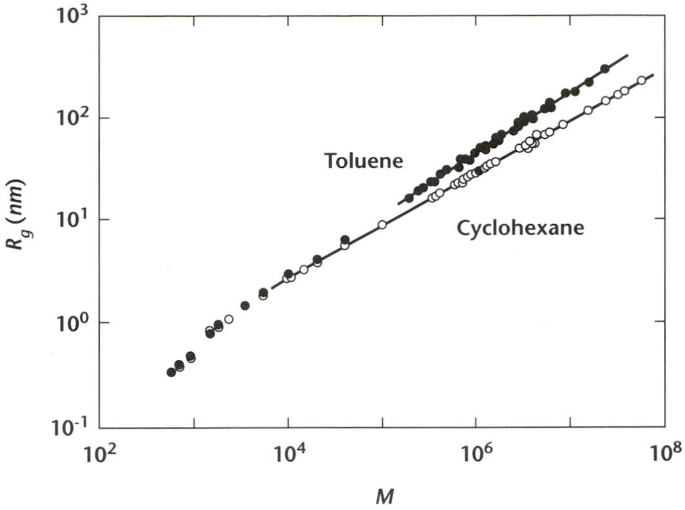

FIGURE 6.1 Radius of gyration as a function of molecular weight for polystyrene in the good solvent toluene and in cyclohexane at the theta condition, $T_\theta = 34.5°C$.

mass is spherically symmetric. The chain units are distributed according to $\rho(S)$, a monotonically decreasing function of distance from the center of mass that falls rapidly toward zero beyond the mean radius of gyration R_g. Let us ignore the details of this distribution and work with averages alone. Take R_g to be the radius of a sphere whose volume defines the average spatial domain permeated by the units of a single polymer molecule, its *pervaded volume:*

$$v_{per} = \frac{4\pi R_g^3}{3} \tag{6.7}$$

The mass of the polymer is M/N_a, and its average mass concentration in its pervaded volume is the *self-concentration:*

$$c_{self} = \frac{M}{N_a v_{per}} \tag{6.8}$$

Apart from an arbitrary multiplying factor of order unity, the self-concentration of polymer molecules in a dilute solution is the same as the overlap concentration c^*, a quantity defined in various ways in the previous chapter (Section 5.1.1, for example). Roughly speaking, overlap is reached when the pervaded volumes of the solute molecules just fill up the space. At that point, $c^* = c_{self}$, and

$$c^* = \frac{3M}{4\pi[R_g(0)]^3 N_a} \tag{6.9}$$

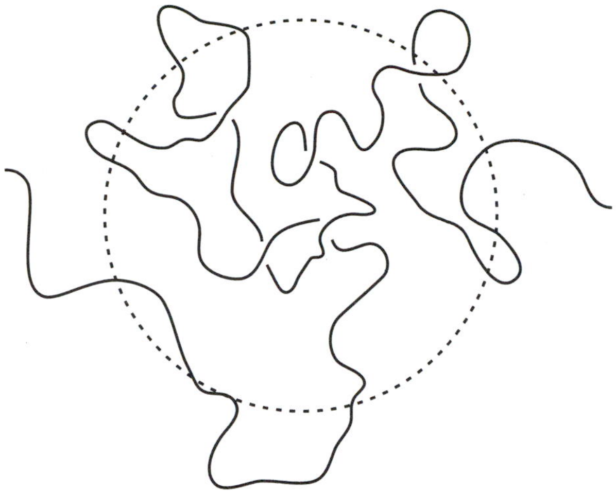

FIGURE 6.2 The pervaded volume of a flexible polymer molecule.

where $R_g(0)$ is the coil size in the dilute limit. The c^* to M relationship for any system with known a known R_g versus M relationship can thus be obtained using Eq. 6.9. Evidently,

$$c^* \propto M^{-p} \tag{6.10}$$

where $p = 1/2$ for random walks and 0.764 for self-avoiding walks ($\gamma = 0.588$). Values of c^* calculated using Eq. 6.9 and $R_g(M)$ for the systems represented in Tables 6.1 through 6.6 agree to within about 30% with those from Eqs. 5.7, 5.59, and 5.65. The values from Eq. 5.38 are smaller by about a factor of 3.

It is sometimes convenient to express self-concentration dimensionlessly, as a volume fraction. Thus, in dilute solutions, to a good approximation:

$$\phi_{self} = \frac{c_{self}}{\rho} = \frac{\upsilon_{occ}}{\upsilon_{per}} \tag{6.11}$$

in which ρ is the undiluted polymer liquid density and υ_{occ} is $M/\rho N_a$. The size–mass relationships in Eq. 6.4 yield the following values for polyisobutylene with molecular weight 815,000:

$$\begin{aligned} \phi_{self} &= 168/M^{0.785} = 0.38 \times 10^{-2} \quad \text{(CHN)} \\ \phi_{self} &= 164/M^{0.746} = 0.64 \times 10^{-2} \quad \text{(HEP)} \\ \phi_{self} &= 24/M^{0.533} = 1.70 \times 10^{-2} \quad \text{(IAV)} \end{aligned} \tag{6.12}$$

The range of a few percent is typical for polymer solutions, although of course ϕ_{self} or c_{self}, like ϕ^* or c^*, vary with chain length, or its surrogate, M in this case.

Physically, ϕ_{self} represents the fraction of the pervaded volume taken up by the molecule itself. The idea of self-concentration has a useful meaning outside the dilute solution regime, and Eq. 6.11 sets its magnitude with R_g at the conditions of interest. The overlap concentration, on the other hand, always refers to the size in dilute solution.

Self-exclusion. All virial coefficients beyond the first depend on the molecular interactions of the system. For solutions, these can be expressed in terms of the *net* or *effective* solute–solute interactions, and for solutions of spherical molecules[5],

$$\begin{aligned} A_2 &= \frac{N_a}{2M^2}\hat{\beta}(T) \\ \hat{\beta}(T) &= \int_0^\infty [1 - \exp(-\psi(r)/kT)]4\pi r^2 dr \end{aligned} \tag{6.13}$$

in which $\hat{\beta}(T)$ is the *binary cluster integral,* and $\psi(r)$ is the potential of mean force for the interactions of solute molecules. Corresponding expressions for the higher virial coefficients are also available[6]. The following expressions apply to hard sphere solutes interacting by volume exclusion alone, Eq. 2.4:

$$\begin{aligned} A_2 &= \frac{4N_a\upsilon}{M^2} \\ A_3 &= \frac{10N_a^2\upsilon^2}{M^3} \end{aligned} \tag{6.14}$$

in which υ is the sphere volume. For hard spheres having van der Waals interactions,

$$\begin{aligned} A_2 &= \frac{4N_a\upsilon}{M^2}F_2(T) \\ A_3 &= \frac{10N_a^2\upsilon^2}{M^3}F_3(T) \end{aligned} \tag{6.15}$$

in which the $F_i(T)$ depends on the form of the potential. For the example of square-well attraction, as in Eq. 2.5,

$$\begin{aligned} F_2(T) &= 1 - 7\Delta \\ F_3(T) &= \frac{5 - 17\Delta + 136\Delta^2 - 162\Delta^3}{5} \end{aligned} \tag{6.16}$$

in which $\Delta = \exp(\varepsilon/kT) - 1$, and ε is the well depth. For simple mixtures (Section 3.4), the effective well depth is the exchange energy. For athermal mixtures, $\varepsilon = 0$: all F_i are unity so all A_i have their hard sphere values, given by Eq. 6.14.

An attractive interaction between solute molecules tends to offset the excluded volume repulsion. The two usually depend differently on the temperature, so at some compensation temperature the two contributions can cancel to yield $A_2 = 0$. (The same compensation of short-range repulsion and longer-range attraction occurs in the PVT behavior of gases: The second virial coefficient of the gas goes to zero at some species-dependent temperature, in that case called the *Boyle temperature*.) As described below, this compensation temperature for polymer solutions is the *Flory temperature* or *theta temperature* T_θ. For the square-well example [Eq. 6.16], F_2 and thus $A_2 = 0$ at $T_\theta = \varepsilon/[k \ln(8/7)]$. At T_θ, $F_3 = 0.975$; so in this case A_3 hardly differs from its hard sphere value.

This principle of excluded volume cancellation plays a central role in the Flory theory of excluded volume.[4,7] Flory suggested that the balance of two factors dictates the size of polymer molecules in the dilute limit:

- a net repulsion of chain units, arising from excluded volume and acting to expand the coils.
- a net attraction of chain units, arising from their connectivity [the spring force Eq. 4.85] and acting to resist coil expansion.

Flory estimated the contribution of each to the free energy per molecule with its average value, then minimized the free energy with respect to size to obtain an equation relating equilibrium coil size to chain length. Because Flory's derivation is rather lengthy, we use here a more compact version to illustrate his idea. A slightly different compact version has been developed by de Gennes[8].

Consider a collection of polymer chains having molecular weight M and pervaded volume υ_{per} in a solution so dilute ($\phi \ll \phi^*$) that the chains do not interact with one another. Envision each molecule as a uniform "microsolution" of mers, confined within the pervaded volume. As represented in Figure 6.3, two countervailing pressures act upon each microsolution. The osmotic pressure π_{op}, governed by mer concentration, tends to expand the microsolution volume. The elastic pressure π_{el}, related to the coil dimensions, resists the expansion. The volume of the microsolution at equilibrium is determined by the zero pressure condition:

$$\pi_{el} + \pi_{op} = 0 \tag{6.17}$$

The elastic pressure from a uniformly expanded or contracted coil can be obtained using Eq. 4.94 for the coil free energy and Eq. 2.11 to relate that value to pressure. Thus, expressing the size expansion factor with respect to $(\upsilon_{per})_o$, the pervaded volume of the coil without excluded volume,

$$\alpha_s^3 = \frac{\upsilon_{per}}{(\upsilon_{per})_o} \tag{6.18}$$

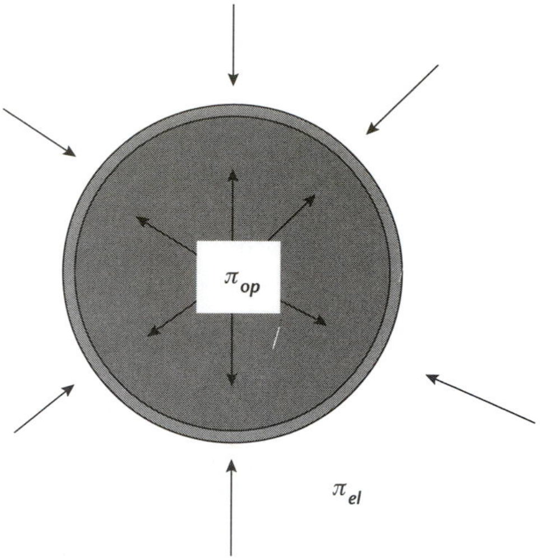

FIGURE 6.3 Modeling coil dimensions of flexible polymer molecules as determined by a balance of osmotic and elastic pressures.

yields:

$$\pi_{el} = -\frac{\partial \Delta A}{\partial \upsilon_{per}} = -\frac{kT}{(\upsilon_{per})_o \alpha_s^2}\left(\alpha_s - \frac{1}{\alpha_s}\right) \tag{6.19}$$

From Eq. 6.17, $\pi_{el} = 0$ if $\pi_{op} = 0$, and hence from Eq. 6.19, $\alpha = 1$ or $\upsilon_{per} = (\upsilon_{per})_o$. The chains simply take up their random coil dimensions, unperturbed by volume exclusion.

For a long chain, the microsolution is dilute, $\phi_{self} \ll 1$, so its osmotic pressure can be represented by the first few terms of a virial expansion [Eq. 5.5]:

$$\pi_{op} = N_a kT[a_2(c_{self})^2 + a_3(c_{self})^3 + \cdots] \tag{6.20}$$

in which the a_i are the virial coefficients for the microsolution. Note that the first virial coefficient (the $1/M$ term) has been omitted from the expression. Each chain is localized in its own microsolution, so the translational entropy operating in a many-chain macroscopic solution, which is represented by that term, plays no role here. Only the interaction of the mers, reflected in the quadratic and higher order terms, is relevant. Equivalently, to evaluate π_{op}, view each chain as a segment of some much longer chain L whose overlap concentration is effectively zero, $c_L^* \ll c_{self}$. Then, from the semidilute ideas developed in Section 7.1.2, π_{op} is the osmotic pressure of a semidilute solution of the longer chains at the concentration $c_L = c_{self}$ for the segment of interest. This second view permits using models for a_2 that are only applicable above the overlap concentration.

Approximate expressions for $a_2, a_3, \ldots$ can be obtained by representing the mers as hard spheres, each with volume υ_m, with square-well attraction. The volume per solvent molecule υ_s is generally different, but that can be handled in the manner of the simple mixture model (Section 3.4), by redefining the chain units to have the same volume as the solvent molecules. Accordingly, a chain with n mers consists of $n\upsilon_m/\upsilon_s$ spheres, each with volume υ_s and molecular weight $M\upsilon_s/n\upsilon_m$. Thus, from Eq. 6.15,

$$a_2 = \frac{4N_a\upsilon_s F_2(T)}{(M\upsilon_s/n\upsilon_m)^2} = \frac{4F_2(T)}{\rho_p^2 V_s} \tag{6.21}$$

$$a_3 = \frac{10(N_a\upsilon_s)^2 F_3(T)}{(M\upsilon_s/n\upsilon_m)^3} = \frac{10F_3(T)}{\rho_p^3 V_s} \tag{6.22}$$

where $\rho_p = M/n\upsilon_m N_a$ is the density of undiluted liquid polymer, and V_s is the solvent molar volume. Expressing c_{self} using Eq. 6.8 then leads to:

$$\pi_{op} = \frac{3RT}{4\pi N_a R_g^3}\left[\frac{3M^2F_2(T)}{\pi\rho_p^2 N_a R_g^3 V_s} + \frac{45M^3F_3(T)}{8\pi^2\rho_p^3 N_a^2 R_g^6 V_s} + \cdots\right] \tag{6.23}$$

The first term in the brackets varies with chain length as $n^{2-3\gamma}$ (approximately $n^{+0.23}$ in good solvents) and the second as $n^{3-6\gamma}$ (approximately $n^{-0.53}$). Thus, the first term dominates for long chains in good solvents, the case of interest here; hence the a_3 and higher terms are omitted.

Substituting Eq. 6.19 and Eq. 6.20 (with only the a_2 term retained) into Eq. 6.17, then using Eqs. 6.7 and 6.18, finally yields:

$$\alpha_s^5 - \alpha_s^3 = \frac{a_2 M^2}{N_a(\upsilon_{per})_o} \tag{6.24}$$

With $(\upsilon_{per})_o = 4\pi(R_g^3)_o/3$ from Eq. 6.7, this result can also be written:

$$\alpha_s^5 - \alpha_s^3 = \left[\frac{3a_2}{4\pi N_a \left(R_g^2/M\right)_o^{3/2}}\right] M^{1/2} \tag{6.25}$$

where the quantity in brackets depends on the polymer–solvent system and temperature, but not on chain length. Apart from committing to an explicit expression for a_2, Eq. 6.25 is the coil expansion formula as first derived by Flory in 1949[4].

According to Eq. 6.25, coil expansion vanishes for all chain lengths at the theta condition: $a_2(T) = 0$ at some temperature $T = T_\theta$, a property that depends on the polymer–solvent system; at the theta temperature, $R_g = (R_g)_o$. Even well away from T_θ, the expansion factor remains near unity in the short chain region, but it grows with chain length. Eventually the relationship between α and n becomes a power law. Thus, for long enough chains ($\alpha^5 \gg \alpha^3$), Eq. 6.25 yields:

$$R_g \propto (R_g)_o n^{1/10} \propto n^{3/5} \tag{6.26}$$

The Flory prediction for the volume-exclusion exponent, $\gamma = 3/5 = 0.600$, agrees well with the observed range for many polymer–solvent systems. The excluded volume exponent for self-avoiding walks is only slightly smaller, $\gamma \sim 0.588$ [Eq. 6.2], and the smaller value does usually fit the data slightly better (see Section 6.2.3). de Gennes has pointed out the danger of mean field methods, such as those used here, to deal with coil self-avoidance[8]. Both chain conformation and pair-wise contacts are strongly fluctuating, correlated properties, so using the averages of each to calculate an average of their net effect could lead to large net errors. In this case, a near cancellation of errors leads to a sensible result. The slight offset of the exponents is probably one of the consequences of noncancellation[8].

The term in brackets in Eq. 6.25 can be written in various ways, depending on the specific form chosen for a_2. Thus, using Eq. 6.21, the hard sphere with square well expression:

$$\frac{\left(\alpha_s^5 - \alpha_s^3\right)}{M^{1/2}} = \frac{3F_2(T)}{\pi \rho_p^2 N_a \left(R_g^2/M\right)_o^{3/2} V_s} \tag{6.27}$$

The value of a_2 deduced from the Flory-Huggins expression for osmotic pressure [see Eq. 7.36] leads to:

$$\frac{\left(\alpha_s^5 - \alpha_s^3\right)}{M^{1/2}} = \frac{3[1/2 - \chi(T)]}{4\pi \rho_p^2 N_a \left(R_g^2/M\right)_o^{3/2} V_s} \tag{6.28}$$

The relationship between expansion factor and chain length has been studied extensively since the Flory formula first appeared. The parameter combination that appears in the various versions has been used to define the *excluded volume parameter*[9]:

$$z = \left(\frac{1}{4\pi \left(R_g^2/M\right)_o}\right)^{3/2} \frac{\hat{\beta}(T) M^{1/2}}{\rho_p^2 V_s^2} \tag{6.29}$$

in which, from Eqs. 6.13 and 6.15, the dependence on the binary cluster integral, $\hat{\beta}(T) = 8 v_s F_2(T)$, has been made explicit. The original Flory formula[4] then becomes:

$$\alpha_s^5 - \alpha_s^3 = 2.60\, z \tag{6.30}$$

whereas the right side is 5.32z if, for example, Eq. 6.27 is used. In 1960, Stockmayer[10] suggested choosing a numerical coefficient for z that forces agreement with the linear term in the perturbation formula for the size (see Section 6.2.2). The main interest here is the observable size measure R_g, so the Stockmayer variant is adopted:

$$\alpha_s^5 - \alpha_s^3 = \frac{134}{105} z \tag{6.31}$$

The Flory theory provided not only an exponent and form for the expansion function, but also a method for establishing the unperturbed chain dimensions. Thus, in dilute solutions and for small enough self-concentration (high enough molecular weight), pair-wise interactions dominate both the *intermolecular* and *intramolecular* excluded volume. Both should therefore vanish at the theta condition. The intramolecular part affects the chain dimensions, as we have just discussed, and the intermolecular part governs the virial coefficients of the solution. Accordingly, the theta condition for a system should be identifiable as the temperature at which the second virial coefficient, which depends on pair-wise interactions alone, passes through zero:

$$A_2(T_\theta) = 0 \tag{6.32}$$

If pair-wise interactions dominate (small-enough c_{self}), then that temperature should be independent of molecular weight. That behavior is demonstrated in Figure 6.4,

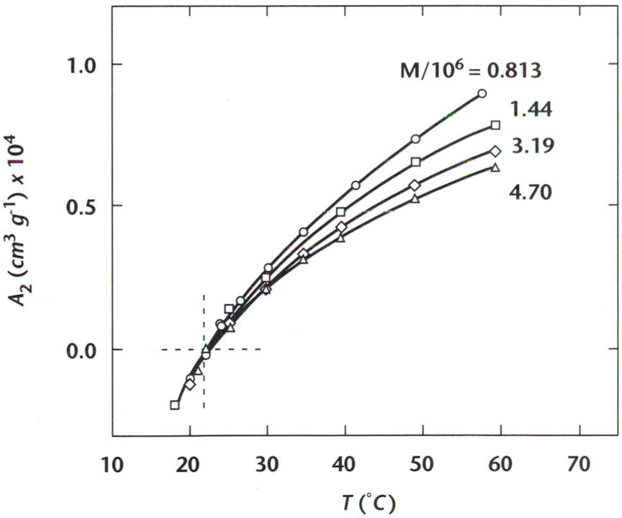

FIGURE 6.4 Second virial coefficient as a function of temperature for polyisobutylene samples in isoamyl-isovalerate. Determination of theta temperature $T_\theta = 22.1°C$. Data from Matsumoto et al[11].

showing A_2 versus T for several samples of polyisobutylene in isoamylvisovalerate[11]. The values of A_2 for different molecular weights cross zero at the same temperature, $22.1°C$, thus establishing that to be T_θ for this system.

Finding theta solvents for a polymer species—those that provide theta temperatures in an experimentally convenient range—is largely a matter of trial-and-error. A systematic screening for mutual solubility is useful, because the theta condition is always near a liquid–liquid phase boundary. (The connection between phase-equilibria and the theta condition is treated in Section 7.5.) If one theta solvent is known for a polymer species, finding others through slight variations in molecular structure is usually rather easy. Theta solvents and temperatures are known for numerous polymer species[12,13].

Many predictions of the Flory theory have been borne out experimentally. Thus, the size–length relationship obtained at the theta condition is consistent with the random coil model: $(R_g)_o/M^{1/2}$, is independent of molecular weight and is essentially a property of the polymer species alone. The value does not vary appreciably when measured for the same species in different theta solvents with similar theta temperatures. Specific solvent effects have been observed, however, and although typically small, they are still large enough to cast doubts on the temperature coefficients of chain dimensions deduced from measurements of $(R_g)_o$ versus T for a series of theta solvents. The values of κ in Table 4.2 were obtained by the more reliable methods described in Chapters 7 and 10. The Flory method is still widely used for estimating unperturbed dimensions, although it is being supplemented by new methods that are applicable to the undiluted state (Chapter 8). Most values of C_∞ in Table 4.2 were obtained by some variant of the Flory method.

Toluene at ambient conditions is a *good solvent* for polystyrene, meaning that the two are not only miscible at all molecular weights but also at conditions very far from those required for phase separation. The phrase "good solvent" itself has also acquired the general meaning that intramolecular excluded volume effects are important in the dilute regime, as shown by the size data for polystyrene–toluene solutions in Figure 6.1. The theta temperature for polystyrene–cyclohexane solutions is 34.5°C, so the sizes for that system, shown in the same figure, supply the unperturbed dimensions needed to calculate expansion factors for polystyrene in toluene. Values of α_s versus M obtained in this way are shown in Figure 6.5. The effect of molecular weight errors was minimized by including only data obtained on the same sample by the same investigators for both solvents. The line drawn as a guide to the eye through the large M data intersects $\alpha_s = 1$ at $M \sim 8000$, which is a rough indicator of the chain lengths required to exhibit size effects for this system. Also shown is the prediction of Eq. 6.31, obtained by treating $z/M^{1/2}$ as an adjustable constant and scaling the result to fit the data for large M. As seen here and examined in detail in Section 6.2, the shape of the full curve is not well described by Eq. 6.31.

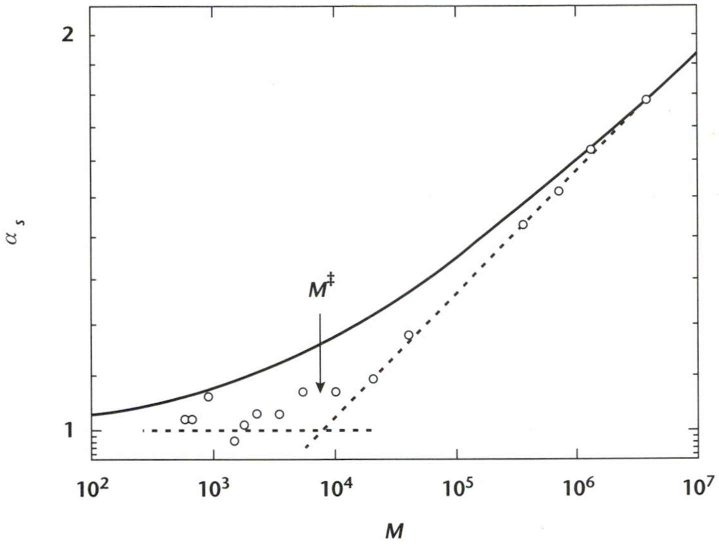

FIGURE 6.5 Size expansion factor for polystyrene in the good solvent toluene over a range of molecular weights, used to establish the crossover value for this polymer–solvent system.

Because the observable measure of chain length is the polymer molecular weight M, it is useful to express the expansion data in dimensionless form, as expansion factor versus reduced molecular weight $M/M^‡$. As exemplified by Figure 6.5, the limiting behavior intersects the line $\alpha = 1$ at some molecular weight $M^‡$ that locates the onset of significant coil expansion. Thus, Eq. 6.31 takes on the form

$$\alpha_s^5 - \alpha_s^3 = \left(M/M_s^‡\right)^{1/2} \tag{6.33}$$

in which $M_s^‡ = M/(134z/105)^2$, or using Eq. 6.29:

$$M_s^‡ = 1220\frac{\left(R_g^2/M\right)_o^3 \rho_p^4 V_s^4}{\hat{\beta}^2} \tag{6.34}$$

In the Flory-Huggins form, from Eq. 6.28:

$$M_s^‡ = 17.5\frac{\left(R_g^2/M\right)_o^3 \rho_p^4 V_s^2}{(1/2 - \chi)^2} \tag{6.35}$$

In this terminology, the crossover molecular weight $M^‡$ plays the role of an excluded volume parameter. It is a property of the polymer–solvent system, and it corresponds to the molecular weight at which z is of order unity. Its direct physical meaning makes $M^{\ddagger}$ a useful intermediary for organizing experimental data (Section 6.2.3).

The vanishing of A_2 at the theta condition means that the pervaded volumes become thermodynamically "transparent" to one another at T_θ. However, well away from T_θ, A_2 is positive, so the coils repel one another and each coil interior is thermodynamically shielded from the others. In the dilute limit, the pervaded volume takes on the character of an *impenetrable sphere,* because only the outermost chain units interact intermolecularly. This manifests itself in an interesting way—coil volume calculated from A_2 [Eq. 6.14] agrees rather well with coil volume calculated from R_g [Eq. 6.7]. Thus, with this A_2-derived volume written as $4\pi(R_t)^3/3$, Eq. 6.15 rearranges to an expression for the *thermodynamic radius* of the coil:

$$R_t = \left(\frac{3A_2M^2}{16\pi N_a}\right)^{1/3} \qquad (6.36)$$

The ratio R_t/R_g is plotted in Figure 6.6 as a function of $M/M^{\ddagger}$ for the samples of several species of linear polymers in a variety of good solvents[3]. A wide range of molecular weights is represented, yet no discernable trend appears, and the spread of values is consistent with the experimental uncertainty. The average of R_t/R_g for all the samples is 0.68. The hard-sphere analogy as a general organizing principle for polymer coil properties is examined in some detail in Section 6.3.3.

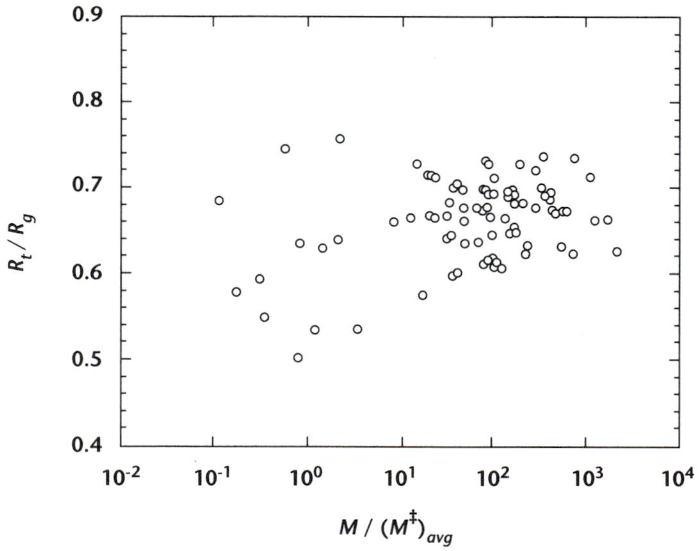

FIGURE 6.6 The ratio of thermodynamic to physical size as a function of reduced molecular weight for several polymer–solvent systems. Data from Hayward and Graessley[3].

6.1.2 Hydrodynamic Interaction

The molecular friction coefficient in the limit of zero concentration, the *Stokes coefficient* Z_o, is the simplest of all dynamical parameters. Defined in Eq. 5.62, Z_o is the proportionality constant in *Stokes law*, which relates an external force acting on solute molecules to their average velocity relative to solvent:

$$\boldsymbol{F} = Z_o \boldsymbol{U} \tag{6.37}$$

At steady state, the force is balanced by an equal and opposite frictional force, and that in turn acts on nearby solvent molecules, dragging them along in the same direction. The friction coefficient for a dilute suspension of hard spheres was calculated by Stokes in 1851[14]:

$$Z_o = 6\pi \eta_s R \tag{6.38}$$

in which η_s is the solvent viscosity and R is the sphere radius.

The frictional force for the polymer case can be expressed microscopically as the sum of contributions from the monomeric units ($i = 1, 2, \ldots, n$). The monomeric force exerted by unit i is $\zeta_o \boldsymbol{u}_i$, the product of its *monomeric friction coefficient* and its velocity relative to the solvent in its neighborhood. Accordingly,

$$\boldsymbol{F} = \zeta_o \sum_{i=1}^{n} \boldsymbol{u}_i \tag{6.39}$$

The local velocity $\boldsymbol{u}_i$ can then be expressed as $\boldsymbol{U} - \Delta\boldsymbol{U}_i$, where $\Delta\boldsymbol{U}_i$ is the change in solvent velocity at the ith unit caused by the drag on the solvent of the other $n-1$ units. Thus, $\boldsymbol{F} = n\zeta_o\boldsymbol{U} - \zeta_o \sum \Delta\boldsymbol{u}_i$, and using Eq. 6.37:

$$Z_o\boldsymbol{U} = n\zeta_o\boldsymbol{U} - \zeta_o \sum_{i=1}^{n} \Delta\boldsymbol{U}_i \tag{6.40}$$

Accordingly, the molecular friction coefficient depends in detail on the solvent velocity distribution induced inside and near the pervaded volume by the monomeric friction forces. The Oseen approximation[15] has been widely used to estimate this velocity field[16–18]. Here, only the two extremes of behavior are considered. In the weak limit, the drag has a negligible effect on the velocity field of the solvent, and $\Delta\boldsymbol{U}_i = 0$ everywhere. In the strong limit, the drag causes the solvent in the coil interior to move with the velocity of the polymer ($\Delta\boldsymbol{U}_i = \boldsymbol{U}$ except for units near the edges of the coil), thus forcing the solvent flow lines into the outermost regions. The streamlines of solvent relative to solute for each limit is illustrated in Figure 6.7.

From Eq. 6.40, in the weak interaction or *free-draining coil* limit, the coil interior is not shielded from the flow, and the Stokes coefficient is $n\zeta_o$. For the approximation

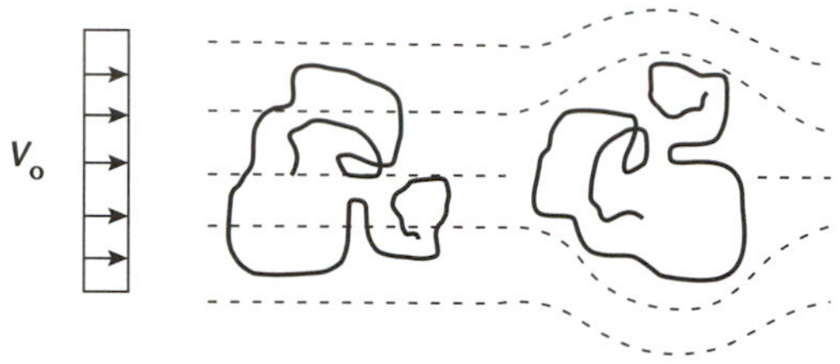

FIGURE 6.7 Streamlines of solvent relative to polymer for the hydrodynamic interaction limits, weak (free-draining) and strong (nonpenetration).

that the monomeric units behave as spheres of some radius r_o, $Z_o = 6\pi\eta_s r_o n$ from Eq. 6.38, so from Eq. 5.53,

$$D_o = \frac{kT}{6\pi\eta_s r_o n} \quad \text{(unshielded)} \tag{6.41}$$

In the strong or *impenetrable coil* limit, the coil interior is shielded from the flow, and the coil behaves hydrodynamically like a hard sphere having some radius R_h. Accordingly, using Eq. 6.38,

$$D_o = \frac{kT}{6\pi\eta_s R_h} \quad \text{(shielded)} \tag{6.42}$$

In the strong limit, the impenetrable volume and the pervaded volume should approximately coincide, so R_h, the *hydrodynamic radius* of the coil, should be of the order of the radius of gyration R_g.

Whether polymeric solutes behave as shielded or unshielded coils or something in between is a question that can be easily settled experimentally. The two limits lead to quite different relationships between the diffusion coefficient and molecular variables. With the plausible assumption for the strong limit that $R_h \propto R_g$, the shielded coil expression predicts that $\eta_s D_o R_g/T$ is a universal constant. The unshielded coil, on the other hand, predicts no direct dependence of D_o on R_g, but instead predicts that the product $\eta_s D_o M/T$ is the same for each polymer species in all solvents. The two models also predict widely differing values for long chains. Thus, the product nr_o is of the order of the contour length of the chain, $L = nl$, so, from Eqs. 6.42 and 6.41,

$$\frac{(D_o)_s}{(D_o)_{us}} \sim \frac{L}{R_g} \tag{6.43}$$

and $L/R_g \gg 1$ for long chains. Values of D_o for long chains, when estimated from Eq. 6.42 having R_h replaced by R_g, agree well with observed values, whereas those from Eq. 6.41 having nr_o approximated by L are much smaller than observed. Beyond this, the experiments quite clearly support the shielded coil model in all tests. Remarkably, the shielded coil picture for dilute solutions was first proposed in the mid-1930s[19].

Relationships based on the hydrodynamically shielded coil are also consistent with the observations on the intrinsic viscosity. The relative viscosity for a dilute suspension of hard spheres depends only on ϕ, the volume fraction of spheres:

$$\eta = \eta_s \left(1 + \frac{5}{2}\phi + 6.2\phi^2 + \cdots\right) \tag{6.44}$$

The linear term was derived by Einstein in 1910[20] and the quadratic term by Batchelor in 1977[21]. The definitions of intrinsic viscosity and the Huggins coefficient in Eq. 5.64, when applied to Eq. 6.44, lead to the following:

$$\begin{aligned} [\eta] &= \frac{5}{2}\left(\frac{\phi}{c}\right) \\ k_H &= 0.99 \end{aligned} \qquad \text{(hard spheres)} \tag{6.45}$$

The ratio ϕ/c for polymer coils is expressed as follows. The volume fraction of coils in the solution is the product of molecular number density and pervaded volume, $\phi = \nu \upsilon_{per}$. The mass concentration is the product of molecular number density and mass per molecule, $c = \nu M/N_a$. Hence, using Eq. 6.45,

$$[\eta] = \frac{5}{2}\frac{N_a \upsilon_{per}}{M} \tag{6.46}$$

and, using υ_{per} from Eq. 6.7,

$$[\eta] \sim \frac{10\pi N_a}{3}\frac{R_g^3}{M} \tag{6.47}$$

This approximate relationship[22], a direct consequence of the shielded coil model, is consistent with the data for many species of flexible polymers. It is supported by what is now a truly vast array of data for various chain sizes, architectures, and solution thermodynamics. The formula correctly predicts the Mark-Houwink exponent of 1/2 for the theta condition. The prediction of 0.764 for the good solvent limit, based on $\gamma = 0.588$, is in reasonable accord with the observed values. It also predicts that, in the long chain limit, the dimensionless parameter $[\eta]M/R_g^3$ is of the order $10\pi N_a/3 = 10.5N_a$ for all polymer–solvent systems. The experimental value for linear polymers[23] is about $6.1N_a$.

The success of the impenetrable coil model implies that the values of D_o and $[\eta]$ for a polymer reflect primarily its molecular dimensions. Those properties in fact can be expressed as sizes—the hydrodynamic radius R_h for D_o and a *viscometric radius* R_υ for $[\eta]$. Thus, from Eq. 6.42:

$$R_h = \frac{kT}{6\pi\,\eta_s\,D_o} \tag{6.48}$$

and, using Eq. 6.46 and $\upsilon_p = 4\pi\,(R_\upsilon)^3/3$,

$$R_\upsilon = \left(\frac{3[\eta]M}{10\pi\,N_a}\right)^{1/3} \tag{6.49}$$

The ratio R_υ/R_g in both good and theta solvents is shown as a function of reduced chain length $M/M^\ddagger$ for several species[3] in Figure 6.8. The ratios in the molecular weight region of most common interest, $M^\ddagger \lesssim M \lesssim 100M^\ddagger$, are all about 0.80. In theta solvents, that value persists out to the highest molecular weights studied, but in good solvents a very weak but clear downward drift occurs with increasing molecular weight in that region, ~3 percent per decade. For shorter chains however, $M \lesssim M^\ddagger$, the behavior changes. Depending on species, the ratio below some chain length begins to rise very rapidly with decreasing chain length. In contrast to the long chain behavior, the onset of the rise and the ratios themselves are the same in both good and theta solvents. The strong rise for small chains may well be an increasing penetration of flow lines into the coil interior, the necessary crossover from shielded to unshielded behavior on approaching the monomeric unit size[23–26]. The origin of the weak trend for long chains in good solvents is still uncertain. Its effect is examined below.

The ratio of dynamic sizes, R_υ/R_h, is plotted as a function of $M/M^\ddagger$ in Figure 6.9 for several species of linear polymers in theta solvents and good solvents[3]. As in Figure 6.6, comparing for R_t and R_g, no trend with molecular weight occurs. Also, no significant difference appears between good and theta solvent ratios. The combination of these results with those in Figure 6.8 demonstrate that the loss of shielding for short chains and the trends for long chains are common features of both dynamic sizes and cancel from their ratios. The average of all R_υ/R_h values is 1.08.

Data for $[\eta]$ and $[\eta]_\theta$ over a wide range of molecular weights are shown for polystyrene in Figure 6.10. As expected from the impenetrable coil model, the behavior parallels that for R_g and $(R_g)_\theta$ in Figure 6.1. Thus, $[\eta]$ begins to depart perceptibly from $[\eta]_\theta$ for $M \gtrsim 10^4$, and it settles into a limiting power law,

$$[\eta](cm^3 g^{-1}) = 0.0102\,M^{0.729} \qquad 0.1 \le 10^{-6}M \le 18 \tag{6.50}$$

The theta solvent values vary as $M^{1/2}$ over virtually the entire range, extending below $M^\ddagger \sim 7000$ and departing in a significant way only for very short chains, $M \lesssim 10^3$, where the random coil model itself begins to break down.

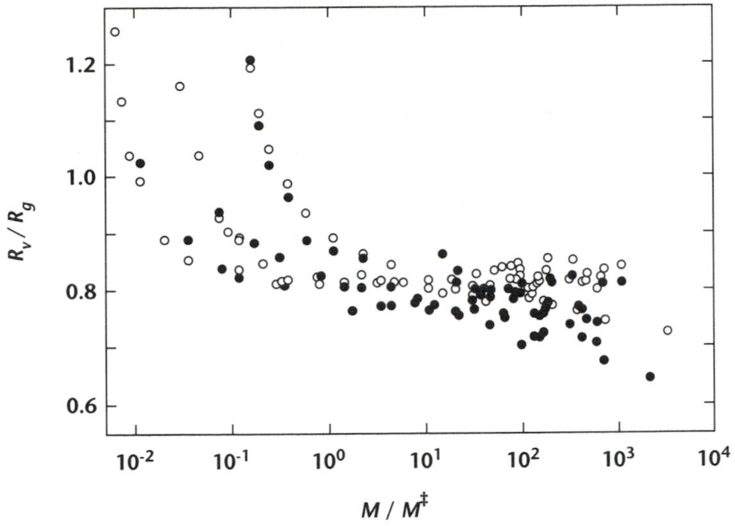

FIGURE 6.8 The ratio of viscometric to physical size as a function of reduced molecular weight for several polymer–solvent systems. Data from Hayward and Graessley[3]: good solvents (●) and theta solvents (O).

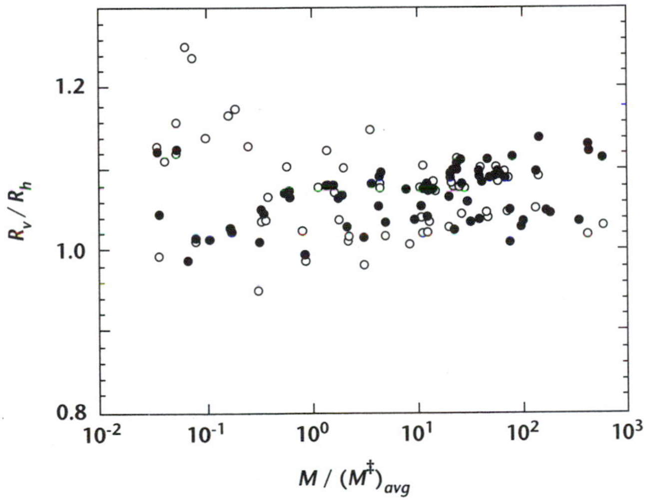

FIGURE 6.9 The ratio of viscometric to hydrodynamic size as a function of reduced molecular weight for several polymer–solvent systems. Data from Hayward and Graessley[3]: good solvents (●) and theta solvents (O).

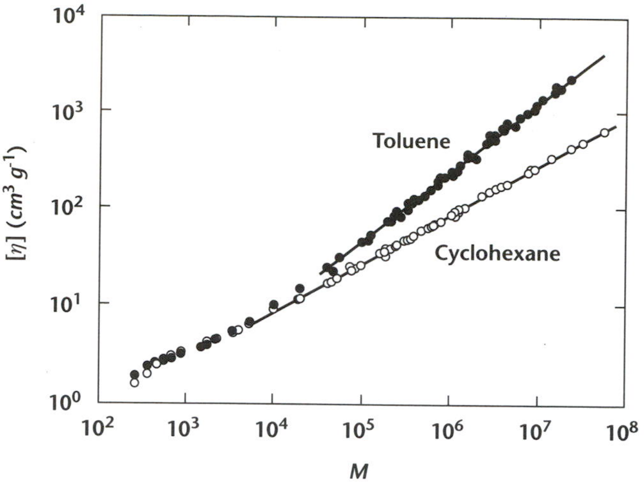

FIGURE 6.10 Intrinsic viscosity as a function of molecular weight for polystyrene in the good solvent toluene and in cyclohexane at the theta condition, $T_\theta = 34.5°C$.

6.2 Linear Polymers

Theories for the dilute solution properties of polymers have been developed and refined extensively in recent years[5,23,24,26,27]. Molecular simulation is now making significant contributions as well[28,29]. With several properties to consider, and a variety of molecular architectures as well, it is helpful to organize the data within some simplified but theoretically neutral framework. The analogy of polymer coil to hard sphere offers that possibility. The second virial coefficient, diffusion coefficient, and intrinsic viscosity can be expressed in terms of equivalent radii through Eqs. 6.36, 6.48, and 6.49. For hard spheres, with the uniform-density radius of gyration $R_g = (3/5)^{1/2}\,R$ included, the various measures of size are related as $R_t = R_h = R_v = 1.291R_g$, or, as size ratios,

$$\begin{aligned} R_t/R_g &= (5/3)^{1/2} = 1.291 \\ R_v/R_g &= (5/3)^{1/2} = 1.291 \qquad \text{(hard spheres)} \qquad (6.51) \\ R_v/R_h &= 1 \end{aligned}$$

Departures from this set of ratios will, of course, reflect departures from hard sphere character for whatever objects are being considered. Theories of A_2, D_o, and $[\eta]$ for various chain architectures (linear chains, three-arm stars, macrocyclic rings, etc.) can thus be judged by their predictions of these ratios in both good and theta solvents. Colloidal objects would have their own characteristic sets of ratios.

Expressing dilute solution properties in terms of coil size is not new. Thus, recasting D_o in terms of R_h goes back many years. Biochemists were already expressing the hydrodynamic and viscometric properties of proteins in terms of molecular size in 1976[30]. Akcasu and Benmouna expressed A_2 in terms of R_t in 1978[31]. Roovers and Toporowski introduced R_v and discussed the ratio R_v/R_h for various molecular architectures in 1980[32]. Burchard, working with both polymeric and colloidal systems, was already inferring structural features through comparisons of various size ratios in 1983[33]. Other dimensionless numbers—the interpenetration function Ψ, the Mandelkern-Flory parameter P, and the Fox-Flory parameter Φ—have been used to convey the same information, and theoretical results are still commonly expressed in those terms[23,24,26]. It is easy to convert such formulations to size ratios and back:

$$\begin{aligned} \Psi &\equiv \frac{A_2 M^2}{4\pi^{3/2} N_a R_g^3} = 0.752(R_t/R_g)^3 \\ P &\equiv \frac{Z_o}{6^{1/2}\eta_s R_g} = 7.70 R_h/R_g \\ \Phi &\equiv \frac{[\eta]M}{6^{3/2} R_g^3} = 0.713 N_a (R_v/R_g)^3 \end{aligned} \tag{6.52}$$

so nothing is lost in either formulation. Comparing the various dilute solution properties on the common basis of equivalent size, however, seems more direct and physically appealing. The discussion here is cast in terms of sizes and size ratios.

6.2.1 Limiting Size Ratios

As noted, the ratios R_t/R_g, R_v/R_g, and R_v/R_h are insensitive to species for long chains (See Figures 6.6, 6.8, and 6.9). Averages and values for a few species are listed in Table 6.1. Recent theoretical predictions for the ratios, obtained using a variety of methods, are also listed. In 1980, Zimm reported the first solution of the Kirkwood-Riseman equations for intrinsic viscosity and molecular friction coefficient in the shielded coil limit that did not employ average distances between chain units to evaluate the effect of hydrodynamic interaction[34]. Assisted by Monte Carlo simulation, he showed that preaveraging had introduced 12 to 13% errors (in opposite directions, remarkably) in earlier predictions of $[\eta]$ and Z_o at the theta condition[17,35]. The corrected values, expressed as $(R_v/R_g)_\theta$ and $(R_v/R_h)_\theta$ in Table 6.1, agree well with the data. (Fluctuations in conformation, not included in the Zimm calculation, partially offset the corrections[36–38], so a rigorous result might well differ slightly[39,40]). Oono used a renormalization procedure to predict limiting values in both good solvents and theta solvents[27,41], and found excellent agreement with the Zimm calculations and with the data. In 1984, Douglas and Freed used renormalization to predict the interpenetration function[42], which led to R_t/R_g in good accord with observations. The Monte Carlo simulations of Friere et al.[28] leading toR_v/R_h, and of Li et al.[1] leading to R_t/R_g, complete a picture of remarkable agreement between theory and experiment for linear chains.

TABLE 6.1 Limiting size ratios for linear polymers in good and theta solvents[3]

	R_t/R_g	R_v/R_g		R_v/R_h	
Polymer	**Good**	**Good[(a)]**	**Theta**	**Good**	**Theta**
Polystyrene	0.67	0.74	0.85	1.10	1.09
Poly(α-methylstyrene)	0.68	0.71	0.82	1.11	1.05
Poly(methylmethacrylate)	0.67	0.76	0.81	1.06	1.03
Poly(vinyl acetate)[(b)]	0.66	0.71	—		—
Polyisobutylene	0.67	0.78	0.82	1.05	1.02
1,4 Polyisoprene	0.68	0.80	0.83	1.11	1.06
1,4 Polybutadiene	0.71	0.84	0.84	1.17	1.13
Average	0.68	0.76	0.83	1.09	1.06
Hard Spheres	1.29	1.29	1.29	1.00	1.00
Renormalization Group[41]	0.66	0.73	0.84	1.14	1.12
Simulation-Theory[39,40]	—	—	0.835	—	1.11
Renormalization Group[42]	0.71	—	—		
Monte Carlo[28]	—	—	1.11	—	
Monte Carlo[1]	0.69	—		—	

(a) R_v/R_g in good solvents depends weakly on chain length owing to slight differences in limiting exponents (see Table 6.4). The values listed apply to the $10^5 \lesssim M \lesssim 10^6$ range.
(b) Calculated from results in ref. 108.

6.2.2 Expansion Factors

The discussion of excluded volume effects has thus far focused on the original theory of Flory (1949) and its application to good solvents and at the theta condition through the size expansion factor $\alpha_s = R_g/(R_g)_\theta$. As seen in Figure 6.5, the Flory formula [Eq. 6.25] does not correctly describe the observed dependence on chain length. Much work has been done since that time, summarized by Yamakawa[9], Casassa and Berry[5], and Fujita[23], seeking to derive a form that fits the data in the good solvent regime. A related aim has been a more precise description of short and intermediate chain length behavior, taking species contributions into account through the helical wormlike chain model[26]. A third aim has been the development of a rigorous analysis of excluded volume effects for long chains, but near the theta condition where the effects are weak. A proper description of short chain properties and a detailed consideration of behavior near T_θ are both beyond the scope of the book. However, the development of perturbation theory, which deals with weak excluded volume interactions, has provided a bridge to good solvents systems, a class of major interest in this book. We therefore consider here, with particular attention, the *two-parameter theory* and attempts to extend it into the realm of relatively strong interactions.

The perturbation theory of volume exclusion appears to have been initiated in 1951 by Teramoto[23]. We begin, however, with the second order perturbation results of Fixman[43] for the expansion of coil size and end-to-end distance. Considering only binary interactions, he obtained:

$$\begin{aligned} \alpha_s^2 &= 1 + \frac{134}{105}z - 2.082z^2 + \cdots \\ \alpha_R^2 &= 1 + \frac{4}{3}z - 2.075z^2 + \cdots \end{aligned} \tag{6.53}$$

in which α_s is $R_g/(R_g)_o$, α_R is $(\langle R^2\rangle/\langle R^2\rangle_o)^{1/2}$, and z is the excluded volume parameter [Eq. 6.29].

In 1960, Stockmayer proposed what has come to be called the two-parameter theory of dilute polymer solutions[10]. He suggested that the coil size of long flexible-chain polymers, reduced by the unperturbed size, is a universal function of the chain length—the latter in reduced form—scaled by the excluded volume parameter. Thus,

$$\begin{aligned} \alpha_s &= f_s(z) \\ \alpha_R &= f_R(z) \end{aligned} \tag{6.54}$$

with the same relationships applying over the entire range of excluded volume strength, from near the theta condition where it is weak, to the good solvent region where it is strong. Stockmayer also extended the idea to the second virial coefficient. When Eq. 6.52 is used to express ψ in terms of size ratio, the proposed form is equivalent to:

$$\frac{R_t}{(R_g)_\theta} = 1.1\,[za(z)]^{1/3} \tag{6.55}$$

A second order perturbation formula for $a(z)$ was published by Tanaka and Solc in 1982[44].

The universality idea was soon applied to size-dominated dynamic properties:

$$\begin{aligned} \alpha_h &= f_h(z) \\ \alpha_v &= f_v(z) \end{aligned} \tag{6.56}$$

where $\alpha_h = R_h/(R_h)_\theta$ and $\alpha_v = R_v/(R_v)_\theta$. First order perturbation expressions for these values were obtained by Stockmayer and Albrecht[45] and by Shimada and Yamakawa[46].

All these calculations are difficult, and the difficulty increases rapidly as more terms are added. Even if extended, however, the series converge slowly and may not converge at all even if all the terms are known. In any case, the sums would still only be applicable for weak excluded volume effects ($z \ll 1$), not in the good-solvent range of interest. To circumvent this apparent dead end, Domb and Barrett[47] proposed an interpolation approach, bridging from the perturbation formulas (small z regime) to the well-developed body of simulation data for self-avoiding walks on lattices (large z regime). Based on the two-parameter model, they required a formula agreeing with the perturbation coefficients in a series expansion, Eq. 6.53, and approaching at large z the lattice walk data, as fitted to the Flory exponent Eq. 6.26. They arrived at:

$$\alpha_R = \left[1 + 10z + \left(\frac{70\pi}{9} + \frac{10}{3}\right)z^2 + 8\pi^{3/2}z^3\right]^{1/15}$$
$$\alpha_s = \left[1 + 9.57z + 24.07z^2 + 26.5z^3\right]^{1/15} \qquad (6.57)$$

In 1984, Barrett proposed interpolation formulas for the dynamic sizes[48,49] based on the first order perturbation results and numerical solutions of the Kirkwood-Riseman shielded coil equations for large excluded volume. He completed the set of interpolation formulas with the development of one governing the second virial coefficient[50]. The Domb-Barrett formulas have the Flory limiting exponent built into them. However, modifications allowing other limiting exponents can be easily made[3].

The validity of the two-parameter theory has been investigated extensively. Based on the specification that all expansion factors depend on the same system variable z, Miyaki and Fujita[23,51] proposed an elegant method for testing the theory independently of the forms chosen to fit the data. The requirements are twofold:

- It must be possible to develop a master curve for each expansion factor from the data on its molecular weight dependence in several polymer–solvent systems merely by permitting a free choice of the binary cluster integral $\hat{\beta}$ for each system; and
- a unique relationship must exist between each pair of expansion factors because they depend upon one another parametrically, through their dependence on z alone.

Some results support the theory. Fujita and Miyaki found, for example, that the α_s versus M relationship can be expressed by a single master curve for a wide range of polymer–solvent systems. However, the α_v versus M data for the same systems do not form a master curve. The relationships between $R_t/(R_g)_\theta$ and α_s, and between α_v and α_s, are also not universal[23]. Even the apparently favorable evidence for size ratio cited above must be viewed cautiously, because the interaction strengths do not cover a wide range, and the overlap of data from systems with different interaction strengths is somewhat limited.

Thus, from the Miyaki-Fujita data analysis, the two-parameter model does not satisfy the crucial experimental test of universality. Oono[27] and Freed[24] have arrived at similar conclusions on theoretical grounds. Li et al.[1] offer an interesting and powerful objection to the two-parameter principle based on the relationship between A_2 and R_g. For α_s near unity—the limit of small z—they distinguish between the behavior of long chains with weak excluded volume (small $\hat{\beta}$) and short chains with strong excluded volume (large $\hat{\beta}$). The natural limit of $R_t/(R_g)_\theta$ in the former is 0 because the excluded volume per monomeric unit goes to 0. However, the limit in the latter should be of the order of 1.29, the ratio R/R_g for hard spheres. As a general principle, therefore, the two-parameter model fails. Despite failure as a general description of dilute solution behavior, the two-parameter idea has been found to apply to the important special case of solutions in good solvents. That aspect is described in the following section.

6.2.3 Good Solvent Master Curves

The expansion factors for many species of linear chains in good solvents obey the Miyake-Fujita criteria for two-parameter behavior[3]. Hayward and Graessley used data for eight species studied extensively during the early 1990s by the Yamakawa-Einaga group, supplementing earlier results for the same species from many other laboratories, to draw that conclusion. Expansion factors for the radius of gyration, hydrodynamic radius, and viscometric radius were calculated as functions of molecular weight in approximately twenty good solvent systems. The unperturbed radii for calculating expansion factors were obtained, whenever available, from measurements at the theta condition for the same sample. When unperturbed radii were not available, unperturbed values for samples of high molecular weight were calculated using the theta solvent expressions in Table 6.2. Calculation of an expansion factor for the second virial coefficient in this manner is not possible, of course, since $(R_t)_\theta$ is zero. To include A_2 data for uniformity of treatment, the data on thermodynamic radius in the good solvents were cast in expansion factor terms using a surrogate, $(R_g)_\theta$, scaled by the average R_t/R_g in good solvents (see Table 6.1), to provide the normalization:

$$\alpha_t = \frac{R_t}{0.68(R_g)_\theta} \tag{6.58}$$

For each polymer–solvent system, the four expansion factors vary with molecular weight in a qualitatively similar manner—values near unity for short chains, power law behavior for long chains, and a smooth transition between. The molecular weight $M^\ddagger$ which locates the transition varies with the polymer-solvent system and, to a somewhat lesser extent, with the property being considered. The power law exponent for each of the four properties varies with the property but is insensitive to the polymer-solvent system. It was thus possible to form a master curve for each property in the Miyake-Fujita manner[23] by choosing appropriate scale factors along the molecular weight axis for each polymer–solvent system. The scale factors were expressed as values of

TABLE 6.2 Coefficients in dilute solution relationships at the theta condition for selected polymer species

Polymer	Theta Condition	$(R_g/M^{1/2})_\theta$(nm)	$([\eta]/M^{1/2})_\theta$(cm^3 g^{-1})	$(R_h/M^{1/2})_\theta$(nm)
polystyrene	cyclohexane (34.5°C)	0.028	0.085	0.022
poly(α-methylstyrene)	cyclohexane (34.5°C)	0.028	0.074	0.021
polymethylmethacrylate	acrylonitrile (44.0°C)	0.025	0.055	0.020
polydimethylsiloxane	bromocyclohexane (29.5°C)	0.0295	0.083	0.022
polyisobutylene	isoamylisovalerate (22.1°C)	0.031	0.110	0.024
1,4 polyisoprene	1,4 dioxane (34.5°C)	0.0335	0.130	0.026
1,4 polybutadiene	1,4 dioxane (26.5°C)	0.0375	0.180	0.028
1,2 polybutadiene	2-octanol (32.8°C)	0.0345	0.120	
polyethylene	biphenyl (127.5°C)	0.0455	0.323	
polypropylene	biphenyl (125.1°C)	0.034	0.152	
polyethylene oxide	methylisobutylketone (50.0°C)	0.0365	0.120	

Values of $(R_g/M^{1/2})_\theta$ from Table 4.2; $([\eta]/M^{1/2})_\theta$ and $(R_h/M^{1/2})_\theta$ for first seven species from ref. 3; $([\eta]/M^{1/2})_\theta$ for 1,2 polybutadiene from ref. 12a $([\eta]/M^{1/2})_\theta$ for polyethylene, polypropylene, and polyethylene oxide from ref. 13.

TABLE 6.3 Characteristic molecular weights for various good solvent systems[3]

Polymer–Solvent System	$M_s^\ddagger$	$M_t^\ddagger$	$M_v^\ddagger$	$M_h^\ddagger$	$n^{\ddagger(b)}$
polystyrene-benzene	4800	3880	8070	(14,000)[(a)]	107
polystyrene-toluene	6420	4200	12,800	(13,200)	176
polystyrene-dichloroethane	12000	8090	14,800	-	223
polystyrene-ethylbenzene	(8300)	7600	19,500	(30,000)	261
polystyrene-tetrahydrofuran	(7700)	2590	10,800	(4900)	129
polystyrene-1,2,4 trichlorobenzene[(1)]			10,900		210
polystyrene-carbon disulfide[(2)]	(3630)		(10,000)		(70)
poly(methyl methacrylate)-acetone	19,900	19,700	33,000	–	484
poly(methyl methacrylate)-nitroethane	4430	3070	6730	–	94
poly(methyl methacrylate)-chloroform	950	420	1010	–	16
poly(methyl methacrylate)-benzene	–	–	3700	–	74
polyisobutylene-heptane	27,700	29,800	33,000	30,000	1080
polyisobutylene-cyclohexane	5790	2940	5180	3220	154
polyethylene-1,2,4 trichlorobenzene[(3)]	5480		5510		393
i-polypropylene-1,2,4 trichlorobenzene[(3)]	4310		3420		181
1,4 polybutadiene-cyclohexane	10,300	5500	5720	8700	563
1,4 polybutadiene-tetrahydrofuran	–	–	4700	–	348
1,4 polybutadiene-toluene			2660		197
poly(ethylene oxide)-methanol[(4)]			2160		147
poly(ethylene oxide)-water[(4)]			1290		88

(a)Values in parentheses are less certain.
(b)Average number of backbone bonds at crossover, calculated with average $M^\ddagger$ for the system with less-certain values omitted: $n^\ddagger = \langle M_x^\ddagger \rangle / m_o$, with molecular weight per backbone bond m_o from Table 4.2.
(1) From data in ref. 12; (2) from data in ref. 86a; (3) from Eqs. 5.75 and 5.76; (4) from data in ref. 87.

$M^\ddagger$, the molecular weight at the power-law intersection with the $\alpha = 1$ axis, as shown in Figure 6.5. The crossover molecular weights obtained for each system and property are given in Table 6.3 and the power law exponent for each property in Table 6.4.

The exponents for the equilibrium properties, α_s and α_t, are the same within the errors, and the shapes of their master curves are indistinguishable. The same is true for

TABLE 6.4 Limiting power-law exponents for expansion factors[3]

Expansion Factor	Exponent
α_s	0.092 ± 0.003
α_t	0.088 ± 0.002
α_v	0.079 ± 0.001
α_h	0.077 ± 0.006

the dynamic properties α_v and α_h. Each pair of master curves was merged to form the equilibrium and dynamic master curves shown in Figures 6.11 and 6.12. The power-law exponents for the merged plots, $p_{eq} = 0.089 \pm 0.001$ and $p_{dyn} = 0.078 \pm 0.001$, correspond to $R_{eq} \propto M^{0.589}$ and $R_{dyn} \propto M^{0.578}$ in the long chain limit. Thus, limiting good-solvent dependence on chain length for the properties of interest is as follows:

$$\begin{aligned} R_g &\propto M^{0.589} \\ A_2 &\propto M^{-0.233} \\ [\eta] &\propto M^{0.734} \\ D_o &\propto M^{-0.578} \end{aligned} \tag{6.59}$$

The chain-length exponent for R_g is clearly smaller than the Flory value of 0.6 and in excellent accord with the renormalization and lattice walk exponent of 0.588 [see Eq. 6.2]. The slightly weaker dependence of R_{dyn} on chain length has been noted

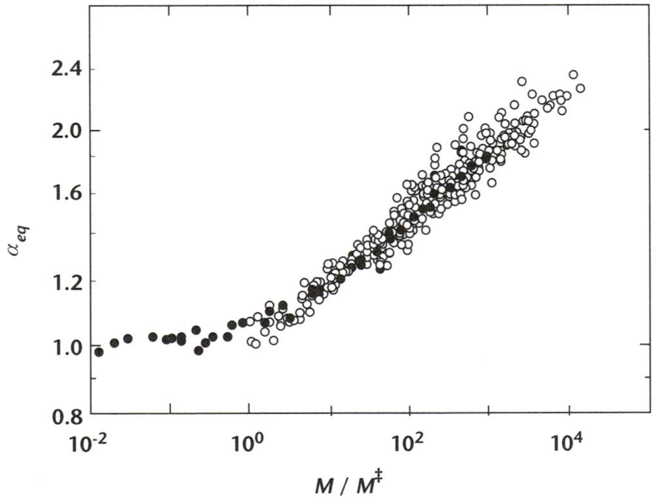

FIGURE 6.11 Expansion factors for physical and thermodynamic sizes: same sample for good and theta data (●), good solvent data with interpolated theta value (O). Data from Hayward and Graessley[3].

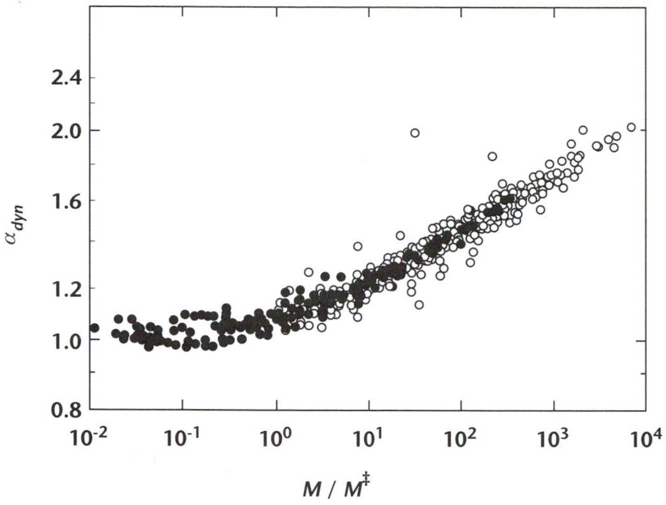

FIGURE 6.12 Expansion factors for viscometric and hydrodynamic sizes: good and theta data measured on the same sample (•), good solvent data with interpolated theta value (O). Data from Hayward and Graessley[3].

previously [9,23,24] and is commonly considered to reflect a species-dependent weakening in hydrodynamic interaction with coil expansion. From the Hayward-Graessley observations, the phenomenon would appear to be general, although possibly with some species-dependent variations.

Two additional features of good solvent systems are of interest: the shape of the master curves—$\alpha_x(M/M_x^{\ddagger})$ for $x = s, t, h, v$—and the relationship between the fitting parameter $M_x^{\ddagger}$ and other system properties. A similarity of master curve shapes for the two equilibrium properties, and for the two dynamic properties, is not in itself surprising. It follows directly from the observed insensitivity of the ratios R_t/R_g and R_v/R_h to both system and chain length (see Figures 6.6 and 6.9). The Domb-Barrett interpolation formulas for equilibrium properties and dynamic properties, also are indistinguishable when expressed as functions of $M/M^{\ddagger}$. All experimental curves, however, have more abrupt transitions between $\alpha_x \sim 1$ and power-law behavior than predicted by any of the theoretically based curves considered here.

Master curve forms. A recent off-lattice Monte Carlo simulation[52] has explored the crossover in the data from $\alpha_s \sim 1$ to power law behavior. The chain was modeled as a string of $n + 1$ spheres with center-to-center distance σ. The spheres interact through a Lennard-Jones potential function (see Eq. 2.3) with a cutoff distance at r_c:

$$\begin{aligned} \psi(r) &= 4\varepsilon\left[(\sigma/r)^{12} - (\sigma/r)^{6} - (\sigma/r_c)^{12} + (\sigma/r_c)^{6}\right] \quad r < r_c \\ \psi(r) &= 0 \qquad r > r_c \end{aligned} \tag{6.60}$$

When r_c is $2^{1/6}\sigma$, the potential is purely repulsive and close to a hard-sphere condition, approximating the *athermal* case. The cutoff was also set at $r_c = 2.5\sigma$, thus adding an attraction to permit the generation of a theta condition and thereby allowing behavior to be studied in the interval between T_θ and the athermal condition. The relative importance of monomer–monomer attraction, and thus solvent goodness, was varied through the temperature $T = \varepsilon/k$. Values of R_g were determined to the limits of available computer power, for lengths up to $n + 1 = 24{,}000$ for the athermal case and for the range $16 \leq n + 1 \leq 3200$ for the various temperatures.

The simulation data were analyzed in the manner of polymer solution results. The theta condition was established, $T_\theta \sim 3.18\varepsilon/k$, by a search to locate the $R_g \propto n^{1/2}$ behavior. The sizes at T_θ were used to compute the characteristic ratio for the chains, $C_\infty = 2.06$, and thereby to permit calculation of the expansion factor for the various lengths and conditions. The values of the crossover length $n^\ddagger$ were established for the athermal chains and several high temperatures, although that grew increasingly problematic at the lower temperatures owing to the weakness of expansion on approaching T_θ.

The form of the $\alpha_s(n/n^\ddagger)$ relationship changes with temperature in the lower range. Thus, the simulation suggests that the two-parameter theory already begins to break down under conditions of mild volume exclusion. However, the high temperature and athermal simulations, corresponding to the good solvent region, are superposable. This indicates that at least some aspects of the two-parameter theory reappear in a region of stronger exclusion; a region that contains most conventional good-solvent systems. The master curve obtained for the high temperature simulations is shown in Figure 6.13. It has the same rapid transition to the power law as the experimental good solvent systems. Indeed, the master curves from simulation and experiment are the same within the errors, as shown by the direct comparison in Figure 6.14.

The Domb-Barrett formula, Eq. 6.57, when adjusted to give the observed power-law exponent, has a more abrupt transition than the Flory formula, but a more gradual transition than either simulation or experiment. With x referring to the property and p_x its limiting exponent, as given in Table 6.4, the following formula describes well the simulation-based size master curve and all four experimental master curves:

$$\alpha_x = \left(1 + M/M_x^\ddagger\right)^{p_x} \tag{6.61}$$

Onset molecular weights. The magnitudes of $M^\ddagger$ for nearly all the good solvent systems investigated were much larger than anticipated from the data on self-avoiding lattice walks. A significant expansion is evident in lattice walks of even a few steps, while many times that number of Kuhn steps are typically required for the polymers in even the best of good solvents. Fujita had pointed out this curious phenomenon earlier[25]. The crossover lengths in the simulation at even the highest temperatures;

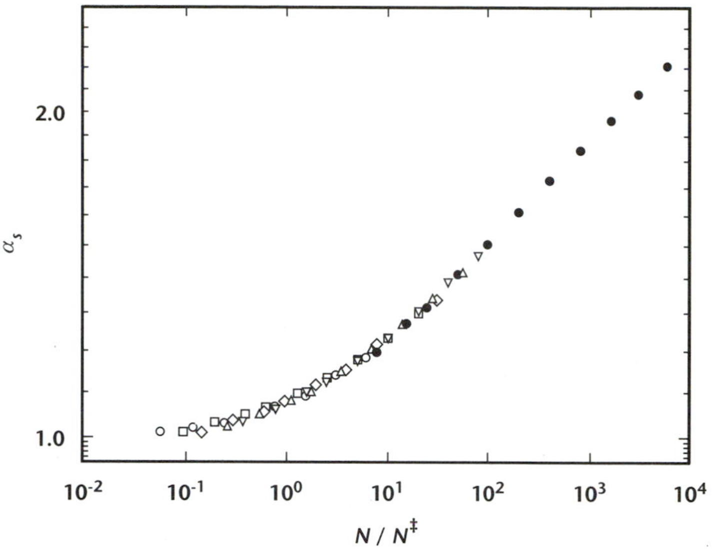

FIGURE 6.13 Size expansion factor as a function of reduced chain length from simulation for athermal interactions and for attractive interactions at several temperatures. Temperature $kT/\varepsilon = 3.6(\bigcirc), 3.8(\square), 4.0(\diamond), 4.5(\triangle), 5.0(\nabla)$, and athermal (●). Data from Graessley et al.[52].

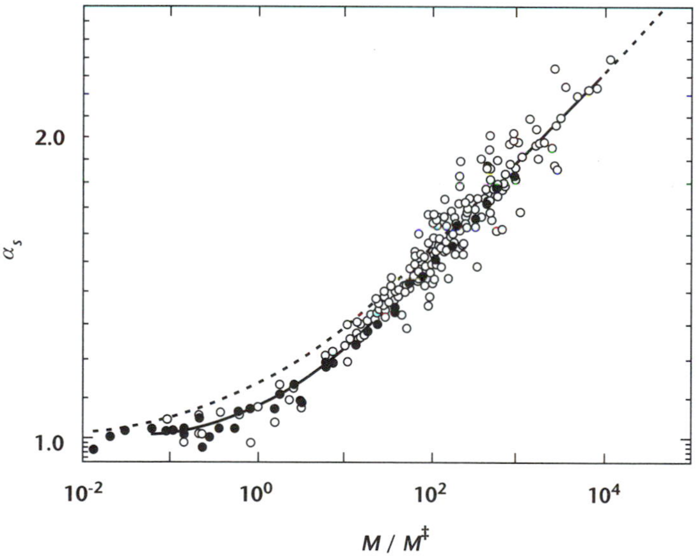

FIGURE 6.14 Comparison of size expansion–reduced length relationships obtained from simulation, adjusted two-parameter theory, and experiment. Experimental with same sample (●), experimental with interpolated theta size (O), simulation (—), and two-parameter theory (– – –). Data from Graessley et al.[52].

TABLE 6.5 Characteristic lengths in Kuhn steps for experiment and simulation[52]

Polymer-Solvent System	$T_{sim}(\varepsilon/k_B)$	$n^{\ddagger(a)}$	$n_K^{\ddagger}$
poly(methyl methacrylate)-chloroform		15.8	1.7
simulation	athermal	4.03	2.0
poly(methyl methacrylate)-benzene		74	8.2
poly(methyl methacrylate)-nitroethane		94	10.4
polystyrene-benzene		148	11
polystyrene-tetrahydrofuran		125	14
poly(ethylene oxide)-water		88	16
polystyrene-toluene		177	18
simulation	5.00	40.3	20
poly(α-methyl styrene)-toluene		241	22
polyisobutylene-cyclohexane		154	23
polystyrene-ethylbenzene		315	24
poly(ethylene oxide)-methanol		147	27
simulation	4.50	57.2	28
polypropylene-trichlorobenzene		183	29
poly(dimethyl siloxane)-cyclohexane		208	31
1,4 polybutadiene-toluene		197	36
1,4 polyisoprene-cyclohexane		200	43
simulation	4.00	105	51
polyethylene-trichlorobenzene		393	53
poly(methyl methacrylate)-acetone		484	53
1,4 polybutadiene-tetrahydrofuran		348	64
simulation	3.80	159	77
poly(dimethyl siloxane)-toluene		440	82
1,4 polybutadiene-cyclohexane		563	104
simulation	3.60	262	127
polyisobutylene-heptane		1080	159
simulation	3.50	460	222

(a) For experimental values, averages from Table 6.3.

that is, for the best "solvents," are also much larger than the athermal value. To facilitate comparison, the crossovers from both experiment and simulation, expressed as numbers of Kuhn steps,

$$n_K^{\ddagger} = n^{\ddagger}/C_{\infty} \tag{6.62}$$

are listed together in Table 6.5.

For the athermal simulation[52], $n_K^{\ddagger} \sim 2$. The data for walks on a simple cubic lattice[1], with C_{∞} for nonreversing walks on a simple cubic lattice estimated to be 3/2 [Eq.(4.27) with $\overline{\cos\theta} = -1/5$], leads to $n_K^{\ddagger} \sim 2.5$. The results from earlier lattice walk studies[53,54] give similar values. However, even for the highest simulation temperature ("best solvent"), $n_K^{\ddagger}$ is about 20, and, with very few exceptions, the values for polymer solutions are 10 or more. The simulations demonstrate that the good solvent anomaly cannot be explained by a mere rescaling of the segment size, a suggestion made by Yamakawa[9]. An explanation for this slow onset of expansion has been suggested[52]: Weak volume exclusion in good solvent systems is a free-volume effect, a characteristic of essentially all polymer solutions. The basis for this suggestion is discussed in Section 7.4.2.

6.3 Branched and Macrocyclic Polymers

The previous section dealt exclusively with linear polymers. Whether homopolymers or statistically uniform copolymers, and with or without side groups of modest size, sufficiently long chains in dilute solution behave similarly, in the flexible linear-chain manner. The same polymeric species, however, can also be formed with nonlinear architectures, most commonly as molecules with long-chain branch points but also as macrocyclic rings. Molecular sizes are smaller and self-concentrations larger than for linear chains of the same molecular mass. The branches of interest are many Kuhn steps in length, and at low branching density, their sizes are not very different from those for linear chains of the same length. However, relationships among the various molecular size measures (R_g, R_υ, etc.) depend on the large-scale architecture. Moreover, at higher branching densities, new colloidlike features appear and eventually dominate the dilute solution behavior. To reflect this difference, the discussion is divided into two parts, one for lightly branched structures (coil-like and relatively self-dilute), the other for heavily branched structures that behave as soft colloids with relatively dense and nonuniform interiors.

Long-chain branches can be formed in many ways. They can be created during polymerization, through multifunctional monomers [55–58], polymer transfer[59–61], or terminal bond incorporation[62,63]. They can also be formed by post-polymerization reactions such as crosslinking[7], grafting[64], or coupling[65]. Some sources of branching, such as crosslinking and grafting are, in principle at least, rather general. Others are species-specific, such as polymer transfer and terminal bond incorporation (polyethylene, polyvinyl acetate). Many branching reactions lead to gelation and the formation of networks (Chapter 9). However, other branching reactions do not, as explained in careful detail by Flory[7].

Long branches can have important effects on the dynamic behavior of polymeric liquids (see Volume 2). These effects are sometimes beneficial in the melt processing of polymers, and for this reason, long branches are occasionally introduced purposely. Long branches can have rheological consequences even at levels that are far too low to detect spectroscopically or by most dilute-solution techniques[66,67]. Sometimes branches are introduced inadvertently, leading to unanticipated properties[68]. For that reason, branching has become an explanation of last resort for unusual flow behavior. Statistical branching of the sorts mentioned above usually leads to a significant broadening of the molecular weight distribution[69]. A heterogeneous mixture of branched structures is also obtained, thus adding significant complexity to the problem of the molecular characterization of statistically branched polymers.

Branched polymers with well-defined architectures are useful for many purposes. Anionic polymerization, the source of many model linear polymers, can also be used to synthesize model branched materials[70–73]. Symmetric star polymers can be reliably synthesized by linking preformed polymeric carbanions, typically with lithium counterions, to a common junction by chlorosilane chemistry. Entire families of stars that are nearly monodisperse with respect to both arm molecular weight M_a and branch-point functionality f have been synthesized and studied. Nearly monodisperse polymeric combs have also been made by this route. Anionic polymerization followed by fractionation has also been used to synthesize nearly monodisperse H-shaped molecules as well as pom-poms and macrocyclic rings.

Two varieties of model branched polymers are emphasized here. Stars permit the exploration of self-concentration effects, ranging from self-dilute to self-concentrated, which encompasses the transition from polymeric to colloidal phenomena. Combs serve as well-defined models for individual species in the polydisperse systems that result from statistical branching processes. The aim is to develop relationships between structure and dilute solution properties that can be applied in the interpretation of data from multidetector SEC analysis.

The dilute solution properties of nonlinear polymers are commonly expressed in relation to those for linear polymers of the same molecular weight. The size ratio g for nonlinear structures was defined in Section 4.3, and the ratios for the virial coefficient, intrinsic viscosity, and hydrodynamic radius are defined analogously. Thus,

$$\begin{aligned} g &= R_g^2/\left(R_g^2\right)_{lin} \\ a &= A_2/(A_2)_{lin} \\ g' &= [\eta]/([\eta])_{lin} \\ h &= R_h/(R_h)_{lin} \end{aligned} \tag{6.63}$$

The symbols g_θ, $g_\theta{}'$, and h_θ refer to measurements at the theta condition; no subscript indicates good solvent data. When the ratios are applied to polydisperse systems, the linear polymer reference is understood to have the same molecular weight distribution as the branched polymer. In the following, *lightly branched* polymers are those for which both self-concentration ϕ_{self} and branch-point functionality f are relatively small. Polymers are *highly branched* if either ϕ_{self} or f is large enough to cause the appearance of new phenomena.

6.3.1 Lightly Branched Model Structures

Four groups of nonlinear chain structures are considered in this section, regular symmetric stars with modest branch-point functionalities, macrocyclic rings, H-shaped species and combs. The rings have neither branch points nor chain ends, the stars have one branch point but several branches, the combs have many branches distributed along a linear backbone, and the H-polymers, having two branch points, serve as a bridge between stars and combs. All were formed by organolithium-based polymerization chemistry and well-established linking reactions, and all are effectively monodisperse. They were chosen for inclusion here because of their availability in sets with systematically varied and thoroughly proven structures, making possible a confident and detailed examination of dilute solution theories and simulations.

Regular stars. Many studies have been made of star polymers in dilute solution. Most of those considered here were formed by coupling polystyrene (PS), 1,4 polyisoprene (PI), or 1,4 polybutadiene (PBD) anions, synthesized by organolithium chemistry, to multifunctional chlorosilanes. Nearly monodisperse stars, with wide and controllable ranges of branch lengths, and with branch-point functionalities (arms per star) that range from 3 to 128, have been made in this way. The studies have shown clearly that, for long enough arms, the exponents in the power laws relating coil dimensions and chain length for linear polymers in both good solvents and theta solvents carry over to stars. Figure 6.15 compares the R_g versus M relationship for linear and star PBD with a wide range of arm numbers in cyclohexane, a good solvent[74]. A similar comparison of $[\eta]$ versus M for linear and star PI in the good solvent toluene[75,76] is

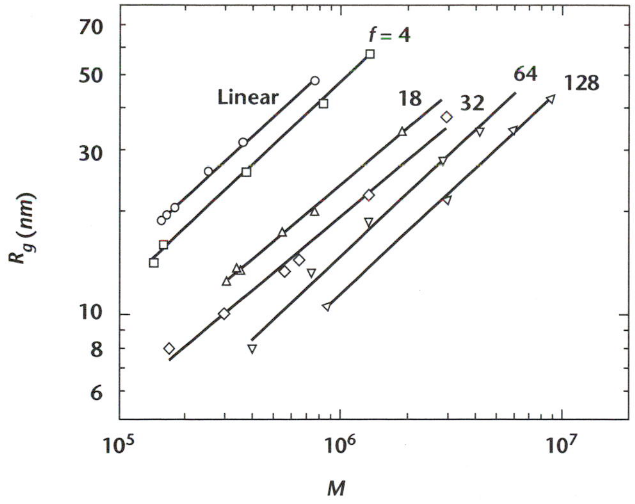

FIGURE 6.15 Size as a function of molecular weight for regular polybutadiene stars with various numbers of arms in cyclohexane. Data from Roovers[74].

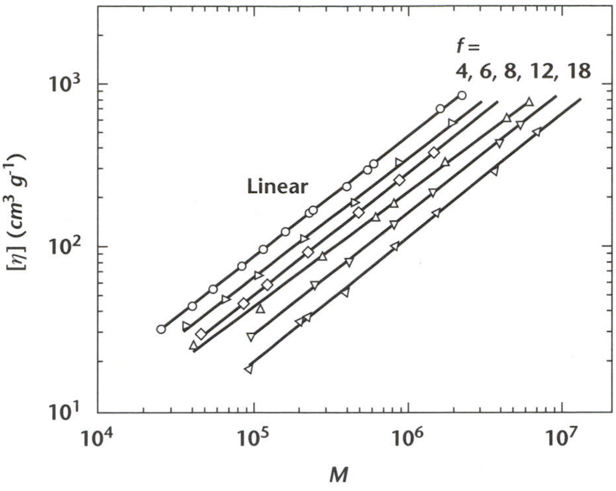

FIGURE 6.16 Intrinsic viscosity as a function of molecular weight for regular polyisoprene stars with various numbers of arms in toluene. Data from Hadjichristidis and Roovers[75] and from Bauer et al[76].

shown in Figure 6.16. The slopes of the lines drawn in the figures correspond to the power-law exponents for the linear chains. Such behavior is also found for the other size measures and at the theta condition as well. Accordingly,

$$R_x = K_x M^{\gamma_x} \tag{6.64}$$

where $x = s, t, h, v$ as before, and the ratios are obtained experimentally as:

$$\begin{aligned} g &= K_s^2/(K_s^2)_{lin} \\ a &= K_t^3/(K_t^3)_{lin} \\ g' &= K_v^3/(K_v^3)_{lin} \\ h &= K_h/(K_h)_{lin} \end{aligned} \tag{6.65}$$

Expressions for g, h, and g' for stars having random walk arms and no volume exclusion were derived respectively by Kramers[77], Stockmayer and Fixman[35], and Zimm and Kilb[78]:

$$\begin{aligned} g_{rw} &= (3f-2)/f^2 \\ h_{rw} &= f^{1/2}/[2-2^{1/2}+(2^{1/2}-1)f] \\ g'_{rw} &= [(2^{3/2}-1)f-2(2^{1/2}-1)]/f^{3/2} \end{aligned} \tag{6.66}$$

Without excluded volume, $a_{rw} = 0$. The Kirkwood-Riseman theory for shielded coils with preaveraged hydrodynamic interactions were used to obtain h_{rw} and g'_{rw}.

Utilizing the Monte Carlo procedure applied earlier for linear chains, Zimm eliminated preaveraging to obtain h and g' for four-arm and six-arm stars, with and without volume exclusion[40]. Rey et al. used similar procedures to obtain values for stars with six, twelve, and eighteen arms[28]. Kremer and Batoulis[79,80], using Monte Carlo lattice simulations, and Grest[81], using molecular dynamics, obtained g and g_θ for wide ranges of f. Douglas and Freed used a renormalization group method to provide information about a for several values of f[42]. Ohno et al.[82,83] used lattice-based Monte Carlo simulation to obtain a as well as g, g', and h for several values of f.

The results to $f = 20$ from these various theoretical predictions are compared, in Tables 6.6 through 6.9, with the averages of experimental values for PS, PBD, PEO, and PI stars[74,76,84–86,87]. Up to $f \sim 6$, the experimental values of g and g_θ agree rather well with the predictions and even with g_{rw}. Beyond that functionality, g decreases more rapidly than g_θ, but in each case agreement is still good between experiment and simulation. As f increases, the values of g_θ fall more slowly than g_{rw} while, curiously, g and g_{rw} continue to agree remarkably well. A similar pattern is found for g', g'_θ, and g'_{rw}, in which experiment and simulation agree well for $f \lesssim 6$; in this case, g'_{rw} decreases less rapidly than either g' or g'_θ. That pattern carries over for h, h_θ, and h_{rw}, although here the departures are already becoming evident at $f = 4$, and h_{rw} decreases more rapidly than either h or h_θ. The rather sparse data on the virial coefficient ratio suggests a similar pattern. The large departure for $f = 12$ is consistent with the Douglas–Freed remark that renormalization works best for $f \lesssim 6$[42]. The discrepancies that grow with increasing f are almost certainly symptoms of crowding near the branch point. The effect of crowding is considered in Section 6.3.3.

Rings. Some biological polymers, such as various versions of DNA, are macrocyclic rings, an observation that has led to a good deal of interest in that architecture[88–90]. Cyclic polydimethylsiloxane, obtained by fractionating equilibrated ring-chain mixtures, formed the basis for the first extensive studies of synthetic rings by Semlyen[91,92]. The molecular weights were low ($M < 30{,}000$), as were those for the polystyrene rings made by self-linking the ends of polymeric dianions [93,94]. High molecular weight rings followed soon afterward[95–97], forming the basis for the major studies of rings in dilute solution[97–100].

Some new phenomena appear in the case of rings. The second virial coefficient does not go to zero at T_θ, as determined with linear chains[98]. Such effects are observed with low molecular weight samples of other architectures (see Section 6.3.3), but not at high molecular weight. Even large rings have nonzero values at T_θ, and $(A_2)_\theta \sim 0.45 \times 10^{-4} cm^3 g^{-1}$ for $23 \lesssim 10^{-3} M \lesssim 440$, with a very weak decrease as M increases.

The origin of this repulsion appears to be a *topological interaction* of macrocyclic rings[101,102]. Chain molecules held together by permanent bonds cannot literally cross through the backbone contours of either their own or other chains. A pair of permanent macrocyclic rings looped around one another but not chemically bound—called

TABLE 6.6 Size ratios for symmetric stars[74,76,84–86,87]

	g_θ				g					g_{rw}
f	**Expt.**	**Refs. 28,40**	**Ref. 81**	**Refs. 79,80**	**Expt.**	**Refs. 28,40**	**Ref. 83**	**Ref. 81**	**Refs. 79,80**	**Eq. 6.66**
1,2	1	1	1	1	1	1	1	1	1	1
3	0.82		0.74	0.79	0.78		0.78	0.72	0.77	0.778
4	0.64	0.59	0.65	0.68	0.63	0.63	0.63	0.68	0.61	0.625
5			0.54	0.55			0.52		0.51	0.520
6	0.46	0.47	0.45	0.48	0.45	0.45	0.45	0.44	0.44	0.444
8	0.42		0.40	0.39	0.35		0.34			0.344
10			0.34					0.29		0.280
12	0.30	0.33		0.28	0.24	0.23	0.24			0.236
18	0.23	0.21			0.17	0.17	(0.16)			0.160
20			0.20					0.14		0.145

TABLE 6.7 Intrinsic viscosity ratios for symmetric stars[74,76,84–86,87]

	g'_θ		g'			g'_{rw}
f	Expt.	Refs. 28,40.	Expt.	Refs. 28,40.	Ref. 83	Eq. 6.66
1,2	1	1	1	1	1	1
3	0.86		0.83		0.84	0.896
4	0.76	0.71	0.73	0.73	0.735	0.811
5	0.74		0.60			0.744
6	0.63	0.58	0.58	0.57	0.56	0.690
8	0.53		0.43			0.610
12	0.42	0.39	0.34	0.37	0.335	0.508
18	0.31	0.38	0.24	0.22	(0.25)	0.420

TABLE 6.8 Hydrodynamic radius ratios for symmetric stars[74,76,84–86,87]

	h_θ		h				h_{rw}
f	Expt.	Refs. 28,40	Expt.	Refs. 28,40	Ref. 83	Refs. 79,80	Eq. 6.66
1,2	1	1	1	1	1	1	1
3	0.93		0.93		0.945	0.96	0.947
4	0.94	0.95	0.92	0.94	0.92	0.91	0.892
5						0.87	0.842
6	0.89	0.89	0.86	0.89	0.875	0.83	0.798
8			0.81		0.82		0.726
12	0.78	0.79	0.73	0.82	0.73		0.623
18	0.72	0.68	0.66	0.76	0.66		0.527

TABLE 6.9 Virial coefficient ratios for symmetric stars[74,76,84–86,87]

	a		
f	Experiment	Monte Carlo Ref. 41	Renorm. Grp. Ref. 42
1,2	1	1	1
3	~ 1	0.98	0.97
4	0.94	0.93	0.92
6	0.86	0.81	0.81
12	0.56		0.34

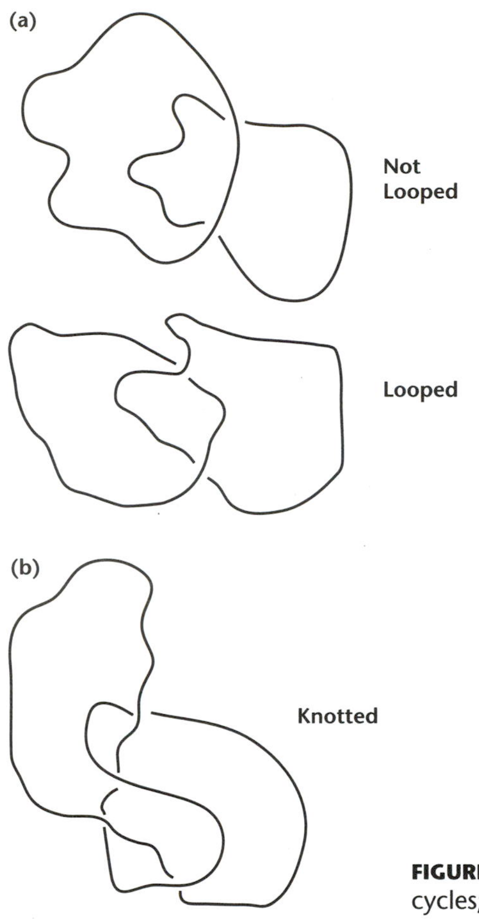

FIGURE 6.17 (a) Looped and not-looped macrocycles; (b) knotted macrocycle.

a *polycatenane molecule*[103,104]—is held together by the *uncrossability constraint*. A pair of independent macrocycles, on the other hand, belongs to a different topological class. In free space, the independent macrocycle pain has more accessible configurations than the corresponding polycatenane, but fewer than a pair of mutually crossable rings. Both configurations illustrated in Figure 6.17(a) are available to phantom (crossable) rings, but only one to permanent macrocycles. Uncrossability resembles volume exclusion, yet also makes a separate contribution that remains in force even in the thermodynamic limit of zero volume exclusion.

Owing to their mutual uncrossability, macrocycle pairs that are not linked topologically undergo a progressive loss of accessible configurations as they approach one another, which translates to a repulsive interaction. Ringless (open chain) polymeric architectures are not subject to this repulsion: All configurations are attainable unless

TABLE 6.10 Summary of structure ratios for macrocyclic rings

Ratio	Good Solvent	Theta Solvent
g	0.52 (0.516$^{(a)}$, 0.563$^{(b,c)}$)	0.53 (0.500$^{(d)}$)
g'	varies with M	0.67 (0.662$^{(d)}$)
h	0.859 (0.899$^{(b)}$)	0.85 (0.848$^{(d)}$)
a	0.90 (0.879$^{(a)}$)	—

Experimental values in good and theta solvents taken from Fujita's analysis of the ring literature in ref. 23. Theory and simulation results are in parentheses: (a) Douglas-Freed, ref. 42.; (b) Chen, ref. 90; (c) Bruns-Naghizadeh, ref. 89; (d) Zimm-Bloomfield, ref. 88.

more than one of the species involved contains macrocycles. Iwata was able to fit the Roovers data on $(A_2)_\theta$ for polystyrene rings, including its weak M dependence, to a theory of the topological interaction[102] by choosing the value of a single adjustable parameter, the reduced interaction per repeat unit, $\bar{\gamma} = 0.0041$.

The $[\eta] - M$ relationship in good solvents is another unusual property of rings. The Mark-Houwink exponent is smaller for rings than for linear chains. It is also different for rings made by different investigators[98]. McKenna et al. suggested that knots in the rings—structures illustrated in Figure 6.17(b)—might somehow be responsible for these differences[100]. *Knots* are topological entities, permanent features of uncrossable closed loops such as the rings. They could well be trapped in at the time of closure, and the various types could have different frequencies depending on the conditions at closure. As the summary in Table 6.10 shows, however, other solution properties of rings are only slightly (if at all) anomalous. It is unclear why the $[\eta] - M$ relationship in good solvents alone would be peculiarly sensitive to the presence of knots. That question and others about macrocycles, well summarized elsewhere[23], remains unresolved.

H-Polymers. The entire body of detailed dilute solution work on the regular H architecture (five strands of equal length joined to form a molecule with two trifunctional branch points) is based on six polystyrene H's, made and studied by Roovers and Toporowski[105]. This model is the simplest two-branch molecule. It is also the simplest having a part—the bridge—that has no free end, a characteristic of considerable interest to flow properties (see Volume 2). Like long-arm polystyrene stars, it has the same dilute solution exponents and the same T_θ as linear polystyrene. The experimental structural ratios are summarized and compared with theory and simulation[42,106,107] in Table 6.11. The results for this series are examined with other members of the comb family of architectures below.

Combs. The discussion to this point has involved families of nonlinear structure that have the same architecture, with the members differing from one another only in size scale. This has permitted tabular comparisons with theory. Combs are not available nicely partitioned into iso-structural families, and graphical comparisons

TABLE 6.11 Structure ratios for H-shaped molecules[105]

Ratio	Good Solvent	Theta Solvent
g	0.69 (0.72$^{(a)}$, 0.71$^{(b)}$)	0.70 (0.712$^{(c)}$)
g′	0.73	0.80
a	0.97 (0.95$^{(a)}$)	

Theory and simulation results are from (a) Douglas-Freed, ref. 42; (b) Lipson et al, ref. 107; (c) Casassa-Berry, ref. 106.

are more efficient. Combs are also better models than stars for the components in polydisperse branched polymers. They are more difficult to synthesize with precisely defined structures, however, and information on their dilute solution properties is rather limited.

In contrast with the literature on polymeric stars, the published work on combs seems not to have been critically reviewed recently. In 1965, Berry et al.[108] provided the first significant body of dilute-solution data on comb polymers, made in that case by free-radical grafting of polyvinyl acetate. Berry[109] also studied two polystyrene combs made using anionic techniques. About that time Noda et al.[110] used the same synthetic method as the basis for the first extensive study of combs. Lacking the guidance of a critical review, the discussion here relies primarily on the more recent results of Roovers[111–113] for polystyrene and polybutadiene combs, and those of Hadjichristidis et al.[73] for polyethylene combs. Data for other architectures with multiple branch points are included as well: the Roovers H-shaped polystyrenes[105] and the Hadjichristidis et al. H-shaped and pom-pom polyethylenes[73].

The Casassa-Berry formulas for the random walk size ratio g_{rw} of regular combs[106] are given in Eq. 4.53. Values of h_{rw} and g'_{rw} have been calculated for selected comb structures by Kurata and Fukatsu[114] and by Osaki et al.[115]. Berry[116] derived an analytical expression for h_{rw} when the branches are randomly placed along the comb backbone. Unfortunately, no analytical expression for g'_{rw} has been derived for any type of regular comb, and the 1974 calculations of Osaki et al.[115] cover only a limited range of rather small structures. Approximating g'_{rw} for combs with some combination of h_{rw} and g_{rw} remains an open possibility[117].

Simulations for comb polymers are relatively rare. Although trifunctional branch points are the main interest, the relevant structural ranges—long branches, large chain distance between branch points, large number of branches—require enormous computing power. The lattice-walk calculations of g_θ by McCrackin and Mazur[118] and of g and h by Lipson[119] involve relatively small structures but are nonetheless consistent with the data, as shown below.

The experimental size ratios for polystyrene combs[111,112] are shown as functions of the random walk ratios in Figure 6.18. Size ratios for the polyethylene combs[73] (not

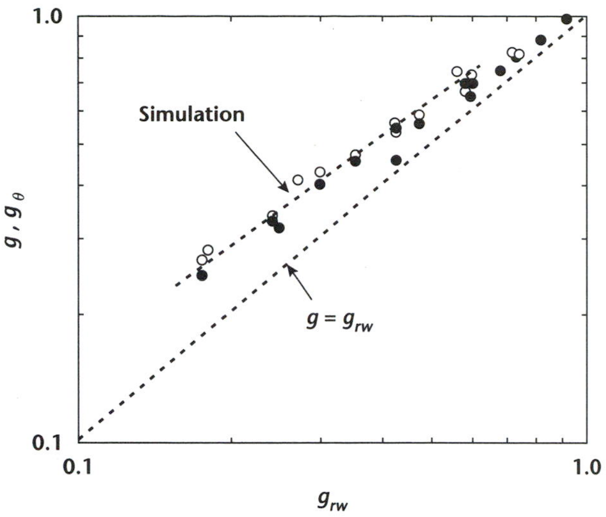

FIGURE 6.18 Comparison of size ratio for polystyrene combs with the random walk calculation. Data in a good solvent toluene (●) and a theta solvent cyclohexane at 34.5°C (O). Data from Roovers[111,112]. The dashed line indicates the Monte Carlo simulation[118].

shown) agree well, and the values of g and g_θ are nearly the same, but with g_θ slightly larger over the entire range. Both g and g_θ, however, are systematically larger than the random walk prediction g_{rw}. The McCrackin-Mazur results for g_θ and the Lipson results for g, represented by the dashed line in Figure 6.18, are consistent with the experiments over the range covered by the simulations and with the observation that g and g_θ are nearly the same. The experimental data, within their uncertainties, are well described by a power law:

$$g \sim g_\theta = g_{rw}^{0.77} \tag{6.67}$$

Results for the hydrodynamic size ratios[32], shown in Figure 6.19, are more scattered. However, the same pattern seems evident; the differences from h_{rw} in this case merely become slightly more exaggerated. The Lipson simulations of h agree very well indeed with the good solvent results.

There being no regular comb expression for g'_{rw}, the corresponding comparison of intrinsic viscosity ratios cannot be made. However, from an examination of the comb data it would seem that g and g_θ are nearly the same as g' and g'_θ. This rather surprising observation is confirmed by the direct comparison of size and intrinsic viscosity ratios for both polystyrene and polyethylene combs shown in Figure 6.20. Results for the lower order members of the comb family—three-arm star, H-shaped, and pom-pom structures—are also included in the plot, as are good solvent and theta

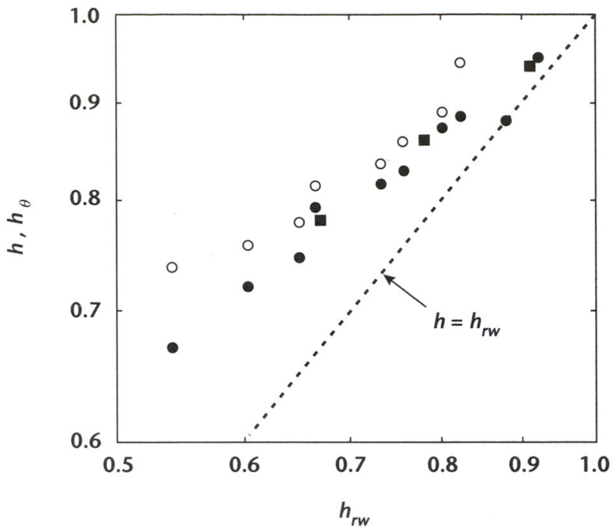

FIGURE 6.19 Comparison of hydrodynamic ratio for polystyrene combs with random walk calculation. Data in a good solvent toluene (●) and a theta solvent cyclohexane at 34.5°C (O). Data from Roovers[111,112].

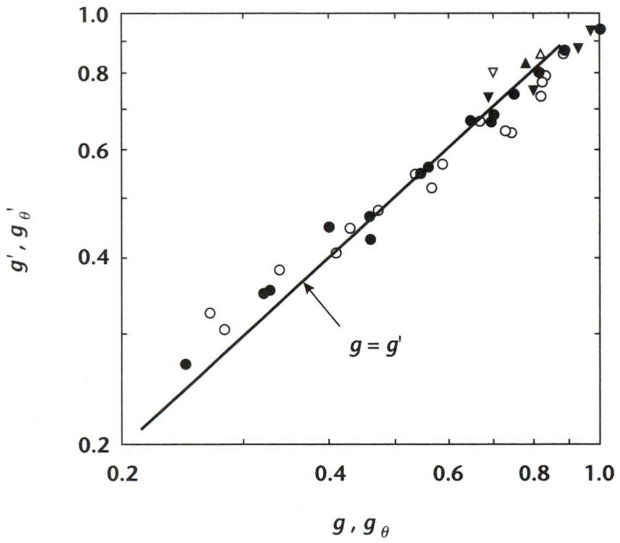

FIGURE 6.20 Comparison of experimental viscometric ratio and size ratio for comb polymers. Filled symbols indicate good solvent conditions; open symbols indicate theta conditions. Data for polystyrene combs in toluene[112] and polyethylene combs in 1,2,4 trichlorobenzene[73] (●); polystyrene combs in cyclohexane[111,112] (O); H-shaped and pom-pom polystyrene[105] and polyethylene[73] (▼, ▽). Values for three-arm stars are from Table 6.6 (▲, △).

condition data. Essentially, a single line fits all the data, suggesting the following approximate relationships for polymeric combs:

$$g \sim g_\theta \sim g' \sim g'_\theta \qquad (6.68)$$

Whether this remarkable coincidence offers some possibilities for characterizing statistically branched polymers, considered briefly in the following Section, remains to be seen.

6.3.2 Lightly Branched Statistical Structures

Statistically branched polymers are, by their nature, polydisperse in both structure and molecular weight, a feature that must be taken into account in their characterization by dilute solution methods. Separation, usually by size-exclusion chromatography (see Section 5.3), is an essential part of the analysis. In-line viscometry and light scattering on the stream eluting from the SEC system offers possibilities for the continuous measurement of absolute molecular weight, intrinsic viscosity, and molecular size over the distribution that the separation affords. Even under idealized circumstances, each slice of the emerging stream, although monodisperse in some measure of molecular size, contains a distribution of architectures[120]—molecular weights, number of branches, branch lengths, and the line. Because of this diversity, it is literally impossible to provide a detailed characterization on a sample of unknown origin. The use of supporting information, such as the branching mechanism together with the analysis of samples obtained from a systematic variation of synthetic conditions, is a good start. Qualitative questions—is sample A more branched than sample B, assuming similar formation mechanisms, or is the branching in sample A more skewed toward the high molecular weight tail—may sometimes be answerable.

Elution liquids are invariably good solvents for the polymer species; otherwise, adsorption on the columns and other complications occur. The properties relevant for in-line detection of branching are therefore g' and g in good solvents. The comb family would appear to provide the most appropriate available models for the irregular structures that populate statistically branched systems. From Figure 6.20, g' and g appear to be indistinguishable and insensitive to both solvent and polymer species for combs and hence perhaps also for the irregular structures of interest. Some support for the comb-irregular structure analogy is offered in Figure 6.21, showing g' and g as functions of molecular weight for the SEC-generated slices from a sample of polyethylene made by free radical polymerization[121]. Such polymers contain both long-chain branches from polymer transfer reactions and short-chain branches from intramolecular transfer[122,123]. The approximation $g = g'$, noted for combs in Eq. 6.68, appears also to describe (over some range at least) the SEC slices of this sample.

Most efforts to characterize statistically branched polymers are based on results from an early paper by Zimm and Stockmayer[55]. Those authors used the Kramers

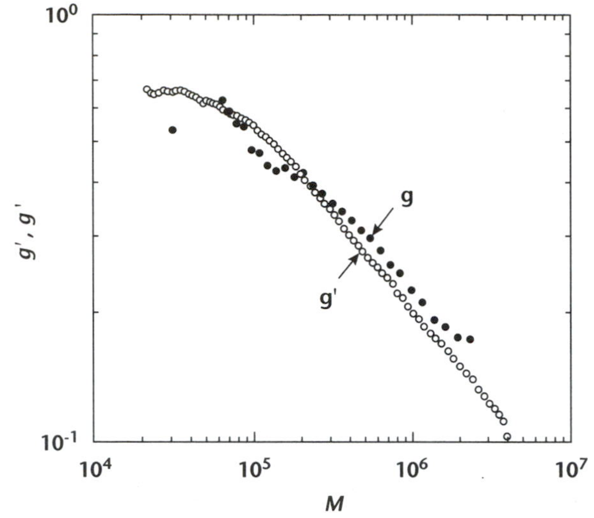

FIGURE 6.21 Comparison of size and viscometric ratios for molecular volume slices obtained for a sample of branched polyethylene by SEC-LALS-VISC and SEC-MALS in 1,2,4 trichlorobenzene at 135°C. Data from Cotts[121].

formula [Eq. 4.46] to calculate the random walk size ratio g_{rw} for several classes of branched architectures, including those generated in random A_2B_f polycondensations. The expressions for various average values of g_{rw} were derived for unfractionated trifunctional ($f = 3$) and tetrafunctional ($f = 4$) branching site polymers and their constant molecular weight fractions. For the trifunctional polymer fractions,

$$\langle g_{rw} \rangle_3 = \frac{1}{[(1 + \lambda M/7)^{1/2} + 4\lambda M/9\pi]^{1/2}} \tag{6.69}$$

in which λM is the average number of branches per molecule in a constant M slice. Unfortunately, SEC does not produce constant M slices, so some formula other than Eq. 6.69 is required, one more tailored to SEC separations.

Figure 6.22 shows the result of an attempt to quantify long-chain branching density with the data in Figure 6.21. It was assumed that Eq. 6.69 could be applied and, hoping for simplicity, that λ is independent of molecular weight. Then, the comb model was assumed to apply, and Eqs. 6.67 and 6.68 were used to convert the values of g', which cover a wider range of molecular weights than the g data, to values of $\langle g_{rw} \rangle_3$. First, however, the values were adjusted for the effect of short chain branching, using $g' = g'_{SCB}\, g'_{LCB}$ and choosing g'_{SCB}—assumed independent of M—such that $g'_{LCB} \sim 1$ for small M. Finally, $\langle g_{rw} \rangle_3 = (g'_{LCB})^{1/0.77}$ was obtained, and the resulting $\langle g_{rw} \rangle_3$ versus M data were fitted to Eq. 6.69, giving the result shown in

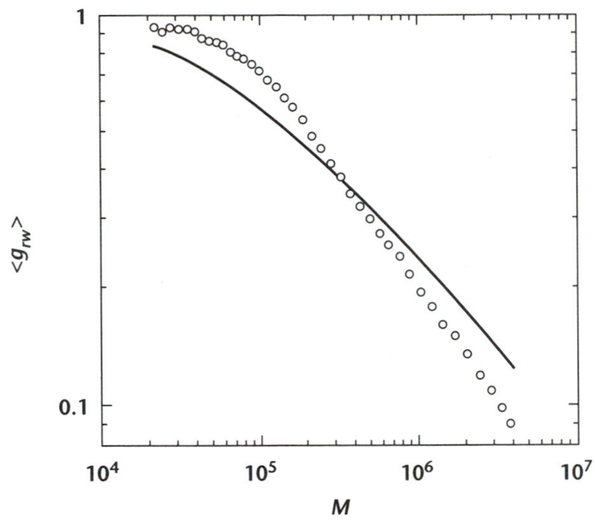

FIGURE 6.22 Fit of viscometric ratios for the SEC slices of a low-density polyethylene sample in 1,2,4 trichlorobenzene at 135°C to the Zimm-Stockmayer expression for constant molecular weight slices of trifunctionally branched polycondensates. Data from Cotts[121].

Figure 6.22 for the choice $\lambda = 1.0 \times 10^{-4}$. The agreement between predicted and observed curves is obviously rather poor.

The exercise above illustrates some of the obstacles to drawing unambiguous conclusions about molecular structure in complex systems even by means of multidetector SEC methods. Among the many imponderables is the applicability of the Zimm-Stockmayer formulas to long-chain branching structures arising from the statistically quite different mechanism of polymer transfer. The recent branching and SEC simulation studies of Tobita[63,124] would seem to open the way for answering such questions. The nonlinear polycondensations whose structural evolution has been studied by Schosseler et al.[125], Weissmüller and Burchard[120], and Lusignan et al.[57,58] are essentially Zimm-Stockmayer systems. These successes demonstrate the significant advantages of multidetector SEC in characterizing the structural distribution of branched systems when the chemistry and the model for analysis are well matched.

6.3.3 Highly Branched Structures

At constant molecular weight, the self-concentration of polymer molecules increases with long-chain branching. Thus, Eqs. 6.8 and 6.11, and the size ratio definition in Eq. 6.63 lead to:

$$\phi_{self} = (\phi_{self})_l / g^{3/2} \tag{6.70}$$

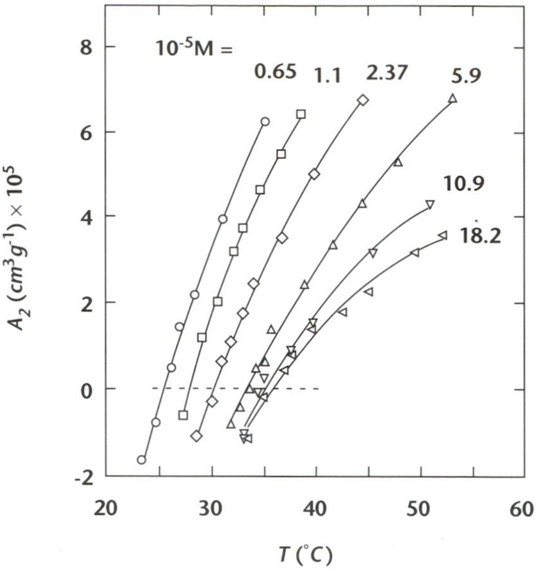

FIGURE 6.23 Second virial coefficient as a function of temperature for six-arm polystyrene stars in cyclohexane. Data from Roovers and Bywater[126].

Increasing internal coil concentration increases the importance of three-body excluded volume interactions. One effect is to shift the temperature at which $A_2 = 0$ for the solution, shown for the case of six-arm polystyrene stars[126] in Figure 6.23. Thus, at $T = T_\theta$, as determined with linear polymers, the branched coils may remain somewhat swollen and may still repel one another, since only the two-body contributions have been cancelled. Three-body effects are undoubtedly present for linear polymers as well, but the shifts for typical molecular weights are negligible. Such shifts only become prominent for the more highly branched architectures, and they become negligible even for those in the long-arm region.

Candau, Rempp, and Benoit[127] introduced in 1972 a theory for three-body excluded volume contributions to coil expansion. They developed formulas by including a ternary term in the Flory treatment of excluded volume. In the analysis developed here, they carried the ternary term in Eq. 6.23 through to the expansion factor expression, with the result:

$$\alpha^5 - \alpha^3 = \frac{l_p^3}{\upsilon_s g_o^{3/2}} \left[\frac{3}{\pi} F_2(T) \frac{(R_g)_{o,L}}{l_p} + \frac{45}{8\pi^2} \frac{F_3(T)}{g_o^{3/2} \alpha^3} \right] \tag{6.71}$$

where the Candau et al. expression has been recast in terms of l_p, the packing length for the species [Eq. 4.41]; g_o, the unperturbed size ratio for the architecture; $(R_g)_{o,L}$, the unperturbed size of linear chains with the same molecular weight; and

$\alpha = R_g/(R_g)_o$, the expansion ratio. At T_θ the binary term is zero, leading to:

$$\alpha_\theta^8 - \alpha_\theta^6 = \frac{45}{8\pi^2}\frac{l_p^3}{\upsilon_s}\frac{F_3(T_\theta)}{g_o^3} \tag{6.72}$$

where $\alpha_\theta = (R_g)_\theta/(R_g)_o$, the ratio of size in a theta solvent to size in the melt state. Thus, the expansion ratio at T_θ is predicted to depend on architecture but not molecular weight. At lower temperatures, the binary term is negative and eventually brings the bracketed quantity in Eq. 6.71 to 0, an event we assume to correspond macroscopically to $A_2 = 0$. Assuming also that F_2 varies linearly with T in that range, the resulting expression can be written as:

$$\Lambda = (T_\theta - T_{A_2})(R_g/l_p)_{o,L}\, g_o^{3/2} \tag{6.73}$$

where Λ may depend on the polymer–solvent system, but not on architecture or molecular weight.

The observed behavior is qualitatively consistent with Eq. 6.73. Thus, $T_\theta - T_{A_2}$ does indeed decrease toward zero with increasing molecular weight and increasing size ratio. The quantitative picture, however, is less satisfactory. Values of Λ calculated with Eq. 6.73 for polystyrene stars[126,128] ($f = 3$, 4, 6, 12) show little variation with f and M and average to $\Lambda = 6.1$. The average for polyisoprene stars[129] ($f = 8$, 12) is $\Lambda = 4.9$. The averages for polystyrene H-molecules[105] and combs[111,112] are larger, $\Lambda = 7.6$ and 15.6. Unfortunately, the evidence may be slightly compromised by a chemical factor: small amounts of butadiene units added to facilitate linking but apparently influencing T_{A_2} in some cases as well[128]. The chemical structure of the linking agent and the large number of end groups may also contribute. Instances in which nonzero $T_\theta - T_{A_2}$ would be expected, but were not found, have also occurred[75].

The situation regarding three-body contributions remained unchanged until a study by Boothroyd and coworkers[130,131]. They used small-angle neutron scattering to determine chain dimensions in the melt state and in a theta solvent to evaluate α_θ for a series of linear and star model polyethylenes ($f = 3$, 4, 12, 18). They found, as expected, that α_θ is independent of molecular weight. They also modified Eq. 6.71 to describe more accurately the elastic pressure resisting expansion for symmetric stars [see Eq. 6.19], the effect being to multiply the right side of Eqs. 6.71 and 6.72 by $(15f - 14)/(3f - 2)^2$. They went on to show that the parameter expressing three-body contributions to intramolecular effects [the modified Eq. 6.72] and to the intermolecular effects [Eq. 6.73] was the same within the uncertainties. They also demonstrated that size in good solvents for these same samples can be understood in terms of two-body interactions alone.

Daoud-Cotton model. The size ratio for symmetric stars is shown as a function of arm number, $3 \le f \le 128$, for good solvents in Figure 6.24 and for the theta condition in Figure 6.25. Data and theory in the lower range, $f \le 20$, were compared in

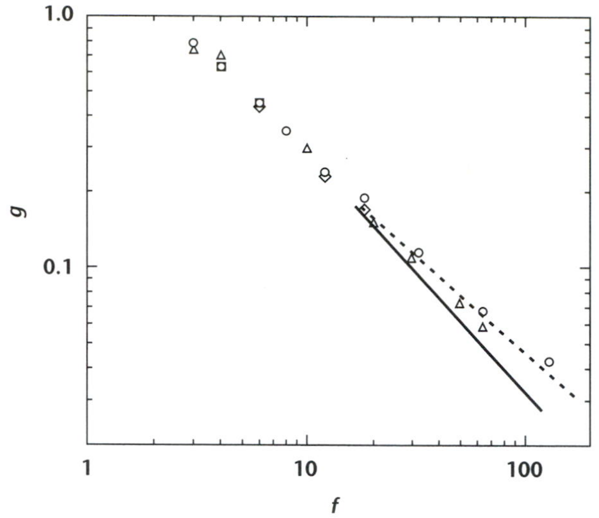

FIGURE 6.24 Size ratio for multiarm stars of various species in good solvents. Results of simulation and theory are from Zimm[40] (□); Rey et al.[28] (◇), and Grest[81] (Δ); experimental data from several sources (○). The solid line indicates values of g_{rw}; the dashed line is drawn with the slope predicted by Daoud and Cotton[132].

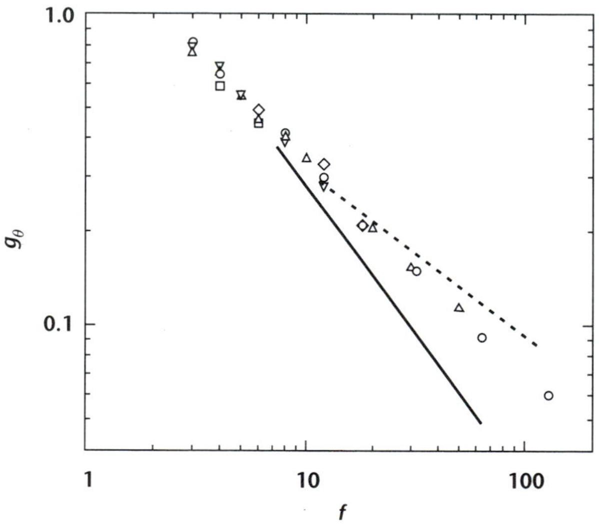

FIGURE 6.25 Size ratio for multiarm stars in theta solvents. Simulation results of Zimm[40] (□), Rey et al.[28] (◇), Batoulis and Kremer[79,80], (∇), and Grest[81] (Δ). Experimental data from several sources (○). The solid line indicates values of g_{rw}; the dashed line is drawn with the slope predicted by Daoud and Cotton[132].

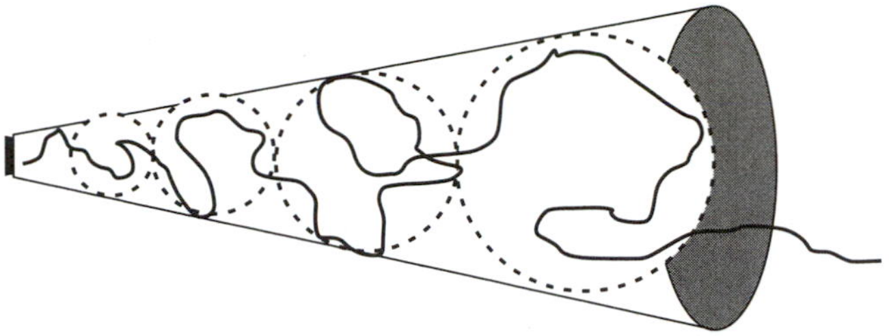

FIGURE 6.26 The Daoud-Cotton picture for one of the arms in a multiarm star.

Section 6.3.1. It was noted that the random walk expression for the size ratio given in Eq. 6.66 describes the data fairly well over that range. The g_{rw} versus f variation for large f is indicated by the solid line in the figures. It departs systematically from both experiments and simulations, which continue to agree rather well, with g_{rw} becoming progressively too small.

The failure of Eq. 6.66 at large f is almost certainly a result of crowding in the coil interior, which forces the arms to extend and thus to expand in overall size as branches are added. The *Daoud-Cotton* theory[132], published in 1982, treats the crowding problem explicitly. Consider a dilute solution of symmetric star molecules with many arms ($f \gg 1$), each arm having a large number n of monomeric units. Represent the star by a sphere of radius R having its total of fn units distributed radially in a manner dictated by the mutual crowding of the arms. A core at the center has some radius r_c, composed of the linking groups and attached arm ends, but containing little solvent. Consider arms long enough to make $r_c/R \sim 0$. Beyond r_c, the arms fan out, but each is assumed to be confined by their mutual volume exclusion to its own conical channel—apex at the center and area of base at the surface $4\pi R^2/f$—as illustrated in Figure 6.26. Accordingly, cone width w is related to r, the axial distance from the apex, as at the surface:

$$w^2 = r^2/f \tag{6.74}$$

The trajectory of each arm, anchored at one end to the core, propagates radially along its conical channel, placing its free end finally on the surface. Because the cone widens with increasing distance from the core, the local concentration of solvent increases; $\nu(r)$, the average number density of arm units, decreases accordingly. The star radius $R(n, f)$ can be obtained if $\nu(r)$ is known, because conservation of monomeric units relates them:

$$fn = \int_0^R 4\pi r^2 \nu(r)\, dr \tag{6.75}$$

An estimate of $\nu(r)$ is made as follows: Well away from the core, the arm trajectories become self-avoiding walks in good solvents and random walks in theta solvents (the three-body interactions are assumed sufficient to provide confinement in the latter). From Eq. 6.3,

$$\sigma = l_\gamma z^\gamma \tag{6.76}$$

in which σ and z are chain size and length respectively, with $\gamma = 0.588$ for good solvents and $\gamma = 1/2$ at theta. Divide the cone into a sequence of sections, Δr_i, $i = 1, 2, \ldots$. Make the sections roughly spherical by choosing Δr_i equal to $w(r_i)$—the cone width at that location—so that the volume of section i is about $[w(r_i)]^3$. If section i is already far from the core, the piece of the arm it contains is self-dilute, and, using Eq. 6.76, the number of arm units in the section is about $(w(r_i)/l_\gamma)^{1/\gamma}$. The average density of units in section i is thus:

$$\nu(r_i) = \frac{[w(r_i)]^{(\frac{1}{\gamma}-3)}}{[l_\gamma]^{\frac{1}{\gamma}}} \tag{6.77}$$

Relating $w(r_i)$ to r_i using Eq. 6.74, then treating $v(r_i)$ as a continuous function and applying Eq. 6.75, yields:

$$fn \propto (R/l_\gamma)^{\frac{1}{\gamma}} f^{\frac{3\gamma-1}{2\gamma}} \tag{6.78}$$

Hence, with the size ratio defined as (R^2/R_{lin}^2), and for the size of linear chains having the same number of units given by $R_{lin} = l_\gamma(fn)^\gamma$, Eq. 6.78 leads to the Daoud-Cotton formula:

$$g \propto f^{1-3\gamma} \tag{6.79}$$

Thus, for many-arm star polymers, Eq. 6.79 predicts $g \propto f^{-0.764}$ in good solvents and $g \propto f^{-0.5}$ at the theta condition. The dashed lines in Figures 6.24 and 6.25, shifted arbitrarily to coincide with the values of g_{rw} at $f = 10$, represent those predictions. Both experiment and simulation at large f agree very well with the good solvent prediction. The large-f prediction for theta solvents, although not quite matching the experiment and simulation trends, is still a vast improvement over g_{rw}. Predictions of the Daoud-Cotton theory about the internal structure of many-arm star polymers in dilute solution have been investigated by neutron scattering[133].

Many-arm star properties. Various dilute solution properties of linear polymers, expressed as limiting size ratios, were discussed in Section 6.2.1. As summarized in Table 6.1, the values of R_t/R_g, R_v/R_g, and R_v/R_h are insensitive to the polymer species and, apart from R_t/R_g, are only mildly dependent on thermodynamic

TABLE 6.12 Size-related properties for symmetric stars in good and theta solvents[76,86,134–136]

	R_t/R_g	R_v/R_g		R_v/R_h		
f	Good	Good	Theta	Good	Theta	k_H
1,2	0.68	0.77	0.83	1.09	1.06	0.31
3	0.77	0.78	0.86	1.04	1.08	0.37
4	0.84	0.91	0.95	0.99	1.03	0.36
5	0.94	—	—	—	—	0.38
6	0.96	0.99	1.11	1.00	1.02	0.37
8	1.00	0.99	1.02	0.99	—	0.52
12	1.09	1.10	1.07	1.01	1.06	0.69
18	1.13	1.11	1.14	1.08	1.03	0.74
32	1.28	1.26	1.25	1.03	1.04	0.76
64	1.40	1.31	1.39	1.04	1.04	0.91
128	1.41	1.36	1.42	1.03	1.03	0.94

interactions. Speculations about the trends with arm number in symmetric stars, based on limited data ($f \leq 18$)[76], were corrected in an important series of works by Roovers and coworkers[134–136]. These researchers used anionic polymerization and the well-tested chlorosilane linking chemistry to synthesize families of star molecules with 18, 32, 64, and 128 arms, and they determined their dilute solution properties using a variety of methods. Table 6.12 contains the size ratios as well as values of the Huggins coefficient for well-defined star polymers in the range $3 \leq f \leq 128$.

The relative sizes change gradually with increasing f: The radius of gyration moves from outside to inside the thermodynamic radius, and from outside to inside the viscometric and hydrodynamic radii in both good and theta solvents. These changes reflect an increasing uniformity of mass density and molecular sphericity, resulting finally in the approach at large f to the size ratios characteristic of hard spheres. In good solvents, the variation with f is shown for R_t/R_g[76,86] in Figure 6.27 and for k_H from previously unpublished work by Roovers[137] in Figure 6.28. The dashed lines correspond to hard sphere values, $(5/3)^{1/2} = 1.291$ for R_t/R_g and 0.99 for k_H^{21}. (For reasons unknown, the Huggins coefficient is difficult to evaluate reliably when f is large, and the situation is worse at T_θ, where the values of k_H even in linear polymers are distressingly variable[137].)

Colloidal analogies. Interpreting many-arm stars as a variety of ultrasoft colloidal particles[138] is a natural consequence of the properties just described. It is therefore of some interest to compare the size ratio trends for many-arm stars with results obtained for dilute colloidal systems of various kinds. Colloids cover a truly vast area of both practical and fundamental interest, so large in fact that we can only include a few representative examples. Static and dynamic light scattering have been used to determine R_g and R_h for latex particles, κ-casein micelles, and microgels, some

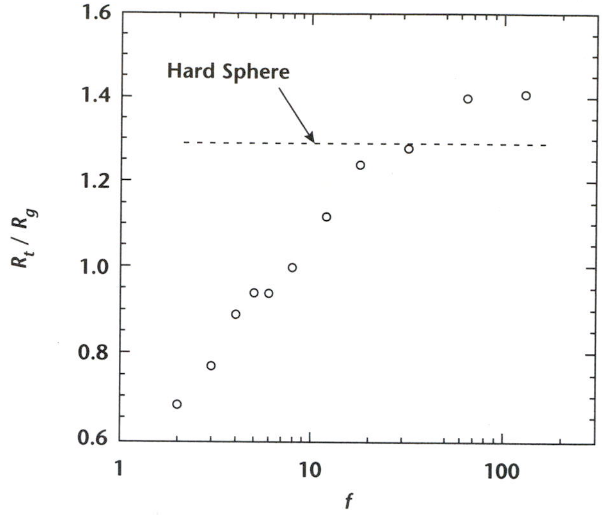

FIGURE 6.27 Ratio of thermodynamic size to physical size for multiarm stars in good solvents. Data from the compilations of Bauer et al.[76] and Grest et al[86].

with chains grafted to the surface[139,140]. For the lattices, $R_h/R_g = 1.29 \pm 0.02$ is found, nicely consistent with expectations for hard spheres. The ratio is smaller for casein micelles: R_h/R_g decreases from ~0.9 to ~0.4 as the concentration is reduced, the value of R_h becoming constant while R_g continues to grow. The ratio for linear polymers, $R_h/R_g = 0.765/1.09 = 0.70$, from the average of values for good solvents

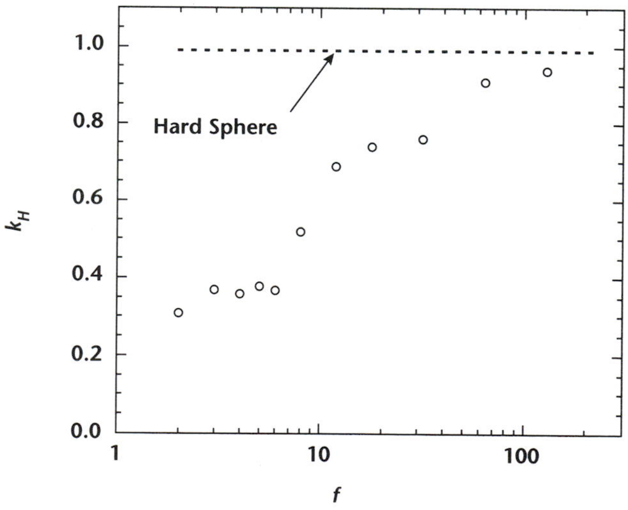

FIGURE 6.28 Huggins coefficient for multiarm stars in good solvents. Data from Roovers[137].

in Table 6.1, falls within the same range, which probably indicates a porous spherelike character for the casein micelles, with significant hydrodynamic penetration. The microgels behave in the opposite fashion, having values of R_h/R_g larger than those for hard spheres—in the range of 1.8 or even more—thus indicating an object with scattering power concentrated rather strongly near its center.

Light scattering has been used to study the aggregates in dilute solution formed by highly asymmetric diblock copolymers in a selective solvent[141]. In this case, the number of molecules per aggregate N was determined from the aggregate molecular weight. The sizes R_g and R_h, were also measured, and, in one aggregate at least, R_t was evaluated as well. One might expect N-arm starlike micelles to form in this case, and indeed the results are consistent with that picture. Thus, $R_h/R_g = 1.25$ was found for seventeen-molecule aggregates, and the value for the corresponding star (estimated from the data in Table 6.12) is about 1.1. Also, $R_h/R_g = 1.35$ was found for $N \sim 100$, whereas the estimated star polymer ratio is about the same. The ratios $R_h/R_g = 1.4$ and $R_t/R_h = 0.9$ were found for $N \sim 80$, with corresponding star polymer ratios of 1.3 and 1.1.

These few comparisons show that many-arm stars provide an interesting intermediate between linear polymers and colloidal aggregates in the dilute regime. How this role evolves at higher concentrations will be examined briefly in Chapter 7, where the liquidlike ordering of highly branched stars is discussed.

Other sorts of molecular or colloidal systems are macromolecular in size and structure. Microgels are formed typically in dilute solution or other dispersed media by the intramolecular crosslinking of polymer chains[142,143]. Microgels, polymacromers[144], dendrimers[145], arborescent (graft-on-graft) species[146], and various biological molecules and aggregates are frequently studied by the methods of dilute solution characterization discussed here; these are vast subjects in their own right and beyond the scope of this book.

6.4 Concentration Regimes

Various means were used in Sections 5.1 and 5.2 to locate c^*, the overlap concentration for polymer solutions. Most were chosen pragmatically, merely by setting the limiting contribution in a concentration expansion to the interactions contribution. The A_2-based Eq. 5.7 and Eq. 5.38, the k_D-based Eq. 5.59, and the $k_H\,[\eta]$-based Eq. 5.65 were obtained in this way. The R_g-based Eq. 6.9 of c^* was obtained differently, as a space-filling concentration for the pervaded volumes. In fact, it is an easy exercise to show all the various definitions are equivalent, within some multiplier of order unity, to various choices of coil size measures R_g, R_t, R_v, or R_h. in Eq. 6.9. Based on Tables 6.1 and 6.12 and on Figures 6.16 through 6.20, and excepting R_t-based values near T_θ, the various estimates of c^* for a given architecture and solvent power do not differ greatly from one another. The choice in any case is arbitrary when dealing with nearly monodisperse polymers, but some differences exist in practical terms, especially for polydisperse systems.

The interaction term from viscometry has been widely used, usually without the Huggins coefficient[147]:

$$c^* = \frac{1}{[\eta]} \tag{6.80}$$

Although based on a dynamic property, this definition has three significant advantages: It is easy to measure accurately, it is relatively insensitive to solute polydispersity, and it does not require the solute molecular weight or even the solute species to be known. As a practical matter, insistence upon always using an R_g-based estimate of c^*, even as a reducing variable for simply organizing concentration-dependent data, is bound to become very inconvenient. It is difficult to measure R_g accurately over wide ranges of size, and its value when obtained properly through the use of a static scattering experiment is extremely sensitive to solute polydispersity. The A_2-based definition of c^* has been used as a reducing variable for osmotic pressure and scattering behavior in good solvents[33,148]. It has the convenience that both A_2 and M are obtainable from the same experiment. When correlating equilibrium properties, $c^* = (A_2 M)^{-1}$ also offers the comfort of a definition based itself on an equilibrium property. In other studies, however, it suffers the disadvantage of being not generally available except from scattering or osmometry, neither an easy experiment, and it is not useful at all near T_θ.

Some time has been devoted here to discussing the overlap concentration because it does more than simply define the extent of the dilute regime for experimental purposes. It also plays an important role as a reducing parameter for the concentration and molecular weight dependence of properties at higher concentrations. Two regions of polymer concentration can be usefully distinguished: moderately concentrated and highly concentrated[149,150]. The moderately concentrated region has also been dubbed the *semidilute regime*[8,151], a useful term adopted here as well. The upper end of the highly concentrated regime is the undiluted polymeric liquid.

Beyond the overlap concentration, the presence of other chains in the pervaded volumes has the effect of screening out the intramolecular excluded volume and hydrodynamic interactions that dominate in the dilute regime. The molecules eventually approach their unperturbed dimensions at some concentration $c^\ddagger$. For purposes here, $c^\ddagger$ locates the crossover from semidilute to concentrated behavior. The dynamic properties of polymer solutions appear to approach unshielded behavior in the same range of concentrations, a matter for consideration in Volume 2. The overlap of pervaded volumes also permits molecular entanglement, an intermolecular interaction arising from the mutual uncrossability of chain backbones and one which strongly affects dynamical properties.

At concentrations beyond $c^{\ddagger}$, chain entanglement and a composition-dependent time scale for local motion increasingly dominate the behavior. Uncrossibility and entanglement, as noted already in the topological interaction of polymeric rings (Section 6.3.1), can contribute substantially to the equilibrium behavior of networks, discussed in Chapters 9 and 10. Chapters 7 and 8 deal with the equilibrium properties of polymeric liquids beyond overlap, including screening and scaling ideas and the phase behavior of polymer solutions and blends.

REFERENCES

1. Li B., N. Madras, and A.D. Sokal. 1995. *J. Stat. Phys.* 80:661.
2. Le Guillou J.C. and J. Zinn Justin. 1989. *J. Physique* 50:1365.
3. Hayward R.C. and W.W. Graessley. 1999. *Macromolecules* 32:3502.
4. Flory P.J. 1949. *J. Chem. Phys.* 17:268.
5. Casassa E.F. and Berry G.C. 1989. *Comprehensive Polymer Science* (Vol. 2), Allen G. and Bevington J.C. (eds). Oxford, UK: Pergamon, p. 71.
6. Hirschfelder J.O., Curtiss C.F., and Bird R.B. 1954. *Molecular Theory of Liquids and Gases,* New York: Wiley.
7. Flory P.J. 1953. *Principles of Polymer Chemistry,* Ithaca, NY: Cornell University Press.
8. de Gennes P.-G. 1979. *Scaling Concepts in Polymer Physics,* Ithaca, NY: Cornell University Press.
9. Yamakawa H. 1971. *Modern Theory of Polymer Solutions,* New York: Harper & Row.
10. Stockmayer W.H. 1960. *Makromol. Chem.* 35:54.
11. Matsumoto T., N. Nishioka, and H. Fujita. 1972. *J. Polym. Sci.: Pt. A-2* 10:23.
12. Kurata M. and Tsunashima Y. 1989. *Polymer Handbook,* 3rd ed., Brandrup J. and Immergut E.H. (eds.). New York: Wiley-Interscience p. VII-1.

12a. Xu Z., N. Hadjichristidis, J.M. Carella, and L.J. Fetters. 1983. *Macromolecules* 16:925.

13. Sundararajan P.R. 1996. *Physical Properties of Polymers Handbook,* Mark J.E. (ed.). Woodbury, NJ: American Institute of Physics, p.197.
14. Batchelor G.K. 1967. *Introduction to Fluid Mechanics,* Cambridge: Cambridge University Press.
15. Oseen C.W. 1927. *Hydrodynamik,* Leipzig: Acad. Publ. Co.
16. Kirkwood J.G. and J. Riseman. 1948. *J. Chem. Phys.* 16:565.
17. Zimm B.H. 1956. *J. Chem. Phys.* 24:269.
18. Doi M. and Edwards S.F. 1986. *The Theory of Polymer Dynamics,* Oxford: Oxford University Press.
19. Kuhn W. 1934. *Kolloid Z.* 68:2.
20. Einstein A. 1906. *Investigations on the Theory of the Brownian Movement,* New York: Dover Publications (1956 publication of 1906 doctoral dissertation, translated and edited by R. Furth , 1926).
21. Batchelor G.K. 1977. *J. Fluid Mech.* 83:97.
22. Fox T.G. and P.J. Flory. 1949. *J. Phys. Colloid Chem.* 53:197.
23. Fujita H. 1990. *Polymer Solutions,* Amsterdam: Elsevier.
24. Freed K. 1987. *Renormalization Group Theory of Macromolecules,* New York: Wiley.
25. Fujita H. 1988. *Macromolecules* 21:179.

26. Yamakawa H. 1997. *Helical Wormlike Chains in Polymer Solutions,* New York: Springer-Verlag.
27. Oono Y. 1985. *Adv. Chem. Phys.* 61:301.
28. Rey A., J.J. Freire, and J.G. de la Torre. 1987. *Macromolecules* 20:342.
29. Sokal A.D. 1995. *Monte Carlo and Molecular Dynamics Simulations in Polymer Science,* Binder K. (ed.). New York: Oxford University Press.
30. Eisenberg H. 1976. *Biological Macromolecules and Polyelectrolytes in Solution,* Oxford, UK: Oxford University Press.
31. Akcasu A.Z. and M. Benmouna. 1978. *Macromolecules* 11:1193.
32. Roovers J. and P.M. Toporowski. 1980. *J. Polym. Sci., Polym. Phys. Ed.* 18:1907.
33. Burchard W. 1983. *Adv.Polym. Sci.* 48:1.
34. Zimm B.H. 1980. *Macromolecules* 13:592.
35. Stockmayer W.H. and M. Fixman. 1953. *Ann. N. Y. Acad. Sci.* 57:334.
36. Fixman M. 1981. *Macromolecules* 14:1706.
37. Fixman M. 1981. *Macromolecules* 14:1710.
38. Fixman M. 1983. *J. Chem. Phys.* 78:1588.
39. Zimm B.H. 1984. *Macromolecules* 17:795.
40. Zimm B.H. 1984. *Macromolecules* 17:2441.
41. Oono Y. 1983. *J. Chem. Phys.* 79:4629.
42. Douglas J.F. and K. Freed. 1984. *Macromolecules* 17:1854.
43. Fixman M. 1955. *J. Chem. Phys.* 23:1656.
44. Tanaka G. and K. Solc. 1982. *Macromolecules* 15:791.
45. Stockmayer W.H. and C.A. Albrecht. 1958. *J. Polym. Sci.* 32:215.
46. Shimada J. and H. Yamakawa. 1978. *J. Polym. Sci.:Polym. Phys. Ed.* 16:1927.
47. Domb C. and A.J. Barrett. 1976. *Polymer* 17:179.
48. Barrett A.J. 1984. *Macromolecules* 17:1561.
49. Barrett A.J. 1984. *Macromolecules* 17:1566.
50. Barrett A.J. 1985. *Macromolecules* 18:196.
51. Miyaki Y. and H. Fujita. 1981. *Macromolecules* 14:742.
52. Graessley W.W., R.C. Hayward, and G.S. Grest. 1999. *Macromolecules* 32:3510.
53. Domb C. 1963. *J. Chem. Phys.* 38:2957.
54. Wall F.T. and F.T. Hioe. 1970. *J. Phys. Chem.* 74:4416.
55. Zimm B. and W.H. Stockmayer. 1949. *J. Chem. Phys.* 17:1301.
56. Cozewith C., W.W. Graessley, and G. Ver Strate. 1979. *Chem. Eng. Sci.* 34:245.
57. Lusignan C.P., T.H. Mourey, J.C. Wilson, and R.H. Colby. 1995. *Phys. Rev. E* 52:6271.
58. Lusignan C.P., T.H. Mourey, J.C. Wilson, and R.H. Colby. 1999. *Phys. Rev. E* 60:5657.
59. Beasley J.K. 1953. *J. Am. Chem. Soc.* 75:6123.
60. Graessley W.W. 1965. *Proc. A.I.CH.E. -I. Chem. E. Joint London Meet.* 3:16.
61. Chatterjee A., W.S. Park, and W.W. Graessley. 1977. *Chem. Eng. Sci.* 32:167.
62. Graessley W.W., H. Mittelhauser, and R. Maramba. 1965. *Makromol. Chem.* 86:129.
63. Tobita H. 1996. *Macromol. Theory Simul.* 5:129.
64. Datta S. and Lohse D.J. 1996. *Polymeric Compatibilizers,* Munich: Hanser Publishers.
65. Macosko C.W. and D.R. Miller. 1976. *Macromolecules* 9:199.
66. Janzen J. and R.H. Colby. 1999. *J. Mol. Struct.* 485–486:569.
67. Shroff R.N. and H. Mavridis. 1999. *Macromolecules* 32:8454.
68. Hogan J.P., C.T. Levett, and R.T. Werkman. 1967. *SPE J.*, 87.
69. Dotson N.A., Galvan R., Laurence R.L., and Tirrell M. 1996. *Polymerization Process Modeling,* New York: VCH.

70. Small P.A. 1975. *Adv. Polym. Sci.* 18:1.
71. Morton M. and L.J. Fetters. 1975. *Rub. Chem. Tech.* 48:359.
72. Roovers J. 1985. *Encycl. Polym. Sci. Eng.* (Vol. 2) Mark H. (ed.). New York: Wiley-Interscience p.478.
73. Hadjichristidis N., et al. 2000. *Macromolecules* 33:2424.
74. Roovers J. 1999. In *Star and Hyperbranched Polymers,* Mishra M.K. and Kobayashi S. (eds.). New York: Marcel Dekker p. 285.
75. Hadjichristidis N. and J.E.L. Roovers. 1974. *J. Polym. Sci. Polym. Phys. Ed.* 12:2521.
76. Bauer B.J., L.J. Fetters, W.W. Graessley, N. Hadjichristidis, and G.F. Quack. 1989. *Macromolecules* 22:2337.
77. Kramers H.A. 1946. *J. Chem. Phys.* 14:415.
78. Zimm B.H. and R.W. Kilb. 1959. *J. Polym. Sci.* 37:19.
79. Batoulis J. and K. Kremer. 1988. *Europhys. Lett.* 7:683.
80. Batoulis J. and K. Kremer. 1989. *Macromolecules* 22:4277.
81. Grest G.S. 1994. *Macromolecules* 27:3493.
82. Ohno K., K. Shida, M. Kimura, and Y. Kawazoe. 1996. *Macromolecules* 29:2269.
83. Shida K., K. Ohno, M. Kimura, Y. Kawazoe, and Y. Nakamura. 1998. *Macromolecules* 31:2343.
84. Roovers J., N. Hadjichristidis, and L.J. Fetters. 1983. *Macromolecules* 16:214.
85. Douglas J.F., J. Roovers, and K.F. Freed. 1990. *Macromolecules* 23:4168.
86. Grest G.S., L.J. Fetters, J.S. Huang, and D. Richter. 1996. *Adv. Chem. Phys.* 94:67.
86a. Chen S.J., G.C. Berry, and D.J. Plazek. 1995. *Macromolecules* 28:6539.
87. Comanita B., B. Noren, and J. Roovers. 1999. *Macromolecules* 32:1069.
88. Bloomfield V.A. and B.H. Zimm. 1966. *J. Chem. Phys.* 44:315.
89. Bruns W. and J. Naghizadeh. 1976. *J. Chem. Phys.* 65:747.
90. Chen Y. 1983. *J. Chem. Phys.* 78:5192.
91. Dodgson K. and J.A. Semlyen. 1971. *Polymer* 18:1265.
92. Semlyen J.A. 1996. *Large Ring Molecules,* New York: Wiley.
93. Geiser D. and H. Höcker. 1980. *Macromolecules* 13:653.
94. Hild G., A. Kohler, and P. Rempp. 1980. *Eur. Polym. J.* 16:525.
95. Roovers J. and P.M. Toporowski. 1983. *Macromolecules* 16:843.
96. Hild G., C. Strazielle, and P. Rempp. 1983. *Eur. Polym. J.* 19:721.
97. Roovers J. and P.M. Toporowski. 1988. *J. Polym. Sci., Part B: Polym. Phys.* 26:1251.
98. Roovers J. 1985. *J. Polym. Sci.: Polym. Phys. Ed.* 23:1117.
99. Hadziioannou G., P.M. Cotts, G. ten Brinke, C.C. Han, P. Lutz, C. Strazielle, P. Rempp, and A.J. Kovacs. 1987. *Macromolecules* 20:493.
100. McKenna G.B., G. Hadziioannou, P. Lutz, G. Hild, C. Strazielle, C. Straupe, P. Rempp, and A.J. Kovacs. 1987. *Macromolecules* 20:498.
101. Frank-Kamenetskii M.D., A.V. Lukashin, and A.V. Vologodskii. 1975. *Nature* 258:398.
102. Iwata K. 1985. *Macromolecules* 18:115.
103. Schill G. 1971. *Catenanes, Rotaxanes and Knots,* New York: Academic Press.
104. Raymo F.M. and J.F. Stoddart. 1999. *Chem. Rev.* 99:1643.
105. Roovers J. and P.M. Toporowski. 1981. *Macromolecules* 14:1174.
106. Casassa E.F. and G.C. Berry. 1966. *J. Polym. Sci. Pt.* A-2 4:881.
107. Lipson J.E.G., D.S. Gaunt, M.K. Wilkinson, and S.G. Whittington. 1987. *Macromolecules* 20:186.
108. Berry G.C., L.M. Hobbs, and V.C. Long. 1965. *Polymer* 6:31.
109. Berry G.C. 1971. *J. Polym. Sci., Part* A-2 9:687.

110. Noda I., T. Horikawa, T. Kato, T. Fujimoto, and M. Nagasawa. 1970. *Macromolecules* 3:795.
111. Roovers J.E.L. 1975. *Polymer* 16:827.
112. Roovers J. 1979. *Polymer* 20:843.
113. Roovers J. and P.M. Toporowski. 1987. *Macromolecules* 20:2300.
114. Kurata M. and M. Fukatsu. 1964. *J. Chem. Phys.* 41:2934.
115. Osaki K., Y. Mitsuda, J.L. Schrag, and J.D. Ferry. 1974. *Trans. Soc. Rheol.* 18:395.
116. Berry G.C. 1968. *J. Polym. Sci.: Pt. A-2* 6:1551.
117. Berry G.C. 1988. *J. Polym. Sci., Part B* 26:1137.
118. McCrackin F.L. and J. Mazur. 1981. *Macromolecules* 14:1214.
119. Lipson J.E.G. 1991. *Macromolecules* 24:1327.
120. Weissmüller M. and W. Burchard. 1997. *Polym. Internat.* 44:380.
121. Cotts P.M. 2001. Private communication.
122. Scholte T.G. 1993. In *Developments in Polymer Characterization-4,* Dawkins D.V. (ed.). Oxford: Applied Science Publishers.
123. Striegel A.M. and M.R. Krejsa. 2000. *J. Polym. Sci.: Pt. B: Polym. Phys.* 38:3120.
124. Tobita H. and N. Hamashima. 2000. *J. Polym. Sci.: Pt. B: Polym. Phys.* 38:2009.
125. Schosseler F., H. Benoit, Z. Grubisic-Gallot, and C. Strazielle. 1989. *Macromolecules* 22:400.
126. Roovers J.E.L. and S. Bywater. 1974. *Macromolecules* 7:443.
127. Candau F., P. Rempp, and H. Benoit. 1972. *Macromolecules* 5:672.
128. Khasat N., R.W. Pennisi, N. Hadjichristidis, and L.J. Fetters. 1988. *Macromolecules* 21:1100.
129. Bauer B.J., N. Hadjichristidis, L.J. Fetters, and J.E.L. Roovers. 1980. *J. Am. Chem. Soc.* 102:2410.
130. Horton J.C., G.L. Squires, A.T. Boothroyd, L.J. Fetters, A.R. Rennie, C.J. Glinka, and R.A. Robinson. 1989. *Macromolecules* 22:681.
131. Boothroyd A.T., G.L. Squires, L.J. Fetters, A.R. Rennie, J.C. Horton, and A.M.B.G. De Vallera. 1989. *Macromolecules* 22:3130.
132. Daoud M. and J.P. Cotton. 1982. *J. Phys.* 43:531.
133. Willner L., O. Jucknischke, D. Richter, J. Roovers, L.L. Zhou, P. M. Toporowski, L.J. Fetters, J.S. Huang, M.Y. Lin, and N. Hadjichristidis. 1994. *Macromolecules* 27:3821.
134. Roovers J., P. Toporowski, and J. Martin. 1989. *Macromolecules* 22:1897.
135. Zhou L.L., N. Hadjichristidis, P.M. Toporowski, and J. Roovers. 1992. *Rub. Chem. Tech.* 65:303.
136. Roovers J., L.L. Zhou, P.M. Toporowski, M. van der Zwan, H. Iatrou, and N. Hadjichristidis. 1993. *Macromolecules,* 26:4324–4331.
137. Roovers J. 1999. Private communication.
138. Likos C.N., H. Löwen, M. Watzlawek, B. Abbas, O. Jucknischke, J. Allgaier, and D. Richter. 1998. *Phys. Rev. Lett.* 80:4450.
139. Kunz D., A. Thurn, and W. Burchard. 1983. *Coll. Polym. Sci.* 261:635.
140. Kunz D. and W. Burchard. 1986. *Coll. Polym. Sci.* 264:498.
141. Vagberg L.J.M., K.A. Cogan, and A.P. Gast. 1991. *Macromolecules* 24:1670.
142. Allen G., J. Burgess, S.F. Edwards, and D.J. Walsh. 1973. *Proc. R. Soc. London* A 334:477.
143. Antonietti M. and C. Rosenauer. 1991. *Macromolecules* 24:3434.
144. Ito K. and S. Kawaguchi. 1999. *Adv. Polym. Sci.* 142:129.
145. Roovers J. and B. Comanita. 1999. *Adv. Polym. Sci.* 142.

146. Hempenius M.A., W. Michelberger, and M. Möller. 1997. *Macromolecules* 30:5602.
147. Frisch H.L. and Simha R. 1956. *Rheology* (Vol. 1). Eirich F.R., New York: Academic Press Ch.14.
148. Adam M., L.J. Fetters, W.W. Graessley, and T.A. Witten. 1991. *Macromolecules* 24:2434.
149. Berry G.C. and T.G. Fox. 1968. *Adv. Polym. Sci.* 5:261.
150. Graessley W.W. 1974. *Adv. Polym. Sci.* 16:1.
151. Daoud M., J.P. Cotton, B. Farnoux, G. Jannink, G. Sarma, H. Benoit, C. Duplessix, C. Picot, and P.G. De Gennes. 1975. *Macromolecules* 8:804.

CHAPTER 7

Polymer Solutions

This chapter begins by considering the changes with concentration in coil size, osmotic pressure, and scattering behavior in the semidilute region. Excluded volume screening and its effect on coil size, leading finally to unperturbed dimensions at the concentrated regime crossover, are discussed together with the attempts to document those changes experimentally. Predictions of osmotic pressure in both good and theta solvents by screening–scaling methods and by the simple mixture models are compared with experiment. Two interpretations of the scattering correlation length are considered, as screening length and as simple mixture consequence, with evidence seeming to favor the latter. Various methods for determining the simple mixture interaction parameter are then introduced; its domination in polymer solutions by free volume effects, including even behavior in the dilute limit, is described. Liquid–liquid phase behavior and scattering properties near the critical region are considered for simple mixtures and also in light of the nonclassical critical behavior of polymer solutions. The chapter closes with an explanation of the Ginzburg criterion for nonclassical critical behavior as applied to simple mixtures, demonstrating that polymer blends, the subject of Chapter 8, can in fact behave classically in the critical region.

Chain conformation and intermolecular interactions play major roles in determining the properties of polymeric liquids. The direct measurement of chain dimensions at concentrations beyond coil overlap became possible in the early 1970s with the advent of small-angle neutron scattering[1–4]. The results cleared away a variety of conflicting ideas about coil size at high concentrations[5]. In 1974, Cotton et al. gave a particularly convincing demonstration that the chains in polymer melts have about the same size as those in dilute solutions at the theta condition[6]. These researchers also demonstrated a more detailed result: that the form factor for chains in the melt is well described by the Debye equation [Eq. 4.101] down to small distances and hence that the distribution of large-scale conformations is Gaussian [Eq. 4.68]. Following a brief discussion of chain dimensions in the melt state, the progressive screening out of intramolecular volume-exclusion effects with increasing concentration beyond c^*, and the effect of screening on chain dimensions, osmotic pressure, and some aspects of scattering behavior, is described in Sections 7.1 through 7.3. Section 7.4 deals with the measurement and interpretation of the interaction parameter χ in polymer solutions—application of Flory-Huggins theory and free volume effects. Section 7.5 treats liquid–liquid phase equilibrium—upper and lower critical behavior scattering near criticality and the Ginzburg criterion.

7.1 Coil Size beyond Overlap

The size of polymer molecules has a direct and very strong influence on the dynamics and rheology of polymeric liquids. In addition, a new type of interaction, molecular entanglement, becomes important as concentration increases and the coils begin to overlap extensively. Though relatively unimportant for equilibrium properties, entanglement dominates dynamic behavior. Knowing how coil dimensions change with concentration is essential information for understanding the impact of entanglement on properties from the coil overlap region to the undiluted state. As described in this section, however, the relationship between coil size and concentration is not a fully settled matter.

7.1.1 Chain Dimensions in Polymer Melts

As an adjunct to his theory of excluded volume in dilute solutions, Flory argued that chain conformations in the melt state would be unaffected by volume exclusion[7]. He noted that the excluded volume parameter z [Eq. 6.29] can be expressed in the form $z = Kn^{1/2}/p$, n being the number of monomeric units per chain, p the ratio of solvent to monomeric unit volume υ_s/υ_m and K a parameter independent of solvent size. Equation (6.30) can thus be expressed as:

$$\alpha^5 - \alpha^3 = Kn^{1/2}/p \tag{7.1}$$

Each chain in the melt state is in a "solvent" of chains, so $p = n$ and $\alpha^5 - \alpha^3 = K/n^{1/2}$. Accordingly, $\alpha = 1$ for $n \gg 1$, and hence $R_g = (R_g)_\theta$ in the melt.

Flory also offered a rather nice physical argument against coil expansion in the melt state. In dilute solutions, the entropy penalty from coil expansion is compensated by a reduction in the excluded volume repulsion because the local concentration of chain units is reduced. In the melt, however, the situation is different. The pervaded volume and surroundings of each chain are filled with chain units. No offset occurs for the coil expansion penalty in melts by reduction in local concentration, because the concentration of chain units is spatially uniform. Hence, no incentive for expansion exists, so the effect of intramolecular excluded volume interactions is nullified.

A reduction in coil dimensions brought about by an increase in the molar volume of solvent, as suggested by Eq. 7.1, is consistent with various observations on dilute

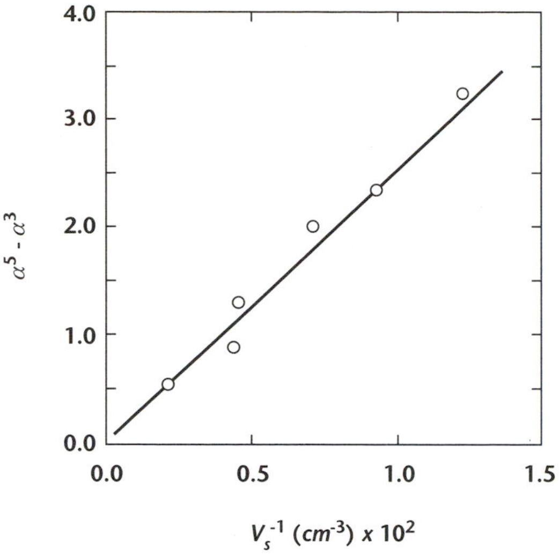

FIGURE 7.1 Relationship between coil expansion and molar volume of the solvent (from Colby et al.[10]).

solutions. For example, Kirste and Lehnen[8] used SANS to measure R_g for long polydimethylsiloxane chains in polydimethylsiloxane diluents of various chain lengths and found trends that are qualitatively consistent with Eq. 7.1. McAdams and Williams[9] measured $[\eta]$ for polystyrene in various mixtures of good solvent with oligomeric polystyrene and found a rapid decrease toward $[\eta]_\theta$ with increasing oligomer size. Colby et al. found that the intrinsic viscosity of 1,4 polybutadiene decreases progressively with υ_s for a series of apparently similar good solvents[10]. Indeed, Figure 7.1 shows that the combination $\alpha^5 - \alpha^3$ is inversely proportional to υ_s/υ_m, as suggested by Eq. 7.1. Landry recently conducted an extensive study of diluent size effects on coil dimensions in polystyrene systems[11], finding some differences from Eq. 7.1 but overall consistency with the predicted trends.

It should be noted that the cancellation of the coil expansion effect of volume exclusion owing to the competition of other chains differs in principle from the thermodynamic cancellation at the theta condition. Only the two-body interactions are cancelled at T_θ, but interactions of all orders are cancelled by intermolecular competition in melts. To retain the distinction, denote size at the theta condition $(R_g)_\theta$ and size at full excluded volume cancellation $(R_g)_0$. The result for linear chains is essentially the same: The two-body term dominates because the self-concentration is small, so $(R_g)_\theta = (R_g)_0$. In general, however, the two can differ—in multiarm star polymers, for example, if the self-concentration is large (see Section 6.3.3).

In the early years of neutron scattering, chain dimensions in the melt were determined in the manner of the light scattering investigations of size in dilute solutions[1–4,12]. Thus, scattering profiles for several dilute solutions of deuterated chains in a hydrogenous chain matrix of the same species (or the reverse) were extrapolated to zero polymer concentration to obtain R_g (Section 5.1.2). It was soon discovered that higher concentrations of labeled species could be used to obtain melt state dimensions, and that extrapolation was for most purposes an unnecessary refinement. Careful experiments at a single concentration suffice if the thermodynamic interactions between labeled and nonlabeled species are weak enough[13]. The other procedure is still used occasionally to obtain R_g, but now mainly as an extra precaution against isotopic interaction effects or when the greatest possible precision is sought[14]. The basis for the new analysis, which is now used as well to determine the thermodynamic interactions of the components in polymer blends, is explained briefly in Section 7.3.2. Values of C_∞ and κ for many species in the undiluted liquid state are now available[14–17], as are comparisons of SANS data with predictions based on the molecular scattering functions for linear chains and stars[18]. Table 7.1 compares values of C_∞ obtained by SANS data for melts and by dilute solution light scattering at T_θ[19]. Considering the disparate nature of the experimental conditions, the agreement is very good indeed.

TABLE 7.1 Comparison of chain dimensions in the undiluted state with values from dilute solution measurements[14–19]

		$R_g/M^{1/2}$ (nm) $\times 10^2$	
Polymer	**T(°C)**	**Dilute Solution**	**Melt State**
polyethylene	140	4.4	4.55
1,4 polybutadiene	25	3.8	3.8
polyethylene oxide	74	2.6–5.2	3.75
1,4 polyisoprene	34	3.3	3.15
poly(ethylene alt-propylene)	27	3.8	3.7
a-polypropylene	38	3.3	3.35
i-polypropylene	183	3.05	3.4
hh-polypropylene	57	3.25	3.4
poly(ethylene alt-1-butene)	25	3.3	3.45
polyisobutylene	25	3.0	3.1
a-poly 1-butene	25	2.8	2.75
a-polyvinylchloride	25	3.5	4.0
Polydimethylsiloxane	25	2.75	2.7
a-polymethylmethacrylate	45	2.35	2.65
i-polymethylmethacrylate	40	2.85	2.9
s-polymethylmethacrylate	72	2.45	2.7
a-polystyrene	35	2.8	2.7
i-polystyrene	25	3.6	2.6
a-polycyclohexylethylene	25	2.35	2.35

7.1.2 Chain Dimensions in Semidilute Solutions

As discussed in Chapter 6, chain dimensions and osmotic pressure in the dilute region depend on particulars of the solution thermodynamics through a species-dependent excluded volume interaction, intramolecular for molecular size and intermolecular for the osmotic pressure. The size of chains in the melt is unperturbed, and the size of dilute chains in a good solvent is expanded. Obviously, the size must contract with increasing concentration. The concentration dependence of polymer size is in fact a property of major concern. The dynamic properties of polymeric liquids at all levels of concentration are strongly dependent on coil size, and it is crucial when considering such properties to know the size and how it varies with concentration. Beyond some region of concentrations, the size will have essentially reached its unperturbed value. To discuss solution dynamics in a meaningful way, it is important to know where that happens.

As introduced in Section 6.4, let $\phi^{\ddagger}$ be the crossover concentration from the *semidilute regime,* where R_g for a polymer may vary in a significant way with concentration, to the *concentrated regime,* where R_g has essentially reached its unperturbed value and no longer varies significantly with concentration:

$$R_g \sim \left(R_g\right)_{\text{o}} \qquad \phi > \phi^{\ddagger} \tag{7.2}$$

Changes of R_g with concentration must have counterparts in other thermodynamic properties such as the osmotic pressure[20], so a general theory of excluded volume effects in semidilute solutions would be quite useful. Unfortunately, theory with the level of rigor possible for infinite dilution becomes prohibitively difficult at higher concentrations[21,22]; indeed, even at concentrations still below the overlap concentration ϕ^*. A more qualitative approach, based on screening ideas and scaling relationships[23–25], was developed in the mid 1970s and has provided a useful framework for organizing the data that pertain to volume exclusion effects. The method and its application to $R_g(M, \phi)$ is explained in the next paragraph. Application to osmotic pressure $\pi(M, \phi)$ is described in Section 7.2.

Excluded volume screening. Consider a solution of long polymer chains that is well above the overlap concentration ($\phi^* \ll \phi$). Select some representative monomeric unit on one of the chains and consider the monomeric units, each of volume υ_m, contained within a sphere of volume δV centered on that unit. Choose the sphere radius to be much smaller than the coil radius in the dilute limit $R_g(0)$ and much larger than the monomer size:

$$\delta V = \frac{4\pi r^3}{3} \qquad \upsilon_m^{1/3} \ll r \ll R_g(0) \tag{7.3}$$

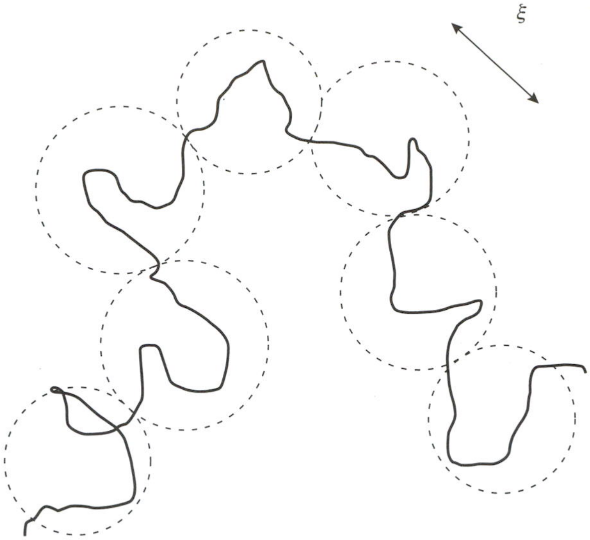

FIGURE 7.2 Correlation volume partitioning for a polymer coil.

As indicated by Figure 7.2, the sphere necessarily contains some length of the chain itself, a segment that has $n_s(r)$ units on average, and a self-concentration as given by Eq. 6.11:

$$\phi_{self} = \frac{n_s v_m}{\delta V} \tag{7.4}$$

The sphere also contains monomeric units from other chains. Estimate the average number of such units from the global concentration of the solution:

$$n_o = \frac{\delta V \phi}{v_m} \tag{7.5}$$

Consider now the situation when the segmental self-concentration is much larger than the global concentration ($\phi_{self} \gg \phi$ or equivalently $n_s \gg n_o$). The environment of the chain segment is about the same as it is in the dilute limit, so the excluded volume interactions among its various chain units are about the same. Accordingly, the relationship between segment length n_s and segment size $\langle r^2 \rangle^{1/2} = r$ is about the same as it would be in dilute solution. Thus, for long enough segments, from Eq. 6.3,

$$r = (n_s)^{\gamma} \ell \tag{7.6}$$

in which $\gamma \sim 3/5$ in the good solvent limit, $\gamma = 1/2$ at the theta condition, and ℓ is some microscopic length that depends only on the polymer–solvent system. Using

Eqs. 7.3 through 7.6, the segmental self-concentration can be expressed in terms of the sphere radius:

$$\phi_{self} = \left(\frac{3\upsilon_m}{4\pi \ell^{1/\gamma}}\right) r^{-(3-1/\gamma)} \tag{7.7}$$

The term in parentheses is independent of r, so:

$$\begin{aligned} &\phi_{self} \propto r^{-4/3} \qquad \text{(good solvent)} \\ &\phi_{self} \propto r^{-1} \qquad \text{(theta condition)} \end{aligned} \tag{7.8}$$

Equations (7.7) and (7.8) apply only when segmental self-concentration is much larger than the solution concentration. Because ϕ_{self} decreases with increasing radius r, a crossover must eventually occur as radius increases, corresponding to $\phi_{self} \sim \phi$ or $n_s \sim n_o$, at some radius ξ. The excluded volume interactions of the chain segment become increasingly intermolecular as the distance scale grows, and the intermolecular contribution dominates for distances much larger than ξ. In the Flory language, the coil-expanding effect of intramolecular excluded volume is progressively cancelled by the increasing multitude of intermolecular contributions with increasing concentration.

The term used to describe this effect, "excluded volume screening," was introduced in 1966 by Edwards[26]. It was used by Daoud et al. in 1975 to describe the concentration dependence of chain dimensions[23]. The distance ξ defines the range over which the excluded volume interaction is effective, or the *screening length* in the solution.

The Debye-Hückel theory of strong electrolyte solutions[27] provides a fully developed example of interaction screening. Consider a dilute aqueous solution of the salt of some monovalent strong acid and strong base such as NaCl. The salt is fully dissociated, giving equal numbers of positive and negative ions. The electrostatic potential energy for any i, j pair is inversely proportional to their separation distance:

$$\psi(r_{ij}) = \Lambda / r_{ij} \tag{7.9}$$

in which $\Lambda = \pm 1.13 \times 10^{11} e^2/\varepsilon$, where e is the electronic charge, -1.6×10^{-19} Coulomb; ε is 78.5, the dielectric constant of water at 25°C; and + or − applies respectively to pairs with the same or opposite charges.

Like charges repel and opposite charges attract, so the neighborhood of any ion tends to have a reduced concentration of like-charged species and an enhanced concentration of oppositely charged species. More generally, the ions distribute themselves spatially in such a way as to minimize the solution free energy, thus compensating the loss of configurational entropy by a reduction of electrostatic potential energy. Debye and Hückel solved the free energy minimization problem for the dilute range, finding that the free energy in such a case is expressed in terms of a *screened potential:*

$$\psi(r_{ij}) = \Lambda \frac{\exp(-r_{ij}/\xi_D)}{r_{ij}} \tag{7.10}$$

in which Λ/r_{ij} is the unscreened or *bare* potential given by Eq. 7.9, and ξ_D is a *screening length,* whose value for 1-1 salts is:

$$\xi_D = \left[\frac{2.82 \times 10^{-12} \varepsilon kT}{e^2[M_s]}\right]^{1/2} \tag{7.11}$$

in which $[M_s]$ is the molar concentration of salt.

The exponential term in Eq. 7.10 describes a crossover from full effect of each pair interaction when the separation is much smaller than ξ_D to no effect when the separation is much larger. For estimation purposes, the exponential can be approximated by a step function:

$$\begin{aligned} \exp(-r_{ij}/\xi_D) &= 1 \qquad r_{ij} < \xi_D \\ \exp(-r_{ij}/\xi_D) &= 0 \qquad r_{ij} > \xi_D \end{aligned} \tag{7.12}$$

For 1-1 electrolytes in water at 25°C, $\xi_D(nm) = 0.304/[M_s]^{1/2}$. Thus, for example, $\xi_D = 2.3\,nm$ for a sodium chloride concentration of 10^{-3} $g\,cm^{-3}$, and $\xi_D = 0.83\,nm$ in physiological saline $[M_s] \sim 0.14$ molar [28].

The competing interactions in the Debye-Hückel example are attractive and repulsive electrostatics. For semidilute polymer solutions, these interactions are intramolecular and intermolecular volume exclusion, but the screening principle—cancellation by mutual compensation—seems to be broadly applicable. Muthukumar and Edwards derived an expression for the excluded volume crossover[29], finding it to be described by the Debye-Hückel form. However, even without a detailed expression for the crossover relationship, predictions about screening effects can be made simply by employing a step cutoff approximation [Eq. 7.12]. Hence, substituting the macroscopic concentration ϕ for ϕ_{self} and the screening length ξ for r in Eq. 7.7 yields an expression for the screening length:

$$\xi = \left(3\upsilon_m/4\pi\ell^{1/\gamma}\right)^{\frac{\gamma}{3\gamma-1}} \phi^{-\frac{\gamma}{3\gamma-1}} \tag{7.13}$$

Accordingly, for a given polymer–solvent system and long enough chains,

$$\xi \propto \phi^{-3/4} \qquad [\text{good solvent} \quad (\gamma = 3/5)] \tag{7.14}$$

$$\xi \propto \phi^{-1} \qquad [\text{theta condition} \quad (\gamma = 1/2)] \tag{7.15}$$

An alternative expression for ξ can be obtained by a *scaling argument*—essentially a smoothness condition—that requires the screening length at the beginning of the semidilute regime, $\phi = \phi^*$, to be equal to the coil radius in the dilute limit. Imposing $\xi = R_g(0)$ at $\phi = \phi^*$ in Eq. 7.13 leads to[23]:

$$\xi = R_g(0)(\phi/\phi^*)^{-\frac{\gamma}{3\gamma-1}} \tag{7.16}$$

Note that the screening length is independent of chain length and size beyond ϕ^*; from Eq. 7.13, the quantity $R_g(0)(\phi^*)^{\frac{\gamma}{3\gamma-1}}$ is independent of global chain properties. (In all arguments of this sort, "equal to" means "of the order of", so $R_g(0)$ could be replaced, for example, by the dilute limit end-to-end distance $\langle R^2\rangle^{1/2}$ to arrive at the equivalent of Eq. 7.16.)

Concentration dependence. A step cutoff approximation for excluded volume means that dilute solution spacings apply for segments with mean end-to-end separation smaller than ξ and that excluded volume has no effect at all on larger average separations. The *blob* model[23,25] turns that approximation into a means for estimating equilibrium properties in the semidilute region. The solution is considered to consist entirely of excluded volume blobs with volume ξ^3. These blobs fill the volume, so that the solution can be pictured as a "melt of blobs," and each chain is a connected sequence of blobs, as illustrated in Figure 7.2. The single-chain behavior at the dilute limit dominates inside the blobs. Excluded volume is cancelled at mean distances larger than ξ, so the sequences have unrestricted random-walk conformations. Thus, each sequence consists of n/n_s steps, each with step length ξ. Coil size R_g depends on both chain length and screening length, according to the formula for unrestricted walks, Eq. 4.20:

$$R_g^2(\xi) = \frac{1}{6}[n/n_s(\xi)]\xi^2 \tag{7.17}$$

From Eq. 7.6, at $r = \xi$,

$$n_s(\xi) = (\xi/\ell)^{1/\gamma} \tag{7.18}$$

Equations (7.16) through (7.18) and the specification that chain size still be comparable to its dilute limit near overlap, $R_g(\xi) = R_g(0)$ at $\phi = \phi^*$, leads finally to the

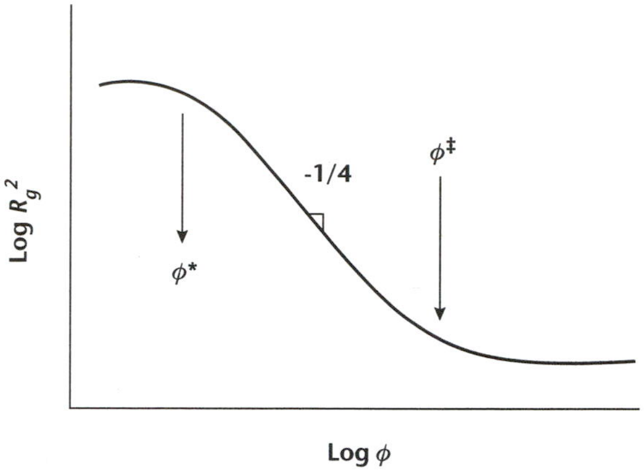

FIGURE 7.3 Coil size as a function of polymer concentration.

semidilute scaling relationship for chain dimensions:

$$R_g^2(\phi) = R_g^2(0)(\phi/\phi^*)^{-\frac{2\gamma-1}{3\gamma-1}} \qquad \phi \gg \phi^* \tag{7.19}$$

In the good solvent limit, with the Flory exponent $\gamma = 3/5$, and at the theta condition $\gamma = 1/2$, for those particular exponents,

$$R_g^2(\phi) = R_g^2(0)(\phi/\phi^*)^{-1/4} \qquad \text{(good solvent)} \tag{7.20}$$

$$R_g^2(\phi) = R_g^2(0) = \left(R_g^2\right)_o \qquad \text{(theta condition)} \tag{7.21}$$

The exponent for good solvents is −0.23 instead of −0.25, if the renormalization group exponent $\gamma = 0.588$ is used. The relationship between polymeric size and concentration in good solvents implied by Eq. 7.20, and including the crossover from the semidilute to concentrated region at $\phi^\ddagger$, as defined by Eq. 7.2, is shown in Figure 7.3.

Semidilute–concentrated crossover. Estimates of $\phi^\ddagger$ have been made using dilute solution data based on scaling and screening ideas[30]. Consider a solution of long chains in a good solvent above ϕ^* and assume that Eq. 7.20 correctly describes $R_g(\phi)$. Obtain the crossover by setting $R_g(\phi)$ equal to $(R_g)_o$, its unperturbed value, at concentration $\phi^\ddagger$:

$$\phi^\ddagger = [R_g(0)/(R_g)_o]^8\phi^* \tag{7.22}$$

Substituting $\phi^* \propto M/(R_g(0))^3$, $R_g(0) \propto M^{3/5}$, and $(R_g)_o \sim (R_g)_\theta \propto M^{1/2}$ into Eq. 7.22 yields:

$$\phi^\ddagger \propto M^0 \tag{7.23}$$

Accordingly, the concentration that separates the semidilute and concentrated regions is independent of the polymer molecular weight.

Values of $\phi^{\ddagger}$ for particular polymer–solvent systems can be estimated from $M^{\ddagger}$, the onset molecular weight for coil expansion in the dilute limit, introduced in Section 6.1.1. The size corresponding to $M^{\ddagger}$, $R_g^{\ddagger} = K_{\theta}(M^{\ddagger})^{1/2}$, defines the screening length, $\xi^{\ddagger} = R_g^{\ddagger}$, which is required to rid the chain dimensions of excluded volume swelling. Thus, when the screening length becomes as small as $\xi^{\ddagger}$, the chain segments inside the blobs have unperturbed dimensions. Thus, the strings of blobs are random walks of random walks and, accordingly, $R_g = (R_g)_o$. Alternatively, $\phi^{\ddagger}$ is simply the overlap concentration for chains having molecular weight $M^{\ddagger}$ in the polymer–solvent system of interest. Thus, $\phi^{\ddagger} = \phi^*$ at $M = M^{\ddagger}$, and using Eqs. 6.9 and 6.11:

$$\phi^{\ddagger} = \frac{3}{4\pi}\frac{M^{\ddagger}}{N_a(R_g^{\ddagger})_o^3\rho} \tag{7.24}$$

Likewise, the crossover range can be estimated using Eq. 6.80:

$$\phi^{\ddagger} = 1/[\eta]^{\ddagger}\rho \tag{7.25}$$

in which $[\eta]^{\ddagger}$ is the intersection of the good and theta solvent Mark-Houwink expressions for the system of interest. Crossover concentrations calculated with values of $M^{\ddagger}$ in Table 6.3 seldom lie outside the range of 0.05 to 0.15, depending on the system and the choice of expression for $\phi^{\ddagger}$. Those based on Eq. 7.25, for several species in various good solvents, are given as examples in Table 7.2. A mid-range estimate from those data is:

$$(\phi^{\ddagger})_{avr} = 0.1 \tag{7.26}$$

TABLE 7.2 Scaling estimates of the semidilute–concentrated crossover range for various polymer species in typical good solvents

Polymer Species	$M^{\ddagger}$	$[\eta]^{\ddagger}$ (cm^{-3} g)	$R_g^{\ddagger}$ (nm)	$\phi^{\ddagger}$	l_K (nm)
PS	13,200	9.8	3.1	0.10	1.48
PαMS	17,000	9.6	3.5	0.10	1.70
PMMA	11,000	5.8	2.8	0.14	1.39
PDMS	13,000	9.4	3.0	0.11	0.92
PIB	19,000	15.0	4.3	0.070	1.04
PI	3,800	8.0	1.9	0.14	0.68
PBD	5,200	13.0	2.8	0.085	0.80
PE	6,500	25.0	3.7	0.050	1.14
iPP	6,600	11.5	2.8	0.11	0.95
PEO	6,700	13.0	3.0	0.070	0.80

Calculated from data in Tables 6.2 and 6.3 as described in the text.

Finally, Noda et al. report departures in osmotic pressure at elevated concentrations from the semidilute power law (see next paragraph) for solutions of polystyrene in toluene[31]. The onset concentration is independent of molecular weight and suggests $\phi^{\ddagger} \sim 0.15 - 0.20$.

Observations. The experimental situation regarding chain dimensions in the semidilute region is somewhat clouded. The determinations of $R_g(c)$ can be done using scattering measurements on ternary systems, mixtures of labeled chains at low concentration in nonlabeled chains, and the solvent. The experiments are technically demanding as well as time-consuming. Four studies have been made—Daoud et al. in 1975[23], King et al. in 1985[32], and Westermann et al. in 2000[33], all using SANS; and Hamada et al. in 1978[34] using SAXS. Even for those experiments, the available q-range was such that only chains of relatively small size could be used—$R_g(0) \lesssim 15nm$—with a correspondingly small difference between $R_g(0)$ and $(R_g)_o$ as shown by extensive data in good solvents for many polymer species (see Section 6.2.3 and Table 6.3). A fifth study, by Kent et al.[35], employed a different three-component technique using dilute polystyrene in the isorefractive pair polymethylmethacrylate-ethyl benzoate, for which the use of light scattering permitted larger chains $[R_g(0) \sim 40nm]$ to be investigated. Unfortunately, the results were contradictory. All but the Westermann et al. studies are based on polystyrene. Hamada et al. furnished the only information on R_g at the theta temperature, confirming the independence of concentration[34], as predicted by Eq. 7.21. The results for polystyrene in good solvents, when normalized by the respective dilute solution sizes and overlap concentrations (Figure 7.4), do not appear to differ greatly from one another. (The results of Hamada et al. were omitted as being too erratic.) Daoud et al. found a power-law relationship for $R_g(c)$ running smoothly from ϕ^* to the undiluted state[23], with an exponent of –0.25—in excellent agreement with the predicted value [Eq. 7.20]. However, Westermann et al. pointed out that the Daoud et al. values of R_g at high concentrations were probably incorrect[33], being significantly smaller than the now known values of $(R_g)_o$ for polystyrene at their molecular weight. King et al. also found a power law running smoothly from $\phi \sim \phi^*$ to $\phi = 1$, but with a much smaller exponent[32], -0.16. The data of Kent et al. (1991), running from well below to well beyond overlap, agree with the predicted exponent and indicate by extrapolation $\phi^{\ddagger} \sim 0.24$.

The study of Westermann et al., employing model polyethylene in nonadecane[33] and covering concentrations from $\phi = 0.25$ to $\phi = 1$, found unperturbed dimensions $[R_g = (R_g)_o]$ throughout that entire range. The results of the four scattering studies [Hamada et al. omitted] having sizes normalized by the melt dimensions to make the high concentration behavior as clear as possible, are shown in Figure 7.5. The inconsistency among investigators leaves unsettled from the standpoint of dynamic properties as well as simple factual knowledge, how R_g varies with concentration beyond ϕ^*. (Indeed, it seems quite remarkable that such an important matter as this

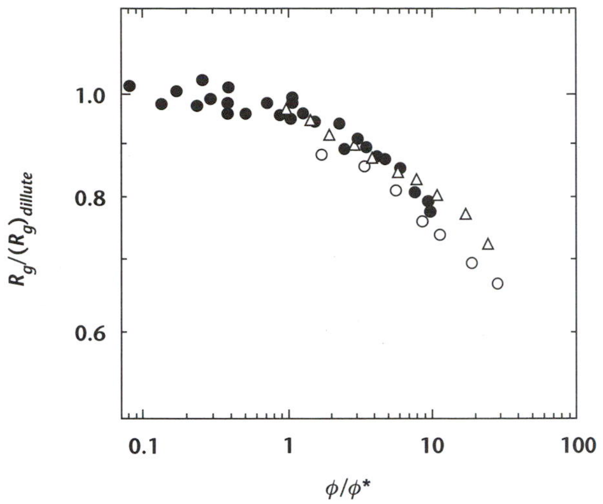

FIGURE 7.4 Reduced coil size as a function of reduced concentration for polystyrene in various good solvents. The symbols represent results in carbon disulfide (○) (Daoud et al.[23]), in toluene (△) (King et al.[32]), and in ethyl benzoate (●) (Kent et al.[35]).

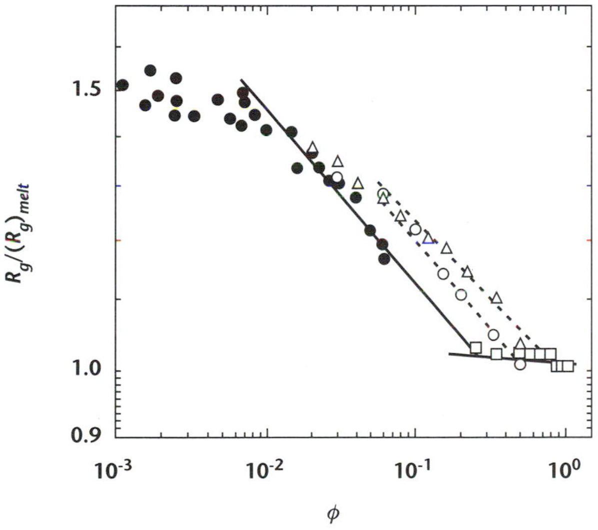

FIGURE 7.5 Coil size normalized by the unperturbed size as a function of concentration for the data in Figure 7.4. The symbol meanings are those in Figure 7.4, except the polyethylene–hexadecane results of Westermann et al.[33] (□) are now included. The lines are merely guides for the eye.

has attracted so little attention over the years.) What is clearly needed is a study likethat of Westermann et al., but using a sample of much longer chains in a solvent known from other evidence to induce large excluded volume swelling and then ranging in concentration from the dilute regime to the undiluted state.

7.2 Osmotic Pressure beyond Overlap

As shown in Eq. 5.1, the osmotic pressure of a solution provides direct information about the component chemical potentials. The osmotic pressure of a dilute polymer solution depends on both molecular weight and concentration [Eq. 5.5]:

$$\frac{\pi(c,M)}{RT} = \frac{c}{M} + A_2(M)c^2 + \cdots \qquad c \lesssim c^* \tag{7.27}$$

The concentration dependence of osmotic pressure increases with increasing concentration, thus reflecting a growing contribution from intermolecular interactions as the coil overlap condition is approached. The molecular weight dependence weakens over the same range, as shown for example by the data for solutions of poly(α–methyl styrene) in toluene[20] in Figure 7.6a. Beyond ϕ^*, the data for samples of different molecular weight merge to become a function of concentration alone:

$$\frac{\pi(c,M)}{RT} = f(c) \qquad c > c^* \tag{7.28}$$

For a polymer solution, the scattering intensity in the forward direction depends on molecular weight and concentration through the osmotic compressibility. In the case of light scattering, from Eq. 5.31:

$$\frac{(\Delta R_\theta)_c}{RT} = \frac{C_{LS}c}{(\partial\pi/\partial c)_{P,T}} \qquad qR_g \ll 1 \tag{7.29}$$

In dilute solutions [from Eq. 5.32],

$$\frac{(\Delta R_\theta)_c}{C_{LS}} = Mc - 2A_2Mc^2 + \cdots \qquad c < c^* \tag{7.30}$$

while in the semidilute region, from Eqs. 7.28 and 7.29,

$$\frac{(\Delta R_\theta)_c}{C_{LS}} = \frac{c}{df/dc} \qquad c > c^* \tag{7.31}$$

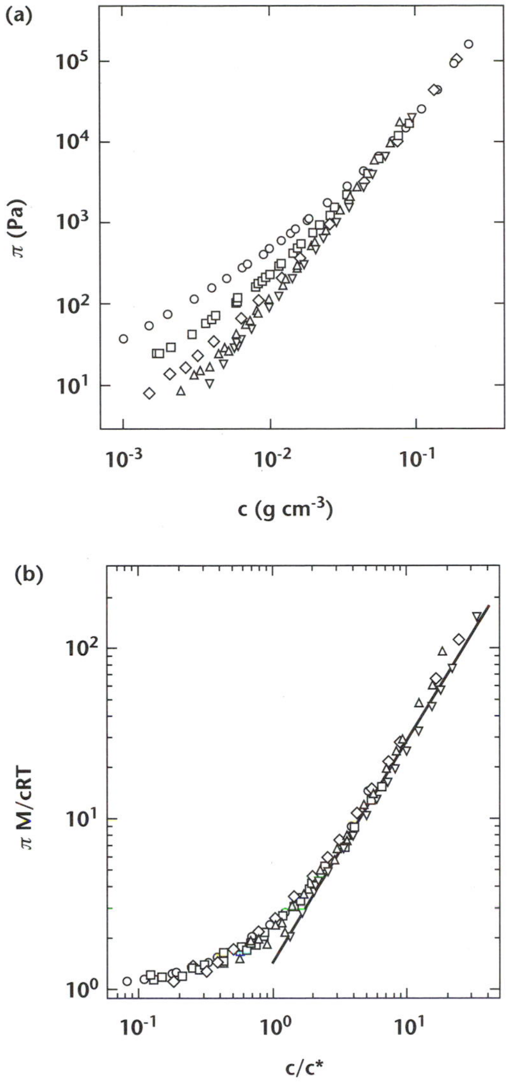

FIGURE 7.6 (a) Osmotic pressure as a function of concentration for several samples of poly(α-methylstyrene) in toluene at 25°C (from Noda et al.[20]). Molecular weights: $10^{-5}M = 0.708$ (○), 2.04 (□), 5.06 (◇), 11.9 (△), 18.2 (▽). (b) Reduced plot of osmotic pressure data in Figure 7.6a.

Thus, even for solutions that are well beyond coil overlap, both osmometry and forward scattering intensity can provide model-free information about the concentration dependence of the component chemical potentials and hence about the intermolecular interactions.

7.2.1 Flory-Huggins Predictions

The history of events leading to an expression for the combinatorial free energy of mixing for components differing in molecular volume was reviewed in Section 3.3. The addition of a Van Laar interaction term led to Eq. 3.67, the free energy of mixing density for a simple mixture. When applied to polymer solutions, this formulation is commonly called the *Flory-Huggins* equation. Using the conventions set forth by Flory and coworkers—ϕ is the volume fraction of polymer; r is $V_2/V_1 = V_p/V_s$, the ratio of polymer to solvent molar volume; and χ is the interaction parameter, defined with respect to the solvent molar volume of solvent as reference, $V_{ref} = V_s$—Eq. 3.67 becomes:

$$g_m(\phi, T) = RT\left[(1-\phi)\ln(1-\phi) + \frac{\phi\ln\phi}{r} + \chi(T)\phi(1-\phi)\right] \tag{7.32}$$

Likewise, solvent and polymer chemical potentials are given by Eqs. 3.69 and 3.70:

$$\frac{\mu_s - \mu_s^{\mathrm{o}}}{RT} = \ln(1-\phi) + \left(1 - \frac{1}{r}\right)\phi + \chi\phi^2 \tag{7.33}$$

$$\frac{\mu_p - \mu_p^{\mathrm{o}}}{RT} = \ln\phi - (r-1)(1-\phi) + r\chi(1-\phi)^2 \tag{7.34}$$

Osmotic pressure is related to solvent chemical potential, and Eq. 5.1, $\pi = -(\mu_s - \mu_s^{\mathrm{o}})/V_s$, is adequate for purposes here. Thus, the Flory-Huggins expression for osmotic pressure is:

$$\pi = -\frac{RT}{V_s}\left[\ln(1-\phi) + \left(1 - \frac{1}{r}\right)\phi + \chi\phi^2\right] \tag{7.35}$$

An important limitation in applying Eqs. 7.32 through 7.35 to polymer solutions must be borne in mind. Its derivation assumes statistical uniformity in the spatial distribution of interacting elements. This is plausible for single-phase mixtures of weakly interacting monomeric molecules, but it is certainly incorrect for polymer solutions in the dilute range of concentrations. Dilute polymer solutions are intrinsically nonuniform because all chain units are at the self-concentration even in the dilute limit. Thus, $\phi \sim \phi^*$ inside the coils, but the coils themselves are separated from one another by regions of pure solvent, where $\phi = 0$. Such gross heterogeneity probably disappears near the overlap concentration, but it certainly must continue to higher concentrations on more fine-grained distance scales. Intrinsic heterogeneity of this sort probably decreases with increasing concentration in parallel with changes in the screening length ξ, and perhaps disappears finally near the $\phi^\ddagger$ crossover. Thus, the application of Eqs. 7.32 through 7.35 is subject to some uncertainty below $\phi^\ddagger$, and increasingly so upon approaching ϕ^*.

To understand better the Flory-Huggins expression for osmotic pressure and its limitations, it is helpful to examine the formulas it provides in the dilute and semidilute ranges. Expanding $\ln(1-\phi)$ in Eq. 7.35 as a power series in ϕ leads to:

$$\frac{\pi V_s}{RT} = \frac{\phi}{r} + \left(\frac{1}{2} - \chi\right)\phi^2 + \frac{\phi^3}{3} + \cdots \tag{7.36}$$

With volume fraction and chain length expressed in terms of polymeric observables, molecular weight M, mass density in solution c, and partial specific volume $\bar{\upsilon}_p$,

$$\begin{aligned} \phi &= c\bar{\upsilon}_p \\ r &= M\bar{\upsilon}_p/V_s \end{aligned} \tag{7.37}$$

Eq.(7.36) becomes:

$$\frac{\pi}{cRT} = \frac{1}{M} + \left(\frac{1}{2} - \chi\right)\frac{\bar{\upsilon}_p^2}{V_s}c + \frac{\bar{\upsilon}_p}{V_s}\sum_{i=2}^{\infty}\frac{1}{i+1}\left(c\bar{\upsilon}_p\right)^i \tag{7.38}$$

The leading term is the ideal solution limit, a necessary consequence of thermodynamic consistency, as discussed in Chapter 3. The second virial coefficient,

$$A_2 = \left(\frac{1}{2} - \chi\right)\frac{\bar{\upsilon}_p^2}{V_s} \tag{7.39}$$

evidently goes to zero when $\chi = 1/2$, thus corresponding to the theta condition for Flory-Huggins solutions:

$$\chi(T_\theta) = 1/2 \tag{7.40}$$

The same expression for $\chi(T_\theta)$ can be deduced from the liquid–liquid phase behavior of Flory-Huggins solutions (see Section 7.5.1).

A Flory-Huggins solution with $\chi = 0$ has no net interactions beyond the mutual volume exclusion reflected in the combinatorial free energy. This interactionless condition, epitomized by terminology such as *athermal* or *ideally good*, has seldom if ever been observed or even approached in polymer–solvent systems. Not an anomaly, this is in fact a natural consequence of an inherent difference in free volume between polymers and monomers in the dense liquid region. The contribution to χ that results from this free volume mismatch is discussed in Section 7.4.2.

According to Eq. 7.38, all virial coefficients beyond the second are independent of χ. This is to be expected, however, because only binary interactions are contained in the interaction model itself. Of greater concern is the prediction that the second and higher virial coefficients are independent of polymer molecular weight. That is contrary to the data for A_2 and A_3[36] and serves as a reminder of the expected qualitative inadequacy of the Flory-Huggins formulation in the dilute range.

Consider now the Flory-Huggins predictions for the semidilute range: $\phi^* < \phi < \phi^\ddagger \ll 1$. Through the use of Eq. 6.9, the overlap concentration $\phi^* = c^*/\rho$ is expressed in terms of the model parameters as follows:

$$\phi^* = \left(\frac{3}{4\pi}\right)\left(\frac{6^{3/2}m_o}{N_a l_s^3 \rho}\right)\frac{1}{r^{1/2}} \tag{7.41}$$

in which m_o is the molecular weight of a monomeric unit, l_s is the statistical segment length for the polymer species, and $\bar{v}_p$ has been harmlessly approximated by the reciprocal of ρ, which is the mass density of the undiluted polymer liquid. The polymer density should be of the order of a few $m_o/N_a l_s^3$, so the entire species-dependent prefactor, $(3/4\pi)(6^{3/2}m_o/N_a l_s^3 \rho)$, is of order unity. Thus, Eq. 7.41 can be written:

$$\phi^* = r^{-1/2} \tag{7.42}$$

Now express Eq. 7.36 as a power series in ϕ/ϕ^* and use Eq. 7.42 to obtain:

$$\frac{\pi V_s}{RT} = \frac{1}{r^{3/2}}\left(\frac{\phi}{\phi^*}\right) + \frac{(1/2-\chi)}{r}\left(\frac{\phi}{\phi^*}\right)^2 + \sum_{j=3}^{\infty}\frac{1}{j\, r^{j/2}}\left(\frac{\phi}{\phi^*}\right)^j \tag{7.43}$$

For long chains ($r \gg 1$) and good solvents ($\chi < 1/2$), the second term dominates, so:

$$\pi = \frac{(1/2-\chi)RT}{V_s}\phi^2 \qquad \text{(good solvent)} \tag{7.44}$$

At the theta condition, $\chi = 1/2$, the second term is ϕ. Now, the third term is much larger than the first, and the leading term in the summation dominates. Thus,

$$\pi = \frac{RT}{3V_s}\phi^3 \qquad \text{(theta condition)} \tag{7.45}$$

Thus, the Flory-Huggins model predicts the osmotic pressure is independent of chain length for semidilute solutions. The model also predicts power-law dependence on concentration, $\pi \propto \phi^2$ for good solvents and $\pi \propto \phi^3$ at T_θ. The scaling analysis described in the next section makes similar predictions.

7.2.2 Scaling Analysis

A scaling argument was employed by des Cloizeaux[24] to predict osmotic pressure behavior in the semidilute range. Thus, in the dilute limit, the variables of interest are related as follows:

$$\frac{\pi}{RT} = \frac{c}{M} \qquad c \ll c^* \tag{7.46}$$

Thermodynamic properties in the semidilute region are dominated by intermolecular interactions. Because the interactions in molecular liquids are short range, osmotic pressure beyond c^* should be a function of polymer concentration and independent of molecular weight. Furthermore, the expression describing its concentration dependence in the semidilute region,

$$\frac{\pi}{RT} = f(c) \qquad c^* \lesssim c \lesssim c^\ddagger \tag{7.47}$$

should join smoothly with the dilute solution expression near the overlap concentration. Thus, using Eq. 7.46,

$$\tilde{\beta}\frac{c^*}{M} = f(c^*) \tag{7.48}$$

in which $\tilde{\beta}$ is some dimensionless numerical coefficient of order unity. Because c^* depends on a power of M, and Eq. 7.48 must be satisfied for all M, $f(c)$ must, for dimensional consistency, be a power-law:

$$f(c) = Bc^p \tag{7.49}$$

The coefficient B is a system-dependent parameter; the exponent p is determined by the condition that π is independent of M beyond c^*. Hence, for smooth joining at c^*,

$$\tilde{\beta}\frac{c^*}{M} = B(c^*)^p \tag{7.50}$$

Substituting $c^* \propto M/(R_g(0))^3 \propto M^{1-3\gamma}$ into Eq. 7.50 and then equating the powers of molecular weight on each side determines the semidilute concentration exponent:

$$\frac{\pi}{RT} \propto c^{3\gamma/(3\gamma-1)} \qquad c^* \lesssim c \lesssim c^\ddagger \tag{7.51}$$

Thus, with γ either 3/5 or 1/2,

$$\frac{\pi}{RT} = Bc^{9/4} = Bc^{2.25} \qquad \text{(good solvent)} \tag{7.52}$$

$$\frac{\pi}{RT} = Bc^3 \qquad \text{(theta condition)} \tag{7.53}$$

For the renormalization group exponent, $\gamma = 0.588$,

$$\frac{\pi}{RT} = Bc^{2.31} \qquad \text{(good solvent)} \tag{7.54}$$

Accordingly, osmotic pressure in the semidilute region varied with ϕ^3 at T_θ from both scaling analysis and the Flory–Huggins model, Eq. 7.45. However, the predictions for good solvent systems are different, $\pi \propto \phi^{2.3}$ from scaling and $\pi \propto \phi^2$ from Flory-Huggins, Eq. 7.44.

The scaling results can also be expressed in terms of excluded volume screening length. Using $B = (c^*)^{1-p}/\tilde{\beta}M$ Eqs. 7.47 through 7.50 leads to:

$$\pi = \tilde{\beta}\frac{c^* RT}{M}\left(\frac{c}{c^*}\right)^{\frac{3\gamma}{3\gamma-1}} \qquad c^* \lesssim c \lesssim c^\ddagger \tag{7.55}$$

Then, using Eq. 7.16, $c/c^* = \phi/\phi^*$, and the overlap concentration defined according to Eq. 6.9, yields

$$\pi = \frac{3\tilde{\beta}}{4\pi}\frac{kT}{\xi^3} \qquad \phi^* < \phi < \phi^\ddagger \tag{7.56}$$

Thus, for good and theta conditions alike, and apart from a dimensionless coefficient of order unity, osmotic pressure in the semidilute region is simply kT multiplied by ξ^{-3}, the number density of screening volumes.

In 1982, Muthukumar and Edwards[29] developed an analytical theory of osmotic pressure and chain dimensions for polymer solutions in good solvents. They obtain expressions for $R_g(\phi)$ and $\pi(\phi)$ and found them to approach the scaling and screening predictions [Eqs. 7.20 and 7.52] for the semidilute long-chain limit. They also predict $\pi \propto \phi^2$ at high concentrations, a result that has not yet been observed experimentally. Finally, they provide an interpolation formula for $\pi(\phi)$ through the dilute to semidilute crossover.

7.2.3 Experimental Observations

Experiments have generally confirmed the scaling predictions about osmotic pressure in semidilute solutions[20,31,37,38]. Extensive data from both osmometry and osmotic compressibility by light scattering [see Eq. 5.31] are available for good solvents. A power-law dependence on concentration is commonly observed. The exponents are almost always greater than 2.0, clustered for the most part in the range $2.1 < p < 2.4$, with an average that is close to the scaling prediction of 2.3. Data at the theta condition are comparatively scarce, but all show exponents larger than 2.7 and, within the errors at least, in reasonable agreement with the predicted $p = 3$.

Osmotic pressure data for poly(α-methyl styrene) in toluene, shown in Figure 7.6a, are plotted in the scaling form, $\pi M/cRT$ versus c/c^*, in Figure 7.6b. The data collapse to a single curve from the dilute range to well beyond overlap. A well-defined power law is found in the semidilute range, and values for the exponent and coefficient have been established[20]:

$$p = 2.325 \quad \text{P}\alpha\text{MS-TOL} \quad \tilde{\beta} = 1.50 \tag{7.57}$$

Similar values are reported for polystyrene in toluene[38]:

$$p = 2.275 \quad \text{PS-TOL} \quad \tilde{\beta} = 2.2 \tag{7.58}$$

The exponents, corresponding to $\gamma = 0.585$ and $\gamma = 0.595$ respectively, agree well with those expected for good solvent systems. The coefficients given above were obtained with c^* defined according to Eq. 6.9.

Light scattering determinations of $(\partial\pi/\partial c)_T$ for linear and star polyisoprene in the good solvent cyclohexane[39] show similar success in reduction to a single power law; the normalizing parameter used in that case is $c^* = (MA_2)^{-1}$. Light scattering data for polystyrene at the theta condition[40] confirmed the predictions based on Eq. 7.55 and, within the uncertainties, yields $\gamma = 1/2$. A comparison of $\pi(\phi)$ for polyisobutylene in a good solvent and at the theta condition[41] is shown in Figure 7.7, the lines indicating power-law exponents of $p = 2.3$ and 3.0. Based on the good

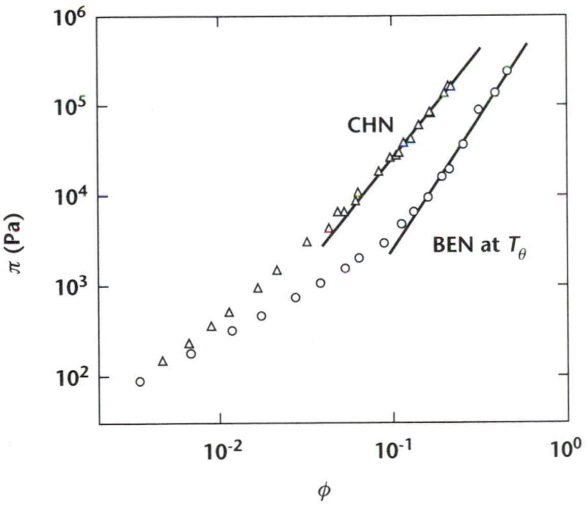

FIGURE 7.7 Osmotic pressure as a function of concentration for solutions of polyisobutylene ($M_n = 90{,}000$) in a good solvent cyclohexane at 30°C and in benzene at the theta condition (24.5°C) (from Flory and Daoust[41]).

solvent results in particular, the des Cloizeaux scaling predictions for osmotic pressure are superior to those from Flory-Huggins.

7.3 Scattering beyond Overlap

Scattering properties depend on pair-wise correlations in scatterer locations. Forward scattering provides osmotic pressure information, as described earlier; the angular dependence of scattering in the low q limit defines a distance, known as the *correlation length*. Beyond separation distances of the order of the correlation length, the relative locations of scattering sites are not correlated and do not contribute to the scattering intensity. In dilute solutions, $\phi \lesssim \phi^*$, the correlation length is the polymer size. Pairs of scattering sites separated by distances much greater than approximately R_g belong to different molecules, and their contribution to coherent scattering is mediated by the intermolecular interactions. In the dilute limit, pairs on different molecules are uncorrelated in relative position, and the sum of scattering amplitudes for such pairs contributes nothing to the coherent scattering intensity. For concentrated solutions, $\phi \gtrsim \phi^\ddagger$, the correlations involve many chains, but the situation simplifies because the chain locations are essentially random. In semidilute solutions $\phi^* \lesssim \phi \lesssim \phi^\ddagger$, the situation is made complicated by screening effects, which introduce correlations that depend on distance and which vary with concentration, as discussed in Sections 7.1 and 7.2.

7.3.1 Correlation Length

Self-concentration and intramolecular pair distribution are related properties. Thus, for r as the distance relative to some reference unit in the chain, $\phi_{self} r^3$ is proportional to the number of chain units in self-volume r^3, and $d(\phi_{self} r^3)/dr^3$ is therefore proportional to $[g(r)]_{intra}$, the mean concentration of chain units on the same chain at distance r. Accordingly, with $\phi_{self}(r)$ from Eq. 7.8, and for distances from well above the mer size r_o to roughly R_g,

$$[g(r)]_{intra} \propto r^{-4/3} \qquad \text{(good solvent)}$$
$$[g(r)]_{intra} \propto r^{-1} \qquad \text{(theta condition)} \tag{7.59}$$

Using Eq. 2.58 and Table 2.4, for coherent scattering in the intermediate range,

$$[S(q)]_{intra} \propto q^{-5/3} \qquad \text{(good solvent)} \tag{7.60}$$

$$[S(q)]_{intra} \propto q^{-2} \qquad \text{(theta condition)} \tag{7.61}$$

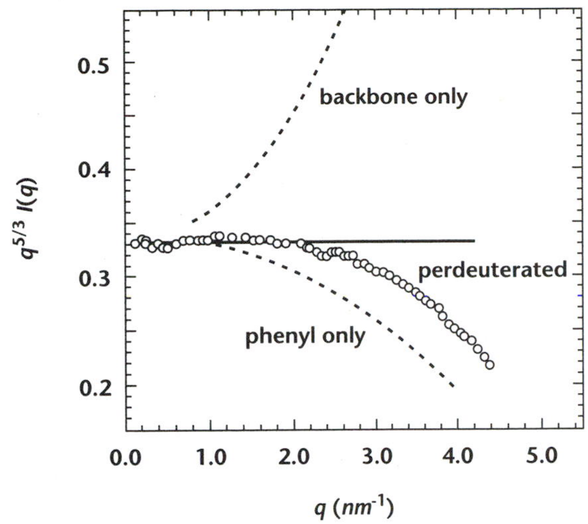

FIGURE 7.8 High q SANS scattering data for dilute CS_2 solutions of labeled polystyrene, backbone deuterated, phenyl groups deuterated, and perdeuterated. The solid line indicates the expected result for all at intermediate q, $I(q) \propto q^{-5/3}$ (from Rawiso et al.[44]).

Equation (7.60) has been verified by Okano et al.[42] using SAXS for polystyrene in good solvents in the dilute limit and by Farnoux et al.[43] and Rawiso et al.[44] using SANS.

In dilute solutions, and assuming self-avoiding walks, one might naturally anticipate $r^{-4/3}$ behavior and hence $I(q) \propto q^{-5/3}$ from distances of a few Kuhn steps to about R_g. The Okano and Farnoux results seem to support that expectation. However, the Rawiso scattering data dispute their lower limit for $r^{-4/3}$ behavior and hence the upper limit for $q^{-5/3}$ scattering. Rawiso et al. reasoned that the details of monomeric unit structure, including the location of deuterium labels on the monomeric unit, should not affect the observations in any true intermediate $q^{-5/3}$ region. These researchers tested this theory with three mer labeling schemes—deuterium substitution on the chain backbone only, on the phenyl groups only, and perdeuteration. The results are shown in Figure 7.8. These results reveal that, despite $q^{-5/3}$ behavior for the perdeuterated sample being observed up to $q \sim 2nm^{-1}$, the contribution of local structure does not disappear until q becomes smaller than about 0.3 nm^{-1}. The other two studies, utilizing only fully hydrogenous PS[42] or perdeuterated PS[43], led to the natural but incorrect conclusion of self-avoiding walk behavior down to much shorter chain distances.

The intramolecular pair distribution for long chains in good solvents is illustrated in Figure 7.9a for the three concentration regimes. The corresponding pair distributions are shown in Figure 7.9b. The r^{-1} region for distances up to the crossover $R_g^{\ddagger}$ in Figure 7.9a refers to the dilute range observation that short chains are Gaussian even in the "best" good solvents (see Table 6.2). The dilute solution scattering behavior illustrated in Figure 7.9b is roughly consistent, in both crossover location and q dependence beyond, with the Rawiso et al. scattering result.

In dilute and semidilute solutions, this r^{-1} region is followed by $r^{-4/3}$ behavior, where volume exclusion is acting, and then, in semidilute solutions, by a second r^{-1} region, where volume exclusion is screened. The two r^{-1} regions merge at the crossover to the concentrated regime, $\phi \sim \phi^{\ddagger}$. Corresponding behavior for intramolecular scattering is shown in Figure 7.9b.

Coherent scattering behavior depends in general on the total pair distribution, not the intramolecular part alone. In the dilute limit, no intermolecular contribution occurs, so $[g(r)]_{intra} = [g(r)]_{total}$. The same applies to the structure factor, so the dilute ranges shown in Figures 7.9a and 7.9b are applicable. The semidilute sketches apply only to a special case—long labeled chains that are dilutely dispersed, $\phi_D \ll \phi^*$, in a mixture of nonlabeled chains and solvent. In broad outline, this is the framework for determining R_g versus $\phi \equiv \phi_D + \phi_H$, as discussed in Section 7.1.2. It can also be used to evaluate the screening length ξ from the crossover between $q^{-5/3}$ and q^{-2} intensity dependence[45]. That method was reported to be less precise, however, than one based on the Ornstein-Zernike correlation length[23], to be discussed later.

A size attributed to the screening length for intramolecular volume exclusion has been calculated from the angular dependence of scattering intensity in semidilute solutions. The scattering behavior is interpreted as that of a one-component system of independent objects—the screening volumes—behaving as if it were a collection of independent scatterers with size ξ. It is argued, in effect, all pair-wise correlations for distances beyond the screening length are nullified. Hence, the screening length, as discussed in connection with the concentration dependence of chain dimensions and osmotic pressure, is the same as the correlation length for scattering. This interpretation ignores the contribution to scattering from the correlations produced by intermolecular excluded volume, as discussed earlier. It is not at all obvious that this is correct or under what conditions it can be safely applied. Nonetheless, the fundamental assumption made for scattering in semidilute solutions is that the excluded volume screening length ξ is equivalent to the Ornstein-Zernike correlation length ξ_{OZ}.

The *Ornstein-Zernike* method is commonly used to extract a correlation length from the angular dependence of scattering intensity. From the Guinier formula [Eq. 4.116], the small-q scattering intensity for any system of noninteracting objects can be written[46]:

$$1/I(q, \phi) = A(\phi) + B(\phi)q^2 + \cdots \tag{7.62}$$

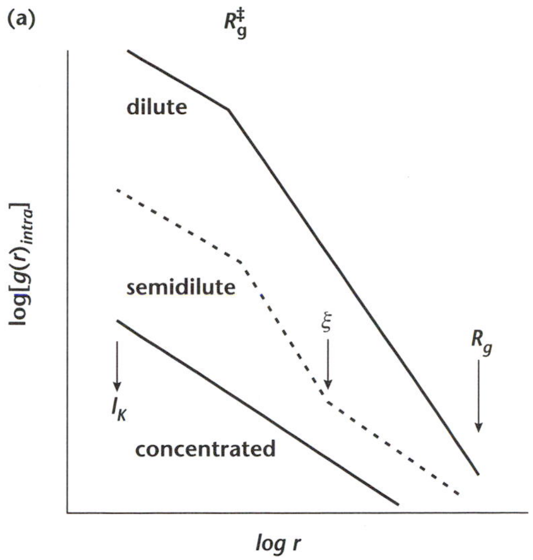

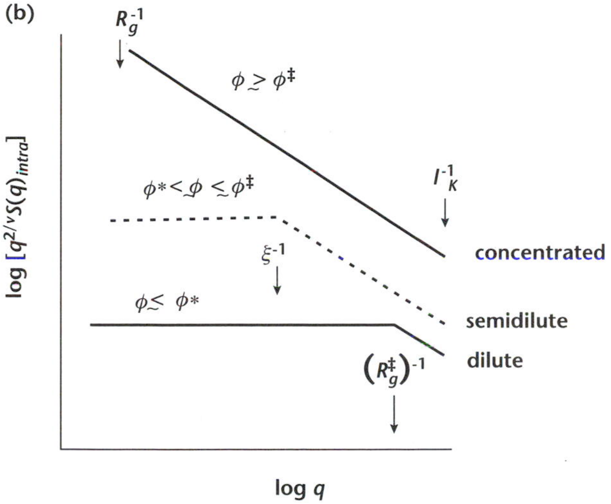

FIGURE 7.9 (a) The intramolecular part of the pair distribution for polymer solutions in the three concentration regimes. Slopes are –1 and –4/3. (b) The intramolecular part of the structure factor for polymer solutions in the three concentration regimes. Slopes are 0 for $S(q) \propto q^{-5/3}$ and –1/3 for $S(q) \propto q^{-2}$.

The correlation length is obtained from the ratio of initial slope-to-intercept for a plot of $1/I$ versus q^2:

$$\xi_{OZ} = \left[\frac{B(\phi)}{A(\phi)}\right]^{1/2} \tag{7.63}$$

This procedure has been used to analyze the scattering behavior for many types of systems. Small-angle neutron scattering and x-ray scattering cover the screening length range of interest for polymer solutions, usually fractions of R_g, typically 2 to 20 nm.

Ornstein-Zernike plots are shown in Figure 7.10a for polystyrene in the good solvent methylene chloride[47]. The concentration dependence of ξ_{OZ} extracted from such plots is shown in Figure 7.10b. A power law describes the data beyond the overlap concentration: $\phi^* \sim 0.02$ in this case. The power-law exponent is about -0.67, somewhat smaller than the predicted value of -0.75 for screening length in good solvents [Eq. 7.14], but not badly off. Power laws with exponents near –0.75 are reported for many good solvent systems[48]. Semidilute solution data at the theta condition are more limited. Some literature results collected by Brown and Nicolai[48] for polystyrene are fairly well described by a power law, and the observed exponent, -1.01, agrees with the screening length prediction [Eq. 7.15].

Despite this pattern of seeming agreement with the supposition of a correspondence between excluded volume screening length and the scattering correlation length, some peculiarities exist. One is the rather sizeable difference between measured osmotic pressure and values predicted by the des Cloizeaux formula, Eq. 7.56, and ξ_{OZ} from the angular dependence of scattering intensity. For example, osmotic pressure for polystyrene–toluene solutions in the semidilute range[31,38] is well described by the formula $\pi(Pa) = 7.8 \times 10^6 \phi^{2.40}$, giving $\pi = 3.1 \times 10^{-2}$ MPa at $\phi = 0.1$, for example. From the SAXS data for semidilute polystyrene–toluene solutions in Figure 7.11a[32,49,50], $\xi_{OZ} = 1.4\,nm$ at $\phi = 0.1$. The values reported for PS-TOL solutions by Wiltzius et al.[51] are very similar. The osmotic pressure calculated using Eq. 7.56, $\xi_{OZ} = 1.4\,nm$ and $\tilde{\beta} = 2.2$ from Eq. 7.58 is 1.4 *MPa,* nearly 50 times larger than the measured value.

The magnitude of ξ_{OZ} relative to the other characteristic sizes listed in Table 7.2 is another peculiarity. Thus, as shown in Figure 7.11a, ξ_{OZ} crosses through $R_g^{\ddagger}$ for polystyrene in typical good solvents at $\phi \sim 0.04$ and becomes smaller than the Kuhn step length of polystyrene beyond $\phi \sim 0.1$. It is unclear what screening could possibly mean in this sub-l_K region. The polystyrene–toluene system is not an isolated example[52] and has been a cause of concern for others[48,53]. Of course, both peculiarities could have the same cause: ξ_{OZ} as a measure of the screening length could simply be abnormally small. Multiplying all values of ξ_{OZ} by a factor of 3 or 4

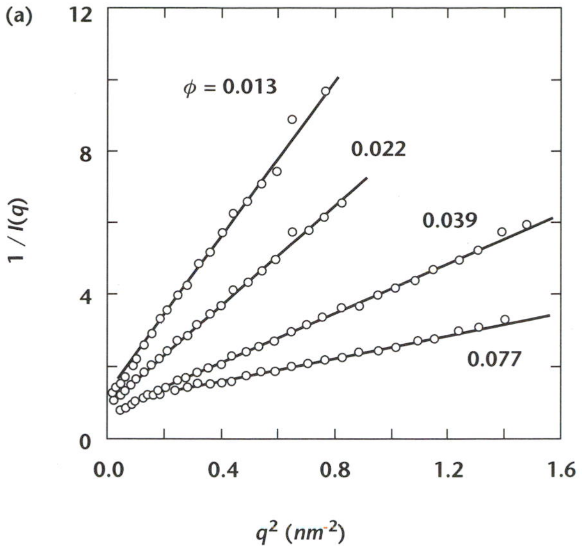

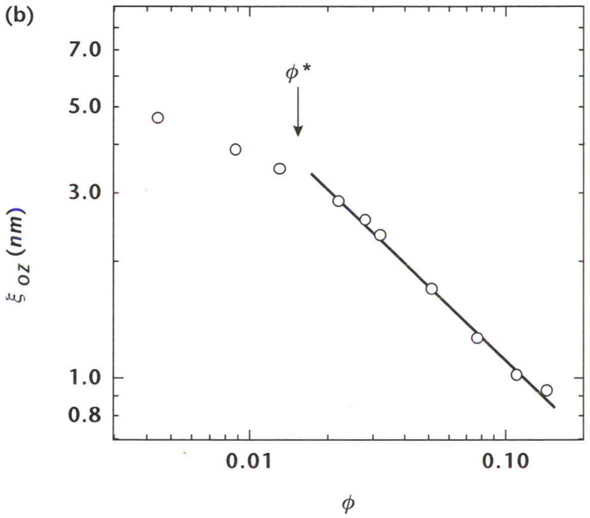

FIGURE 7.10 (a) Ornstein-Zernike plot of neutron scattering intensity for polystyrene ($M = 1 \times 10^5$) at various concentrations in a good solvent dichloromethylene at 25°C (from Brown and Mortensen[47]). (b) Ornstein-Zernike correlation length as a function of concentration for polystyrene–dichloromethylene solutions (from Brown and Mortensen[47]). A power-law exponent, 0.67, is obtained for the data beyond ϕ^*.

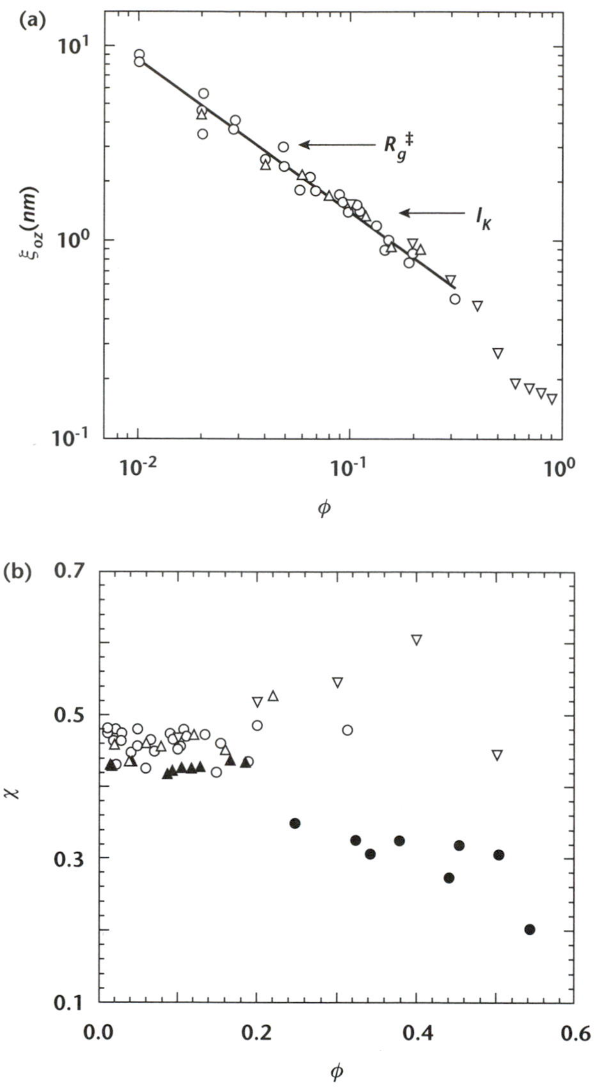

FIGURE 7.11 (a) Ornstein-Zernike correlation length as a function of concentration for polystyrene–toluene solutions ($M > 10^5$, $\phi > \phi^*$) at 25°C [Hamada et al. (○)[49], King et al. (△)[32], Brown et al. (▽)[50]]. (b) Interaction parameter for the polystyrene–toluene system obtained from the scattering data in Figure 7.11a with Eq. 7.53. Open symbols indicate χ_{scat} data (see Figure 7.11a), filled symbols indicate χ data from vapor sorption [Noda et al. (●)[31]] and osmometry [Kuwahara et al. (▲)[56]].

would in fact remove both concerns; Huang and Witten have deduced $\xi_{OZ}/\xi_{\pi} = 3.81$ for good solvents and 2.97 at T_{θ} from the reported data[54]. Refining the definition of ξ_{OZ} to reflect a "true" screening length—including the Guinier factor of 1/3, and so on—might possibly supply a factor of that magnitude. Nonetheless, it seems evident that the association of ξ_{OZ} with the intramolecular excluded volume screening length must be viewed with at least some measure of caution. An alternative interpretation is offered in the next section.

7.3.2 The Random Phase Approximation

The screening idea supposes that the total pair correlation vanishes for distances beyond ξ. That situation may perhaps apply to a semidilute solution of infinitely long chains ($\phi^* \to 0$; $\phi \to 0$). However, correlations beyond ξ that produce coherent scattering are developed in the concentrated region as a result of intermolecular excluded volume, and it is not at all evident whether or where they vanish as the concentration is reduced. This is still an active subject of debate within the theoretical community[55], and it is perhaps worthy of concern when interpreting scattering data even in the semidilute regime.

Equations (4.143) and (4.145) describe the scattering intensity for a simple mixture with no interactions beyond volume exclusion; that is, for $\chi = 0$. The reduced coherent scattering intensity $I_r(q)$, the contrast factor C_{sc}, and the structure factor for the mixture are related through

$$\frac{C_{sc}}{I_r(q)} = \frac{1}{\phi_1 V_1 P_1(q)} + \frac{1}{\phi_2 V_2 P_2(q)} \tag{7.64}$$

where the ϕ_i, V_i, $P_i(q)$ are the component volume fractions, molar volumes, and scattering form factors. Consider now Eqs. 5.27 and (5.31), which combine to form a relationship between the forward scattering intensity ($q = 0$) and the free energy of mixing density:

$$\frac{C_{sc}}{I_r(0)} = \frac{1}{RT}\left(\frac{\partial^2 g_m}{\partial \phi^2}\right)_T \tag{7.65}$$

When the simple mixture expression for g_m [Eq. 3.67] is applied,

$$\frac{C_{sc}}{I_r(0)} = \frac{1}{V_1\phi_1} + \frac{1}{V_2\phi_2} - 2\frac{\chi}{V_{ref}} \tag{7.66}$$

where χ is the interaction parameter and V_{ref} its reference volume [see Eq. 3.66]. Neither Eq. 7.64, which has no interaction term, nor Eq. 7.66, which has no q

dependence, is adequate for analyzing the scattering data of concentrated mixtures. These values can be combined to form a relationship that contains both by simply adding the interaction term in Eq. 7.66 to Eq. 7.64, as suggested by Higgins and Benoit[13]. The result is the *de Gennes equation*[25]:

$$\frac{C_{sc}}{I_r(q)} = \frac{1}{\phi_1 V_1 P_1(q)} + \frac{1}{\phi_2 V_2 P_2(q)} - 2\frac{\chi}{V_{ref}} \tag{7.67}$$

de Gennes derived this equation by a different procedure, the *random phase approximation,* which makes use of relationships that would take the analysis too far afield for purposes here. The Higgins-Benoit procedure, which uses direct addition of an interaction term, is a natural one to use, being the same as used to develop the simple mixture model itself. Thus, as discussed in Section 3.3, a configurational entropy of mixing, taking into account only intermolecular volume exclusion, was derived first. An interaction contribution was then simply appended to that result, the assumption being that the energy change is small enough to leave the structural arrangements unchanged. It would appear that nothing deeper is involved in going from Eq. 7.64 to Eq. 7.67 via the Higgins-Benoit route.

The de Gennes equation has been used primarily in the determination of chain dimensions and interaction strength for polymer blends (Chapter 8). It has the limitations of the simple mixture model. Accordingly, the assumption of spatial uniformity and local randomness makes it inappropriate for dilute solutions. There may well be problems in applying the equation to scattering in the semidilute range, in which properties such as R_g vary with concentration, and pair correlations depend on r in some complicated way. The nature of those problems and their importance in semidilute solution scattering is a matter for exploration. There seems no special reason, however, to reject its use for solutions in good solvents above approximately $\phi^{\ddagger}$ and even at the theta condition above approximately ϕ^*. In the usual terminology for polymer solutions ($\phi = \phi_p$, the polymer volume fraction; $V_2 = V_p$; $V_1 = V_s = V_{ref}$), Eq. 7.67 becomes:

$$\frac{C_{sc} V_s}{I_r(q)} = \frac{1}{1-\phi} + \frac{V_s}{\phi V_p P(q)} - 2\chi \tag{7.68}$$

in which $P(q)$ is the polymeric form factor or the Debye equation [Eq. 4.108] for linear chains.

At $q = 0$, the polymeric form factor is unity, making the second term on the right of Eq. 7.68 of order $V_s/V_p \sim 10^{-3}$ or less, unless the polymer volume fraction ϕ is very small. The polymeric contribution to forward scattering intensity is therefore negligible compared to the other two terms beyond the dilute range:

$$\frac{C_{sc}V_s}{I_r(0)} \sim \frac{1}{1-\phi} - 2\chi \tag{7.69}$$

The polymer contribution becomes significant, however, in the region of intermediate $q(R_g^{-1} < q < r_K^{-1})$. There, $P(q) = 2/(qR_g)^2$ for random coils [Eq. 4.119], and Eq. 7.68 becomes:

$$\frac{C_{sc}V_s}{I_r(q)} = \frac{1}{1-\phi} - 2\chi + \frac{1}{2\phi}\left(\frac{M_s\rho_p}{\rho_s}\right)\left(\frac{R_g^2}{M}\right)_p q^2 \tag{7.70}$$

where ρ is pure component liquid density. Accordingly, the scattering behavior beyond $\phi^\ddagger$, ~ 0.1 from Eq. 7.26, depends on two physical properties, χ and the unperturbed chain dimensions, R_g^2/M. Equation (7.70) reflects only local interactions and, for $\phi > \phi^\ddagger$, only the unperturbed chain dimensions. The Ornstein-Zernike correlation length derived from Eq. 7.70 by applying Eq. 7.63 has nothing to do with an excluded volume screening length.

7.3.3 Interpretation Alternatives

The modified de Gennes expression, Eq. 7.70, accounts well for reported scattering results. Power laws are not obtained, but the curves calculated with unperturbed chain dimensions of the species and known or estimated values of χ are quantitatively consistent with the reported correlation lengths[52]. In this interpretation, the angular dependence of scattering reflects nothing more than the interplay of unperturbed chain dimensions, polymer concentration, and local interactions as described, at least approximately, by Eq. 7.70. An expression for converting the reported values of ξ_{OZ} to values of χ is readily obtained from Eqs. 7.63 and (7.70):

$$\chi = \frac{1}{2}\left[\frac{1}{1-\phi} - \frac{1}{\xi_{OZ}^2\phi}\left(\frac{M_s\rho_p}{2\rho_s}\right)\left(\frac{R_g^2}{M}\right)_p\right] \tag{7.71}$$

The values of χ calculated through Eq. 7.71 using $(R_g^2/M)_o$ for polystyrene (Table 4.2) and the polystyrene–toluene ξ_{OZ} data in Figure 7.11a are compared in Figure 7.11b with results obtained by measurements of relative chemical potentials[31,56] (see Section 7.4.1). Although the trends with concentration do seem different, values from the two sources are certainly in the same range. The consistency with independently determined results even extends below $\phi^\ddagger$ and seemingly without major problems right into the ϕ^* range. More examples demonstrating similar agreement, and further discussion of this method of evaluating χ, are given in Section 7.4.1.

Based on these observations, the assumption that the correlation length from scattering gives reliable information on excluded volume screening length seems untenable. This is not to say that the concept of intramolecular excluded volume screening, with some attendant length scale related to concentration, should be rejected. On the contrary, the analyses of osmotic pressure and coil dimensions that flow from that idea lead to useful and nonobvious results. It simply means that other correlations exist, related to intermolecular interactions and detected by scattering, that are adequately described by the de Gennes equation and that, it would appear, overwhelm the screening contributions to scattering. Perhaps the original procedure—identifying the screening length with the $q^{-5/3}$ to q^{-2} crossover in scattering behavior[45]—offers the best chance after all.

7.3.4 Branch-Induced Ordering

The scaling and screening behavior of lightly branched polymers in semidilute solution should parallel closely that of linear polymers, as has indeed been found for symmetric star polymers with relatively few arms[39]. However, some new features appear for stars with many arms, primarily the development of an osmotic pressure jump and of liquidlike ordering near the overlap concentration. Both features are associated with the increased self-concentration of many-arm stars and the resulting resistance to molecular interpenetration as ϕ^* is approached[57].

As in monomeric liquids (Section 2.2.1), ordering can be detected by scattering methods. It is indicated by the presence of a broad peak at finite q in the q-dependence of scattering intensity. Huber et al.[58] observed the development of a SANS intensity peak with increasing concentration for twelve-arm polystyrene stars. Adam et al.[39] detected a very broad light scattering intensity maximum for an eighteen-arm polyisoprene star ($M_a = 378,000$) near ϕ^*. The SANS results of Willner et al.[59] for a much smaller eighteen-arm polyisoprene ($M_a = 7800$) are shown in Figure 7.12. The peak is sharpest near ϕ^*, about 0.1 in this case, and gradually subsides at both higher and lower concentrations. Further examples of ordering have been demonstrated with f-arm polyisoprene and polybutadiene stars for $8 \leq f \leq 128$[60], and even ordering in the melt state for 132-arm stars[61]. The osmotic jump, a rapid rise in osmotic pressure over a relatively narrow range of concentrations, reflects the resistance to coil interpenetration. It was seen first as only slight displacements from linear chain behavior beyond ϕ^* for eight-arm and eighteen-arm stars[39]. It was later demonstrated in quite spectacular fashion for stars with 32, 64, and 128 arms[62]. The jump magnitude was found to be proportional to $f^{3/2}$, as predicted earlier by Witten et al.[57].

The properties of star polymers in the semidilute region, like the dilute solution examples in Section 6.3.3, demonstrate a smooth evolution in behavior, from linear chain to colloidal particle, with increasing numbers of arms. Self-concentration

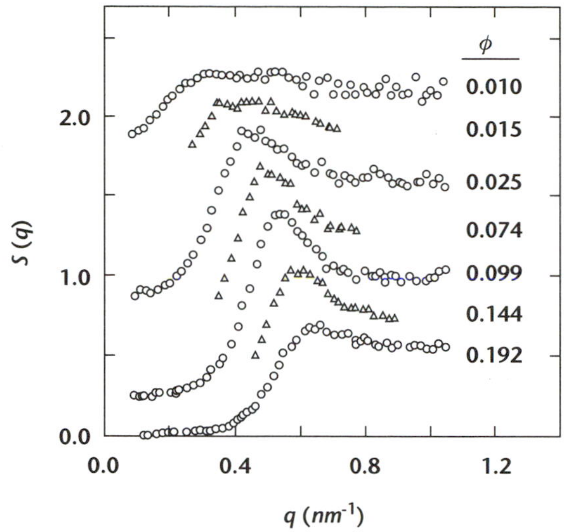

FIGURE 7.12 Structure factor for an eighteen-arm 1,4-polyisoprene star ($M_a = 7{,}500$) for several concentrations in a good solvent methylcyclohexane (from Willner et al.[59]). The units of $S(q)$ are arbitrary; $c^* \sim 0.1$; the data have been shifted vertically by various amounts for clarity.

appears to be the controlling variable, with relatively large values evidently required to evoke colloidlike properties. In many cases of practical interest, however, the self-concentrations of branched polymers, like those of linear polymers, are small enough to permit easy coil interpenetration and solution uniformity at all concentrations beyond overlap.

7.4 The Interaction Parameter

The limitations of the Flory-Huggins model were considered briefly in Section 7.2. Such limitations notwithstanding, Eqs. 7.32 through 7.35 provide a very useful framework for organizing the thermodynamic data on polymer solutions. The interaction parameter is defined operationally as:

$$\chi \equiv \frac{(\mu_s - \mu_s^{\rm o}) - (\mu_s - \mu_s^{\rm o})_{com}}{\phi^2 RT} \tag{7.72}$$

where $(\mu_s - \mu_s^{\rm o})$ is the relative chemical potential of solvent as determined experimentally and $(\mu_s - \mu_s^{\rm o})_{com}$ is the combinatorial contribution $RT[\ln(1-\phi) + (1 - V_s/V_p)\phi]$. As discussed briefly in Section 3.6.2, this is the Flory definition of χ, which permits it to depend on ϕ as well as T when χ reflects short range interactions, not on chain length for long chains.

7.4.1 Evaluation of χ

Three χ evaluation methods—vapor sorption, osmometry, and inverse chromatography—provide the relative chemical potential of solvent directly. The fourth, a scattering method, provides χ_{scat}, which differs slightly from χ as defined by Eq. 7.72 but is easily converted to it. In evaluating χ, the effects of volume change with mixing are ignored, and r, a measure of the number of mers per chain, is taken to be large. The interpretation of the interaction parameters so obtained, for example, the effect of differences in component cohesive energy density and free volume, is described in Section 7.4.2.

Vapor composition and sorption. As discussed briefly in Section 3.3, the magnitude of departures from Raoult's law at ambient pressures ($P \sim 0.1\ MPa$) is an especially striking characteristic of polymer solutions. Raoult's law relates the vapor and liquid compositions at equilibrium through p_i^o, the pure component vapor pressures. With y_i and x_i the vapor and liquid phase mole fractions,

$$y_i = \left(\frac{p_i^o}{P}\right) x_i \qquad \text{(Raoult's law)} \tag{7.73}$$

The vapor pressure of polymers is undetectably small—the mole fraction in the vapor phase is effectively zero at all compositions. For typical solvents, however, the mole fraction in the gas phase is readily measurable and, over a wide range of solution composition, y_1 is much larger than values calculated with Eq. 7.73.

Large departures from Raoult's law are quite general for polymer solutions. They increase with polymer molecular weight, are insensitive to temperature, and are found to similar degree for both polar and nonpolar constituents[7]. The main features of such behavior are readily understood from Eq. 7.33. Thus, at modest pressures, the chemical potential of solvent can be expressed in terms of gas-phase properties[63]:

$$\mu_s - \mu_s^o = RT \ln \frac{y_s P}{p_s^o} \tag{7.74}$$

From Eq.(7.33),

$$\ln \frac{y_s P}{p_s^o} = \ln(1-\phi) + \left(1 - \frac{1}{r}\right)\phi + \chi\phi^2 \tag{7.75}$$

The vapor composition at equilibrium is related mainly to ϕ and χ. Beyond ϕ^*, y_s is insensitive to r if r is large (see below). Fitting vapor composition data to Eq. 7.75 was the first method used for determining χ in polymer–solvent systems[7]. It is easiest to apply and most sensitive to χ in the concentrated region. Experimental difficulties increasingly plague the vapor composition-sorption methods as concentration is decreased. Such methods are seldom useful at concentrations below about 30% polymer.

Osmometry. Osmotic pressure measurements offer another source of information about the chemical potential of solvent in polymer solutions. From Eq. 7.35,

$$-\frac{\pi V_s}{RT} = \ln(1-\phi) + \left(1-\frac{1}{r}\right)\phi + \chi\phi^2 \tag{7.76}$$

Osmotic pressure at concentrations above ϕ^* has been obtained for several polymer–solvent systems[41,64–71]. Osmotic compressibility $(\partial\pi/\partial\phi)^{-1}$, obtained from scattering intensity data extrapolated to zero angle by applying Eq. 7.29, offers another source of osmotic pressure information[39,40]. Like vapor composition, osmotic pressure is expected to depend mainly on ϕ and to be insensitive to chain length, except at low concentrations. That is indeed found true, as was discussed in Section 7.2.3.

The measurement of osmotic pressure becomes increasingly difficult at higher polymer concentrations. The pressures eventually become large enough to present serious mechanical problems in membrane support, and the equilibration times become impracticably long as well. Solution clarification for forward scattering determinations also becomes difficult at elevated concentrations. Osmometry is accordingly useful for evaluations of χ over a lower range of concentrations than vapor sorption. Sorption and osmometry are in fact complimentary methods. Figure 7.13 compares values of χ obtained by high-pressure osmometry (shown as osmotic pressure

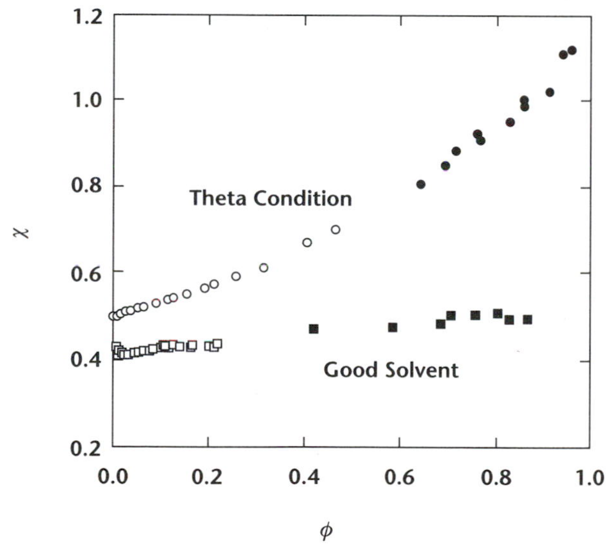

FIGURE 7.13 Interaction parameter as a function of concentration for polyisobutylene in a good solvent cyclohexane at 30°C and in benzene at the theta condition (24.5°C) (from Eichinger and Flory[65,66]). Filled symbols indicate values from vapor sorption and open symbols from osmometry.

in Figure 7.8) and vapor sorption for polyisobutylene solutions in the good solvent cyclohexane and at the theta condition in benzene. Although the ranges covered by the two methods do not quite overlap, it is clear enough that the results they provide are mutually consistent.

Inverse chromatography. In this method—called inverse because it supplies information about stationary phase properties—an aliquot of the solvent is injected as a small pulse into an inert carrier gas passing continuously over the polymeric liquid (the stationary phase in a chromatographic column), at some steady volumetric flow rate $\dot{q}$. The time lapse from injection to the emergence of the pulse is related to the elution volume $\upsilon = \dot{q}\,\Delta t$. With suitable calibration and a small enough pulses, υ can be related to the relative chemical potential of solvent at equilibrium with the polymer phase for low solvent concentrations[72]. The method, combined with the Flory-Huggins model and a good estimate of the second virial coefficient for solvent in the gas phase, provides values of χ at $\phi \sim 1$[73,74].

Inverse chromatography is readily applied at elevated temperatures, an experimentally awkward region for osmometry and conventional vapor sorption. It is thus especially valuable for polymers that are crystalline at room temperature[75–77]. The chromatographic method has the disadvantage that only values of χ near $\phi = 1$ are obtainable, but as shown below, the interaction parameter is not very sensitive to ϕ for good solvents, the usual case of interest.

Values of χ obtained by inverse chromatography for linear polyethylene in several n-alkane solvents—C_8, C_9, C_{10}, and C_{12} at 149°C[75]–are shown in Figure 7.14a. These values form a trend with a reciprocal carbon number that is qualitatively consistent with the expectation of $\chi = 0$ at $1/n = 0$; that is, for a mixture of polyethylene in polyethylene. As shown in Figure 7.14, they also agree fairly well with other data for the n-alkane polyethylene system obtained by small-angle scattering, discussed next, and from PVT-based information, discussed in Section 7.2.3. The significance of non-zero χ for solutions of polymers in their own oligomers is considered briefly in Section 7.4.2.

Scattering. The value of χ obtained from the angular dependence of scattering intensity through applying Eq. 7.71 depends on the derivative of chemical potential. Thus, from Eq. 7.72,

$$RT\frac{\partial \phi^2 \chi}{\partial \phi} = \left(\frac{\partial \mu_s}{\partial \phi}\right) - \left(\frac{\partial \mu_s}{\partial \phi}\right)_{com} \tag{7.77}$$

Equations (3.31) and (7.69) then lead to a relationship between the interaction parameter values obtained from scattering experiments and those from relative chemical potential:

$$\chi_{scat} = \frac{1}{2\phi}\frac{\partial \phi^2 \chi}{\partial \phi} \tag{7.78}$$

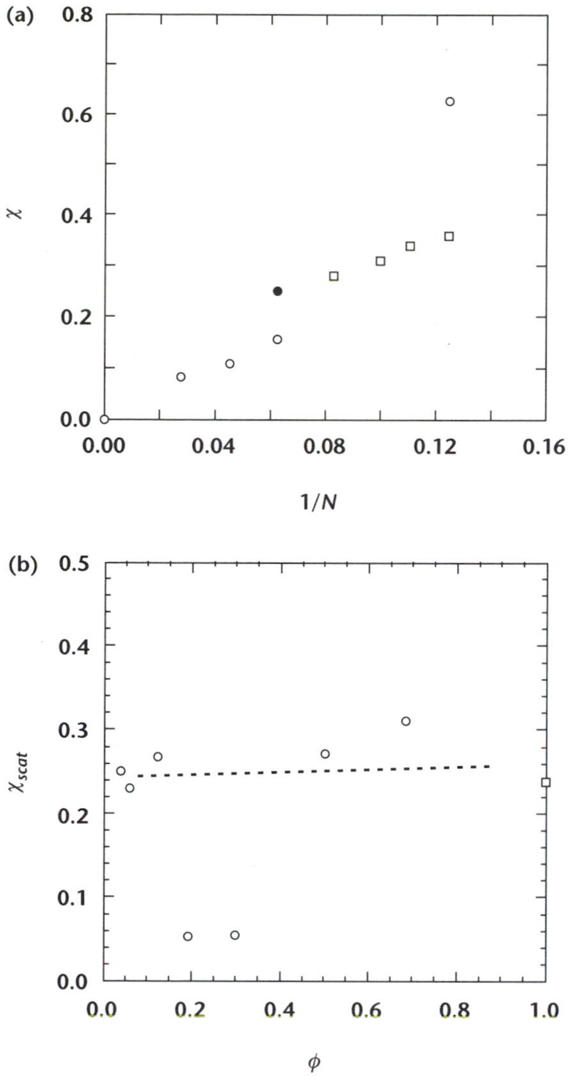

FIGURE 7.14 (a) Interaction parameters from gas chromatography for solutions of linear polyethylene in various n-alkanes at 149°C. The values (□)(from Schreiber et al.[75]) are plotted against the reciprocal of carbon number. Also shown are values based on scattering data from Tao et al.[53] (●) and on FOV parameters and simulations (O) as explained in Section 7.2.3. (b) Interaction parameters from scattering for polyethylene in n-hexadecane at 140°C. A value of $\chi(1)$ obtained by extrapolation the chromatography-based values in Figure 7.16a to $n = 16$ is also shown (□).

If χ does not depend on ϕ, then $\chi_{scat} = \chi$, but if a dependence exists, then Eq. 7.78 provides a conversion formula from one definition to the other. Conversion formulas to the interaction parameter then appears in the expression for free energy of mixing density itself can also be derived. As noted in Section 3.4, however, there is not much point to applying them, because all observable properties of mixtures depend on the compositional derivatives of g_m, not on g_m itself. The definition given by Eq. 7.72 is used here. Finally, if the composition dependence of χ_{scat} is expressed as a power series, the coefficients for χ_{scat} and χ are related through Eq. 7.78:

$$\begin{aligned} \chi_{scat} &= a_o + a_1\phi + a_2\phi^2 + \cdots \\ \chi &= a_o + 2a_1\phi/3 + a_2\phi^2/2 + \cdots \end{aligned} \tag{7.79}$$

Values of χ_{scat} obtained from ξ_{OZ} using Eq. 7.71 are shown for various systems in Figures 7.14b through 7.16. Only scattering data obtained for concentrations beyond ϕ^* are used. For the good solvent system polyethylene–hexadecane at 140°C, shown in Figure 7.14b[53], two of the points depart considerably; the other points indicate $\chi_{scat} \sim 0.25$ and insensitivity to composition over a wide range. The results are also consistent, perhaps fortuitously, to $\chi(1) = 0.24$ for PE-C_{16}, obtained by extrapolating the correlation of chromatographic results in Figure 7.14a.

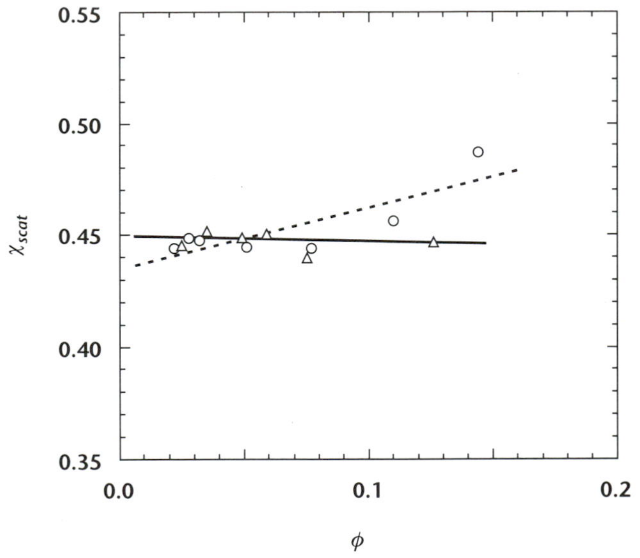

FIGURE 7.15 Interaction parameters calculated from scattering data for polystyrene in good solvents. Scattering data in methylene chloride (O) and tetrahydrofuran (Δ) from Brown et al.[47] were used (note Figure 7.10b).

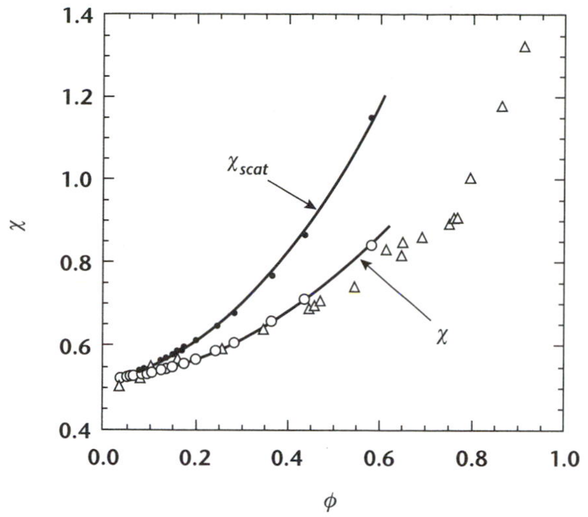

FIGURE 7.16 Interaction parameters from scattering for polystyrene at the theta condition in cyclohexane. The filled circles (●) are χ_{scat} values obtained from scattering data, as reported by Brown and Nicolai[48]; the open circles (○) are those from converting χ_{scat} to χ; the open triangles (Δ) are values of χ obtained with osmometry and vapor sorption measurements by Höcker et al.[70].

The results typical of good solvents, $0.40 \lesssim \chi \lesssim 0.48$ with relatively weak composition dependence, are shown in Figure 7.15 for solutions of polystyrene in methylene chloride and tetrahydrofuran[47]. A value of $\chi\,(1) = 0.34$ for PS-MEC at 40°C and the range $-0.16 \leq \chi\,(1) \leq 0.7$ for PS-THF at elevated temperatures has also been reported[78].

Figure 7.16 compares χ_{scat} and χ for polystyrene in cyclohexane at the theta condition. The values of χ_{scat} were obtained with ξ_{OZ} data from several sources[48]; those of χ were obtained from osmotic pressure and vapor sorption measurements[70]. Although both approach 0.5 at low concentrations, they progressively diverge with increasing concentration, with observable differences beginning at about $\phi = 0.15$. If χ varies significantly with concentration, as is the case here, departures between χ_{scat} and χ with increasing concentration are to be expected according to Eq. 7.79. The values from scattering fit a quadratic equation rather well: $\chi_{scat} = 0.520 + 0.131\phi + 1.61\phi^2$, yielding from Eq. 7.79 $\chi = 0.520 + 0.087\phi + 0.805\phi^2$. Each value of χ_{scat} was converted to χ with this equation and shown in Figure 7.16. Although not perfect, the agreement is good enough to support the use of the scattering method. Thus, the relatively few test cases available show reasonable consistency between the scattering method and more traditional means for determining values of the interaction parameter.

7.4.2 Interpretation of χ

The interaction parameters for two good solvent systems—natural rubber-benzene[64] and polyisobutylene-cyclohexane[66]—are shown for a wide range of concentrations in Figure 7.17a. The results are remarkably similar: The intercept $\chi(0)$ is approximately 0.40 for both, $\chi(\phi)$ is essentially linear in ϕ over the entire range, and the slopes, $(\partial\chi/\partial\phi)_o \sim 0.1$, are about the same. Interaction parameters for polymers at the theta condition—polyisobutylene-benzene at 24.5°C[65], polystyrene-cyclohexane at 34°C[70], polydimethylsiloxane-methylethylketone at 25°C[79]—are shown in Figure 7.17b. As in the good solvent comparison, the behavior of $\chi(\phi)$ for all three systems is remarkably similar over the entire range. Not surprisingly, all three intercepts are near the theta solvent value of 0.5 (see Section 7.3.1). However, the initial slopes are also about the same, as is the curvature at higher concentrations. The concentration dependence of χ is much stronger at the theta condition than for good solvents. The initial slope is larger, $(\partial\chi/\partial\phi)_o \sim 0.25$, and that, together with the curvature, means that χ at high concentrations is much larger for theta solvents.

It is obviously dangerous to infer general behavior from the observations for only five systems. Many others have been studied, and compilations with data sources are available[78,80,81]. The properties for a selection of well-studied systems are given in Table 7.3. The first four are theta systems $[\chi(0) = 0.5]$. The next three—polymethylmethacrylate–acetone, polyisobutylene–pentane, and polystyrene–methylethylketone—are poor solvent systems $[0.47 \lesssim \chi(0) \lesssim 0.49]$, sometimes called *marginal solvent systems*, but in any case between good and theta systems in solvent quality. The values of χ increase significantly with concentration, but not as rapidly as they do for the theta systems. The others in the table are good solvent systems. The typical range for good solvents is $0.40 \lesssim \chi(0) \lesssim 0.46$, and the concentration dependence is typically weak, $(\partial\chi/\partial\phi)_o \lesssim 0.1$, and sometimes even negative. Occasionally, values less than 0.40 are found, polymethylmethacrylate–chloroform and polyethylene oxide–benzene being clear examples. Other examples include the polyethylene-n-alkane series (Figures 7.14a and 7.14b). The most striking observation about all these data is the absence of systems with interaction parameters anywhere near $\chi(0) = 0$: what might be called the interactionless Flory-Huggins solution. These observations are considered in more detail later, where the effect of component free volume differences is considered.

When Eqs. 3.72 and 3.73 are applied to χ versus T data for polymer solutions, it is found that χ_h is small compared with χ_s: The experimental χ is much less sensitive to temperature than $1/T$ behavior. Indeed, for many good solvent systems, the values are essentially independent of temperature. Such behavior appears to be general for polymer solutions. Moreover, the enthalpic contribution to χ is typically positive, reflecting a net repulsion between polymer and solvent and thus unfavorable to mixing. The entropic contribution seems always to be positive, and thus is also

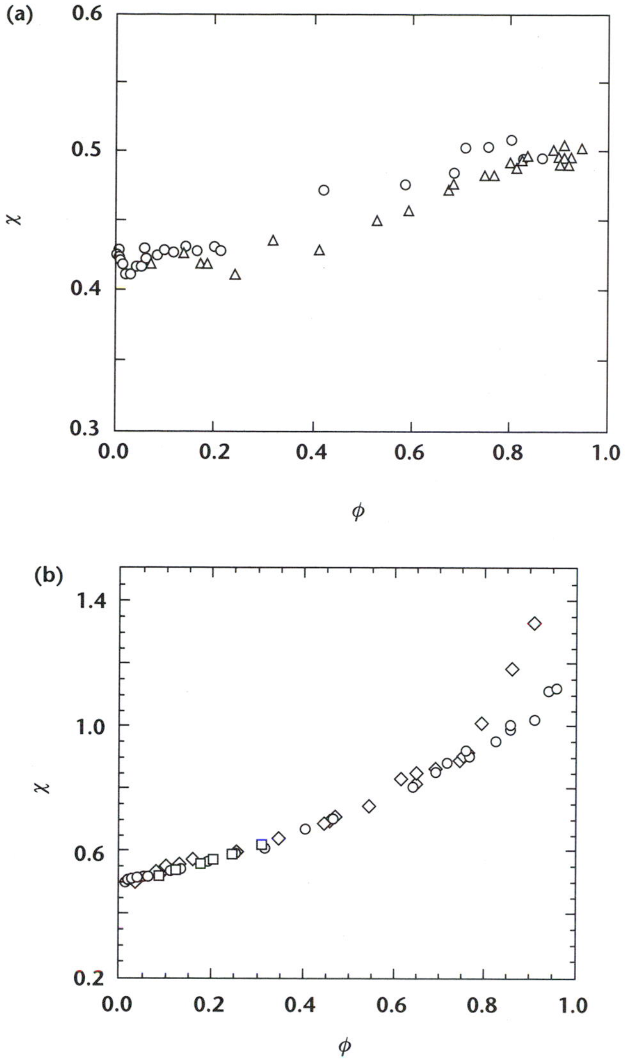

FIGURE 7.17 (a) Interaction parameter as a function of concentration for solutions in good solvents: polyisobutylene–cyclohexane (○) from Eichinger and Flory[66], and polyisoprene (natural rubber)–benzene (Δ) from Eichinger and Flory[64]. (b) Interaction parameter as a function of concentration for solutions at the theta condition: polystyrene-cyclohexane (○) from Höcker et al.[70], polyisobutylene–benzene (Δ) from Eichinger and Flory[65], and polydimethylsiloxane–methyl ethyl ketone from Shiomi et al.[79].

TABLE 7.3 Interaction parameters near the dilute limit for selected polymer/solvent systems

	Experimental Values[a]				Calculated Values[b]	
System	$\chi(0)$	$(\partial\chi/\partial\phi)$	$\chi_h^{(c)}$	$\chi_s^{(c)}$	χ_{EXC}	χ_{FV}
PS/cyclohexane	0.50	0.4	0.3	0.2	0.03	0.42
PDMS/benzene	0.50	0.3	-	-	0.65	0.05
PDMS/2-butanone	0.50	0.4	-	-	0.69	0.10
PIB/benzene	0.50	0.4	0.26	0.24	0.16	0.31
PMMA/acetone	0.48	-	0.03	0.45	0.00	0.48
PIB/n-pentane	0.48	0.3	-	-	0.01	0.58
PS/2-butanone	0.47	0.4	−0.03	0.50	0.01	0.50
PIB/n-octane	0.44	0.2	−0.17	0.63	0.00	0.45
PIB/cyclohexane	0.41	0.1	-	-	0.03	0.34
PI/benzene	0.42	0.1	0.1	0.3	0.01	0.16
PS/chloroform	0.45	-	-	-	0.03	0.41
PS/toluene	0.44	−0.1	-	-	0.00	0.38
PS/ethylbenzene	0.39	0.3	−0.02	0.42	0.00	0.41
PDMS/cyclohexane	0.42	0.19	0.19	0.23	0.39	0.05
PDMS/n-octane	0.40	-	-	-	0.27	0.07
PVAc/benzene	0.42	-	-	-	0.02	0.16
PVAc/acetone	0.40	-	-	-	0	0.28
PMMA/benzene	0.45	-	0.03	0.42	0.01	0.34
PMMA/toluene	0.45	-	0.03	0.45	0.02	0.33
PMMA/chloroform	0.36	-	−0.08	0.44	0.00	0.37
PEO/benzene	0.26	−0.10	-	-	0.02	0.25
PPO/benzene	0.28	−0.21	-	-	~0	-

[a] Selected from the values in refs. 78, 80, 81, 89, and 90.
[b] Estimated with Eqs. 7.81 and 7.87 as described in the text.
[c] Obtained from the temperature dependence of $\chi(0)$ see Eq. 3.72.

unfavorable to mixing. Values of χ_s and χ_h for several polymer–solvent systems are given in Table 7.3. These various observations are considered here.

Free volume effects. As shown by the examples Figures 7.14 through 7.17 and Table 7.3 but found quite generally, the interaction parameters for solutions of non-polar or weakly polar polymers are always positive and seldom smaller than about 0.40. Positive values are consistent with the solubility parameter formalism for χ (Section 3.5), but the absence of *small* positive values is surprising. For regular solutions [Eq. 3.93] and solvents with the typical molar volume, $V_1 = 10^2\, cm^3$, the available miscible range, $0.4 \lesssim \chi \lesssim 0.5$, corresponds to a difference in solubility parameters in the range $3.2 < |\delta_1 - \delta_2| < 3.5$ in units of $(J\, cm^{-3})^{1/2}$ or $MPa^{1/2}$. Because polymeric liquids are nonvolatile, their solubility parameters cannot be obtained in the usual way, using the internal energy of evaporation. Nonetheless, solubility parameters are local properties, and those for polymers originate from the same kinds of functional groups and associated intermolecular forces that govern the solubility parameters of their solvents. The solubility parameter values for nonpolar and weakly polar molecular liquids in Table 2.6 span a range of only about $5(J cm^{-3})^{1/2}$. It is quite inconceivable that, among the many polymer–solvent systems studied, essentially all just happen to have solubility–parameter differences of less than about $3.5(J\ cm^{-3})^{1/2}$ and yet none have differences of less than about $3(J cm^{-3})^{1/2}$. Evidently, something besides solubility parameter differences contributes significantly to the interactions in polymer solutions.

As discussed in Section 3.6, interactions can arise in mixtures from a difference in component free volumes. In fact, a large systematic difference occurs between the free volume fraction in polymeric liquids and monomeric liquids. From Eq. 3.108, free volume fraction f_{υ} is directly related to reduced volume $\tilde{\upsilon}$ at the temperature of interest. That quantity, as well as the characteristic pressure and temperature P^* and T^*, can be obtained with the thermal expansion coefficient α and isothermal compressibility β for the liquid of interest, using the FOV-based Eqs. 3.118 through 3.120. The values of α and β for selected monomeric liquids are given in Table 2.2. The values for selected polymeric liquids[82] are given in Table 7.4. The values of $\tilde{\upsilon}$, P^*, and T^* derived therefrom are listed in Tables 7.5 and 7.6. The values of the free volume fractions at a common temperature of 20°C, calculated using Eq. 3.108, are also given in Tables 7.5 and 7.6. For the monomeric liquids, the average $\bar{f}_{\upsilon}$ is about 0.22—about 0.12 for the polymers—with no overlap from the two populations. The free volume fraction and PVT properties are thus related, and from Eqs. 3.109 and 3.118, the difference in free volume fraction is evidently dictated by the difference in thermal expansion coefficients.

TABLE 7.4 Thermal expansion coefficient and isothermal compressibility for selected polymeric liquids[82]

Species	$T(°C)$	$\rho(g\ cm^{-3})$	$\alpha(K^{-1}) \times 10^4$	$\beta(MPa^{-1}) \times 10^4$
polyisobutylene	75	0.892	5.64	6.2
polystyrene	175	0.974	5.83	7.3
poly(methylmethacrylate)	175	1.116	6.3	4.1
1,4 cis polyisoprene	20	0.916	6.6	5.0
i-polypropylene	200	0.754	6.6	13.8
i-poly(1-butene)	180	0.776	6.7	12.1
poly(vinylmethylether)	125	0.984	7.07	9.3
poly(ethylene oxide)	150	1.022	7.1	7.8
poly(vinyl acetate)	65	1.155	7.18	5.9
polyethylene	175	0.765	7.22	11.7
poly(dimethylsiloxane)	50	0.948	9.14	13.9

TABLE 7.5 FOV parameters at 20°C for selected monomeric liquids

Liquid	$T^*(K)$	$P^*(MPa)$	$v^*(cm^3\ g^{-1})$	$\tilde{v}$	f_v
n-tetradecane	5624	415	1.077	1.217	0.178
n-decane	5134	427	1.098	1.248	0.199
ethylbenzene	5134	541	0.924	1.248	0.199
n-hexanol	5107	572	0.980	1.250	0.200
toluene	5053	540	0.920	1.254	0.203
acetic acid	4977	553	0.759	1.26	0.206
carbon disulfide	4882	562	0.624	1.268	0.211
benzene	4838	557	0.894	1.272	0.214
carbon tetrachloride	4838	515	0.493	1.272	0.214
cyclohexane	4816	484	1.009	1.274	0.215
n-octane	4794	432	1.089	1.276	0.216
chloroform	4693	588	0.523	1.285	0.222
2-butanone	4547	570	0.956	1.300	0.231
ethyl acetate	4450	602	0.847	1.311	0.237
ethanol	4376	638	0.960	1.320	0.242
n-hexane	4361	432	1.147	1.322	0.243
acetone	4293	601	0.951	1.331	0.248
methanol	4255	644	0.946	1.336	0.251
n-pentane	4085	408	1.173	1.361	0.265

Obtained with Eqs. 3.108 and 3.118–20 from the data in Table 2.2.

From Tables 2.2 and 7.4, α is consistently larger for solvents than for polymers. The average of those listed is $10^4\alpha(K^{-1}) = 11.9 \pm 1.8$ for the solvents and 6.86 ± 0.9 for the polymers, with essentially no overlap. [In contrast, the average of the isothermal compressibilities from the same tables, $10^4\beta(MPa) = 10.8 \pm 2.0$ for solvents and 9.1 ± 3.2 for polymers, is much closer with extensive overlap.] A systematic free volume contribution to the net interactions in polymer solutions can therefore be expected.

Equations (3.125) through (3.127) provide an approximate expression of the free volume contribution to the interaction parameter, according to a modified FOV model. For polymer solutions,

$$\chi_{FV}(\phi) = \chi_{FV}(0)\left[1 + \frac{2\left(3 - 4\alpha_s T - 4\alpha_s^2 T^2\right)}{9}\left(\frac{T_p^* - T_s^*}{T_p^*}\right)\phi + \cdots\right] \tag{7.80}$$

in which:

$$\chi_{FV}(0) = \frac{(\Pi_{CED} V\alpha)_s}{2R}\left(\frac{T_p^* - T_s^*}{T_p^*}\right)^2 \tag{7.81}$$

TABLE 7.6 FOV parameters for selected polymeric liquids

Liquid	$T(°C)$	$T^*(K)$	$P^*(MPa)$	$v^*(cm^3\ g^{-1})$	$\tilde{v}$	f_v	$\tilde{v}(20°C)$	$f_v(20°C)$
polyisobutylene	75	7874	436	0.956	1.173	0.148	1.139	0.121
polystyrene	175	8476	534	0.840	1.222	0.181	1.127	0.113
poly(methylmethacrylate)	175	8105	583	0.725	1.237	0.191	1.134	0.118
1,4 cis polyisoprene	20	6695	530	0.932	1.171	0.146	1.171	0.146
polypropylene	230	8267	351	1.062	1.274	0.215	1.130	0.115
poly(1-butene)	180	7871	393	1.030	1.251	0.201	1.139	0.122
poly(vinylmethylether)	125	7213	462	0.822	1.236	0.191	1.155	0.134
poly(ethylene oxide)	150	7392	601	0.783	1.249	0.199	1.150	0.130
poly(vinyl acetate)	65	6682	601	0.717	1.208	0.172	1.171	0.146
polyethylene	175	7522	442	1.033	1.265	0.209	1.147	0.128
poly(dimethylsiloxane)	50	5698	330	0.847	1.246	0.197	1.213	0.176

Obtained with Eqs. 3.108 and 3.118–20 from the data in Table 7.4.

Equation 7.81 is the *Patterson formula*[83] as applied to polymer solutions. The interaction parameters estimated using Eqs. 7.80 and 7.81 are in the range of observed values for polymer–solvent systems. With typical property magnitudes for nonpolar or mildly polar monomeric liquids—$\Pi_{CED} \sim 340\, Jcm^{-3}$ (average from Table 2.6 with the last three entries omitted), $V_s \sim 10^2\, cm^3, \bar{\alpha} = 1.2 \times 10^{-3}\, K^{-1}$—the average free volume contributions become:

$$\chi_{FV}(0) \sim 2.4\left(1 - T_s^*/T_p^*\right)^2 \tag{7.82}$$

$$\left(\frac{\partial \chi_{FV}}{\partial \phi}\right)_o \sim 0.24\left(\frac{T_p^* - T_s^*}{T_p^*}\right)\chi_{FV}(0) \tag{7.83}$$

At ambient pressures, the following expression, easily obtainable from Eqs. 3.118 and 3.119, relates the characteristic temperature to the thermal expansion coefficient:

$$T^* = \frac{3}{\alpha}\frac{(1 + 4\alpha T/3)^4}{(1 + \alpha T)^3} \tag{7.84}$$

Working with these averages, and using $T = 300K$ for the solvents and $T = 400K$ for the polymers (the temperature ranges in which most data for the two sets were obtained), Eq. 7.84 gives $T_s^* = 4790K$ and $T_p^* = 7360K$, or an average $T_s^*/T_p^* \sim 0.65$ for polymer solutions. Thus leading, with Eqs. 7.82 and 7.83, to a rough global average for polymer solutions:

$$\begin{aligned} \chi_{FV}(0) &\sim 0.30 \\ (\partial \chi_{FV}/\partial \phi)_o &\sim 0.025 \end{aligned} \tag{7.85}$$

The estimate of χ at low concentrations agrees well with the proposal of Blanks and Prausnitz[84] for good solvents, $\chi_s = 0.34$. It is certainly in the range of observed values for good solvents in Table 7.3. The estimate of concentration dependence has the typically observed sign, but it is smaller in magnitude than that found for most good solvent systems. Assuming for simplicity that the contributions of cohesive energy mismatch and free volume mismatch are additive, the interaction parameter can be written, as in Eq. 3.130,

$$\chi = \chi_{FV} + \chi_{CED} \tag{7.86}$$

in which χ_{FV} can be estimated using Eq. 7.81, and χ_{CED} can be expressed in terms of solubility parameter differences through Eq. 3.93 and then estimated with δ_s and

δ_p from the internal pressures of the components using Eq. 3.120:

$$\chi_{CED} \sim \chi_{IP} = \frac{V_s}{RT}\left[\left(P_s^*/\tilde{v}_s^2\right)^{1/2} - \left(P_p^*/\tilde{v}_p^2\right)^{1/2}\right]^2 \tag{7.87}$$

which is an alternative form of Eq. 3.129. The following is equivalent to Eq. 7.86, but expressed entirely in terms of pure component properties of solvent and polymer that can be obtained from fits of PVT data to the FOV equation of state:

$$\frac{\chi\, RT}{V_s} = \left[\left(P_s^*/\tilde{v}_s^2\right)^{1/2} - \left(P_p^*/\tilde{v}_p^2\right)^{1/2}\right]^2 + \frac{\alpha_s T\, P_s^*}{2\tilde{v}_s^2}\left[1 - T_s^*/T_p^*\right]^2 \tag{7.88}$$

The values of χ_{CED} and χ_{FV} for several of the well-characterized systems discussed previously and estimated using Eq. 7.88 and the FOV-based data in Tables 7.5 and 7.6, are listed in Table 7.3. Given the wide choice of solvents available, it is relatively easy to match the solubility parameters to arrive as near zero as desired. Thus, the values of P^* extend over a range for both solvents and polymers, and the ranges overlap extensively. Even if solubility parameters are not available, systematic experimentation with solvent choice should suffice to reduce the first term on the right of Eq. 7.88 to zero. The values of T^* are systematically larger for polymeric liquids: The ranges for polymers and solvents either do not overlap, or, if the values of T^* do coincide, then the values of P^* differ enough to make the exchange interaction large. Those for χ_{CED} are small in nearly all cases; the values of χ_{FV} dominate the observed magnitude of $\chi = \chi_{FV} + \chi_{CED}$.

Solutions of polymers in their oligomers appear to be exceptions to this interpretation. Thus, as shown in Figure 7.14a, chromatographic values of χ for polyethylene in n-alkanes[75] decrease systematically toward zero as the diluent chain length increases and reach values well below 0.4 even for rather modest lengths. However, free volume ideas not only predict this trend but also supply a reasonable estimate for the values observed. Orwoll and Flory[85] measured and also compiled from other sources a body of accurate data on liquid densities, expansion coefficients, and thermal pressures for linear polyethylene and several n-alkanes—C_6,C_8,C_{16},C_{22},C_{36}. These data permit the evaluation of molar volume, internal pressure Π_{IP} from Eq. 2.113, and characteristic pressure T^* from Eq. 7.84. The simulation-based relationship between cohesive energy and internal pressure for n-alkanes[86] was then used to obtain Π_{CED}. These values are sufficient to evaluate χ_{FV} using Eq. 7.81 and χ_{CED} using Eqs. 3.93 and 3.94 to finally obtain χ using Eq. 7.86. The values (C_6 omitted) are plotted in Figure 7.14a. The agreement with results from other methods is reasonable. For the three longest chain solvents, χ_{FV} is 2 to 4 times larger than χ_{CED}.

The systematic difference in thermal expansion coefficient, and hence in T^*, between monomeric and polymeric liquids under comparable conditions of intermolecular potentials and distances is not a complete surprise. For polymeric liquids, covalent bond lengths determine distances between adjacent intrachain units, and these are essentially independent of temperature. Only the distances between interchain neighbors, those which are normal to the local chain trajectory direction, are free to change with temperature through the competition between van der Waals attraction and repulsion that governs thermal expansion in monomeric liquids. From this, one might guess $\alpha_p/\alpha_m = 2/3$, a ratio not very different from that observed for averages quoted above, $\bar{\alpha}_p/\bar{\alpha}_m = 0.58$. One might also suppose that the Prigogine parameter c, the so-called ratio of external to total degrees of freedom, would reflect this difference as well, and that c_p/c_s would have a value of about 2/3 as well. According to Eq. 3.110, if $P_s^* = P_p^*$ then $c_p/c_s = T_s^*/T_p^*$, and the observed ratio of averages, $\bar{T}_s^*/\bar{T}_p^* = 0.65$, is remarkably close to the guess of 1/2. In that sense at least, polymeric liquids could perhaps be said to have a kind of two-dimensional character in their thermodynamic properties that has a significant impact on their solution properties.

Impact on dilute solutions. In Chapter 6 it was noted that, for even the best good solvents, the expansion of polymer chain dimensions in dilute solutions is much smaller than the expansion of self-avoiding walks in simulations[87,88]. That observation is the intramolecular counterpart of the intermolecular interaction behavior discussed early in this section: Namely, that the interaction parameter is not only positive but also bounded well away from zero in virtually all polymer solutions. Both indicate the presence of a net attraction between chain units that offsets a significant part of the hard-core excluded-volume repulsion. As shown above, the inherent difference of free volume density in monomeric and polymeric liquids produces a net repulsion of about the right magnitude to explain the intermolecular effect. It thus seems reasonable to attribute the excluded volume weakening in dilute solutions to the same cause.

The trends and even the magnitudes of the intramolecular and intermolecular effects are consistent with one another. From Eqs. 6.31 and 6.33, the excluded volume parameter z is related to the characteristic molecular weight $M^\ddagger$. From Eq. 6.35, $M^\ddagger$ and hence $n_K^\ddagger$, the characteristic number of Kuhn steps, then vary as $1/(1/2 - \chi)^2$ for a given set of polymer and solvent properties with different interaction strengths. The athermal case is $\chi = 0$, so $n_K^\ddagger/(n_K^\ddagger)_{athermal} = (1/4)/(1/2 - \chi)^2$. The athermal case corresponds to the strictly self-avoiding walk, and from Table 6.5, $n_K^\ddagger \sim 2$. Hence,

$$n_K^\ddagger \sim \frac{2}{(1 - 2\chi)^2} \tag{7.89}$$

From Table 7.3, based on intermolecular interaction measurements, it would appear that $\chi \sim 0.38$ is about the minimum interaction parameter for the typical good solvent range. For that case, the estimate from Eq. 7.89 is $n_K^\ddagger \sim 30$. That value overstates the effect to some extent, as judged by the values for good solvent systems in Table 6.5, but it is certainly in the right range.

Athermal solutions. The phrase "athermal polymer solutions" is widely used in the literature. Sometimes mere insensitivity of volume exclusion to temperature is meant. Sometimes, however, this term indicates $\chi = 0$, a literal absence of interactions aside from hard core repulsion, as in self-avoiding walk simulations. Many athermal examples of the first kind exist. In that sense, the majority of conventional good solvent systems are approximately athermal. However, the relative paucity of polymer–solvent systems that even approach athermal behavior of the second kind, where $\chi = 0$, has been noted at various places in this and previous chapters. Such systems might be interesting and useful, in that they have much larger coil sizes at modest chain lengths than ordinary good solvent systems, so how one might achieve that state is considered briefly here.

The use of oligomers of the polymer species of interest as solvents offers one route to small values of χ, as discussed above and shown for polyethylene in its oligomers in Figure 7.14a. However, from Eqs. 6.28 and 6.31, for a given polymer species the excluded volume parameter is proportional to the ratio $(1/2 - \chi)/V_s$. Hence, raising oligomer size is a race between increasing z by decrease of χ and decreasing z by an increase of solvent molar volume. From the polyethylene data in Figure 7.14a, no optimum exists: The ratio $(1/2 - \chi)/V_s$ simply decreases monotonically with increasing V_s.

The only feasible route to athermal-like coil expansions would thus appear to be choosing solvents with small molar volume that match the polymer in free volume fraction. Polydimethylsiloxane has an unusually large free volume (last column, Table 7.6), one that actually approaches solvent free volumes (Table 7.5). However, its solutions—at least for the PDMS solvents represented in Table 7.3—do not show abnormally small values of χ.

Only three carefully studied systems having chemically distinct solvent and polymer microstructure and $\chi < 0.39$ have been identified (Table 7.3): polyethylene oxide–benzene (PEO-BEN), polypropylene oxide–benzene (PPO-BEN), and polymethylmethacrylate–chloroform (PMMA-CHL). (There may be more; the search was not exhaustive.) For POE-BEN[89], $\chi(0) = 0.26$ and $(\partial \chi / \partial \phi)_o = -0.10$; for PPO-BEN[90], $\chi(0) = 0.28$ and $(\partial \chi / \partial \phi)_o = -0.21$, both from vapor composition measurements over a range of concentrations and temperatures. The authors, Booth and Devoy, inferred from their results the existence of a specific attractive interaction between the components. The strong negative concentration dependence of χ is unusual and also consistent with specific attraction between the components. Attractive interactions between aromatic rings and ether linkages—the likely candidates—have in fact been observed spectroscopically in another system, polystyrene–polyvinylmethylether[91], which also supports the Booth-Devoy inference. Polymer–solvent attraction of course implies polymer–polymer repulsion, so the POE-BEN and PPO-BEN results point a possible way toward highly expanded coil dimensions.

The interaction parameter for the PMMA-CHL system[81], $\chi = 0.36$, is unusual but much less extreme than the POE-BEN and PPO-BEN examples. Osmotic pressure data[92] for PMMA in several solvents clearly places CHL as the best solvent among them, a position also consistent with dilute solution data on PMMA chain dimensions and second virial coefficient in various solvents[93,94]. Interestingly, its position in Table 6.5 based on $n_K^{\ddagger}$ places it directly in the self-avoiding walk range. Although both polymer and solvent are polar substances, making association a possibility, no objective evidence exists of specific polymer–solvent attraction in the PMMA-CHL system. Its estimated value of $n_K^{\ddagger}$ not withstanding, PMMA-CHL appears to be simply a naturally occurring very good solvent system.

The polystyrene–carbon disulfide system (PS-CS_2), although not widely used, is valuable for solution studies using SANS because CS_2 lacks protons and hence generates very little incoherent scattering. The work of Cotton et al. establishes it to be a good solvent system[6], and the estimate of $M_s^{\ddagger}$ in Table 6.3 is based on their data. Intrinsic viscosity data for PS in several solvents[95], leading to the assignment of $M_v^{\ddagger}$, suggest that CS_2 is in fact a slightly better solvent for PS than TOL, but not at all unusual. That CS_2 is a representative good solvent for PS, but clearly not athermal, is significant because that system played a central role in establishing the polymeric correlation length characteristics in the dilute range[43,44] (see Section 7.3.2).

7.5 Liquid–Liquid Phase Behavior

Following the pioneering work of the 1950s[7,96], the phase behavior of polymer solutions remained a relatively inactive area until the mid-1970s. The critical phenomena revolution took place during the interim. The realization that critical region behavior, difficult to study in any case, is also highly sensitive to polymeric polydispersity caused interest to shift towards monomeric systems[97].

With the growing availability of nearly monodisperse polymers from anionic polymerization, interest in the phase behavior of solutions experienced a revival. Systematic studies of various types were made during this later period, and these researchers provide the bulk of the material in this section.

7.5.1 Upper Critical Behavior

Phase diagrams and spinodal curves for solutions of two nearly monodisperse polyisoprenes in dioxane ($T_\theta = 34°C$)[98] are shown in Figure 7.18. The binodal data were obtained by cloud-point determinations. The spinodal points were based on light scattering data in the single-phase state, as described in Section 7.5.3. Phase diagrams for methylcyclohexane (MCH) solutions of nearly monodisperse polystyrenes (PS) ($T_\theta = 70.6°C$) having a wide range molecular weights[99] are shown in Figure 7.19. Section 3.2 describes the method for predicting phase diagrams and spinodal curves classically, using a free energy of mixing expression.

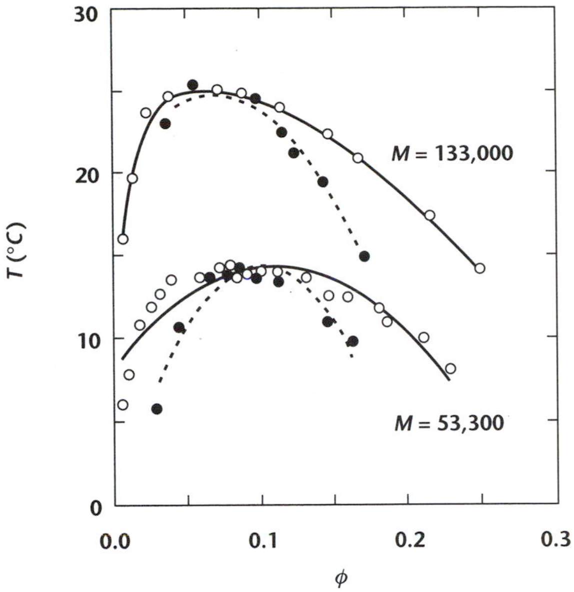

FIGURE 7.18 Phase diagrams and spinodal curves for solutions of 1,4-polyisoprene samples in 1,4 dioxane (data from Takano et al.[98]).

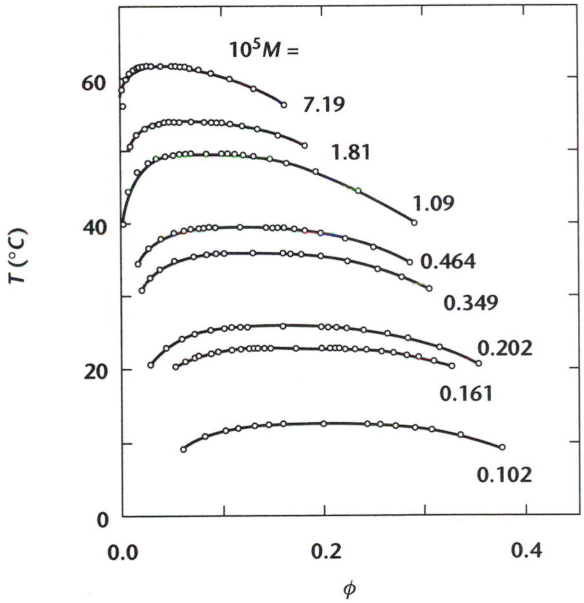

FIGURE 7.19 Liquid–liquid phase diagrams for solutions of polystyrene with various molecular weights in methylcyclohexane (data from Dobashi et al.[99]).

The simple mixture model provides the basis for understanding how properties such as the critical composition and temperature vary with molecular weight for a polymer–solvent system. Thus, from Eqs. 3.80 and 3.81, for solutions of long chains, ($r = V_p/V_s \gg 1$),

$$\phi_c = \frac{1}{r^{1/2}} \qquad r \gg 1 \tag{7.90}$$

$$\chi_c = \frac{1}{2} + \frac{1}{r^{1/2}} \qquad r \gg 1 \tag{7.91}$$

The influence of polymeric polydispersity on phase behavior is at least partially understood. For simple mixtures and long chains, the critical conditions depend on only a single polydispersity ratio[100]:

$$\phi_c = \frac{(r_z/r_w)^{1/2}}{r_w^{1/2}} \qquad r_w \gg 1 \tag{7.92}$$

$$\chi(T_c) = \frac{1}{2} + \frac{\left[(r_w/r_z)^{1/2} + (r_z/r_w)^{1/2}\right]}{2r_w^{1/2}} \qquad r_w \gg 1 \tag{7.93}$$

Thus, ϕ_c shifts to higher concentration with increasing polydispersity and some effect χ_c is somewhat affected as well. The predicted binodals and spinodals also widen with polydispersity[101]. The greater complication of polydispersity for phase behavior is that the system is no longer a true binary. This introduces many new phenomena, including the possibility of fractional separation according to chain length[7]. Such subjects are beyond the scope of this book; they are discussed in careful detail elsewhere[36].

The interaction parameter at the theta condition is 1/2, so Eq. 7.91 gives:

$$\chi_c - \chi_\theta = \frac{1}{r^{1/2}} \qquad r \gg 1 \tag{7.94}$$

From Eq. 7.90, the critical composition for solutions moves to lower polymer concentrations with increasing chain length. The critical composition is also comparable to the overlap concentration ϕ^* when the chains are long [see Eq. 7.42]. For UCT systems ($\partial\chi/\partial T < 0$), the critical temperature rises with increasing chain length at a rate that depends on both chain length and $\partial\chi/\partial T$, reaching T_θ in the long chain limit.

Decreasing ϕ_c and rising T_c with increasing chain length, predicted features for simple mixtures, are seen for PS-MCH solutions in Figure 7.19. The rise in critical temperature usually differs from that predicted by Eq. 7.91 for a purely enthalpic interaction ($\chi \propto 1/T$). However, the interaction parameter for UCT systems commonly varies as $A + B/T$ and can be expressed for purposes here as $\chi_s + \chi_h T_\theta/T$ [see Eq. 3.75]. Thus, for $r \gg 1$, Eq. 7.91 can be rearranged to read:

$$\frac{1}{T_c} = \frac{1}{T_\theta} + \frac{1}{\chi_h T_\theta r^{1/2}} \tag{7.95}$$

Accordingly, a plot of T_c^{-1} versus $r^{-1/2}$ should be linear, with a slope and intercept that permit evaluation of T_θ and χ_h, and hence, from the latter, using Eq. 7.91, the value of χ_s. Equation (7.95) can also be expressed:

$$\left|\frac{T_\theta - T_c}{T_c}\right| \sim \frac{2}{f}\left(\frac{V_s}{V_p}\right)^{1/2} \tag{7.96}$$

where:

$$f \equiv \left|\frac{T_c}{\chi_c}\left(\frac{\partial \chi}{\partial T}\right)_c\right| = \left|\frac{\chi_h}{\chi_s + \chi_h}\right|_c \tag{7.97}$$

The tops of the binodals in Figure 7.19 are rather broad and flat, making accurate evaluations of ϕ_c difficult. Assignment of T_c is easier. A plot of T_c^{-1} versus $r^{-1/2}$ for the polystyrene–methylcyclohexane system[102] is shown in Figure 7.20a. A straight line is obtained, yielding $T_\theta = 69.5°C$. The interaction components at T_θ are found to be $\chi_s = -0.06$ and $\chi_h = 0.56$. Theta temperatures obtained in this way, and those from the vanishing of A_2, have been found for several systems to agree to within approximately 1°C. Little $\chi(T)$ information exists for the comparison of data from other methods, such as osmometry, for near theta systems and in the relevant range of concentrations. The values of χ_s and χ_h for two theta systems, PS-CHN and PIB-BEN, are given in Table 7.3. Only PS-CHN offers a direct comparison, and the result is not encouraging: $f = \chi_h/(\chi_s + \chi_h)$ is 0.6 by osmometry and 3.1 from the T_c^{-1} versus $r^{-1/2}$ plot. Inconsistencies of this sort are one aspect of nonclassical behavior in the critical region caused here by composition fluctuations.

The observed critical concentrations do not agree very well with the predictions of Eq. 7.90. The values of ϕ_c for the PS-MCH system[102], estimated from averages of ϕ_+ and ϕ_- in Figure 7.19, are shown as a function of chain length in Figure 7.20b. These values are not proportional to $r^{-1/2}$, and they are 2 to 3 times larger than predicted by Eq. 7.90. A power law fits the data rather well, however. The least-squares exponent in the figure is 0.77, leading to the observation:

$$\phi_c \propto r^{-0.375} \tag{7.98}$$

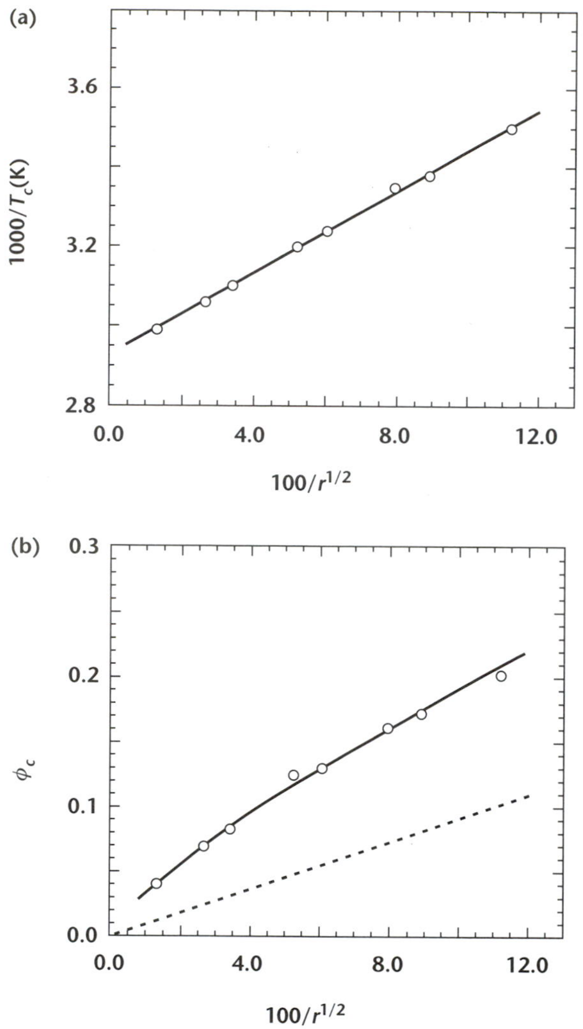

FIGURE 7.20 (a) Molecular weight dependence of critical temperature for polystyrene–methylcyclohexane solutions (data from Dobashi et al.[102]). (b) Molecular weight dependence of critical composition for polystyrene–methylcyclohexane solutions (data from Dobashi et al.[102]). The dashed line indicates the Flory-Huggins prediction.

This is another aspect of fluctuation-driven departures from classical behavior near criticality.

Dobashi et al.[99] compare two binodals from Figure 7.19 with predictions based on $\chi(\phi, T)$, as obtained by fitting $\phi_c(M)$ and $T_c(M)$ for the PS-MCH system. The fit to the data is reasonable for the high molecular weight sample, except very near the critical point, but it is not even qualitatively correct for the other. The curves in

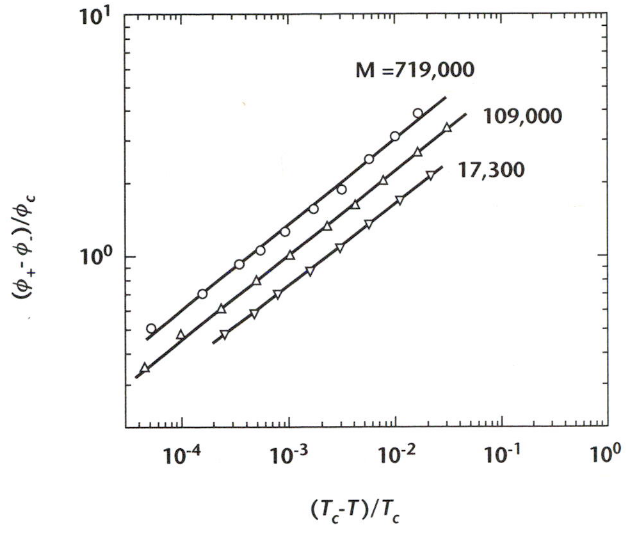

FIGURE 7.21 Phase behavior in the critical region for polystyrene–methylcyclohexane solutions (data from Dobashi et al.[102]). The lines shown correspond to a power-law exponent 0.33.

Figure 7.18 are predicted spinodals and binodals, and they fit the data well. However, obtaining these fits required interaction parameters to depend not only on temperature and composition but also on chain length. Requiring a chain-length dependence for χ conflicts with the assumed local character of the interactions, still another symptom of critical fluctuation complications.

The inconsistency of thermodynamic data near the critical region and the peculiarly flattened shape of the phase diagrams in the critical region, especially evident for the low molecular weight samples in Figure 7.19, are general features of phase behavior in polymer solutions. The flattened shape is another aspect of nonclassical behavior near the critical point. The flattened top is typical behavior for both monomeric liquid mixtures—see Figures 3.3 and 3.5 for example—and polymer solutions. The PS-MCH data in Figure 7.21 demonstrate that aspect more quantitatively[102]. The difference between coexisting phase compositions and distance from the critical point obeys a power law:

$$|\phi_+ - \phi_-| = K\,|T - T_c|^{0.33} \tag{7.99}$$

The experimental exponent differs significantly from the classical exponent of 0.5 [Eq. 3.45]. Within the uncertainties, the same experimental exponent applies to both monomeric mixtures and polymer solutions.

TABLE 7.7 Power-law exponents for various properties in the critical region[104]

Property	Relationship	Classical Exponent	Observed Exponent
Scattering Intensity	$I_r(0) \propto \lvert T - T_c\rvert^{\gamma}$	1	1.24
Binodal Span	$\lvert\phi_+ - \phi_-\rvert \propto \lvert T - T_c\rvert^{\beta}$	1/2	0.33
Scattering Size	$\xi_0 \propto M^{\alpha}$	1/4	0.28
Critical Composition	$\phi_c \propto M^{-\nu}$	1/2	0.39

The behavior outlined above forms a general pattern of departures from the classical theory of critical phenomena. For liquid mixtures, the departures reflect more than merely some inappropriate choice of mathematical function to describe the free energy of mixing. Any choice of analytical expression for mixing free energy, for example, invariably predicts the exponent in Eq. 7.99 to be 0.5. The various departures arise from the nonanalytic character of thermodynamic properties themselves near critical points. Inconsistencies in classical expectations are the result of the composition fluctuations, the effects of which grow very large in the critical region. Power laws with rational exponents are predicted by analytical functions, and power laws with irrational exponents describe the observations.

Of most interest for polymeric systems are the relationships of coexisting composition differences to distance from the critical temperature [Eqs. 7.99 and 3.48], the molecular weight dependence of critical composition [Eqs. 7.98 and 7.90], and scattering behavior in the critical region (Section 7.5.3). The corresponding classical and observed power-law exponents[103,104] are summarized in Table 7.7. The region affected by these nonclassical effects is not necessarily large. Their importance for various classes of mixtures can be estimated by a procedure explained in Section 7.5.4. It seems safe to conclude that nonclassical behavior near liquid–liquid critical points is the source of difficulty in modeling the phase compositions in polymer solutions using simple mixture formalism and nothing more complicated than an arbitrary choice of $\chi(\phi, T)$.

7.5.2 Lower Critical Behavior

Freeman and Rowlinson reported the first example of liquid–liquid phase separation exhibiting lower critical behavior for polymer solutions in 1960[105]. It was soon realized that LCT is a rather general phenomenon for polymer solutions, and an explanation by Patterson and coworkers[83,106] followed shortly. The free volume difference between monomeric and polymeric liquids increases with increasing temperature and the net solvent–polymer interaction becomes increasingly repulsive (eventually exceeding χ_c) and thereby producing liquid–liquid phase separation.

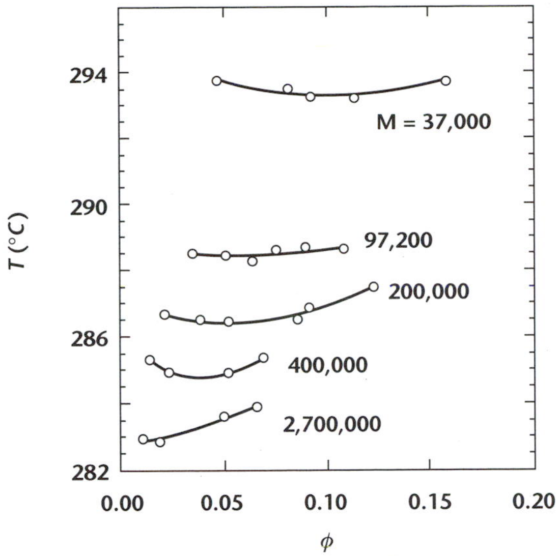

FIGURE 7.22 Lower critical solution behavior for the polystyrene–toluene system (data from Saeki et al.[107]).

Elevated pressures are necessary to suppress vapor phase formation. For polymers in a good solvent, the lower critical temperature is typically not far below $(T_c)_{V-L}$, the vapor–liquid critical temperature for the pure solvent. The upper critical temperature is unrelated to the solvent $(T_c)_{V-L}$ for systems that also undergo phase separation at low temperatures—the UCT type discussed so far. The phase diagrams in such cases are essentially mirror images, reflected along the temperature axis. The phase boundaries are commonly determined by cloud-point measurements for solutions sealed under vacuum in glass tubes. Some pressure beyond atmospheric levels develops at the elevated temperatures, which no doubt influences the phase boundary locations to some extent, but the conditions are otherwise unexceptional.

Consider the phase behavior at elevated temperatures for the polystyrene–toluene system[107], shown in Figure 7.22. At room temperature, toluene is a good solvent for polystyrene, as noted repeatedly throughout this book. As temperature is increased and pressure applied to prevent formation of a vapor phase, the good solvent quality decreases until finally liquid–liquid phase separation occurs. The system demonstrates LCT behavior with $T_\theta = 277°C$, about 40°C below $(T_c)_{V-L}$ for pure toluene.

The polystyrene–methylcyclohexane system, whose UCT behavior is shown in Figure 7.19, also has an LCT at elevated temperatures[107], again not far below the pure solvent $(T_c)_{V-L}$. In the UCT region, the critical temperatures increase with increasing

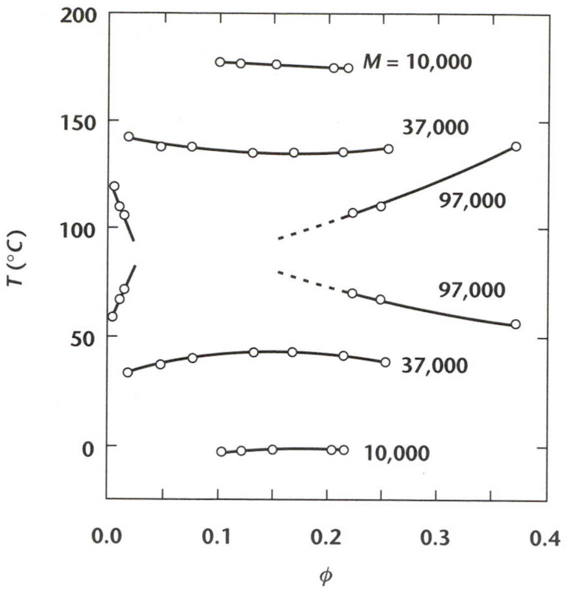

FIGURE 7.23 Upper and lower critical solution behavior for the polystyrene–ethylformate system (data from Konno et al.[109]).

molecular weight, whereas the order is reversed at high temperatures. This system has two theta temperatures under the conditions: 71°C and 219°C. Similar behavior and a similar outcome were found for the polystyrene–cyclopentane system by Berry et al.[108]. That study is especially notable because chain dimensions were also determined at the two theta conditions, 19.6°C and 154.5°C, and $(R_g)_\theta$ was found to be essentially the same despite the ~135°C difference in temperature. In some systems, the range of miscibility is so limited that the high and low temperature regions overlap. The phase diagrams for solutions of polystyrene in ethyl formate[109], shown in Figure 7.23, illustrate that behavior. Beyond $M \sim 90{,}000$, the phase diagrams have "hourglass" shapes and no critical points.

The free volume explanation of universal LCT behavior for polymer solutions is consistent with the observation that phase separation is suppressed by the application of sufficiently high pressures. Increased pressure, like decreased temperature, favors the single-phase state because it increases the density and thereby reduces the free volume difference between polymer and solvent. Indeed, polymers form solutions readily in the supercritical region if the components are energetically compatible and the pressure is high enough to compress the fluid into the dense liquid range[110]. The densities of interest lie below the typical dense liquid range, and the Sanchez-Lacombe[111] and Chen-Radosz[112] models have been used extensively as frameworks for data organization. Precipitation from supercritical solutions has turned out to be an efficient method for separating polydisperse polymers into homogeneous fractions[113]. The practical advantages of pressure change over temperature change as the control variable for inducing phase separation more than offsets the trouble of high-pressure operations. It also permits the use of relatively benign solvents, such as carbon dioxide, for conducting the fractionations. Both universal LCT behavior and supercritical fractionation, major topics in their own right, are beyond the scope of this book.

7.5.3 Scattering and Criticality

From Eqs. 3.36 and 5.27, the structure factor of a binary mixture at $q = 0$ diverges to infinity on the spinodal curve. Consider the scattering intensity $I(0)$ for some mixture as a function of temperature, representing $(\partial\mu_1/\partial\phi)_T$ with the first term of an expansion from a point on the spinodal:

$$\left(-\frac{\partial\mu_1}{\partial\phi}\right)_{T,\phi} = -\left(\frac{\partial^2\mu_1}{\partial T\,\partial\phi}\right)_{T_{sp},\phi}(T - T_{sp}) + \cdots \quad (7.100)$$

From this, it follows that reciprocal scattering intensity at small q is proportional to distance from the spinodal:

$$\frac{1}{I(0)} \propto |T - T_{sp}| \quad (7.101)$$

Plots of $1/I(0)$ versus T, extrapolated to $1/I(0) = 0$, yield the spinodal temperature for that mixture composition, obtaining thereby a point on the spinodal curve. This is the main method for generating spinodal curves. It is applicable only to scattering data from the single-phase region and, as indicated in Figure 7.24, longer and hence more uncertain extrapolations are needed as the distance from ϕ_c increases.

Fluctuations in composition extend over large distances in liquid mixtures near liquid–liquid critical points. Departures from the average composition—the fluctuation amplitudes—remain small as always, but the spatial range over which the composition departures are correlated becomes very large[114]. The resulting light scattering intensities are large enough to be seen unaided, as a kind of iridescent cloudiness called *critical opalescence*. From light scattering measurements, Zimm showed in 1950 that the correlation length is large enough to cause an angular dependence in the scattered intensity[115]. The amplitude of the fluctuations actually grows smaller on approaching ϕ_c, T_c, but their spatial range grows without bound (see Section 7.5.4). The correlation in composition extends over distances much larger than molecular dimensions and finally diverges to infinity at the critical point. The fluctuations near T_c,ϕ_c are important enough to influence macroscopic properties such as the phase diagram shape, noted already in Section 7.5.1.

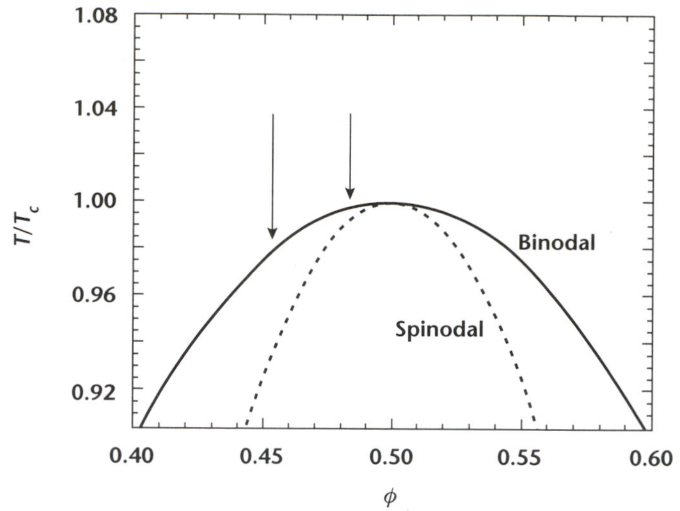

FIGURE 7.24 Constant-composition temperature sweeps for determining spinodal temperatures by scattering near and far from the critical composition.

Light scattering behavior in the critical region can be interpreted by considering the composition fluctuations themselves to be a collection of independent scattering objects. From Eqs. 4.143 and 5.26, and applying the Guinier relationship [Eq. 4.116] to characterize the low-q features of the fluctuations form factor,

$$\frac{C_{sc}}{I_r(q)} = \frac{\bar{\upsilon}_p^2}{\phi V_s RT}\left(-\frac{\partial \mu_s}{\partial \phi}\right)_T (1 + q^2\xi^2/3) \tag{7.102}$$

where ξ is the correlation length. From Eq. 3.41, $(\partial \mu_s/\partial \phi)_T$ goes to zero at the critical point, so using its expansion for small $T - T_c$ along a $\phi = \phi_c$ path is a natural way to represent it. Thus,

$$\left(-\frac{\partial \mu_s}{\partial \phi}\right)_{T,\phi_c} = -a_{11}(T - T_c) + \cdots \tag{7.103}$$

where $a_{11} = \left(\partial^2 \mu_s/\partial T \partial \phi\right)_c$ [see Eq. 3.46] is some system-dependent property at T_c, ϕ_c. When substituted into Eq. 7.102, with properties evaluated at T_c, ϕ_c,

$$\frac{C_{sc} V_s \phi_c}{\bar{\upsilon}_p^2 I_r(q)} = \left|\frac{T - T_c}{T_o}\right| + \frac{q^2\xi^2}{3}\left|\frac{T - T_c}{T_o}\right| \tag{7.104}$$

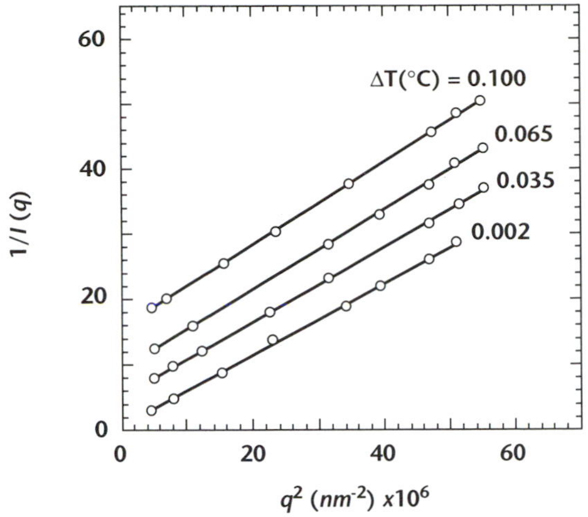

FIGURE 7.25 Ornstein-Zernike plot for single-phase mixtures of n-decane and β, β'-dichlorodiethylether at the critical composition (0.394 mass fraction decane) for several temperatures just above $T_c = 26.5_0\,°C$ (from Kao and Chu[116]).

where $T_o \equiv RT_c/|a_{11}|$. [The left side of Eq. 7.104 is always positive; using absolute values on the right saves the trouble of keeping track of signs.]

Before proceeding further, it is necessary to appreciate some general aspects of scattering behavior in the critical region. A nonpolymeric example from Kao and Chu[116] is shown in Figure 7.25. The system is a solution of n-decane and dichlorodiethylether at the critical concentration, 39.4 wt % n-decane; Ornstein-Zernike plots, are shown at four temperatures in the single-phase region just above $T_c = 26.23°C$. Note that from highest to lowest temperature, the forward scattering intensity $I_r(0)$ changes by more than a order of magnitude, whereas the slope dI^{-1}/dq^2 changes hardly at all. This general feature implies that the product $(T - T_c)\xi^2$ remains essentially constant in the critical region. Accordingly, the correlation length must vary approximately as:

$$\xi = \xi_o \left| \frac{T_o}{T - T_c} \right|^{1/2} \tag{7.105}$$

in which ξ_o is some parameter with the dimensions of length.

Replacing the correlation length in Eq. 7.104 with Eq. 7.105 leads to:

$$\frac{C_{sc} V_s \phi_c}{\bar{v}_p^2 I_r(q)} = \left| \frac{T - T_c}{T_o} \right| + \frac{q^2 \xi_o^2}{3} \tag{7.106}$$

Evidently, ξ_o and T_o can be evaluated from the slope and intercept of an Ornstein-Zernike plot at fixed $T - T_c$. The surmise that ξ_o is of the order of some average molecular size is a natural one: No other length scale exists in the system. The value of ξ_o obtained for the n-decane and dichlorodiethylether system is $0.96nm$[116], a somewhat large but still not unreasonable approximation to the physical size of the component molecules.

During the 1960s, a time of intense activity in critical phenomena, significant practical difficulties made polymer solutions unattractive for critical scattering studies[97], primarily coil size uncertainties, dust removal, polydispersity, and multiple scattering effects. The issues of major interest, such as whether the correlation length exponent in Eq. 7.105 is in fact $-1/2$, as inferred here, are quite general ones. The matter was pursued in other types of critical systems, including liquid–liquid binaries with nonpolymeric components, such as the Figure 7.25 example. However, Debye and coworkers[117,118] did gather some critical scattering data for solutions of nearly monodisperse polystyrene in cyclohexane. The angular dependence of scattering intensity was used to evaluate ξ_o for a wide range of polystyrene molecular weights. The molecular weights, critical conditions, and the sizes obtained by applying Eq. 7.106, are listed in Table 7.8. Values of R_g for polystyrene at T_θ are also shown; it is evident that the polymer size is much larger than ξ_o and increasingly so for longer chains.

The simple mixture model provides an explicit formula, Eq. 7.67, for analyzing scattering behavior. It can also generate a scattering-based molecular size for any simple mixture, including those applied to polymer solutions[119,120]. Thus, from Eq. 7.67 with low-q form factors represented by the Guinier formula, Eq. 4.116,

$$\frac{C_{sc}}{I_r(q)} = \frac{1+q^2\left(R_g^2\right)_1/3}{\phi_1 V_1} + \frac{1+q^2\left(R_g^2\right)_2/3}{\phi_2 V_2} - 2\frac{\chi(T)}{V_{ref}} \tag{7.107}$$

Consider first the case well away from critical points, where $2\chi/V_{ref}$ is negligible relative to the other terms in Eq. 7.107. Define a size for negligible interaction case:

$$\xi_{sm}^2 \equiv \left[\frac{\left(R_g^2\right)_1}{\phi_1 V_1} + \frac{\left(R_g^2\right)_2}{\phi_2 V_2}\right] \Big/ \left[\frac{1}{\phi_1 V_1} + \frac{1}{\phi_2 V_2}\right] \tag{7.108}$$

Now consider the temperature dependence of scattering intensity along a $\phi = \phi_c$ path.

Writing separate terms for the $q = 0$ and q-dependent parts of Eq. 7.107, then applying Eqs. 3.80 and 3.81, leads to the near-critical scattering expression:

$$\frac{C_{sc}}{I_r(q)} = 2\frac{(\chi_h)_c}{V_{ref}}\left|\frac{T-T_c}{T_c}\right| + \left(\frac{1}{V_1^{1/2}} + \frac{1}{V_2^{1/2}}\right)^2 \left(\xi_{sm}^2\right)_{\phi_c} q^2/3 \tag{7.109}$$

TABLE 7.8 Molecular size from critical scattering for polystyrene–cyclohexane solutions[117,118]

$M \times 10^{-4}$	$(V_s/V_p)^{1/2}$	$T_c(°C)$	ϕ_c	$(R_g)_\theta^{(a)}(nm)$	$\xi_o^{(b)}(nm)$	$\xi_{sm}^{(c)}(nm)$	$\xi_{sm}^{(d)}(nm)$
8.0	0.0380	19.24	0.068	7.9	1.68	1.55	1.15
12.3	0.0307	21.82	0.056	9.8	1.88	1.72	1.26
15.3	0.0275	23.19	0.050	11.0	2.01	1.82	1.34
23.9	0.0220	24.95	0.046	13.7	2.12	2.03	1.40
25.3	0.0214	25.09	0.043	16.2	2.35	2.06	1.45
56.9	0.0143	27.79	0.028	21.1	2.98	2.52	1.79
119	0.0099	29.00	0.020	30.5	3.54	3.03	2.12
350	0.0057	29.18	0.0125	52.4	4.67	3.97	2.69

a) Calculated from *M* with PS-CHN formula in Table 6.2.
b) Experimental size from scattering, from refs 117 and 118.
c) Calculated with Eq. 7.110.
d) Calculated with Eq. 7.108 and observed values of ϕ_c.

where:

$$\left(\xi_{sm}^2\right)_{\phi_c} = \left[\frac{\left(R_g^2\right)_1}{V_1^{1/2}} + \frac{\left(R_g^2\right)_2}{V_2^{1/2}}\right] \Big/ \left[\frac{1}{V_1^{1/2}} + \frac{1}{V_2^{1/2}}\right] \qquad (7.110)$$

Thus, $(\xi_{sm}^2)_{\phi_c}$ defines the molecular size parameter for critical scattering by simple mixtures. For the particular case of polymer solutions ($V_2 = V_p \gg V_1 = V_s$, $V_{ref} = V_s$, $(R_g^2)_p/V_p^{1/2} \gg (R_g^2)_s/V_s^{1/2}$), Eqs. 7.109 and 7.110 become:

$$\frac{C_{sc}V_{ref}}{I_r(q)} = 2\left(\chi_h\right)_c \left|\frac{T - T_c}{T_c}\right| + \left(\xi_{sm}^2\right)_{\phi_c} q^2/3 \qquad (7.111)$$

and, with the usual approximation, $V_p = M/\rho$,

$$\left(\xi_{sm}^2\right)_{\phi_c} = \left(\rho R_g^2/M\right)_p \left(V_s V_p\right)^{1/2} \qquad (7.112)$$

where ρ is the density of the undiluted liquid polymer. Equation (7.111) is equivalent to Eq. 7.106 if T_o (in this case) is identified as $T_c/2(\chi_h)_c$ and the Debye size ξ_o as $(\xi_{sm})_{\phi_c}$.

For the near-critical polystyrene–cyclohexane system, $(R_g^2)_{PS}$ should be approximately $(R_g^2)_\theta$. This is justified on two grounds: In a polymer solution approaching the liquid–liquid critical point, the chains go from weakly self-repulsive ($T > T_\theta$), through zero ($T \sim T_\theta$), and finally to weakly self-attractive ($T_\theta < T < T_c$). Excluded volume interactions at the near-coil overlap critical concentrations might accordingly have little effect on the intramolecular correlations. Second, careful experiments by Melnichenko and coworkers have recently demonstrated no significant change in R_g through this range[121].

Molecular sizes were calculated using Eq. 7.108 from $(R_g)_\theta$ for polystyrene, the observed values of ϕ_c, and $(R_g^2)_s = (R_g^2)_{CHN} = 0.1\,nm^2$ for cyclohexane. The results are listed in Table 7.8. Also shown are sizes calculated with the simple mixture predictions for ϕ_c [Eq. 7.110] instead of the experimental values. The calculations in both cases are insensitive to $(R_g^2)_{CHN}$, and the values agree with the data in both magnitude and trend with molecular weight. Those obtained with calculated values of ϕ_c are closer, within about 15% of the experimental sizes. The average size near criticality, experimental or calculated, varies approximately as $\xi_o \propto M^{1/4}$, a behavior that follows directly from Eq. 7.112.

Thus, based on this admittedly limited base of critical scattering data for polymer solutions, the simple mixture model seems to work reasonably well as an estimator of ξ_o. However, other aspects of near-critical behavior for solutions are inconsistent with the simple mixture model, or indeed any analytical mixing model. As shown in Table 7.7, the classical forms, Eqs. 7.106 and 7.112, are not precisely correct for polymer solutions[104]. Some aspects of nonclassical behavior are considered in Section 7.5.4.

7.5.4 The Ginzburg Criterion

Composition fluctuations near the critical point can become large enough to influence the shape of the liquid–liquid binodal and spinodal curves. Indeed, it is probably true that sufficiently near T_c, ϕ_c departures always occur from the phase behavior predicted by any analytical expression for the mixing free energy. The pertinent question to consider is therefore the size of this region of nonclassical behavior. If only a few degrees, it would be of little practical consequence. If significantly larger, however, then the nonclassical complexities would need consideration.

In 1960, Ginzburg[122] suggested a method based on classically calculated quantities for gauging the size of the nonclassical region near a critical point. The criterion involves a comparison of two properties at the critical composition as a function of distance from the critical temperature. For a liquid–liquid critical point, one of the properties is $|\phi_+ - \phi_-|$, the difference in composition of coexisting phases in the two-phase region. The other is $\langle|\delta\phi|\rangle$, the average magnitude of composition fluctuations for the same system in the single-phase state. If $|\phi_+ - \phi_-| > \langle|\delta\phi|\rangle$, the two-phase state is favored; if $|\phi_+ - \phi_-| < \langle|\delta\phi|\rangle$, the single phase is favored. Each component is weakly self-attracting in the critical region, and fluctuations and phase separation offer competing ways of minimizing free energy by reducing the frequency of intercomponent contacts. As shown below, the fluctuation magnitude is always larger for sufficiently small $|\Delta T| = |T - T_c|$, so the single-phase region is always expanded somewhat. The difference in phase composition grows more rapidly with $|\Delta T|$, however, so a crossover occurs at some $|\Delta T^*|$, which thus defines the size of the nonclassical region. The qualitative effect of single-phase stabilization is illustrated in Figure 7.26 for a UCT system.

The nonclassical nature of monomeric phase diagrams was well established by 1970[123], but the effect of polymeric constituents was still unclear. In 1977, de Gennes applied the Ginzburg criterion to polymer solutions and blends[124]. The development below follows in broad outline the similar procedure of Joanny[125].

The classical formula for phase compositions on the near-critical binodal curve was developed in Section 3.1. From Eq. 3.45,

$$(\phi_+ - \phi_-)^2 = 24\frac{\left|\partial^2\mu_1/\partial\phi\partial T\right|_c}{\left|\partial^3\mu_1/\partial\phi^3\right|_c}\Delta T \tag{7.113}$$

The magnitude of the composition fluctuations is estimated using the Einstein formula [Eq. 4.123]. In the terminology used here,

$$\left\langle(\delta\phi)^2\right\rangle = \frac{\phi V_1 kT}{\delta V\,(-\partial\mu_1/\partial\phi)} \tag{7.114}$$

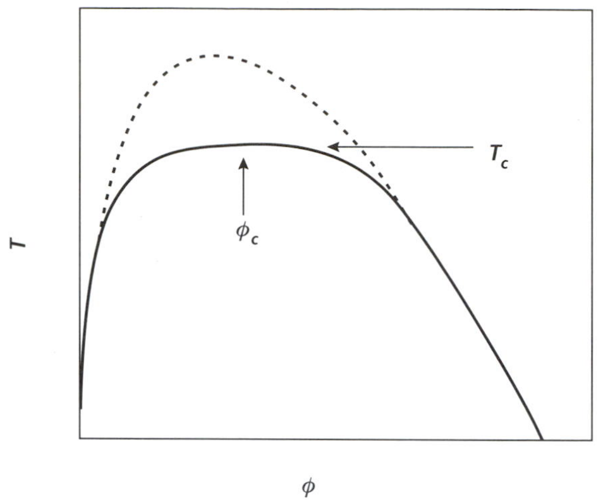

FIGURE 7.26 Illustration of the single-phase stabilization by composition fluctuations in two-component liquids.

in which δV is the correlation volume ξ^3. The correlation length in the critical region is related to the temperature and a molecular size ξ_o by Eq. 7.106. Equation (7.103) describes $\partial\mu_1/\partial\phi$ near T_c, ϕ_c, along the path $\phi = \phi_c$. Thus,

$$\left\langle(\delta\phi)^2\right\rangle = \frac{R V_1 \phi_c}{N_a \xi_o^3 \left|\partial^2\mu_1/\partial\phi\partial T\right|_c} \left|\frac{\Delta T}{T_c}\right|^{1/2} \tag{7.115}$$

The derivatives in Eqs. 7.113 and 7.115 have values that depend on the particular choice of mixing model, but the power laws, $\langle(\delta\phi)^2\rangle \propto |\Delta T|^{1/2}$ and $(\phi_+ - \phi_-)^2 \propto |\Delta T|$, are general forms, arising solely from the assumed analytical nature of the mixing free energy expression. Because of the different exponents, $(\phi_+ - \phi_-)^2$ is always smaller than $\langle(\delta\phi)^2\rangle$ at small enough $|\Delta T|$, but it grows more rapidly with increasing $|\Delta T|$, thus producing a crossover. The expressions intersect at $|\Delta T_{ginz}|$:

$$\left|\frac{\Delta T_{ginz}}{T_c}\right|^{1/2} = \frac{R V_1 \phi_c \left|\partial^3\mu_1/\partial\phi^3\right|_c}{24 T_c N_a \xi_o^3 \left|\partial^2\mu_1/\partial\phi\partial T\right|_c^2} \tag{7.116}$$

and the single-phase state is favored thermodynamically for $|\Delta T| < \left|\Delta T_{ginz}\right|$.

Consider now the Ginzburg criterion as applied to the simple mixture model. A simple mixture value for ξ_o can be obtained with a modified version of the correlation length formula Eq. 7.110:

$$\xi^2 = \left(\xi_{sm}^2\right)_c \left(\frac{T_x - T_c}{T - T_c}\right) \tag{7.117}$$

where T_x is the temperature where $\chi(T) = 0$. Thus, $\xi^2 = (\xi^2_{sm})_c$ at $T = T_x$, where there are no interactions, and then builds as usual as T approaches T_c. Representing $\chi(T)$ with Eq. 3.75 leads to $T_x = -(\chi_h/\chi_s)T_c$ and hence to:

$$\xi^2 = (\xi^2_{sm})_c \left(\frac{T_c}{T - T_c}\right)\left(\frac{\chi_s + \chi_h}{-\chi_s}\right) \tag{7.118}$$

Accordingly, $\xi_o^2 = (\xi^2_{sm})_c/|1 - f|$, where f is a system-specific parameter defined by Eq. 7.97. Simple mixture expressions for $a_{11} = (\partial^2\mu_1/\partial\phi\partial T)_c$ and $a_{30} = (\partial^3\mu_1/\partial\phi^3)_c$ are already available from Eqs. 3.84 and 3.85, so substitutions into Eq. 7.116 then lead to:

$$\left|\frac{\Delta T_{ginz}}{T_c}\right|^{1/2} = \frac{(V_1V_2)^{1/2}}{12N_a\,(\xi_{sm})^3_{\phi_c}}\frac{|1 - f|^{3/2}}{f^2} \tag{7.119}$$

where Eq. 3.80 has been used to relate ϕ_c to the molar volumes; $(\xi_{sm})_{\phi_c}$ is given by Eq. 7.110; and Eq. 7.97 has been used again.

Three categories of binary systems of are interest—monomeric pairs ($V_1 \sim V_2$, $(R_g^2)_1 \sim (R_g^2)_2$, $V_i \sim N_a(R_g^3)_i$); monomeric-polymeric pairs ($V_1 \ll V_2$, $(R_g^2)_1/V_1^{1/2} \ll (R_g^2)_2/V_2^{1/2}$); and polymeric pairs ($V_1 \sim V_2$, $(R_g^2)_1 \sim (R_g^2)_2$, $V_i \ll N_a(R_g^3)_i$). The following estimation formulas are obtained from Eq. 7.119:

$$\begin{aligned}
\left|\Delta T_{ginz}/T_c\right|^{1/2} &= \frac{V}{12\,N_a R_g^3}\frac{|1 - f|^{3/2}}{f^2} && \text{(monomer mixtures)}\\
\left|\Delta T_{ginz}/T_c\right|^{1/2} &= \frac{\left(V_2^{5/4}/V_1^{1/4}\right)}{12\,N_a (R_g)_2^3}\frac{|1 - f|^{3/2}}{f^2} && \text{(polymer solutions)}\\
\left|\Delta T_{ginz}/T_c\right|^{1/2} &= \frac{V}{12\,N_a R_g^3}\frac{|1 - f|^{3/2}}{f^2} && \text{(polymer blends)}
\end{aligned} \tag{7.120}$$

For monomer mixtures, $R_g^3 \propto V$, so the size scale drops out. For polymer solutions and blends, $R_g^3 \propto V^{3/2}$ and size has an effect. Thus,

$$\begin{aligned}
\left|\Delta T_{ginz}/T_c\right|^{1/2} &\propto V^0 && \text{(monomer mixtures)}\\
\left|\Delta T_{ginz}/T_c\right|^{1/2} &\propto V^{-1/4} && \text{(polymer solutions)}\\
\left|\Delta T_{ginz}/T_c\right|^{1/2} &\propto V^{-1/2} && \text{(polymer blends)}
\end{aligned} \tag{7.121}$$

It would appear, therefore, that the size of the nonclassical region would shrink with increasing chain length in both polymer solutions and blends. However, an additional feature must be considered for polymer solutions. Polymeric size is unperturbed by excluded volume at T_θ, so the correlation length must go from molecular dimensions at T_θ to infinity at T_c. Moreover, the gap between T_θ and T_c depends on the polymeric chain length, shrinking to zero in the long chain limit. The following version of Eq. 7.117 offers a plausible interpolation formula:

$$\xi^2 = \left(\xi_{sm}^2\right)_c \left|\frac{T_\theta - T_c}{T - T_c}\right| = \left(\xi_{sm}^2\right)_c \left|\frac{T_\theta - T_c}{T_c}\right| \left|\frac{T_c}{T - T_c}\right| \tag{7.122}$$

Accordingly, a factor $|(T_\theta - T_c)/T_c|^{3/2}$ replaces $1/(1-f)^{3/2}$ in the denominator of Eq. 7.119 when polymer solutions are considered. From Eq. 7.97, $|(T_\theta - T_c)/T_c|^{3/2} = |2/f|^{3/2}(V_1/V_2)^{3/4}$, so modified results replace the equations above:

$$|\Delta T_{ginz}/T_c|^{1/2} = \frac{(V_2^2/V_1)}{24N_a(R_g)_2^3}\frac{1}{|2f|^{1/2}}$$

or: (polymer solutions) (7.123)

$$|\Delta T_{ginz}/T_c|^{1/2} \propto V_2^{1/2}$$

The nonclassical region thus grows with increasing chain length for polymer solutions. Classical behavior near the critical point is apparently attainable only in polymer blends.

Equations (7.119) and (7.123) can be expressed in more quantitative fashion as well. Consider first the ratio $V/N_a R_g^3$ for monomeric pairs. Representing monomeric components as hard spheres leads to a sphere volume $(4\pi/3)(5/3)^{3/2}R_g^3$. Using 0.5, a typical ratio of occupied volume to molar volume in dense liquids, then leads to $V/N_a R_g^3 \sim 18$. In the evaluation of $V/N_a R_g^3$ for polymeric pairs, unperturbed coil sizes apply at all temperatures, because excluded volume is screened out. Accordingly, $R_g^3 \sim [(R_g^2/M)_\theta V\rho]^{3/2}$, where ρ is the mass density of the undiluted polymer liquid, and therefore $V/N_a R_g^3 = 1/N_a[(R_g^2/M)_\theta \rho]^{3/2} V^{1/2}$. Values of $(R_g^2/M)_\theta^{1/2}$ for common species are listed in Table 4.2. Neither $(R_g^2/M)_\theta$ nor ρ vary greatly with species; $(R_g^2/M)_\theta^{1/2} = 0.03\,nm$ and $\rho = 0.8\,g\,cm^{-3}$ are adequate for estimation purposes. A typical polymeric molar volume V is $10^2 N\,cm^3$, where N is the degree of polymerization. With these substitutions,

$$|\Delta T_{ginz}| \sim \frac{2.2\,|1-f|^3}{f^4}T_c \quad \text{(monomer mixtures)} \tag{7.124}$$

$$|\Delta T_{ginz}| \sim \frac{0.063N}{f}T_c \quad \text{(polymer solutions)} \tag{7.125}$$

$$|\Delta T_{ginz}| \sim \frac{0.51\,|1-f|^3}{f^4 N}T_c \quad \text{(polymer blends)} \tag{7.126}$$

Little is known about the ratio $f = |\chi_h/(\chi_s + \chi_h)|$ at T_c for monomeric mixtures. In polymer blends, f can, in principle, range considerably; but for unknown reasons, χ_s and χ_h tend to have opposite signs and, for UCT systems at least, f is commonly about 1.5 to 2.0. The relatively few available values for polymer solutions suggest similar properties. For LCT systems, f can be quite large. Nevertheless, taking $f = 1.7$ as an average, and applying that to all categories, leads to the following estimates for $N = 10^3$ and $T_c = 400K$:

$$
\begin{aligned}
&|\Delta T_{ginz}| \sim 35\,°C && \text{(monomeric mixtures)} \\
&|\Delta T_{ginz}| \sim 16000\,°C && \text{(polymer solutions)} \\
&|\Delta T_{ginz}| \sim 0.008°C && \text{(polymer blends)}
\end{aligned}
\tag{7.127}
$$

From these estimates, critical fluctuations compete strongly in the critical region for monomeric mixtures, and nonclassical critical behavior is expected. Moreover, polymer solutions should also exhibit nonclassical critical behavior. Numerous examples supporting the de Gennes prediction are now known[104]. Polymer blends, conversely, should behave classically except for a small region near T_c, ϕ_c or unless the chains are relatively short. Several studies were undertaken to investigate this prediction using a variety of techniques, mainly searching for a crossover from classical to fluctuations-dominated behavior in blends of various species and chain lengths.

Chu and coworkers used binodal curve determinations and SAXS to compare the near-critical behavior for a blend of long chains, nearly monodisperse polystyrene and 2-chlorostyrene, in the presence and absence of a monomeric diluent, dibutylphthalate[126]. With 22% of the diluent present, a small but definite departure from linearity in the single-phase $I(0)^{-1}$ versus T^{-1} relationship near T_c was observed. Coexisting compositions in the two-phase region near the critical point also varied nonclassically, $|\Delta\phi| \propto |T - T_c|^{0.33}$, (see Table 7.8). Without the diluent, the near-critical behavior in both these respects was classical. This difference between systems is consistent with expectations based on the Ginzburg criterion for a two-polymer, one-solvent ternary system and a two-polymer binary system[127].

Three other tests involved direct tests of Eq. 7.126. Schwann and coworkers used SANS to investigate PS/PVME blends of long chains[128,129]. Short-chain PEP/PI blends were investigated using both SANS and dynamic light scattering by Bates and coworkers[130,131]. Hair et al.[132] used SANS with short-chain PS/PBD blends. All polymers except PVME had narrow distributions; both UCT and LCT systems are represented. Rather complete supporting information was supplied with the PEP/PI data, and its crossover evidence is also particularly clear. Figure 7.27 shows the temperature dependence of the SANS intensity at $q = 0$ from $\sim 100°C$ to the UCT = 38.2°C (from cloud-point data) for a blend having the critical composition. Down to approximately 70°C $I(0)^{-1}$ versus T^{-1} is linear, indicating classical behavior

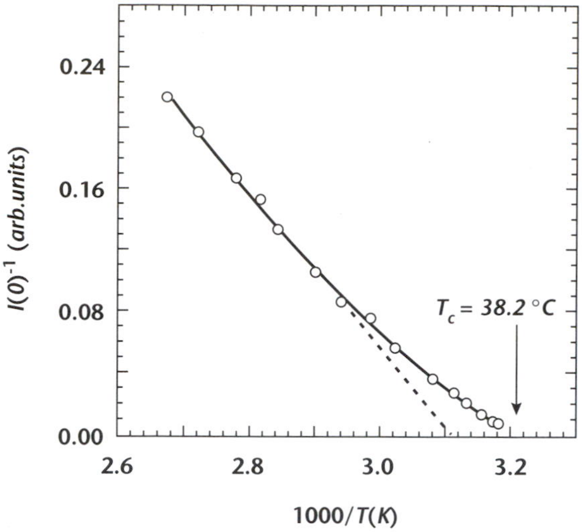

FIGURE 7.27 Crossover from classical behavior at high temperatures to nonclassical behavior near the critical point for a polyisoprene-poly(ethylene-*alt*-propylene) mixture at its critical composition (Bates et al.[130] and Stepanek et al.[131]).

and extrapolating to an apparent $T_c = 48°C$. Below 70°C, the $I(0)^{-1}$ versus T^{-1} path is curved, signaling crossover to the fluctuations-dominated regime. Depending on how that extrapolation is managed, but not in a sensitive way, $I(0)^{-1}$ appears to reach zero within 1°C or so of the independently established cloud-point value of 38.2°C. The pattern of behavior is similar for the PBD/PS and PS/PVME systems, although the span of nonclassical behavior is smaller in both cases.

Results for the three studies are summarized in Table 7.9. Two measures of nonclassical range are recorded: $|\Delta T|_{range}$, the difference between temperature at first departure from classical behavior and observed critical temperature; and $|\Delta T|_c$, the difference between observed critical temperature and critical temperature as extrapolated from the classical region.

It is disappointing to find that $|\Delta T_{ginz}|$, the value predicted by Eq. 7.126, is smaller than either $|\Delta T|_{range}$ or $|\Delta T|_c$ by at least two orders of magnitude. A precise agreement cannot be expected of course, but the magnitude of the discrepancy is disconcerting even when taking the simplicity of the analysis into account. An answer to this apparent dilemma appeared a few years later through the work of Meier, Schwann, Mortensen, and Janssen[133]. Rather than using the intersection method based on simple mixture formulas, they employed a crossover function derived from a renormalization group analysis of the near-critical single-phase region to analyze the data[134].

TABLE 7.9 Observed classical to nonclassical crossover behavior in the critical region and estimates based on the Ginzburg criterion

System	Form	f	$\lvert\Delta T\rvert_{range}$ (°C)	$\lvert\Delta T_c\rvert$ (°C)	$\lvert\Delta T_{ginz}\rvert$ (°C)
PEP/PI(Refs. 130-1)	UCT	1.8	32	10	0.05
PBD/PS (Ref. 132)	UCT	1.7	11	3.3	0.086
PS/PVME (Refs.128-9)	LCT	−69	2.4	0.4	0.0015

Meier et al. found the classical to nonclassical transition to be much more gradual than the intersection method suggests. They also concluded that the simple mixture model underestimates fluctuation sizes near criticality by about a factor of 10, making $|\Delta T_{ginz}|$ from Eq. 7.126 too small by about a factor of 100, and thus accounting for the discrepancy. They went on to suggest, based on the existing data, that the N dependence for blends is much more complicated than the N^{-1} behavior in Eq. 7.126[135]. Thus, there appears to be little change in $|\Delta T_{ginz}|$ between $N \sim 1$ (monomeric mixtures) and $N \sim 30$, but then an N^{-2} dependence up at least to $N \sim 400$. The findings that the crossover is very broad and that the simple mixture grossly underestimates the critical fluctuation size nicely vindicates the simulation results of Sariban and Binder[136,137]. Their conclusion was that very long chains are required for polymer blends to approach fully classical critical behavior.

Chapter 8 deals with polymer blends, and the analysis of phase equilibrium data employs the classical formulation throughout.

REFERENCES

1. Cotton J.P., G. Farnoux, G. Jannink, J. Mons, and C. Picot. 1972. *Comp. Rend.* C275:175.
2. Kirste R.G., W.A. Kruse, and J. Shelten. 1972. *Makromol. Chem.* 162:299.
3. Ballard D.G.H., G.D. Wignall, and J. Shelten. 1973. *Eur. Polym. J.* 9:965.
4. Shelten J., W.A. Kruse, and R.G. Kirste. 1973. *Kolloid-Z. Polym.* Z. 251:919.
5. Graessley W.W. 1974. *Adv. Polym. Sci.* 16:1.
6. Cotton J.P., D. Decker, H. Benoit, B. Farnoux, J. Higgins, G. Jannink, R. Ober, C. Picot, and J. Des Cloizeaux. 1974. *Macromolecules* 7:863.
7. Flory P.J. 1953. *Principles of Polymer Chemistry,* Ithaca, NY: Cornell University Press.
8. Kirste R.G. and B.R. Lehnen. 1976. *Makromol. Chem.* 177:1137.
9. McAdams J.C. and M.C. Williams. 1980. *Macromolecules* 13:858.
10. Colby R.H., L.J. Fetters, W.G. Funk, and W.W. Graessley. 1991. *Macromolecules* 24:3873.
11. Landry M.R. 1997. *Macromolecules* 30:7500.
12. Kirste R.G., Kruse W.A., and Abel K. 1975. *Polymer* 16:120.
13. Higgins J.S. and Benoit H.C. 1994. *Polymers and Neutron Scattering,* New York: Oxford University Press.

14. Zirkel A., V. Urban, D. Richter, L.J. Fetters, J.S. Huang, R. Kampmann, and N. Hadjichristidis. 1992. *Macromolecules* 25:6148.
15. Boothroyd A.T. and L.J. Fetters. 1991. *Macromolecules* 24:5215.
16. Smith G.D., D.Y. Yoon, R.L. Jaffe, R.H. Colby, R. Krishnamoorti, and L.J. Fetters. 1996. *Macromolecules* 29:3462.
17. Fetters L.J., W.W. Graessley, R. Krishnamoorti, and D.J. Lohse. 1997. *Macromolecules* 30:4973.
18. Squires G.L., Boothroyd A.T., Horten J.C., Fetters L.J., Rennie A.R., Glinka C.J., Robinson R.A., and de Vallera A.M.B.G. 1989. In *Molecular Basis of Polymer Networks.* A. Baumgartner and C. Picot (eds.). Heidelberg: Springer-Verlag, p. 200.
19. Fetters L.J. 2001. Private communication.
20. Noda I., N. Kato, T. Kitano, and M. Nagasawa. 1981. *Macromolecules* 14:668.
21. Casassa E.F. and Berry G.C. 1989. In *Comprehensive Polymer Science*. Vol. 2. G. Allen and J.C. Bevington (eds.). Oxford, UK: Pergamon Press, p. 71.
22. Berry G.C. 1994. *Adv. Polym. Sci.* 114:233.
23. Daoud M., J.P. Cotton, B. Farnoux, G. Jannink, G. Sarma, H. Benoit, C. Duplessix, C. Picot, and P.G. De Gennes. 1975. *Macromolecules* 8:804.
24. des Cloizeaux J. 1975. *J. Phys.* 36:281.
25. de Gennes P.-G. 1979. *Scaling Concepts in Polymer Physics,* Ithaca, NY: Cornell University Press.
26. Edwards S.F. 1966. *Proc. Roy. Phys. Lond.* 88:265.
27. McQuarrie D.A. 1976. *Statistical Mechanics,* New York: HarperCollins Publishers.
28. Parsegian J.L. 1975. In *Annual Review of Biophysics and Bioengineering.* Mullins L.J. (ed.). New York: Academic Press, p. 221.
29. Muthukumar M. and S.F. Edwards. 1982. *J. Chem. Phys.* 76:1982.
30. Graessley W.W. 1980. *Polymer* 21:258.
31. Noda I., Y. Higo, N. Ueno, and T. Fujimoto. 1984. *Macromolecules* 17:1055.
32. King J.S., W. Boyer, G.D. Wignall, and R. Ullman. 1985. *Macromolecules* 18:709.
33. Westermann S., L. Willner, D. Richter, and L.J. Fetters. 2000. *Macromol. Chem. Phys.* 201:500.
34. Hamada F., H. Hayashi, and A. Nakajima. 1978. *J. Appl. Crystallogr.* 11:514.
35. Kent M.S., M. Tirrell, and T.P. Lodge. 1991. *Polymer* 32:314.
36. Fujita H. 1990. *Polymer Solutions,* New York: Elsevier.
37. Roots J. and B. Nystrom. 1979. *Polymer* 20:148.
38. Higo Y., N. Ueno, and I. Noda. 1983. *Polym. J.* 15:367.
39. Adam M., L.J. Fetters, W.W. Graessley, and T.A. Witten. 1991. *Macromolecules* 24:2434.
40. Stepanek P., R. Perzynski, M. Delsanti, and M. Adam. 1984. *Macromolecules* 17:2340.
41. Flory P.J. and H. Daoust. 1957. *J. Polym. Sci.* 25:429.
42. Okano K., E. Wada, and H. Hiramatsu. 1974. *Prog. Polym. Sci. Japan* 17:145.
43. Farnoux B., M. Daoud, J.P. Cotton, G. Jannink, M. Nierlich, and F. Boue. 1978. *J. Appl. Crystallogr.* 11:511.
44. Rawiso M., C. Duplessix, and C. Picot. 1987. *Macromolecules,* 630.
45. Farnoux B. 1976. *Ann. Phys. t* 1:73.
46. Stanley H.E. 1971. *Introduction to Phase Transitions and Critical Phenomena,* New York: Oxford University Press.
47. Brown W. and K. Mortensen. 1988. *Macromolecules* 21:420.
48. Brown W. and T. Nicolai. 1990. *Col. Polym. Sci.* 268:977.

49. Hamada F., S. Kinugasa, H. Hayashi, and A. Nakajima. 1985. *Macromolecules* 18:2290.
50. Brown W., K. Mortensen, and G. Floudas. 1992. *Macromolecules* 25:6904.
51. Wiltzius P., H.R. Haller, D.S. Cannell, and D.W. Schaefer. 1983. *Phys. Rev. Lett.* 51:1183.
52. Graessley W.W. 2002. *Macromolecules* 35:3184.
53. Tao H., C.-I. Huang, and T.P. Lodge. 1999. *Macromolecules* 32:1212.
54. Huang J.-R. and T.A. Witten. 2002. *Macromolecules* 35:10225.
55. Müller M., K. Binder, and L. Schäfer. 2000. *Macromolecules* 33:4568.
56. Kuwahara N., T. Okazawa, and M. Kaneko. 1967. *J. Chem. Phys.* 47:3357.
57. Witten T.A., P.A. Pincus, and M.E. Cates. 1986. *Europhys. Lett.* 2:137.
58. Huber K., S. Bantle, W. Burchard, and L.J. Fetters. 1986. *Macromolecules* 19:1404.
59. Willner L., O. Jucknischke, D. Richter, B. Farago, L.J. Fetters, and J.S. Huang. 1992. *Europhys. Lett.* 19:297.
60. Richter D., O. Jucknischke, L. Willner, L.J. Fetters, M. Lin, J.S. Huang, J. Roovers, C. Toporovski, and L.L. Zhou. 1993. *J. Phys. IV* 3: 3.
61. Vlassopoulos D., T. Pakula, G. Fytas, J. Roovers, K. Karatasos, and N. Hadjichristidis. 1997. *Europhys. Lett.* 39:617.
62. Roovers J., P.M. Toporowski, and J. Douglas. 1995. *Macromolecules* 28:7064.
63. Prausnitz J.M., R.N. Lichtenthaler, and E. Gomez de Azevedo. 1986. *Molecular Thermodynamics of Fluid-Phase Equilibria.* 2nd ed. Englewood Cliffs, NJ: Prentice-Hall.
64. Eichinger B.E. and Flory P.J. 1968. *Trans. Faraday Soc.* 64:2035.
65. Eichinger B.E. and P.J. Flory. 1968. *Trans. Faraday Soc.* 64:2053.
66. Eichinger B.E. and P.J. Flory. 1968. *Trans. Faraday Soc.* 64:2061.
67. Eichinger B.E. and P.J. Flory. *Trans. Faraday Soc.* 64:2066.
68. Flory P.J. and H. Höcker. 1971. *Trans. Faraday Soc.* 67:2258.
69. Höcker H. and P.J. Flory. 1971. *Trans. Faraday Soc.* 67:2270.
70. Höcker H., H. Shih, and P.J. Flory. 1971. *Trans. Faraday Soc.* 67:2275.
71. Flory P.J. and H. Shih. 1972. *Macromolecules* 5:761.
72. Braun J.-M. and J.E. Guillet. 1976. *Adv. Polym. Sci.* 21:108.
73. Patterson D., Y.B. Tewari, H.P. Schreiber, and J.E. Guillet. 1971. *Macromolecules* 4:356.
74. Leung Y.-K. and B.E. Eichinger. 1974. *J. Phys. Chem* 78:60.
75. Schreiber H.P., Y.B. Tewari, and D. Patterson. 1973. *J. Polym. Sci.* 11:15.
76. Su C.S. and D. Patterson. 1977. *Macromolecules* 10:708.
77. Du Q., P. Hattam, and P. Munk. 1990. *J. Chem. Eng. Data* 35:367.
78. Orwoll R.A. and Arnold P.A. 1996. In *Physical Properties of Polymers Handbook.* Mark J.E. (ed.). Woodbury, N.Y.: AIP Press, pp. 177–196.
79. Shiomi T., Z. Izumi, F. Hamada, and A. Nakajima. 1980. *Macromolecules* 13:1149
80. Flory P.J. 1970. *Discuss. Faraday Soc.* 49:7.
81. Casassa E.F. 1976. *J. Polym. Sci., Polym. Symp.* 54:53.
82. Orwoll R.A. 1996. In *Physical Properties of Polymers Handbook.* Mark J.E. (ed.). Woodbury, N.Y.: AIP Press, pp. 81–89.
83. Patterson D. 1968. *J. Polym. Sci., Part C* 16:3379.
84. Blanks R.F. and J.M. Prausnitz. 1968. *Ind. Eng. Chem. Fund.* 3:1.
85. Orwoll R.A. and P.J. Flory. 1967. *J. Am. Chem. Soc.* 89:6814.
86. Maranas J.K., M. Mondello, G.S. Grest, S.K. Kumar, P.G. Debenedetti, and W.W. Graessley. 1998. *Macromolecules* 31:6991.

87. Fujita H. 1988. *Macromolecules* 21:179.
88. Graessley W.W., R.C. Hayward, and G.S. Grest. 1999. *Macromolecules* 32:3510.
89. Booth C. and C.L. Devoy. 1971. *Polymer* 12:309.
90. Booth C. and C.L. Devoy. 1971. *Polymer* 12:320.
91. Mirau P.A. and F.A. Bovey. 1990. *Macromolecules* 23:4548.
92. Schulz G.V. and H. Doll. 1952. *Z. Elektrochem.* 56:248.
93. Abe F., K. Horita, Y. Einaga, and H. Yamakawa. 1994. *Macromolecules* 27:725.
94. Abe F., Y. Einaga, and H. Yamakawa. 1994. *Macromolecules* 27:3262.
95. Chen S.J., G.C. Berry, and D.J. Plazek. 1995. *Macromolecules* 28:6539.
96. Shultz A.R. and P.J. Flory. 1952. *J. Am. Chem. Soc.* 74:4760.
97. Chu B. 2000. Private communication.
98. Takano N., Y. Einaga, and H. Fujita. 1985. *Polym. J.* 17:1123.
99. Dobashi T., M. Nakata, and M. Kaneko. 1980. *J. Chem. Phys.* 72:6692.
100. Stockmayer W.J. 1949. *J. Chem. Phys.* 17:588.
101. Scholte T.G. 1972. *J. Polym. Sci.: Pt. C* 39:281.
102. Dobashi T., M. Nakata, and M. Kaneko. 1980. *J. Chem. Phys.* 72:6685.
103. Shinozaki K., F. Hamada, and T. Nose. 1982. *J. Chem. Phys.* 77:4734.
104. Sanchez I.C. 1989. *J. Phys. Chem* 93:6983.
105. Freeman P.I. and J.S. Rowlinson. 1960. *Polymer* 1:20.
106. Delmas D., D. Patterson, and T. Somcynsky. 1962. *J. Polym. Sci.* 57:1962.
107. Saeki S., N. Kuwahara, S. Konno, and M. Kaneko. 1973. *Macromolecules* 6:246.
108. Berry G.C., E.F. Casassa, and P.-Y. Liu. 1987. *J. Polym. Sci.: Pt. B, Polym. Phys.* 25:673.
109. Konno S., S. Saeki, N. Kuwahara, M. Nakata, and M. Kaneko. 1975. *Macromolecules* 8:799.
110. Whaley P.D., H.H. Winter, and P. Ehrlich. 1997. *Macromolecules* 30:4882.
111. Sanchez I.C. and R.H. Lacombe. 1978. *Macromolecules* 11:1145.
112. Chen S.J. and M. Radosz. 1992. *Macromolecules* 25:3089.
113. Kirby C.F. and M.A. McHugh. 1999. *Chem. Rev.* 99:565.
114. Einstein A. 1993. *The Collected Papers of Albert Einstein (English Translation Supplement).* Vol. 3. Klein M.J. and others (eds.). Princeton, NJ: Princeton University Press.
115. Zimm B. 1950. *J. Phys. Colloid Chem.* 54:1306.
116. Kao W.P. and B. Chu. 1969. *J. Chem. Phys.* 50:3986.
117. Debye P., H. Coll, and D. Woermann. 1960. *J. Chem. Phys.* 33:1746.
118. Debye P., B. Chu, and D. Woermann. 1962. *J. Chem. Phys.* 36:1803.
119. de Gennes P.G. 1968. *Phys. Lett.* A26:313.
120. Vrij A. and M.W.J. van der Esker 1972. *Faraday Soc. Trans.* 68:513.
121. Melnichenko Y.B., Wignall G.D., Brown W., Kiran E., Cochran H.D., Salaniwal S., Heath K., Van Hook W.A., and Stamm M. 2000. In *Computational Studies, Nanotechnology, and Thermodynamic Studies of Polymer Systems.* Dadnum M.D. (ed.). New York: Kluwer, p. 15.
122. Ginzburg V.L. 1960. *Sov. Phys. Solid State* 2:1824.
123. Chu B. 1972. *Ber. Bunsenges. Phys. Chem.* 76:202.
124. de Gennes P.G. 1977. *J. Phys. Lett.* 38:L441.
125. Joanny J.F. 1978. *J. Phys. A: Math. Gen.* 11:L117.
126. Chu B., Q. Ying, K. Linliu, P. Xie, T. Gao, Y. Li, T. Nose, and M. Okada. 1992. *Macromolecules* 25:7382.
127. Broseta D., L. Leibler, and J.-F. Joanny. 1987. *Macromolecules* 20:1935.

128. Schwahn D., K. Mortensen, and H. Yee-Madeira. 1987. *Phys. Rev. Lett.* 58:1544.
129. Schwahn D., K. Mortensen, T. Springer, H. Yee-Madeira, and R. Thomas. 1987. *J. Chem. Phys.* 87:6078.
130. Bates F.B., J.H. Rosedale, P. Stepanek, T.P. Lodge, P. Wiltzius, G.H. Fredrickson, and R.P. Hjelm. 1990. *Phys. Rev. Lett.* 65:1893.
131. Stepanek P., T.P. Lodge, C. Kendrowski, and F.B. Bates. 1991. *J. Chem. Phys.* 94:8289.
132. Hair D.W., E.K. Hobbie, A.I. Nakatani, and C.C. Han. 1992. *J. Chem. Phys.* 96:9133.
133. Meier G., D. Schwahn, K. Mortensen, and S. Janssen. 1993. *Europhysics Letters* 22:577.
134. Anisimov M.A., S.B. Kiselev, J.V. Sengers, and S. Tang. 1992. *Physica A* 188:487.
135. Schwahn D., G. Meier, K. Mortensen, and S. Janssen. 1994. *J. Phys. II* 4:837.
136. Sariban A. and K. Binder. 1987. *J. Chem. Phys.* 86:5859.
137. Binder K. 1995. In *Monte Carlo and Molecular Dynamics Simulations in Polymer Science.* Binder K. (ed.). New York: Oxford University Press, p. 356.

CHAPTER 8

Polymer Blends

This chapter begins by reviewing the various sources of net interactions in molecular mixtures, emphasizing scattering methods, PVT data, the simple mixture model, and the special issues that apply to polymer blends. First considered are isotopic blends, mixtures of ordinary and deuterated versions of the same polymer species, which afford simple and carefully studied systems with relatively well understood interactions. Considered next are polyolefin blends, many based on model species. These furnish simple systems for testing ideas such as regular mixing and the solubility parameter scheme, experimental techniques such as PVT-based estimates of the interaction parameter, and molecular simulations for exploring the relationship for polymeric liquids between cohesive energy density and internal pressure. Blends of hydrocarbon polymers more generally, including polydienes and polystyrene, are then considered, with the interaction information being supplied by scattering, pure component PVT data, and the order–disorder transition temperatures of block copolymers. The chapter ends with the consideration of statistical copolymers blends—emphasizing observations on systems studied in detail—and blends of components with specific attractions, in this case the particularly well-studied example of polystyrene–polyvinylmethylether blends.

It is unusual to find two polymer species that have significant mutual solubility. The "like-dissolves-like" homily works well for monomeric pairs. It works only slightly less well for polymers and monomers; but it takes on a much stricter meaning if both liquids are polymeric. Thus, polymer species of similar polarity, whose monomers and oligomers are *miscible*—soluble in all proportions—usually form separate liquid phases. Even adding modest amounts of a common solvent makes little difference. Consider the example of two polymeric alkanes, polyethylene and polypropylene. Both are saturated hydrocarbons and have the same empirical formula, yet for typical molecular weights ($M \sim 10^5$), their mutual liquid-state solubilities are undetectably small. Even copolymers of ethylene and propylene differing in composition by more

TABLE 8.1 Chain length effects on the critical conditions for liquid mixtures

System	ϕ_c (1)	χ_c (2)	$\lvert\delta_2 - \delta_1\rvert_c^{(3)} (J\,cm^{-3})^{1/2}$	$\lvert 1 - T_1^*/T_2^*\rvert_c^{(4)}$
monomeric mixtures ($r_1 = r_2 = 1$)	1/2	2	~8	~1
polymer solutions ($r_1 = 1, r_2 = 10^3$)	0.03	1/2	~4	~0.5
polymer blends ($r_1 = r_2 = 10^3$)	1/2	0.002	~0.25	~0.03

(1) From Eq. 3.80.
(2) From Eq. 3.81.
(3) Calculated from χ_c and Eq. 3.93, with $V_{ref} = 10^2 cm^3$ and $T_c = 100°C$.
(4) Calculated from χ_c and Eq. 7.88 for solutions; Eq. 8.4, with $V_{ref} = 10^2\,cm^3$ and $T_c = 170°C$ for blends.

than about 10% form separate phases throughout the observable liquid range. In 1949, Scott gave the explanation for such observations[1], as the inherently small combinatorial entropy of mixing density when both components are polymeric. The critical conditions for monomeric mixtures, polymer solutions, and polymer blends are compared in Table 8.1. For miscibility, the net intermolecular interactions must be either attractive, which is uncommon, or very weak, if the net interactions are repulsive, as is usually the case.

This nearly universal immiscibility of polymeric liquids is far from the end of the story, however. Polymer blends have, over the years, become important articles of commerce[2]. They are almost always multiphase systems, and their utility arises from the combinations of properties that single-component systems cannot readily achieve. The usefulness of blends depends in turn on the amounts and properties of their individual phases and the strength of their phase interfaces. Both phase composition and interfacial thickness underlie these aspects of blend performance in the solid state, and both are in large part determined by the thermodynamic interactions that govern liquid-state phase behavior.

Polymeric phase interfaces[3–5] and the related topic of polymeric compatibilization[6] are active research subjects at the present time. Block copolymers are also of great current interest in their own right[7], but also in part because of their potential for interfacial activity[8]. These timely topics are still evolving rapidly. Except for occasional usage and comments made in passing, interfacial effects and block copolymers are not considered in this book. The present chapter deals with the bulk thermodynamics of polymer blends, but even that subject is far too broad to be covered in full detail. Only some fundamental aspects of single-phase, two-component liquid mixtures are considered. A brief discussion of the relevant interactions introduces a selected set of categories—isotopic blends, polyolefin blends, general hydrocarbon blends, blends with specific interactions, and blends of copolymers.

8.1 Molecular Interactions

Composition and net interaction strength at liquid–liquid criticality are compared in Table 8.1 for monomeric mixtures, polymer solutions, and polymer blends. The critical composition for polymer blends—like that for monomeric mixtures and unlike that for polymer solutions—lies in the midrange. Critical interaction strength for blends, relative to that for both monomeric mixtures and polymer solutions, is drastically reduced. Two sources of net component interactions were considered for polymer solutions (Chapter 7), those arising from cohesive energy differences and those from free volume differences. In their simplest realizations, the solubility-parameter and corresponding-states formulations, the net interaction density is always repulsive. Miscibility limitations from those sources, expressed as differences in solubility-parameters and characteristic temperatures, are given in Table 8.1. The limits are much smaller for blends, a consequence of their small entropy of mixing density and illustrating again the severity of restrictions on polymeric mutual solubility.

Obtaining polymeric solubility parameters by the monomeric liquid method of measuring evaporation energy [Eqs. 2.106 and 3.94] is of course impossible. Solubility parameters have been assigned to various polymer species by indirect means[9,10]. Some carefully tested estimation procedures, based on the group-contribution concept and on small-molecule data, have also been developed[11]. An assumed equality of cohesive energy density and the internal pressure, the latter obtainable by PVT measurements, has also been employed[12], as described in Section 8.3. Although useful for estimation purposes, such methods are not precise enough to permit dependable quantitative predictions. Section 8.3 also describes a method of assigning self-consistent relative solubility parameters to the species within structurally similar polymer families, based on interaction data for their blends. When such data are not available, the interaction density from cohesive energy mismatch can instead be estimated using PVT-based solubility parameters. Thus, using the Eq. 2.111 approximation that cohesive energy density and internal pressure are equal,

$$\delta_{\mathrm{PVT}} = (\Pi_{IP})^{1/2} = (T\alpha/\beta)^{1/2} \tag{8.1}$$

The estimate of interaction density coefficient $X = RT\chi/V_{ref}$ (Eq. 3.66) for cohesive energy mismatch is therefore [from (Eq. 3.96)],

$$X_{CED} = (\delta_1 - \delta_2)^2_{\mathrm{PVT}} \tag{8.2}$$

The free volume density of polymers and solvents differ from one another systematically, as discussed in Section 7.4.2. The effect of those differences is large enough to make free volume mismatches a universally important phenomenon in polymer solutions. The differences in free volume among polymer species are smaller, but the tolerable difference for miscibility in polymer blends is also smaller. The free volume contributions can be estimated from PVT data. The FOV-based Eq. 3.126, the Patterson equation, is easily extended to blends. Thus, in terms of the interaction density coefficient,

$$X_{FV} = \frac{\overline{\Pi}_{CED}\bar{\alpha}T}{2}\left(\frac{T_1^* - T_2^*}{\overline{T}^*}\right)^2 \qquad (8.3)$$

where component averages of Π_{CED}, α, and T^* are used; the differences are always small for the cases of interest. For estimation purposes, the values $\Pi_{CED} = 300\ MPa$, $\alpha = 7 \times 10^{-4}(K^{-1})$, and $T = 167°C\ (440K)$ are typical for polymers in the liquid state. Using these values, Eq. 8.3 becomes:

$$X_{FV}(MPa) = 46\left(\frac{T_1^* - T_2^*}{\overline{T}^*}\right)^2 \qquad (8.4)$$

Equation 7.84 relates characteristic temperature for the FOV model to the thermal expansion coefficient. It can be expanded for small differences in α, and for $\alpha T = 0.31$ from the typical α and T given above, the result is:

$$\frac{T_1^* - T_2^*}{\overline{T}^*} = -0.54\left(\frac{\alpha_1 - \alpha_2}{\bar{\alpha}}\right) \qquad (8.5)$$

Equation 8.4 then becomes:

$$X_{FV}(MPa) = 14\left(\frac{\alpha_1 - \alpha_2}{\bar{\alpha}}\right)^2 \qquad (8.6)$$

Other sources of interaction can be important for blends. The *specific association* of functional groups may produce miscibility by overcoming the normally repulsive cohesive and free volume contributions. The extreme example of acid–base reactions between otherwise immiscible species can be too strongly attractive, however, and result in the formation of intractable network-like structures. A relatively mild association, such as by hydrogen bonding, can be useful and can also supply spectroscopic signatures for association complex formation[11]. Specific association may be too weak to furnish an unambiguous spectroscopic signature, but it can still produce miscibility. Thus, for example, evidence from NMR has been suggested to demonstrate the existence of specific associations in the blends of polydienes[13], thus accounting for their otherwise surprising miscibility. The signature is subtle, however, and others have registered contrary opinions[14]. Infrared spectroscopy has revealed no trace of such associations[15]. Spectroscopic methods are constantly improving[16,17], so perhaps in time issues such as this one will be resolved. Association effects weaken with increasing temperature and are finally overcome by the underlying net repulsion, thus leading to LCT phase separation.

A fourth type of interaction, based on *local packing differences*, may require some consideration in polymer blends. As noted in Section 2.2.1, the interactions governing cohesive energy density depend on the intermolecular distribution of interacting sites, on average the intermolecular pair distribution $[g(r)]_{inter}$. Like the total pair distribution $g(r)$ from scattering experiments, $[g(r)]_{inter}$, at the distances relevant to cohesive energy, is mainly set by the specie's chemical structure. Accordingly, $[g(r)]_{inter}$ should be insensitive to environment and, in the absence of other considerations, not changed appreciably by mixing. This should offer no particular problem in the formation of polymer solutions—solvent positioning can accommodate any necessary adjustments—although perhaps with some net free energy change. Mixing two polymeric species is another matter, however, because the species must now compete to satisfy their respective $[g(r)]_{inter}$ constraints. Given the minute energy densities that govern phase behavior in blends, the net contribution to free energy change from this configurational competition—the local packing effects—might well be important. The contribution could be either positive or negative, depending on the local packing compromises that exist in the pure component states. Unfortunately, no technique is available at present for observing local packing changes directly.

Scattering in the single-phase region, as interpreted by using the de Gennes equation [Eq. 7.67], often called the random phase approximation or RPA, has become the primary means for measuring interaction strength in polymer blends. Because the de Gennes equation plays such a dominant role, it is well to be aware of its limitations. Two sorts of observations suggest prudence in applying Eq. 7.67. One has to do with the angular dependence of scattering in the low q range. It is very rare to observe a true Guinier range in the scattering behavior. Fitting the data for blends of nearly monodisperse components to Eq. 7.67 using component Debye equations [Eq. 4.108] works well in the intermediate q-range and even below that to some extent, but invariably—or seemingly so—the observed intensities eventually drift above the fit in the still lower q range. Of course, explanations can be made based on contamination by particulates, which is a common problem in light scattering for example, but the effect seems different, being seen to nearly the same extent below approximately $q = 1/R_g$ in short chains and long chains alike. Correlations appear to exist in blend composition at separation distances extending well beyond the component values of R_g that are not encompassed by Eq. 7.67.

A second sort of unsettling observation is the composition dependence of the interaction parameters extracted from the scattering data using Eq. 7.67. The values of χ are commonly found to vary in a roughly parabolic fashion around the critical composition, $\phi_c \sim 1/2$, typically with a minimum in that region (see Figures 8.4 and 8.17). The values in the "wings", $0 \lesssim \phi \lesssim 0.2$ and $0.8 \lesssim \phi \lesssim 1$, are typically higher than those in the midrange, and the midrange values, $0.25 \lesssim \phi \lesssim 0.75$, are typically fairly constant. Stated differently, scattering in the wings is more intense than expected from the value of χ obtained with midrange data, hence the larger inferred χ from Eq. 7.67. The pattern is not a clear one, hence the "usually" and "typically" usage. The parabolic form is not typical of blends with net attractions (see Figure 8.23 below), for example. Various explanations have been offered (see Section 8.3.4).

Whether the two types of departures somehow arise from the approximate nature of Eq. 7.67 or whether they are related to the recently expressed more general concerns about composition correlations in polymeric mixtures[18] is not clear. In any case, the values of χ and R_g, extracted through Eq. 7.67 in the composition midrange and by fits that ignore the small-q drifts, would appear to be valid. That is, χ and R_g obtained in this way agree with such observables as phase separation temperatures and sizes obtained by other means.

Small-angle neutron scattering (SANS) by blends having component labeled with deuterium is the main method for determining χ in blends. The value of isotopic labeling to generate scattering contrast with a minimal disturbance of component interactions was discussed briefly in Section 5.1.2. Small-angle x-ray scattering has also been used occasionally[19,20]; the natural difference in electron density between components supplies the contrast in this case. Labeling to increase x-ray contrast is possible, but the labels must be chosen carefully to avoid disturbances. Light scattering is not useful for determining the interaction density in blends, because the natural contrast from refractive index difference is too small. Labeling to increase light scattering contrast is not an option: both the scattering event and the system interactions strongly involve the outer electrons, so disturbance is inevitable.

It must not be concluded from this discussion that the thermodynamic consequences of labeling polymers with deuterium can be ignored. In 1975, Strazielle and Benoit demonstrated an isotope effect on the theta temperature[21], showing that T_θ for the polystyrene–cyclohexane system increases about 4°C when perdeuterated cyclohexane is used, decreases about 4°C when perdeuterated polystyrene is used, and yet hardly changes at all when both components are perdeuterated. They attributed these changes to a reduction in solubility parameter when deuterium replaces hydrogen. In 1982, Atkin et al. reported a significant shift in the critical point of polystyrene–polybutadiene blends when perdeuterated polybutadiene was used[22]. Yang et al. in 1983 demonstrated clearly the effects of deuteration in the phase behavior of polystyrene–polyvinylmethylether blends[23]. In 1985, Lapp et al. obtained a non-zero interaction parameter for a blend of ordinary and perdeuterated polydimethylsiloxane[24]. These observations and others led to the careful studies of isotopic blends described in Section 8.2.

Sections 8.2 and 8.3 describe the extensive use of specially made saturated hydrocarbon polymers, the *model polyolefins*. These are derived from nearly monodisperse polydienes, made by organolithium-initiated anionic polymerization[25,26], with

TABLE 8.2 Nomenclature for model polyolefin species

Monomer	Polymerization	Model Polyolefin	Other Names	Comments
butadiene	unmodified	EB08	H08	statistical ethylene-1-butene copolymers
	modified	EB08-EB90	H08-H90	statistical ethylene-1-butene copolymers
	modified	EB97-EB100	PEE, PB	statistical ethylene-1-butene copolymers
isoprene	unmodified	PEP	HPI, PM	alternating ethylene-propylene copolymer. ~7% 3,4 incorporation.
	modified	50HPI, 75HPI		~50 and 75% 3,4 incorporation
2-ethylbutadiene	unmodified	PEB	PE	alternating ethylene-1-butene copolymer. ~7% 3,4 incorporation.
	modified	50PEB		~50% 3,4 incorporation.
2-methylpentadiene	unmodified	PP		~1% other isomers
2,3-dimethylbutadiene	unmodified	hhPP		~3% other isomers
styrene	unmodified	PCHE	PVCH	poly(cyclohexylethylene)

and without the presence of microstructural modifiers. The polydienes are converted to model polyolefins by addition of H_2 (or D_2 for labeling purposes), quantitatively and without architectural alteration, to their double bonds. A number of methods for accomplishing this step have been found[27–31]. A variety of acronyms for these model polyolefin materials are used in the literature. The terminology to be used here is given in Table 8.2.

8.2 Isotopic Blends

As discussed in previous chapters, isotopic labeling—the substitution of deuterium for part or all of the hydrogen in a molecule—finds many uses in the polymer field. The scattering cross section for slow neutrons is much different for protons and deuterons, but the properties are affected only slightly. Thus, deuterium substitution supplies SANS contrast and makes information on the structure and interactions of a system accessible without significant alteration. Changes always occur, however, and they are not always negligible, especially for polymer blends. Prudence thus dictates the need for experimental crosschecks and some understanding of the origin of isotopic interactions to correct for them. The following estimate of their magnitude for hydrocarbon substances, based on the regular mixture model (Section 3.5), is a simplified version of the previous analyses[32,33].

8.2.1 Isotopic Interactions

It has been known since the 1930s that deuterium substitution reduces both molar volume V and molecular polarizability α[34]. Both V and α influence Π_{CED}, the cohesive energy density of the liquid. Using Eq. 3.88 and $\varepsilon_{ii} = -w_{ii}r^{-6}$, the energy per pairwise contact in the pure liquid state for species i, ε_{ii}, is proportional to α_i^2/r_{ii}^6, where r_{ii} is the intermolecular distance between polarizable bonds. Assuming r_{ii} is proportional to $V_i^{1/3}$ leads to $\varepsilon_{ii} \propto \alpha_i^2/V_i^2$, or from Eq. 3.92 for the energy density,

$$(\Pi_{CED})_i \propto \frac{\alpha_i^2}{V_i^3} \tag{8.7}$$

where the proportionality constant depends in general on the various features of species structure that are relatively insensitive to small changes in V and α.

In 1942, Bell measured the effect of deuterium substitution on polarizability of the C-H bonds in methane gas[35], finding $\alpha_{CH}/\alpha_{CD} = 1.0145$. Molecular calculations by Bates et al. led to similar values[33], an average of 1.0165 with a range from 1.015 to 1.018, for several hydrocarbon liquids. Measurements of density[33,34] for several hydrocarbon liquids, including polymers, led to an average of $V_H/V_D = 1.003$ with a range from 1.002 to 1.004. Thus, using Eq. 8.7,

$$\frac{(\Pi_{CED})_H - (\Pi_{CED})_D}{(\Pi_{CED})_H} \sim 0.022 \tag{8.8}$$

The effects on polarizability and volume changes tend to opposite directions, but polarizability rather clearly dominates. For typical hydrocarbons, the cohesive energy density decreases by about 2% with perdeuteration.

Using Eq. 3.94 and Eq. 8.8, perdeuteration reduces the solubility parameter by about 1%:

$$\frac{\delta_H - \delta_D}{\delta_H} \sim 0.011 \tag{8.9}$$

According to Eqs. 3.93 and 3.96, having a typical $\delta_H = 16\ J^{1/2}cm^{-3/2}$ for nonaromatic hydrocarbon liquids (Table 2.6) and having a typical molar volume $V = 10^2\ cm^3$ as V_{ref}, a typical isotopic interaction parameter at 25°C is:

$$\chi_{HD} \sim 1.2 \times 10^{-3} \tag{8.10}$$

Only polymeric values of χ_{HD} are available for confirmation. For 1,2 polybutadiene and its fully saturated analog at 23°C, obtained from neutron scattering as discussed below and adjusted to $V_{ref} = 10^2 cm^3$,

$$\begin{aligned} \chi_{HD} &= 1.17 \times 10^{-3} \quad \text{(1,2 polybutadiene)} \\ \chi_{HD} &= 1.42 \times 10^{-3} \quad \text{(atactic poly(1-butene))} \end{aligned} \tag{8.11}$$

Agreement between estimates and observed values is good, nicely supporting the interpretation. The values also demonstrate that deuterium substitution can have a significant influence on χ for blends in the polymeric range of interest (note typical values in Table 8.1).

Experimental aspects. The determination of interaction density by SANS experiments is best conducted with samples as nearly monodisperse as possible. In any case, the samples should be thoroughly characterized in molecular weight distribution and, for the labeled sample, in the level of deuterium as well. It is necessary to use absolute intensity $I_r(q)$, so careful attention to instrument calibration is also required. The SANS contrast factor C_{SANS} is commonly defined in terms of the polymeric repeating unit [see Eq. 4.144], so it is convenient to express the de Gennes equation, Eq. 7.67, in the corresponding manner. Thus, for the case of isotopic blends,

$$\frac{C_{SANS}}{I_r(q)} = \frac{1}{\phi_H N_H V_H P_H(q)} + \frac{1}{\phi_D N_D V_D P_D(q)} - \frac{2\chi_{HD}}{V_{ref}} \tag{8.12}$$

where V_{ref} is the reference volume, and the ϕ, N, V, and $P(q)$ terms are the component volume fraction, mers per chain M/m_o, molar volume per mer, and form factor.

The form factor for monodisperse random coils is the Debye equation [Eq. 4.108], $P(u) = 2(\exp(-u) - 1 + u)/u^2$, where $u = q^2 R_g^2$ or $q^2 l_N^2 N/6$, and l_N is the mer-based statistical length [Eq. 4.34]. Like V, l_N is assumed to be insensitive to labeling. For V_H as the reference volume and assuming $V_D = V_H$, Eq. 8.12 becomes:

$$\frac{C_{SANS} V_H}{I_r(q)} = \frac{1}{\phi_H N_H P(u_H)} + \frac{1}{\phi_D N_D P(u_D)} - 2\chi_{HD} \tag{8.13}$$

where $u_H = q^2 l_N^2 N_H/6$ and $u_D = q^2 l_N^2 N_D/6$.

Aside from the interaction parameter, all quantities in Eq. 8.13 are known or separately measurable—N_H, N_D, and l_N by dilute solution methods; C_{SANS}, V_H, and $\phi_H = 1 - \phi_D$ from the sample compositions; $I_r(q)$ from SANS. Thus, χ_{HD} can be determined by choosing a value that produces the best fit of the SANS data to the equation. Example fits to Eq. 8.13 for scattering by isotopic 1,4-polybutadiene (PBD) blends are shown in Figures 8.1 and 8.2. The result of fitting $I_r(q)$ versus q at 23°C for two blends of the same H and D samples, $\phi_D = 0.38$ and 0.65—the first near ϕ_c, the other off critical—is shown in Figure 8.1. A reasonable fit is achieved for both blends with $\chi_{HD} = 8.7 \times 10^{-4}$, indicating that the interaction parameter is insensitive to blend composition, as expected for simple mixtures.

The results for the $\phi_D = 0.38$ blend at temperatures from −26°C to +87.5°C, plotted in the I_r^{-1} versus q^2 (Ornstein-Zernike) manner, are shown in Figure 8.2.

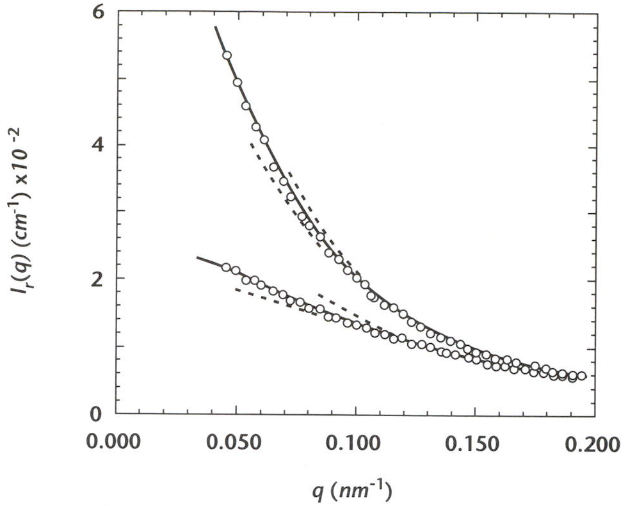

FIGURE 8.1 Angular dependence of reduced SANS scattering intensity for isotopic 1,4 polybutadiene blends at 23°C. The compositions are $\phi_D = 0.31$ (near the critical composition) for the upper curve and $\phi_D = 0.65$ for the lower curve. The fitted curves for both data sets correspond to $10^4\chi = 8.7$, with the dashed curves indicating ± 0.5 and ± 1.3 of that value for the upper and lower curves respectively. Data from Bates et al.[39].

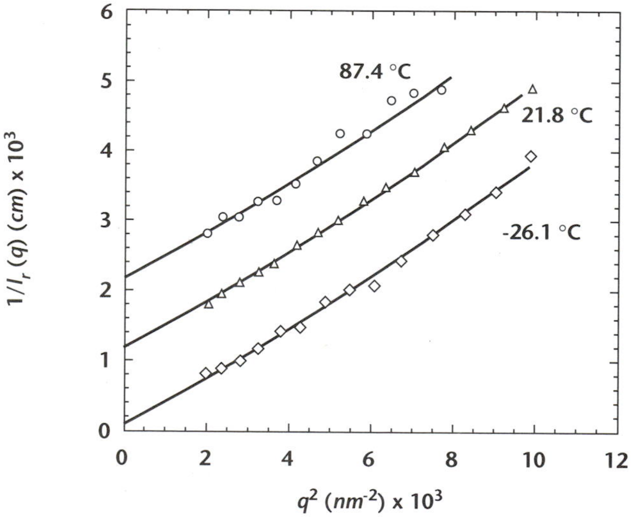

FIGURE 8.2 Ornstein-Zernike plots of the scattering data at selected temperatures for an isotopic 1,4-polybutadiene blend ($\phi_D = 0.31$). Data from Bates et al.[39].

Despite the linear appearance, the fits to the data are actually slightly curved. The nearly linear character is nonetheless helpful in judging whether the system is in the single-phase region. Thus, the $q = 0$ intercept is positive for single phases, zero on the spinodal curve, and negative beyond that local stability limit. From Figure 8.2, the blend is apparently single phase at all the measurement temperatures, but, as viewed from the single phase, it is approaching a spinodal of the UCT type near −26°C.

Some general comments on the labeled-chain SANS procedures for extracting interaction parameters are in order. The labeling level is usually obtained or confirmed by component density measurements; the very slight change in molar volume with labeling is ignored. The fits themselves usually cover about an order of magnitude in scattering angle, beginning from $qR_g \sim 1$ and extending well into the intermediate range. Unlike light scattering data, SANS data seldom extends into the Guinier range, and $I_r(0)$ is obtained by the extrapolation of a fitted curve. Full-range fitting, making use of all the scattering data, is the most common practice. Bates and Wignall[36] determined χ_{HD} as a function of temperature for isotopic polystyrene mixtures. They then used the solubility parameter idea to demonstrate a remarkable quantitative consistency between their data on isotopic blends of undiluted polystyrene at 160°C and the effect of deuteration on T_θ for the polystyrene–cyclohexane system [21].

Having values of χ_{HD} available is obviously beneficial for the practical problem of anticipating the magnitude of undesirable complications that could be caused by isotopic labeling. However, Bates and Wignall evidently recognized as well the value of isotopic blends as model systems that permit tests of mixing theories to be made under the simplest possible circumstances. With nearly monodisperse samples, typically synthesized by anionic methods, they and colleagues used SANS and the de Gennes equation to develop a body of data on isotopic blends for several polymeric species and to use such results in the exploration of several fundamental issues involving blends in general. The interaction data are considered first, then the method, and finally the dependence of the interactions on temperature, labeling level, composition, and chain length dependence.

Interaction parameter data for isotopic blends, hydrogenous and perdeuterated components of the same polymer species[33,37–41], are shown in Table 8.3. The temperature T_m is the midpoint of the range of temperatures covered in the experiments, [see Eq. 3.75], and the interaction parameter χ_m is the value at T_m for midrange compositions, $\phi_H \sim \phi_D$. For each species, the interaction parameters are defined with respect to mer reference volumes, $V_{ref} = m_0/\rho$. From these data, an early dispute about how seriously deuterium labeling affects polymer blend experiments is easily understood. In 1985, Bates et al.[37] found strong interaction, verging on phase separation ($\chi_{HD} \sim \chi_c$), in an isotopic 1,4 polybutadiene blend with a critical interaction parameter [Eq. 3.81], $\chi_c \sim 11 \times 10^{-3}$. A year later, Yang et al.[42] found negligible interaction in an isotopic polystyrene blend with only slightly smaller critical interaction parameter, $\chi_c \sim 8.5 \times 10^{-3}$. From Table 8.3, the interaction parameter χ_m is

TABLE 8.3 Interactions for hydrogenous and perdeuterated polymer blends

Species	$T_m(K)$	$\alpha_m(K^{-1}) \times 10^4$	$\chi_m \times 10^4$	$\chi_h \times 10^4$	$\chi_s \times 10^4$	$-\alpha_m T_m \chi_m \times 10^4$
Polyethylene	443	7.80	4.0	5.0	−1.0	−1.4
poly(ethylene-*alt*-propylene)	419	7.21	7.07	13.6	−6.5	−2.1
poly(1-butene)	336	7.07	7.9	8.3	−0.4	−1.9
1,4 polybutadiene	304	6.35	8.4	10.7	−2.3	−1.6
1,2 polybutadiene	336	6.24	6.36	5.2	+1.15	−1.3
Polystyrene	460	5.83	1.45	4.35	−2.9	−0.4

larger for the 1,4-polybutadiene blend by a factor of nearly 6. Thus, both groups were correct: As found, deuteration effects should indeed have been weak in the Yang et al. experiment and strong in the Bates et al. experiment. The example does illustrate, however, the difficulty of assessing the likely effects of labeling. The chemical nature of C-H and C-D bonds can go well beyond merely the dispersive interaction contribution. Polymer species containing polar or complexing substituents might well respond differently.

8.2.2 Temperature Dependence of the Interactions

How should temperature influence interaction strength for a blend of components obeying the simple liquid model (Section 2.3.5) and the regular mixture model (Section 3.5)? That combination offers the simplest physically meaningful description of dense liquid behavior. The purpose is to investigate how well, if at all, the observed temperature dependence of χ can be accounted for under the most favorable circumstances. The answer is of interest for the polyolefin blends in Section 8.3 as well as for the the isotopic blends considered here.

The cohesive energy density Π_{CED} of a simple liquid is quadratic in the mass density ρ, Eq. 2.87 or Eq. 2.107, and Π_{CED} is equal to the internal pressure Π_{IP}, Eq. 2.111:

$$\Pi_{CED} = (\Pi_{CED}/\rho^2)_{rcp}\rho^2 \tag{8.14}$$

$$\Pi_{CED} = \Pi_{IP} \tag{8.15}$$

The proportionality constant $(\Pi_{CED}/\rho^2)_{rcp}$ is a temperature and pressure insensitive property of the species. The internal pressure, a mechanical property, depends on temperature and pressure only in relation to α and β, the thermal expansion coefficient and isothermal compressibility, Eq. 2.113:

$$\Pi_{IP} = \alpha T/\beta \tag{8.16}$$

For regular mixing, the interaction parameter is related to the difference in component solubility parameters, Eq. 3.93:

$$\chi = \frac{V_{ref}}{RT}(\delta_1 - \delta_2)^2 \tag{8.17}$$

and, from Eq. 3.94, the solubility parameters are square roots of the component cohesive energy densities:

$$\delta_i = (\Pi_{CED})_i^{1/2} \tag{8.18}$$

Thus, from Eq. 8.14, the component solubility parameter is directly proportional to component mass density:

$$\delta_i = (\delta_i/\rho_i)_{rcp}\rho_i \tag{8.19}$$

Finally, the reference volume, as conventionally defined, varies inversely as some average component density $\bar{\rho}$:

$$V_{ref} = (\overline{V_{ref}\rho})_{rcp}/\bar{\rho} \tag{8.20}$$

In the simple-liquid, regular-mixture formulation, the product χT has no explicit dependence on either temperature or pressure. It depends on P and T only through the effects of those variables on the component densities, as determined by the component values of β and α. Thus, from Eqs. 8.19 and 8.20, the solubility parameter difference and reference volume relative to that at an ambient pressure $P_a \sim 0.1MPa$ and some temperature of convenience T_m, are given by:

$$\delta_1(P,T) - \delta_2(P,T) = \delta_1(P_a,T_m)\frac{\rho_1(P,T)}{\rho_1(P_a,T_m)} - \delta_2(P_a,T_m)\frac{\rho_2(P,T)}{\rho_2(P_a,T_m)} \tag{8.21}$$

$$V_{ref}(P,T) = V_{ref}(P_a,T_m)\frac{\bar{\rho}(P_a,T_m)}{\bar{\rho}(P,T)} \tag{8.22}$$

Using Eqs. 3.72 and 3.73, the interaction parameter can be written as the sum of entropic and enthalpic contributions. If χ varies with temperature as $A + B/T$ over the range of interest, then Eq. 3.75 can be used to express the interaction parameter:

$$\chi(T) = \chi_s + \frac{\chi_h T_m}{T} \tag{8.23}$$

in which χ_s and χ_h are the entropic and enthalpic contributions at some temperature of convenience T_m.

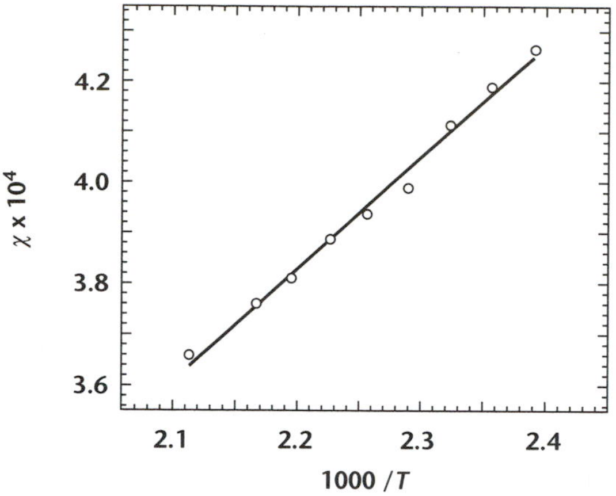

FIGURE 8.3 Temperature dependence of the interaction parameter for an isotopic polyethylene blend. Data from Londono et al.[41].

The form $\chi = A + B/T$ describes the temperature dependence of isotopic blends. Figure 8.3 for polyethylene is typical. Table 8.3 lists values of χ_h and χ_s for hydrogenous/perdeuterated blends of six polymeric hydrocarbon species along with the values of χ_m and T_m already discussed in Section 8.2.1. Now assume that isotopic isomers have essentially the same thermal expansion coefficient α. The density ratios are then equal and can be factored out of the terms in Eq. 8.17 to give:

$$\chi(T) = \chi_m \frac{\rho(T_m)}{\rho(T)} \frac{T_m}{T} \tag{8.24}$$

Approximating the density ratio as:

$$\frac{\rho(T_m)}{\rho(T)} = 1 - \alpha(T - T_m) \tag{8.25}$$

leads to:

$$\chi(T) = \chi_m \left[\frac{T_m}{T}(1 + \alpha T_m) - \alpha T_m \right] \tag{8.26}$$

Thus, by comparing with Eq. 8.23,

$$\begin{aligned} \chi_s/\chi_m &= -\alpha T_m \\ \chi_h/\chi_m &= 1 + \alpha T_m \end{aligned} \tag{8.27}$$

Values of α for the six species and corresponding values of χ_s calculated using Eq. 8.27 are shown in the last column of Table 8.3. The density-based contributions are always negative, which is consistent with five of the six experimental values. Quantitatively, the results are equivocal. Evaluations of χ_s are based on rather long extrapolations, and hence are subject to considerable uncertainty. Nevertheless, from these results alone it is impossible to rule out other sources of contribution to the interactions. This question is examined again in Section 8.3.4, where χ versus T data for polyolefin blends are considered.

8.2.3 Some Properties of χ_{HD}

Other features of relevance for polymer blends were first investigated definitively by the Bates-Wignall group using isotropic blends. The assumption that the interaction parameter is independent of chain length for long-enough chains having low-enough self-concentrations is implicit in the simple mixture model. Indeed, the fundamental idea that interaction is a locally determined property in dense liquids becomes suspect if it is not true. The molecular weight dependence of χ_{HD} was examined with data in the critical region for blends having nearly matched N_H and N_D[40]. The prediction for simple mixtures of this composition is $\chi_c \propto N^{-1}$ [Eq. 3.81 for $V_1 = V_2$], and, for data covering nearly a factor of 5 in N, that prediction was confirmed.

The same study also examined the *random copolymer equation*, a relationship based on local randomness of mixing that connects the interaction parameter for mixtures of copolymers to their chemical compositions (Section 8.5.2). For mixtures of partially deuterated components, the prediction for partially deuterated components is a particular case of Eq. 8.62:

$$\chi_{HD} = (f_1 - f_2)^2 (\chi_{HD})_o \tag{8.28}$$

where $(\chi_{HD})_o$ refers to mixtures of fully hydrogenous and fully deuterated components of the same species, and f_1 and f_2 are the fractions of their $D \Rightarrow H$ substitutions. Equation 8.28 was confirmed, providing the first detailed quantitative test of the random copolymer equation. It was in fact that result that permitted the use of partially deuterated components to cover the large range of chain lengths in the test for N dependence[40].

Another assumption of the simple mixture model—that χ is independent of blend composition—has been tested extensively with isotopic mixtures[33,41,43]. As mentioned in Section 8.1, dependence of a very strange kind is found. Contrary to the mild monotonic variation with ϕ for polymer solutions, χ_{HD} is relatively constant in the midrange, $0.25 \lesssim \phi \lesssim 0.75$, but then changes rapidly and symmetrically at the extremes. The value at the extremes decrease for PS[43] but increase for PE[41], 1,2PBD, and EB99[33], the latter being the fully saturated derivative of 1,2PBD. The variations are not small, as shown by data for 1,2PBD and EB99 in Figure 8.4. This "parabolic" behavior is also found, but in varying degrees, for polyolefin blends[44,45]. Its origin is considered further in Section 8.3.4.

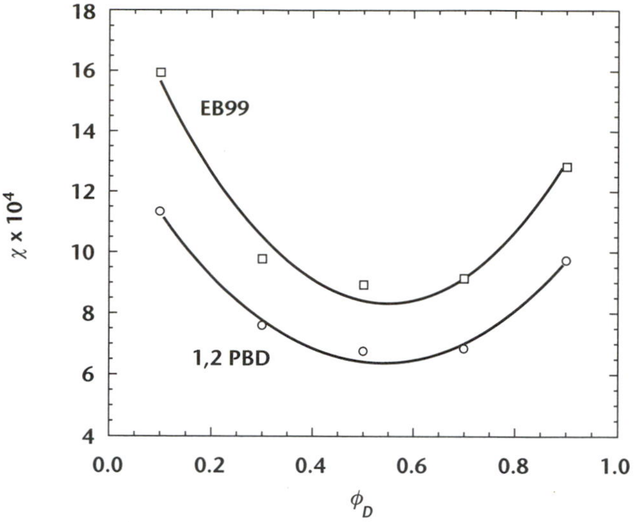

FIGURE 8.4 Composition dependence of the interaction parameter for isotopic blends of 1,2 polybutadiene (37°C) and of its saturated product (26°C). The curves are quadratic fits to guide the eye. Data from Bates et al.[142].

8.3 Polyolefin Blends

The net interactions that govern phase behavior in saturated hydrocarbon polymer blends, like those in isotopic mixtures, originate from induced dipole dispersion forces. Thus, in the simplest instance, phase separation should be driven by mismatches in cohesive energy density. However, the variety of possibilities is inherently much broader for hydrocarbons because, unlike isotopic pairs, hydrocarbons can differ widely from one another in molecular shape. As mentioned earlier in this chapter, shape differences introduce the possibility of local packing effects and attendant energy changes upon mixing. Free volume mismatches can influence the mixing energy as well. How such added complexities of hydrocarbon mixtures play against one another, and whether sensible generalizations can be found, is still a matter of debate.

Polyolefins such as polyethylene, polyisobutylene, polypropylene, and their various copolymers belong to the class of saturated hydrocarbon substances. They also comprise by far the largest family of commercial polymers and are often used in the form of blends. Their melt state phase behavior is thus of interest for practical as well as fundamental reasons. *Cloud point determination* is commonly used to locate phase boundaries in polymer blends. However, that method depends on *optical contrast*, a difference in refractive index between coexisting phases. Although valuable for supplementing or confirming the results from other methods, cloud point measurements for polyolefin blends are inherently difficult, owing to the similarities in refractive index among polyolefin species.

Labeling with deuterium (taking advantage of its various differences from hydrogen while minimizing the chemical disturbance) is a natural course to consider. Lohse conducted the first SANS-based study of the interactions in single-phase polyolefin blends in 1986[46], demonstrating that atactic and isotactic polypropylene are miscible in the melt state. As synthesized commercially, polyolefins have broad microstructural and molecular weight distributions, thus complicating the interpretation of SANS data to quantify the interactions that underlie phase behavior. This led to interest in *model polyolefins*, materials mimicking polyolefins in local structure but having simple well-defined large-scale architectures and easy to label with deuterium.

The synthesis of model polyolefins by saturating the double bonds of nearly monodisperse polydienes is described briefly in Section 8.1. Separate aliquots of the parent polydiene sample are saturated with hydrogen and with deuterium, providing well-characterized *matched pairs*—nonlabeled and labeled samples with the same chemical structure. Crist and coworkers reported the first SANS studies of polyolefin mixing behavior with such materials in 1990[47]. Others joined the effort, and within a relatively few years an extensive body of SANS-based data had been generated on polyolefin interactions, about which essentially nothing of a quantitative nature had been known previously. The labeled model polyethylenes proved useful for other purposes, in particular for the study of polyolefin phase interfaces by *nuclear reaction analysis* (NRA), an isotopic scattering technique for the spatial profiling of composition[48]. Coexisting compositions in the two-phase region have been determined through NRA[48,49], and these data complement SANS-based interaction data on the same systems.

8.3.1 Measurements

Major advances in understanding polyolefin interactions have come from SANS measurements and their interpretation through the de Gennes equation [Eq. 7.67]. The procedure typically begins with SANS measurements at several temperatures on matched pair mixtures, usually $\phi_H \sim 0.5$. The data are analyzed by performing full q-range fits to Eq. 8.13, the isotopic pair expression. The chemical structures of the components are identical, so $N_H = N_D$ and $V_H = V_D$, and $u_H = u_D$ are reasonable approximations. For matched pairs, the scattering equation thus simplifies to:

$$\frac{C_{SANS} V_H}{I_r(q)} = \frac{1}{\phi_H \phi_D P_H(u_H)} - 2\chi_{HD} \tag{8.29}$$

where $P_H(u)$ is the Debye equation, Eq. 4.108, the samples being nearly monodisperse. The fitting parameters are the mer-based statistical length for the species l_N and the isotopic interaction parameter χ_{HD}.

Blends of labeled and nonlabeled components with different chemical structures are then prepared from the resulting pool of well-characterized samples. SANS measurements are conducted at several temperatures, and the results at each temperature are analyzed with another version of the de Gennes equation:

$$\frac{C_{SANS}}{I_r(q)} = \frac{1}{\phi_1 N_1 V_1 P(u_1)} + \frac{1}{\phi_2 N_2 V_2 P(u_2)} - 2\frac{\chi}{V_{ref}} \quad (8.30)$$

The subscript 1 and 2 refer to the components; χ is the interaction parameter for the pair; V_{ref} is the reference volume for χ, commonly chosen as $(V_1 V_2)^{1/2}$; and $u_i = q^2 (l_N^2)_i N_i/6$. The values of N_i, $(l_N)_i$, ϕ_i, and V_i are known at each temperature from either molecular characterization or the matched pair results. The quantity of interest, χ for the component pair, is the only fitting parameter.

All this works well if the fits have small enough error bars, the values of $(l_N)_i$ are unchanged by blending, and the effect of isotopic labeling on χ is negligible. The first two seem to be well-satisfied[50], but labeling effects turn out to be large. Rhee and Crist showed that "swap" experiments, reversing which component is labeled, led to significantly different values of χ[51]. Several explanations were offered[51–53], all coming to a similar result. Solubility parameter ideas and the isotopic blend results, as shown by the following example calculation, provided one of them.

Owing to exchange and addition during saturation, the labeled polybutadiene-based models contain about 3 D and 5 H atoms per mer, corresponding to fractional substitution $f \sim 0.375$ and thus, from Eqs. 8.9 and 8.28, $\chi_{HD} \sim 1.0 \times 10^{-4}$ for a mer reference volume of 60 cm^3 at 23°C. This is the observed range[52]; it is on the border of detectability. Consider now the effect of labeling at that level on the values of χ for a pair of model ethylene-butene copolymers (EB copolymers) made by saturating polybutadienes of different 1,2 contents. Two pairings have SANS contrast, component 1 labeled, and its swap. Expressions for solubility parameter differences are easily obtained:

$$\begin{aligned} \delta_{1D} - \delta_{2H} &= \delta_{1H} - \delta_{2H} - f_1 g_1 \delta_{1H} \\ \delta_{1H} - \delta_{2D} &= \delta_{1H} - \delta_{2H} + f_2 g_2 \delta_{2H} \end{aligned} \quad (8.31)$$

where $g_i = (\delta_H - \delta_D)_i/(\delta_H)_i$ is the fractional change in solubility parameter with perdeuteration for species i, and $f_i g_i$ is the fractional change with partial substitution. For $\delta_{i_D} < \delta_{iH}$, the correct term $f_i g_i \delta_{iH}$ is always a positive quantity, algebraically decreasing $\delta_{1H} - \delta_{2H}$ when species 1 is labeled and increasing it for the swap.

The magnitude of $f_i g_i \delta_{iH}$ can be estimated using data already at hand. From Eq. 8.9, g_i is about 0.01 for full deuterium labeling for both species; so $f_i g_i$, from $f \sim 0.375$ above, is about 0.00375. For saturated hydrocarbons, δ_i is about $16(J\ cm^{-3})^{1/2}$, so $f_i g_i \delta_{iH} \sim 0.056\ (J\ cm^{-3})^{1/2}$. For a blend with typical $\chi \sim 10^{-3}$ and $V_{ref} \sim 60\ cm^3$ at 23°C for unlabeled species, the case of interest here, $|\delta_{1H} - \delta_{2H}| \sim 0.2\ J^{1/2} cm^{-3/2}$ from Eq. 8.17. For the case $\delta_{1H} > \delta_{2H}$, $\delta_{1H} - \delta_{2H} = +0.2\ (J cm^{-3})^{1/2}$. Using these numbers and Eq. 8.17, the following interaction parameters are obtained for the four combinations:

$$\begin{aligned} \chi_{HH} &= \chi_{DD} = 0.98 \times 10^{-3} \\ \chi_{DH} &= 0.51 \times 10^{-3} \\ \chi_{HD} &= 1.60 \times 10^{-3} \end{aligned} \tag{8.32}$$

Note that a simple average of the singly labeled results gives a value near the unlabeled result of interest, $(\chi_{DH} + \chi_{HD})/2 = 1.06 \times 10^{-3}$. If $f_i g_i$ is the same for both components, the unlabeled result, $\chi_{HH} = 0.98 \times 10^{-3}$, is given exactly by the average of the square roots:

$$\chi_{HH} = \left(\frac{\chi_{DH}^{1/2} + \chi_{HD}^{1/2}}{2} \right)^2 \tag{8.33}$$

Some useful conclusions can be drawn from this exercise. When deuteration lowers the cohesive energy density, which is seemingly true for saturated hydrocarbons generally, labeling the species with higher cohesive energy density lowers χ relative to its value for nonlabeled species. Labeling the component with lower cohesive energy density, on the other hand, elevates χ. *The swap test thus identifies which of two saturated hydrocarbon components has the higher cohesive energy density.* This attribute of labeling has been examined extensively and has proved to be reliable.

The effects of labeling can be large even though, in these cases of partial labeling, $\chi_{HD} \sim 10^{-4}$ for the matched pairs is much smaller than $\chi \sim 10^{-3}$ for the blend. If χ is available for both varieties of singly labeled blends, then Eq. 8.33 provides a useful estimate for the nonlabeled blend. If data for only one variety are available, then the results must be supplemented by other means, such as cloud point determinations. That alternative was followed in the case of polyisobutylene (PIB) blends[54] because a substantial optical contrast was available. The corrections for labeling can also be estimated with Eq. 8.31. That equation relies rather heavily on the solubility parameter principle, however, and as shown in Section 8.3.2, that principle does not apply to all polyolefin blends.

8.3.2 Observations

The values of χ for many pairs of model polyolefins were obtained as outlined above: SANS for matched pairs to determine l_N for the species; then SANS for two singly labeled polyolefin blend pairs to determine χ_{HD} and χ_{DH}; and finally the estimation of χ_{HH}, the parameter of interest for each pair, with the square root average [Eq. 8.33]. Usually χ was determined in the composition midrange, $\phi \sim 0.5$, and at several temperatures, typically $27 \le T(°C) \le 167$ unless phase separation or crystallization intervened. Examples are given in Figures 8.5 through 8.9, including the H-H behavior according to Eq. 8.33. In several cases, the critical temperature for

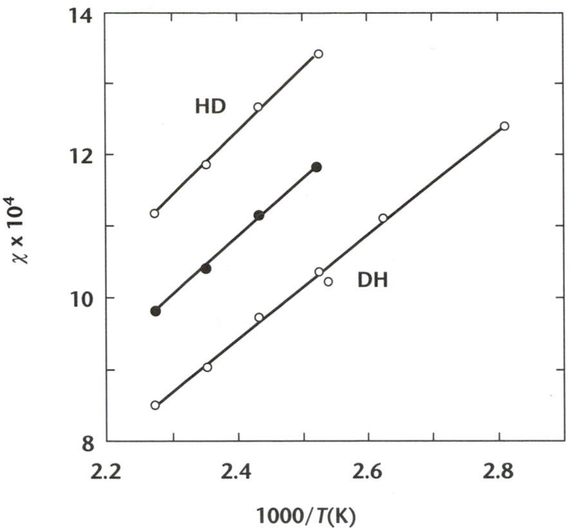

FIGURE 8.5 Temperature dependence of the interaction parameter for singly labeled EB88–EB78 blends. Interaction parameters for nonlabeled (H-H) blends as inferred from Eq. 8.33 are indicated by the filled symbols.

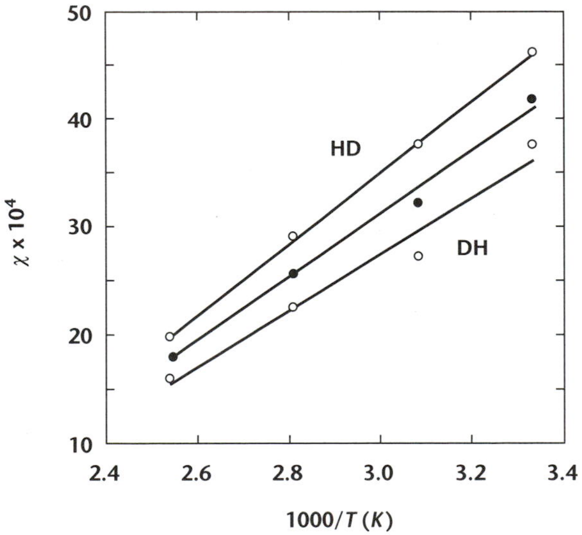

FIGURE 8.6 Temperature dependence of the interaction parameter for singly labeled PP/hhPP blends. Interaction parameters for nonlabeled (H-H) blends as inferred from Eq. 8.33 are indicated by the filled symbols.

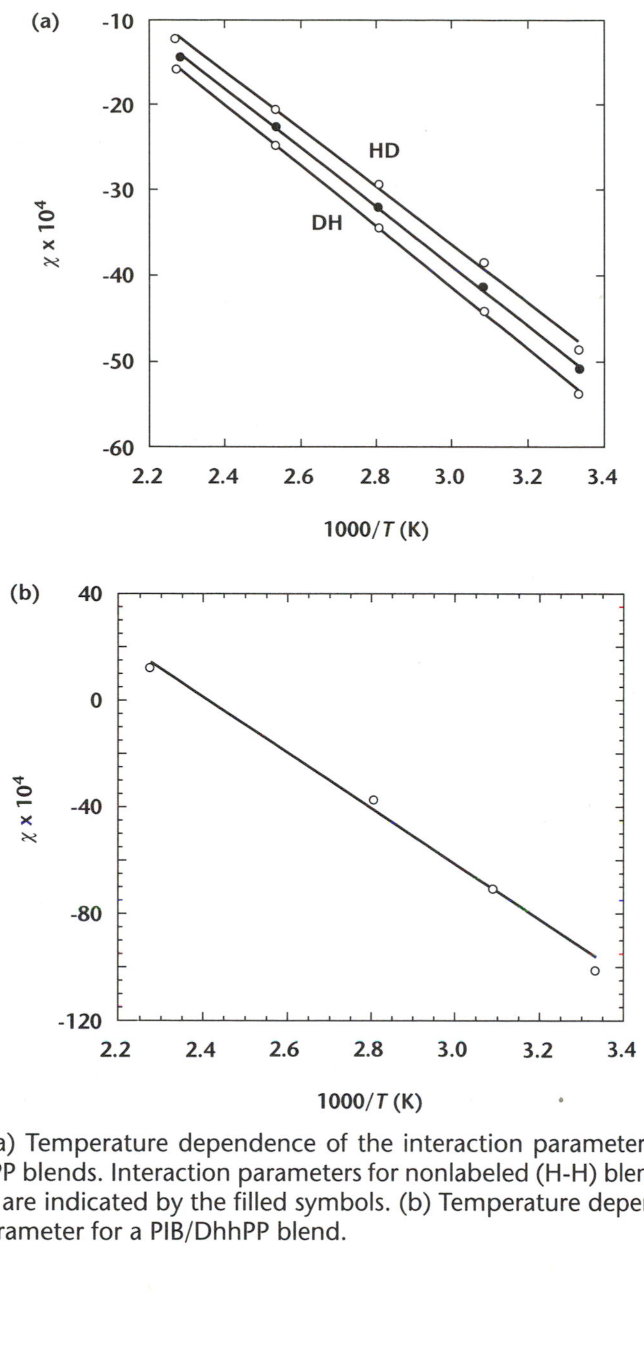

FIGURE 8.7 (a) Temperature dependence of the interaction parameter for singly labeled 50PEB/PP blends. Interaction parameters for nonlabeled (H-H) blends as inferred from Eq. 8.33 are indicated by the filled symbols. (b) Temperature dependence of the interaction parameter for a PIB/DhhPP blend.

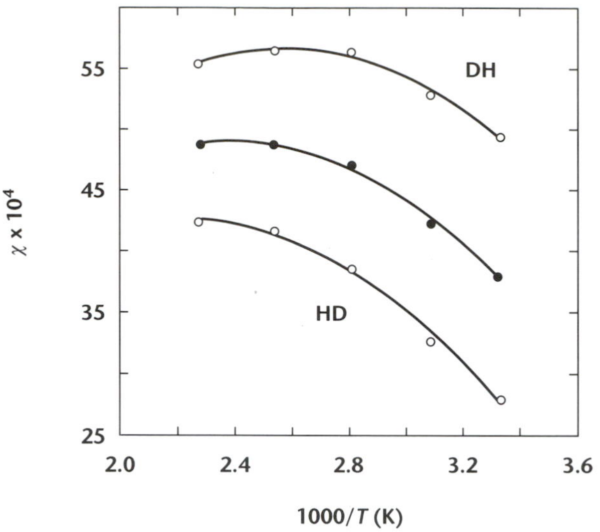

FIGURE 8.8 Temperature dependence of the interaction parameter for singly labeled PEP/PP blends. Interaction parameters for nonlabeled (H-H) blends as inferred from Eq. 8.33 are indicated by the filled symbols.

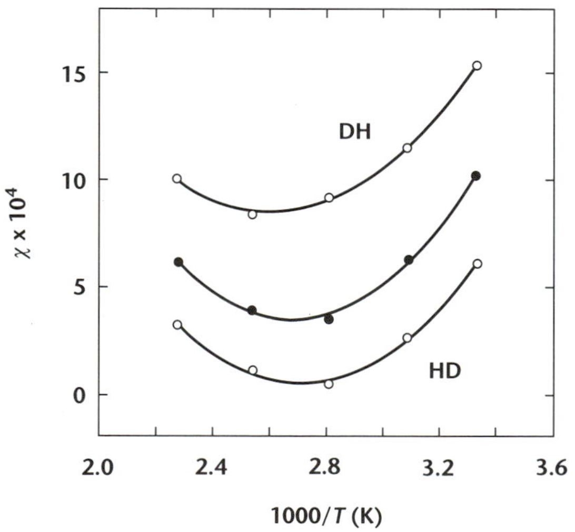

FIGURE 8.9 Temperature dependence of the interaction parameter for singly labeled PEP/hhPP blends. Interaction parameters for nonlabeled (H-H) blends as inferred from Eq. 8.33 are indicated by the filled symbols.

TABLE 8.4 Critical temperatures for model polyolefin blends with various labeling combinations

Components	Labeling	T_c(°C)WALS	T_c(°C)SANS	T_c(°C)NRA
EB88, EB78	DH	123±3[(a)]	115[(a)]	127[(d)]
	DD	98±3		
	HH	95±5		
	HD	68±3	60	62
EB66, EB52	DH	175±5[(b)]	180[(b)]	204[(d)]
	DD	133±3		
	HH	148±3		
	HD	108±3	90	88
EB38, EB06	DH	225±5[(c)]		
	DD	180±10		
	HH	180±10		
	HD	160±10		
PEP, EB38	DH	90±5[(a)]		
	DD	60±5		
	HH	52±3		
	HD	<40*		

* $T_m = 42°C$ for EB38
(a) Krishnamoorti, ref 55; (b) Krishnamoorti et al., ref. 44;
(c) Rhee and Crist, ref 51; (d) Scheffold et al., ref. 48

all four H-D blend combinations was estimated by phase separation at $\phi \sim \phi_c$, as judged by changes in light scattering intensity[50,55], by post-quenching morphology[51], or by nuclear reaction analysis[48]. The observations given in Table 8.4 show that, even for only partial deuteration, the labeling effects are significant. These effects also follow the pattern expected from the solubility parameter argument. According to the same principle, both figures and table give clear predictions of which nonlabeled component has the lower cohesive energy density.

The figures also illustrate the varieties of temperature dependence for χ in polyolefin blends. Figures 8.5 and 8.6 show the conventional $A + B/T$ form found in all the isotopic blends and which corresponds to UCT phase behavior. Like the EB78/EB88 data in Figure 8.5, all blends of EB copolymer pairs show the $A + B/T$ behavior. The PP/hhPP data in Figure 8.6 are of interest because the components are geometrical isomers, differing only in the relative orientation of PP mer insertions along the chain, yet they show a large net repulsion and immiscibility even for relatively short chains.

Figures 8.7a and 8.8b obey the $A + B/T$ form, but the interaction parameters are negative. Negative χ, a net attraction between blend components, is strange and indeed contrary to all expectations for nonpolar species. It is also contrary to the solubility parameter scheme. These are not isolated examples: Several blends with either PP[56] or polyisobutylene[54] as a component exhibit large negative values of χ over extended ranges of temperature. In all cases, the magnitude decreases with increasing temperature, and sometimes χ crosses zero to become positive in the observable range. Phase separation is the ultimate result and corresponds to LCT behavior. Note in Figure 3.6 the comparable monomeric LCT system of water and a tertiary amine.

The form $A + B/T$ is a common one from polyolefin blends, but other forms are also observed. Figures 8.8 and 8.9 exhibit behavior only rarely seen even in monomeric mixtures—the normal repulsive net interaction—but one that passes through a maximum or minimum with temperature. The PP/PEP data in Figure 8.8[57] suggest the occurrence of a maximum, corresponding to the possibility of a two-phase region. Over some range of intermediate temperatures, a behavior like that for the nicotine–water system (Figure 3.7). The hhPP/PP data in Figure 8.9 show a clear minimum, corresponding to the possible existence of a single-phase region over some range of intermediate temperatures, an hourglass form for which no monomeric example could be located. Again, these unexpected properties of liquids composed of only saturated hydrocarbon components are not isolated incidents.

8.3.3 Solubility Parameters

Among the variety of systems studied are what might be called *redundant blends*, those systems with components, for example, whose blends with, some common third component have also been studied: one of the three blends, A–B, A–C, and B–C, is redundant. Redundancy is irrelevant unless some scheme relates the interactions between two components to each of their interactions with a common third component. The regular mixture model does that. From Eq. 8.17,

$$(\delta_i - \delta_j)^2 = \frac{\chi RT}{V_{ref}} \tag{8.34}$$

where δ_i and δ_j are the solubility parameters of components i and j. Thus, for the redundant blend example, the Lin-Roe relations should apply[58],

$$\pm(\delta_A - \delta_B) = \pm(\delta_A - \delta_C) + \pm(\delta_B - \delta_C) \tag{8.35}$$

or

$$\pm\chi_{AB}^{1/2} = \pm\chi_{AC}^{1/2} + \pm\chi_{BC}^{1/2} \tag{8.36}$$

where the $\pm$'s reflect the square root sign ambiguity in applying Eq. 8.34. Note that Eq. 8.33 is equivalent to the earlier Lin-Roe Eq. 8.36.

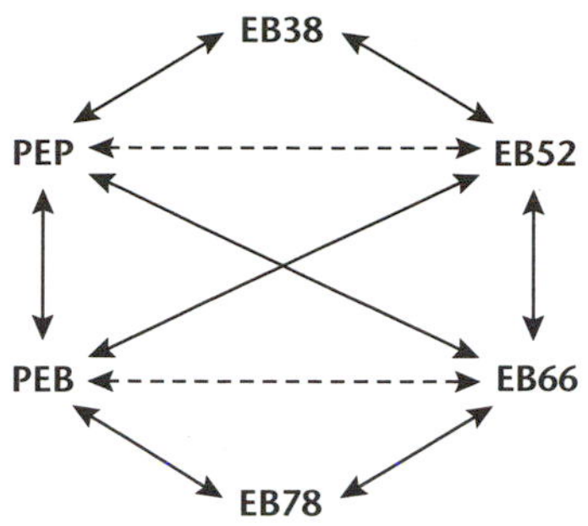

FIGURE 8.10 Diagram of blend combinations used to establish relative solubility parameters for the model polyolefins shown. The solid lines denote blends for which the swap effect gives a clear indication of ordering; those with less ordering are denoted by the dashed lines. Data from Krishnamoorti et al.[60].

Although the sign uncertainties are bothersome, the Lin-Roe relationships permit tests of consistency with the regular mixture idea without requiring the individual evaluation of the solubility parameters. The interaction data for model polyolefin blends offer two advantages in applying such tests. Many more redundancies are available, including the various blends of both labeled and nonlabeled species, and hence more cross checks than in the above AB, AC, BC example. Information is also available from the swap effect, thus ordering the solubility parameters in a blend and thereby helping to eliminate the sign ambiguities[59]. The data for redundant networks of polyolefin blends, such as that depicted in Figure 8.10[60], reveal that the net interactions in more than 80% of the more than 150 blends studied[12,56,61] are consistent with the regular mixture model.

The following criterion was used for regularity of mixing in the polyolefin studies[57]:

$$|X_{ij} - (\delta_i - \delta_j)^2| < 0.0032\, MPa \tag{8.37}$$

in which X_{ij} is the experimental interaction density coefficient for an i, j mixture, δ_i and δ_j are their mutually consistent solubility parameters from sets of redundant blends, and the inequality refers to the average over the temperature range covered. This span is roughly the experimental uncertainty of the interaction density coefficients themselves. All species except PIB formed enough regular mixtures with other components to allow such individual assignment. Only relative solubility parameters can be obtained in this way, however, so one of the species, EB97, was chosen as a reference for all the others. The values of $\delta - \delta_{ref}$ for unlabeled versions of the model polyolefins at five temperatures are shown in Table 8.5.

Blends of polyolefin species made by direct copolymerization of ethylene and various α-olefins have also been studied[61]. The products of single-site catalyst systems, loosely the metallocene-based polyolefins, were natural candidates. These systems have statistically uniform chemical microstructures and only moderate polydispersity: $M_w/M_n \sim 2$. The metallocene-derived polymers are not labeled, but their melt

TABLE 8.5 SANS-based relative solubility parameters* for model polyolefins, assigned by self-consistency tests of their interactions in blends[12,56]

	$[\delta(T) - \delta_{ref}(T)](MPa^{1/2})$				
Species	**27°C**	**51°C**	**83°C**	**121°C**	**167°C**
75SPI	−0.17	−0.17	−0.17	−0.16	−0.16
EB97	0	0	0	0	0
EB90	0.19	0.18	0.18	0.17	0.15
EB88	0.23	0.22	0.21	0.20	0.18
PP	0.22	0.22	0.25	0.26	0.25
EB78	0.49	0.48	0.46	0.44	0.41
50SPI	0.44	0.43	0.41	0.39	0.36
50PEB	0.49	0.48	0.47	0.46	0.42
EB66	0.73	0.72	0.69	0.66	0.62
PEB	0.72	0.72	0.71	0.69	0.66
hhPP	0.82	0.79	0.76	0.72	0.67
PEP	0.92	0.91	0.90	0.89	0.88
EB52	1.01	0.98	0.95	0.91	0.86
EB38		1.20	1.16	1.11	1.05
EB35			1.17	1.13	1.07
EB32			1.28	1.22	1.15
EB25			1.37	1.31	1.23
EB17				1.37	1.30
EB08				1.54	1.43
EB00 (HDPE)				1.60	1.48

*Reference Species is EB97.

state interactions can still be determined from SANS measurements on their blends with the labeled model polyolefins. Two methods were employed. The interactions in a binary blend of nonlabeled and labeled components provide a relative solubility parameter in the usual way, if the components mix regularly. The interaction data for a ternary blend—two nonlabeled and one labeled species—can provide information on the interactions between the nonlabeled species, if interaction parameters for the two nonlabeled with labeled binaries are known from independent experiments[62–64]. Unlike the binary method, the ternary method requires no a priori assumption of mixing regularity.

The ternary method[63] was tested on both regular and irregular model polyolefin blends and found to be satisfactory[64]. Both methods were applied to metallocene-derived polymers, copolymers of ethylene with propylene, 1-butene, 1-hexene, and 1-octene. The interactions of metallocene-derived species obtained through the ternary method were found to be consistent with the regular mixing model, and the relative solubility parameters so obtained agreed well with those from the binary

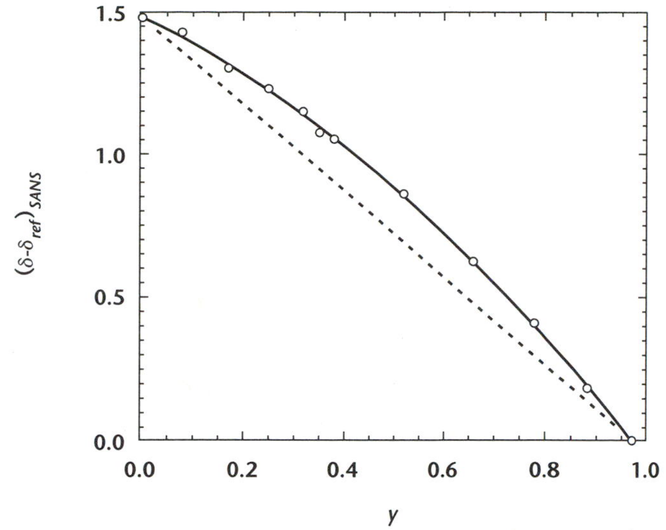

FIGURE 8.11 SANS-based relative solubility parameters for the model EB copolymers at 167°C. The dashed line indicates the linear behavior predicted by the copolymer equation; the curve is a quadratic fit of the data, $\delta - \delta_{ref} = 1.48 - 0.862y - 0.686y^2$, where y is the mole fraction of branched C4 units in the polymer (butene mers).

blend results. Moreover, solubility parameters for the model ethylene-butene copolymers, (the EB series), shown in Figure 8.11 as a function of y, the fraction of branched C4 units, are in essential agreement with those for the corresponding metallocene-derived materials. In addition, the values obtained for all the various ethylene copolymers are essentially the same function of copolymer composition when expressed as weight or volume fraction of comonomer. The comparison is shown in Figure 8.12. The same data fall along different lines when plotted as a function of mole fraction.

Molecular considerations. Relative solubility parameters provide a useful organizing principle for the melt state interactions in many polyolefin blends. As indicated in Table 8.6, they are especially useful for copolymer blends, and these account for a sizable portion of all the regular mixtures encountered. Curiously, these parameters are less reliable for predicting interactions when one or both components have rather uniform structures along the chain—homopolymers, such as PIB, PP, PB, and hhPP, or pseudo-homopolymers, such as PEB and PEP. Indeed, two such structurally uniform species seem almost always to mix irregularly. Irregularity in mixing is considered in greater detail in Section 8.3.5.

Other puzzling features are present. One is the very existence of regular mixtures, especially for polymeric systems. Regularity depends crucially on Eq. 3.90,

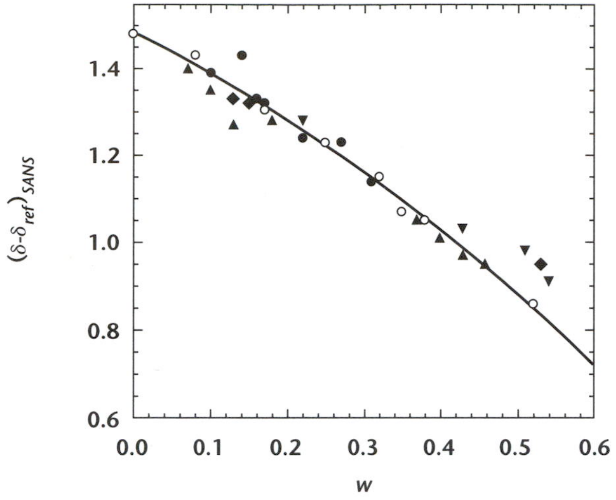

FIGURE 8.12 SANS-based relative solubility parameters at 167°C for copolymers of ethylene and various α-olefins, made with metallocene catalysts. The filled symbols indicate the metallocene-derived materials for the comonomers propylene(♦), 1-butene(•), 1-hexene(▲), and 1-octene(▼). The open symbols and the curve refer to the model ethylene-butene copolymer data in Figure 8.11.

the geometrical mean rule relating the interaction energies, $\varepsilon_{ij} = -(\varepsilon_{ii}\varepsilon_{jj})^{1/2}$. The interaction energies are so large and the net interaction so small—for single-phase polymeric mixtures at least—that the geometric mean must apply within an incredibly small tolerance[57]. Thus, for example, consider proportionality rather than equality for the cross term: $\varepsilon_{ij} = -\lambda(\varepsilon_{ii}\varepsilon_{jj})^{1/2}$. Substitution into Eq. 8.37 then leads to:

$$|2(\lambda - 1)\delta_i\delta_j| < 0.0032\ MPa \tag{8.38}$$

With $\delta_i\delta_j \sim \Pi_{CED} \sim 250\ MPa$ for typical polymer species (see below), the result is $|(\lambda - 1)| \lesssim 5 \times 10^{-6}$, meaning that the geometric mean must be satisfied to a few parts per million for regularity to be satisfied. It seems truly a wonder that any regular mixtures exist at all, yet the majority of polyolefins seem to mix that way.

TABLE 8.6 Number of model Polyolefin blends and component types with regular and irregular mixing

Components	Regular	Irregular
homopolymer/homopolymer	1	9
homopolymer/copolymer	~23	7
copolymer/copolymer	~130	1

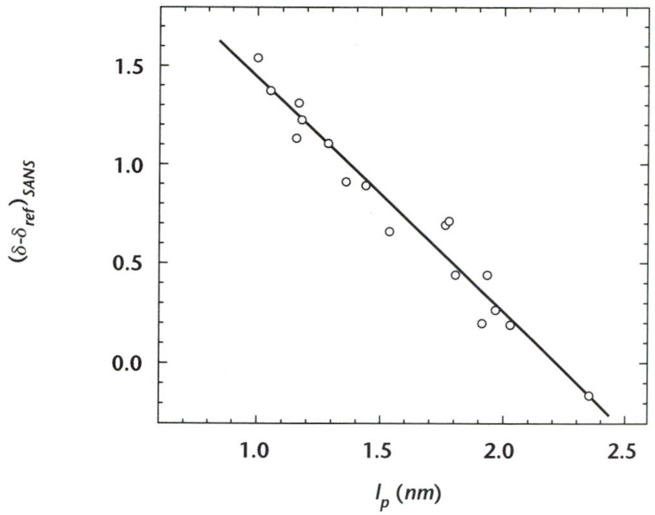

FIGURE 8.13 Relationship between relative solubility parameters and the packing lengths for model polyolefins at 121°C. The least-squares line is $(\delta - \delta_{ref})_{SANS} = 2.60 - 1.17\ l_p$.

An apparently close relationship between the assigned solubility parameter and packing length for the species is another systematic feature of the data. The relationship is rather convincingly a linear one, as shown by Figure 8.13: values of l_p were calculated with the mer-based segment length from the matched-pair SANS data. It suggests the possibility of a geometrical origin for the net interaction, something like the local packing effects discussed earlier or packing on some larger scale as suggested in 1992 by Bates et al.[65] and developed further by Fredrickson et al.[66,67]. Conversely, working with PRISM theory in the threadlike approximation, Schweizer and Singh[68] find that cohesive energy densities, and hence conventional solubility parameters, should also correlate closely with packing length, as shown here.

PVT measurements. The measurement of pure component PVT properties was undertaken in hopes of clarifying the origins of both $(\delta - \delta_{ref})_{SANS}$ and the departures from regular mixing behavior. From Eqs. 8.15 and 8.16, the cohesive energy density and the internal pressure of simple liquids are numerically identical; Π_{IP} can be calculated from α and β, the thermal expansion coefficient and isothermal compressibility of the liquid. Specific volume was measured as a function of pressure and temperature for most of the model polyolefins and analyzed over small ranges to obtain the ambient pressure values of α and β at several temperatures[12]. The results are given in Table 8.7 along with PVT-based values of the solubility parameter from Eq. 8.18:

According to Eqs. 8.14 and 8.15, both δ_{CED} and δ_{PVT} are directly proportional to mass density for a simple liquid. A test of that condition for δ_{PVT} is shown in

TABLE 8.7 Physical properties related to mixing for selected polyolefin species[12]

Sample	$T(°C)$	$\alpha(K^{-1}) \times 10^4$	$\beta(MPa^{-1}) \times 10^4$	$\delta_{PVT}(MPa^{1/2})$
EB100 (iPB)	167	6.92	12.41	15.66
EB97	27	6.94	6.31	18.17
	51	7.12	7.32	17.76
	83	6.99	8.52	17.08
	121	7.14	10.09	16.69
	167	7.25	12.68	15.86
EB88	27	6.81	6.16	18.22
	51	7.08	7.14	17.93
	83	6.98	8.25	17.35
	121	7.16	9.81	16.96
	167	7.37	12.33	16.22
EB66	27	6.92	5.90	18.76
	51	7.24	6.90	18.44
	83	7.17	7.87	18.00
	121	7.30	9.36	17.53
	167	7.43	11.77	16.66
EB52	27	6.97	5.84	18.93
	51	7.25	6.83	18.56
	83	7.12	7.96	17.83
	121	7.35	9.47	17.49
	167	7.50	11.90	16.65
EB34	27	6.23	5.23	18.91
	51	6.48	6.07	18.61
	83	6.70	7.23	18.16
	121	6.93	8.72	17.69
	167	7.47	10.89	17.37
EB32	51	7.02	6.38	18.88
	83	7.30	7.65	18.42
	121	7.41	9.16	17.84
	167	7.71	11.45	17.21
EB16	121	7.09	8.78	17.83
	167	7.65	11.09	17.42
EB08	121	7.45	8.87	18.19
	167	7.65	10.94	17.54
EB08(D)	121	7.42	8.73	18.30
	167	7.62	11.03	17.44
EB00 (HDPE)	167	7.80	10.99	17.67
EP100 (aPP)	27	7.28	6.47	18.37
	51	7.55	7.54	18.02
	83	7.31	8.73	17.25
	121	7.30	10.44	16.60
	167	7.52	13.14	15.87

(Continued)

TABLE 8.7 *(Continued)*

Sample	$T(°C)$	$\alpha(K^{-1}) \times 10^4$	$\beta(MPa^{-1}) \times 10^4$	$\delta_{PVT}(MPa^{1/2})$
EP84	27	6.97	6.01	18.65
	51	7.06	6.90	18.22
	83	6.97	8.06	17.54
	121	7.16	9.67	17.08
	167	7.60	12.02	16.68
EP76	27	6.70	5.98	18.34
	51	6.98	6.90	18.11
	83	6.97	8.06	17.54
	121	7.23	9.61	17.21
	167	7.59	12.08	16.62
EP57	27	6.77	5.89	18.57
	51	7.06	6.88	18.24
	83	7.12	7.87	17.94
	121	7.41	9.28	17.73
	167	8.10	11.49	17.61
EP23	121	8.36	8.90	19.23
	167	7.45	11.07	17.21
PEB	27	7.09	5.80	19.15
	51	6.92	6.59	18.45
	83	6.94	7.70	17.90
	121	7.18	9.30	17.44
	167	7.45	11.78	16.68
PEP	27	6.60	5.82	18.45
	51	6.78	6.74	18.06
	83	6.69	7.87	17.39
	121	6.88	9.30	17.07
	167	7.33	11.50	16.74
50SPI	27	6.89	5.97	18.62
	51	7.12	6.94	18.24
	83	6.92	8.07	17.46
	121	7.10	9.68	17.00
	167	7.45	11.97	16.54
75 SPI	27	6.50	5.57	18.72
	51	6.52	6.50	18.04
	83	6.41	7.80	17.12
	121	6.65	9.39	16.71
	167	6.98	11.79	16.14
50PEB	27	7.55	5.87	19.64
	51	7.20	6.82	18.50
	83	6.81	7.90	17.52
	121	7.05	9.61	17.01
	167	7.45	12.22	16.39

(Continued)

TABLE 8.7 *(Continued)*

Sample	$T(°C)$	$\alpha(K^{-1}) \times 10^4$	$\beta(MPa^{-1}) \times 10^4$	$\delta_{PVT}(MPa^{1/2})$
hhPP	27	6.70	5.69	18.80
	51	6.92	6.60	18.45
	83	6.81	7.60	17.85
	121	7.02	9.01	17.52
	167	7.15	11.3	16.68
PIB	27	5.52	4.8	18.58
	51	5.72	5.52	18.33
	83	5.53	6.29	17.68
	121	5.68	7.41	17.38
	167	6.08	8.85	17.38

Table 8.8. For each model species, the value δ_{PVT} at the five tabulation temperatures normalized to the value at 83°C was calculated, and the species-average of those results was recorded, giving a rough indication of how rapidly δ_{PVT} varies with temperature. The same procedure was repeated for the ratio δ_{PVT}/ρ, a quantity that is independent of temperature for simple liquids. The result indicates that the data are consistent with the simple liquid model.

The PVT-based values of δ_{PVT}, relative to those for the reference species EB97, $(\delta - \delta_{ref})_{PVT}$, are compared with $(\delta - \delta_{ref})_{SANS}$ for the model polyolefins in Figure 8.14. The data are somewhat scattered—the relative PVT-based values are near the precision limits of the instrument used for the measurements[69]. The least-squares fit, $(\delta - \delta_{ref})_{SANS} = 0.014 + 0.83(\delta - \delta_{ref})_{PVT}$, is shown by the solid line. Assuming that $(\delta - \delta_{ref})_{SANS}$ corresponds to the conventional definition $(\delta - \delta_{ref})_{CED}$, the result indicates a close adherence to expectations based on the simple liquid model.

Simulations. Molecular dynamics studies were undertaken to supply information on the relationship between cohesive energy density and internal pressure in dense liquid hydrocarbons[70,71]. The united atom model with the potentials of Jorgensen et al.[72] were used to evaluate Π_{CED} and Π_{IP} for three groups of liquids—various heptane

TABLE 8.8 Consistency test of solubility parameters derived from internal pressure data with the simple liquid model

$T(°C)$	$(\delta_{PVT})_T/(\delta_{PVT})_{83°C}$	$(\delta_{PVT}\rho^{-1})_T/(\delta_{PVT}\rho^{-1})_{83°C}$
27	1.058	1.018
51	1.035	1.013
83	1.000	1.000
121	0.976	1.002
167	0.946	1.002

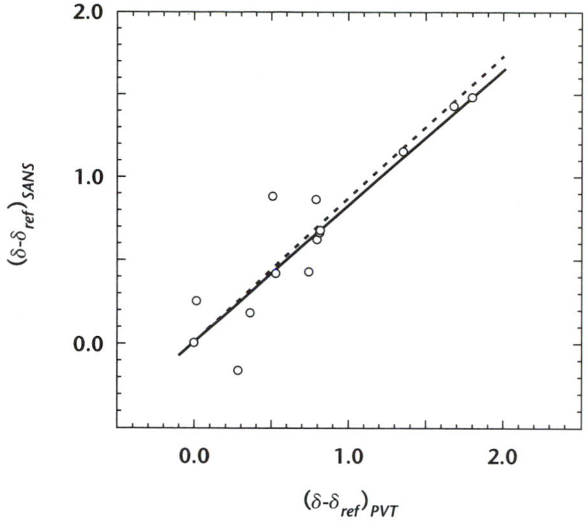

FIGURE 8.14 Comparison of relative solubility parameters at 167°C derived from SANS and from PVT data. The solid line is a least-squares fit, $(\delta - \delta_{ref})_{SANS} = 0.14 + 0.83(\delta - \delta_{ref})_{PVT}$; the dashed line corresponds to $(\delta - \delta_{ref})_{SANS} = 0.87(\delta - \delta_{ref})_{PVT}$, per Eq. 8.41 obtained from simulations as discussed in the text.

(C7) isomers, oligomeric (C30) versions of several model polyolefins, and some polymeric values obtained by extrapolation of length variations to the polymeric range. Realistic values of mass density at ambient pressure, either experimental (C7 at 20°C and polymeric at 150°C) or estimated (C30 at 150°C), were imposed on the simulations. Cohesive energy density was calculated according to its definition as the ratio of average intermolecular potential energy to the simulation volume:

$$\Pi_{CED} = \left(\frac{-\langle \Psi_{inter} \rangle}{V} \right)_{N,T} \tag{8.39}$$

where N is the number of molecules in the simulation. Internal pressure was also calculated according to its definition as the derivative of total energy with respect to volume. The kinetic energy being independent of volume according to equipartition, the total potential energy was evaluated for several volumes near the volume of interest to obtain the slope:

$$\Pi_{IP} = \left(\frac{-\partial \langle \Psi_{total} \rangle}{\partial V} \right)_{N,T} \tag{8.40}$$

The values obtained along with the model polyolefins and their nearest C7 analogs are shown in Table 8.9. For the C7 analogs, the simulation results for Π_{CED} follow the

TABLE 8.9 Simulation results for the cohesive energy density and internal pressure of selected polyolefin species and their small-molecule analogs[70,71]

		$\Pi_{CED}(MPa)$		Π_{CED}/Π_{IP}			
Species	**C7 Analog**	**C7 (exp)**	**C7 (sim)**	**C7 (exp)**	**C7 (sim)**	**C30 (sim)**	**Polymer (sim)**
PE	n-hexane	235	247	0.92	1.02	0.85	0.77
	2-methylhexane	221	230	0.89			
PEB	3-ethylpentane	231	245	0.87	0.94	0.80	0.77
PEP	3-methylhexane	226	237	0.90	0.96	0.81	0.75
PIB	2,2-dimethylpentane	203	208	0.87		0.88	
hhPP	2,3-dimethylpentane	222	230	0.86	0.97	0.82	0.74
PP	2,4-dimethylpentane	206	212	0.87	0.89	0.79	0.75
PB						0.81	0.76

trend of the experimental values, although running 2 to 6% higher. The simulation values of Π_{IP} (not shown) are systematically smaller and more scattered, whereas the ratio Π_{CED}/Π_{IP} runs about 10% higher than the experimental ratio and is close to unity in most cases. Experimental values of Π_{CED}/Π_{IP} are not available for the C30 and polymeric structures. The simulations show the ratio to vary systematically from about unity for C7 down to a polymeric value that is approximately 0.75 for the six species studied. If that limit were to hold true for all polyolefin species, it would mean $\delta_{CED}/\delta_{IP} = (\Pi_{CED}/\Pi_{IP})^{1/2} \sim 0.87$. The relative solubility parameters would be related the same way, such that (assuming the SANS-based values reflect cohesive energy densities),

$$(\delta - \delta_{ref})_{\text{SANS}} = 0.87(\delta - \delta_{ref})_{\text{PVT}} \tag{8.41}$$

Equation 8.41 is shown as a dashed line in Figure 8.14. It fits the data about as well as the least-squares line, but the data scatter is too great to permit a decision.

What is the significance of this drift of Π_{CED}/Π_{IP}, from approximately unity for monomeric liquids to about 0.75 for polymers? It could be related to changes in the pair distribution function. Thus, Π_{CED} depends on an integral over $[g(r)]_{inter}$, whereas Π_{IP}, being a mechanical property and reflecting the transmission of forces, probably involves some corresponding integral over $[g(r)]_{total}$. On going from monomeric to polymeric structure, these two distributions drift away from one another, such that finding $\Pi_{CED} < \Pi_{IP}$ for polymers is not surprising. Finding the ratio to be about the same for all species would be surprising, however, because the difference, $[g(r)]_{intra} = [g(r)]_{total} - [g(r)]_{inter}$, should be highly sensitive to species structure.

Alternatively, the drift might be related to the same factors that cause the thermal expansion coefficients of polymeric liquids to be systematically smaller than those for monomeric liquids. For example, if the drift stems somehow from a polymer having only two directions available for thermal expansion, whereas a monomeric liquid has three, the effect could be universal or nearly so. Now merely speculation, this is a possible avenue for future exploration.

8.3.4 Interaction Properties

This section deals with several properties of the interaction parameter in polyolefin blends. The effects of temperature, pressure, chain length, and composition are examined, extending topics introduced in Section 8.2 for isotopic blends including applicability and limitations of the simple liquid and simple mixture models in polyolefin systems.

Temperature dependence. The approach used for isotopic blends can also be applied to polyolefin blends, but the situation differs somewhat. The majority of polyolefin blends exhibit $A + B/T$ dependence, but many do not. Even for those with $A + B/T$ dependence—the only ones considered here—the factorability of density ratios used for isotopic blends is usually not an acceptable approximation. The calculation must allow for differences between components in properties such as α and β. For temperature variations alone, Eq. 8.21 can be expressed:

$$\delta_2(T) - \delta_1(T) = \delta_2(T_m) - \delta_1(T_m) + \delta_2(T_m)f_2 - \delta_1(T_m)f_1 \tag{8.42}$$

where

$$f_i(T, T_m) = \frac{\rho_i(T)}{\rho_i(T_m)} - 1 \tag{8.43}$$

and T_m is an arbitrary temperature of convenience [Eq. 8.23]. With rearrangement and the definitions $\Delta\delta = \delta_2(T_m) - \delta_1(T_m)$, $\Delta f = f_2(T_m) - f_1(T_m)$, $\bar{\delta} = (\delta_1 + \delta_2)/2$, and $\bar{f} = (f_1 + f_2)/2$, Eq. 8.42 becomes:

$$\delta_2(T) - \delta_1(T) = \Delta\delta\left[1 + \bar{f}\left(1 + \frac{\Delta f}{\bar{f}}\frac{\bar{\delta}}{\Delta\delta}\right)\right] \tag{8.44}$$

For some range of temperature, $\Delta T = T - T_m$,

$$f_i(T, T_m) = -\alpha_i(T_m)\Delta T \tag{8.45}$$

The temperature-dependent solubility parameter difference and interaction parameter can thus be written:

$$\delta_2(T) - \delta_1(T) = \Delta\delta[1 - \bar{\alpha}(1 + \xi)\Delta T] \tag{8.46}$$

$$\chi(T, T_m) = \frac{V_{ref}(T)(\Delta\delta)^2}{RT}[1 - \bar{\alpha}(1 + \xi)\Delta T]^2 \tag{8.47}$$

TABLE 8.10 Temperature dependence of χ for selected polyolefin blends

Blend Pair: 1/2	$(\delta_2 - \delta_1)/\bar{\delta}$	$(\alpha_2 - \alpha_1)/\bar{\alpha}$	$X_E (J\ cm^{-3})$(a)	$\chi_m \times 10^4$	$\chi_h \times 10^4$	$\chi_s \times 10^4$	$\chi_s \times 10^4$ calc
EB97/EB88	0.012	<0.005	~0	10.3	16.7	−6.4	−4.6
EB88/EB78	0.014	<0.005	~0	14.2	22.8	−8.6	−6.0
EB78/EB66	0013	~0.006	~0	12.4	17.2	−4.8	−5.9
EB66/EB52	0.014	~0.008	~0	15.2	24.2	−9.0	−8.0
EB52/EB38	0.012	~0.009	~0	9.7	18.8	−9.1	−6.0
50SPI/hhPP	0.016	0.015	~0	24.0	47.3	−23.5	−16.6
PP/hhPP	0.028	−0.07	−0.196	25.6	82.6	−56.9	+39.1

(a) The excess interaction coefficient: $X_E = X - (\delta_2 - \delta_1)^2$.

in which $\xi = (\Delta\alpha/\bar{\alpha})/(\Delta\delta/\bar{\delta})$. Expressing the temperature-dependent reference volume in Eq. 8.22 as $V_{ref}(T_m)/(1 - \bar{\alpha}\Delta T)$ and applying Eq. 8.17 at T_m leads to:

$$\chi(T, T_m) = \chi_m \frac{T_m}{T} \frac{(1 - \bar{\alpha}(1+\xi)\Delta T)^2}{(1 - \bar{\alpha}\Delta T)} \tag{8.48}$$

For $\bar{\alpha}\Delta T \ll 1$, a condition well satisfied here, Eq. 8.48 leads to the following expressions for the residual temperature dependence, written in terms of entropic and enthalpic contributions to the interaction parameter:

$$\begin{aligned} \chi_s/\chi_m &= -\bar{\alpha}T_m(1 + 2\xi) \\ \chi_h/\chi_m &= 1 + \bar{\alpha}T_m(1 + 2\xi) \end{aligned} \tag{8.49}$$

The data for a selection of seven nonlabeled polyolefin blends[57,73] with $\chi = A + B/T$ behavior, expressed according to Eq. 8.23 with midrange temperature $T_m = 83°C$, are given in Table 8.10. The first five are EB blends, and all show weak to moderate entropic contributions. All five form regular mixtures, as indicated by the last column X_E, the excess interaction energy coefficient at T_m, the portion not accounted for by the solubility parameter. The sixth blend, 50SPI/hhPP, is also a regular mixture ($X_E \sim 0$), but one with a relatively large entropic contribution. The seventh, PP/hhPP, exhibits an even larger entropic contribution, and it also departs significantly from regular mixing behavior. Included in the table are values of $\Delta\delta/\bar{\delta}$ and $\Delta\alpha/\bar{\alpha}$ for each blend at 83°C[12], the former with $\bar{\delta} = 18\ MPa^{1/2}$ and the latter with $\bar{\alpha} = 0.7 \times 10^{-4}(K^{-1})$.

These data were used to calculate ξ and hence χ_s with Eq. 8.49. The calculated values of χ_s are recorded in the last column of the table. The agreement in magnitude and sign for the six regular blends is reasonable, especially considering the uncertainty in $\alpha_2 - \alpha_1$, which is barely one significant figure in most cases. Even the sign is wrong for the irregular blend PP/hhPP, thus indicating contributions beyond density-derived effects to the interaction coefficient. The consistency of observed and calculated values of χ_s for the others lends further support to the assumption of density dependence alone and its corollary: that the net interactions in regular polyolefin are primarily enthalpic in origin.

Pressure dependence. Equation 8.21 is readily extended to express the P and T dependence of solubility parameter differences if the component density ratios differ. Over modest ranges, $\Delta T = T - T_m$ and $\Delta P = P - P_a$,

$$f_i(T) = -\alpha_i(P_a, T_m)\Delta T + \beta_i(P_a, T_m)\Delta P \tag{8.50}$$

which then leads to:

$$\delta_2(P,T) - \delta_1(P,T) = \Delta\delta\left[1 - (\bar{\alpha}\Delta T - \bar{\beta}\Delta P)\left(1 + \frac{\bar{\delta}}{\Delta\delta}\,\frac{\Delta\alpha\Delta T - \Delta\beta\Delta P}{\bar{\alpha}\Delta T - \bar{\beta}\Delta P}\right)\right] \tag{8.51}$$

The P and T dependence of the interaction parameter then follows from Eq. 8.17. For isothermal variations with P,

$$\chi(P,T) = \chi(P_a,T)\frac{[1 + \bar{\beta}(1+\xi)\Delta P]^2}{[1 + \bar{\beta}\Delta P]} \tag{8.52}$$

where $\xi = (\bar{\delta}/\Delta\delta)/(\Delta\beta/\bar{\beta})$.

The predictions of Eq. 8.52 were compared with data on the pressure dependence of χ for two saturated hydrocarbon copolymer blends, an ethylene-*alt*-propylene (PEP) and ethylene-*alt*-butene (PEB) pair[74]and a pair of statistical ethylene copolymers[75], one containing 18.4 wt % hexene-1, the other 32 wt % butene-1. Values of $\bar{\delta}$ and $\bar{\beta}$ at the temperatures of interest were estimated from PVT data[12,76]. Values of $\Delta\delta$ were obtained or estimated from SANS-based assignments[12,61]. Values of $\Delta\beta$ are by far the most uncertain. The component compressibilities themselves, difficult to measure accurately under any circumstances (see Section 3.6.3), had similar magnitudes in both blends; they also vary relatively rapidly with temperature. Even with careful smoothing and interpolation, no more than about one significant figure precision was likely achieved. The resulting values of ξ for both blends at the temperatures of interest are listed in Table 8.11.

A comparison between measured values of $\chi(P)$ and those calculated with Eq. 8.52 is shown for the PEP/PEB blends at four temperatures in Figure 8.15. Except for the data at the highest temperature, the agreement is very good indeed. The predictions for the EH18.4/EB32 blend, however, are less satisfactory. Experimentally, the interaction coefficient $X = (\delta_1 - \delta_2)^2$ at each temperature roughly doubles on going from ambient pressure to 100 MPa (see Figure 1 in reference[75]). The predicted effect is smaller, the factor being about 1.35 at 121°C and 1.7 at 166°C.

TABLE 8.11 Pressure dependence parameter for selected blends and temperatures

Blend	T(°C)	ξ
PEP/PEB	59	2.0
PEP/PEB	89	1.7
PEP/PEB	123	0.26
PEP/PEB	167	−2.6
EH18.4/EB32	121	0.9
EH18.4/EB32	166	1.8

Molecular weight dependence. The interaction parameter for many pairs of polyolefin components was determined with only one sample, and hence one molecular weight M, of each. In some cases, however, two or more molecular weights have been used, and the results therefore provide tests of whether a dependence on M exists. As noted regarding isotopic blends, a significant M-dependence for reasonably long chains compromises the view that the interactions are local in origin. Fortunately, the findings for polyolefin blends[55] indicate clearly that χ or X are insensitive to M, so, as concluded also for isotopic blends, independence of M is an excellent approximation. The results of one such test[59], analogous to the test for M-dependence in isotopic blends[40] are shown in Figure 8.16. In this case blends of linear polyethylene fractions—standard reference materials SRM-1482, SRM-1483, and SRM-1484 supplied by the U.S. National Institutes of Standards and Technology—with various

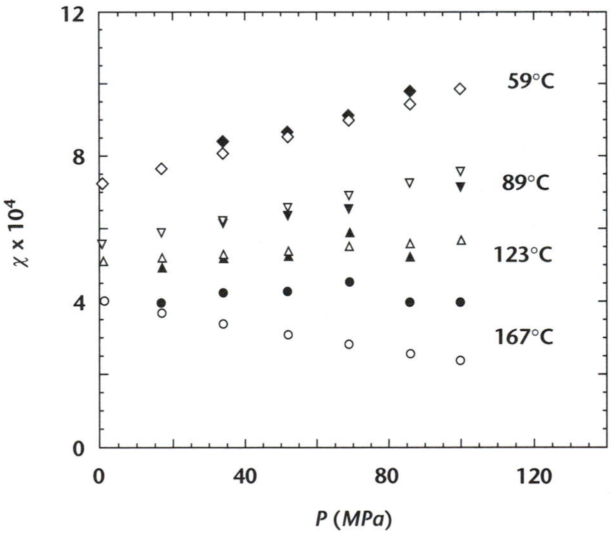

FIGURE 8.15 Calculated and observed effects of pressure on the interaction parameter of PEP/PEB blends. Filled symbols are observed values, open symbols are calculated.

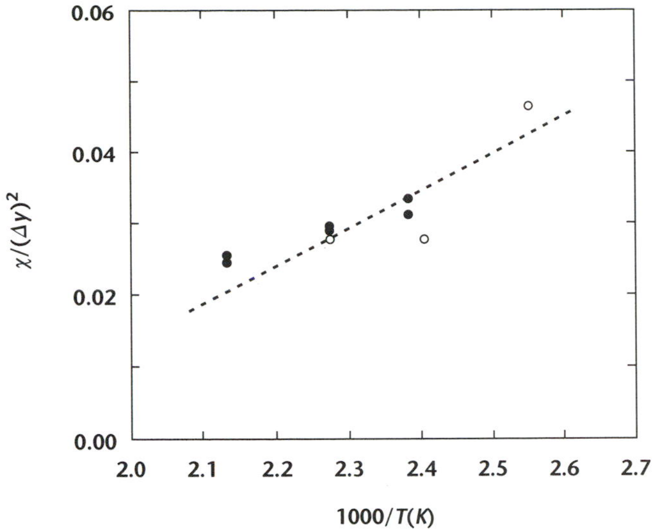

FIGURE 8.16 Test of molecular weight independence for the interaction parameter based on the copolymer equation and data for blends of polyethylene and ethylene-rich model EB copolymers. Filled symbols refer to data for blends of PE standards with various labeled EBs; open symbols refer to a blend of EB08 with labeled EB25. Data from Graessley et al.[59]

labeled EB copolymers are used. The PE components cover a factor of nine in chain length, and the interaction parameters for their blends, corrected for differences in composition with the copolymer equation [see Eq. 8.62], lie along a common line when plotted as a function of temperature.

Composition dependence. Essentially all the results for polyolefin blends discussed so far have been obtained in the composition midrange, $0.25 \lesssim \phi \lesssim 0.75$. In that range, the values of χ are constant or change only slowly with ϕ. However, the strange composition dependence discovered in SANS studies of isotopic mixtures outside that range—a roughly symmetric and fairly rapid increase in χ at the composition extremes—is also found in polyolefin blends[44]. The example of blends of labeled EB52 and nonlabeled EB66 is shown at two temperatures in Figure 8.17a. The upturn becomes less severe with increasing temperature and, based on limited data, may possibly be connected with location relative to the critical region. Not all polyolefin blends exhibit the upturn. The PEP–PEB blend system[45], for example, shows no sign of composition dependence in SANS-derived χ up to $\phi_{PEB} \sim 0.9$.

Also shown in Figure 8.17a are interaction parameters for the same pair of blend components obtained by Klein and coworkers with coexistence curve data using NRA [48,49]. The coexistence data including SANS-derived cloud-point values[44] are shown in Figure 8.17b. After exploring simpler forms, Scheffold et al. used their

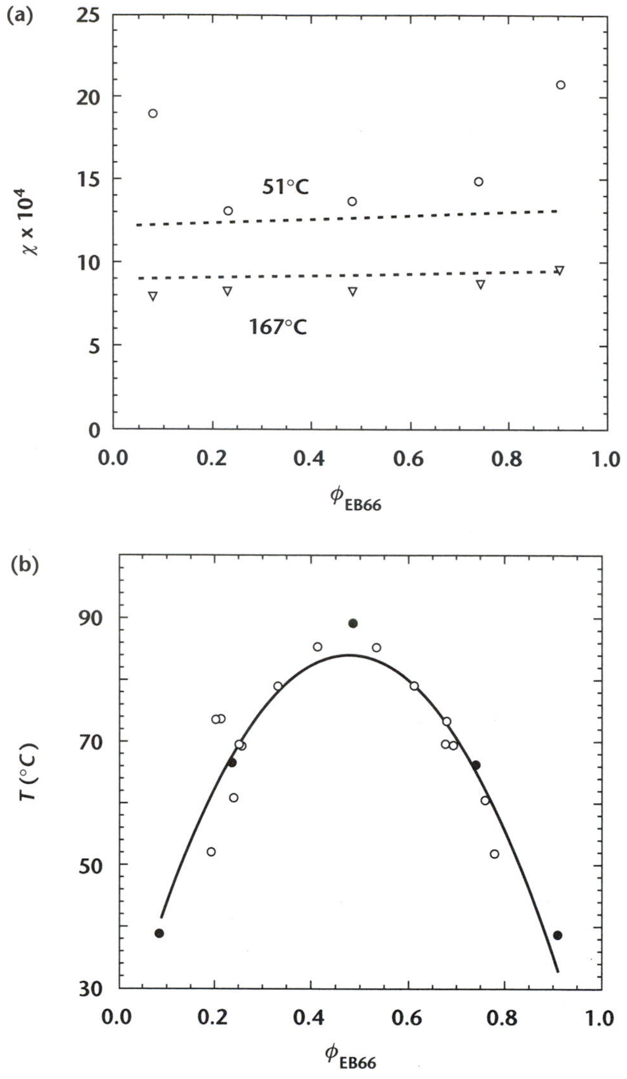

FIGURE 8.17 (a) Composition dependence of the interaction parameter for EB66(H)/EB52(D) blends. The symbols indicate data obtained by SANS (Krishnamoorti et al.[44]); the dashed lines indicate values inferred from direct determination of coexisting phase compositions. Data from Scheffold et al.[48]; Klein et al.[49]. The same samples were used in both studies. (b) Comparison of phase boundary locations as obtained in the studies described in (a). The open circles indicate coexisting phase compositions. Data from Klein and coworkers[48,49]; closed circles are SANS cloud-point determinations from Krishnamoorti and coworkers[44].

coexistence data and the free energy expression for simple mixtures, Eq. 3.67, to obtain values of A, B, and E in the following form for the interaction parameter:

$$\chi_{sm}(\phi, T) = (A + B/T)(1 + E\phi) \tag{8.53}$$

Phase coexistence depends on the equality of chemical potentials, so the appropriate interaction parameter is χ as defined in the Flory manner, Eq. 7.72, now adapted for blends:

$$\chi \equiv \frac{(\mu_1 - \mu_1^{\circ}) - (\mu_1 - \mu_1^{\circ})_{com}}{r_1\phi^2 RT} \tag{8.54}$$

in which $\phi = \phi_2$ for consistency with Eq. 8.53. The interaction parameter from scattering is related to the chemical potential-based expression by Eq. 7.78. Thus, the following formulas based on Eq. 8.53 are relevant to blends:

$$\begin{aligned} \chi &= (A/T + B)(1 - E + 2E\phi) \\ \chi_{scat} &= (A/T + B)(1 - E + 3E\phi) \end{aligned} \tag{8.55}$$

The lines shown in Figure 8.17a are NRA-based predictions of χ_{scat} for the same blends, calculated by Eq. 8.55 with the fitted values of A, B, and E[49]. The agreement between values of χ_{scat} from the two sources, derived through very different experiments, is quite acceptable in the composition midrange but progressively poorer in the extremes.

Similar NRA-SANS comparisons have been made for other polyolefin blends[48,49] with similar results. The composition coefficient E is frequently larger than that obtained for the blend discussed above, and it is sometimes negative. Unfortunately, no clear pattern emerges; attempts to explain the observations are well summarized elsewhere[77,78]. As it turns out, the upturn in χ from SANS contributes very little to the observable parts of the phase diagram, so little can be concluded about the physical reality of the upturns. Whether artifact or real, it makes little practical difference in the systems so far investigated, but it does represent a puzzle that should be sorted out.

8.3.5 Irregular Blends

Positive values of χ and UCT phase diagrams are the conventionally expected behavior for nonpolar mixtures, such as polyolefin blends. Invariably, blends showing other behavior turn out to be *irregular blends* that depart from the prediction based on assigned solubility parameters (Table 8.5). Some blends with positive χ, and even some of those with the conventional $A + B/T$ temperature dependence—PP–hhPP in Table 8.10, for example—turn out to be irregular. The discrepancy in net

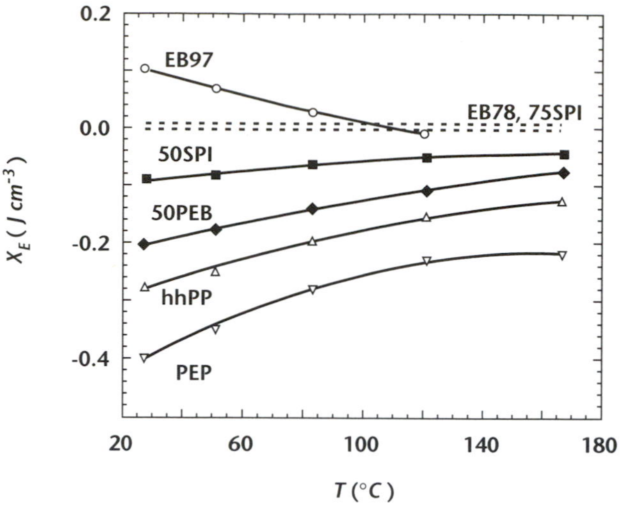

FIGURE 8.18 Excess interaction density coefficient for blends of polypropylene with various second components. The dashed lines indicate the range of X_E for blends deemed to mix regularly. The filled symbols correspond to blends with negative interaction parameters; the others are blends with positive interaction parameters.

interaction density between observation and prediction can be expressed as:

$$X_E = X - (\delta_1 - \delta_2)^2 \tag{8.56}$$

where X is the measured interaction density coefficient for a blend, and X_E is the excess, which can be either positive or negative.

Most remarkable are those blends with values of X that are negative, a feature found in several blends with either PIB or PP as a component. In other blends, X itself is positive but X_E is significant and sometimes even larger than the values for some blends with negative X. The values of X_E for PP blends, shown in Figure 8.18, illustrate the point. The filled symbols indicate blends with negative X; the others have positive X. It further happens that only a few of the irregular blends have positive X_E, and that even for those, the magnitudes are not especially large. Most irregular blends have negativeX_E, and for many, its magnitude is very large.

As already shown in Table 8.6, nearly all blends that mix irregularly share the curious feature that at least one of the components is a homopolymer. Most species studied are statistical copolymers, either the product of a metallocene-mediated copolymerization or derived from a mixed microstructure polydiene. With only one exception, all copolymer–copolymer blends studied mix in a regular fashion. In

TABLE 8.12 Excess energy density coefficients for blends of homopolymers

	X_E (MPa) × 10^2				
Blend	**27°C**	**51°C**	**83°C**	**121°C**	**167°C**
PP/PB	10.2	6.8	2.7	−1.0	
PEB/hhPP	4.2	4.1	3.8	3.4	3.5
PEP/hhPP	3.0	1.8	0.8	0.3	−0.1
PEP/PEB	0.8	0.6	0.2	−0.2	−0.9
PP/hhPP	−26	−24	−19	−16	
PP/PEP	−40	−35	−28	−23	−22
PEP/PE	−45	−43	−42	−40	−40
PEP/PE				−16	−14
PE/PB				−136	−102
PIB/hhPP	−91	−67	−56		−43

contrast, homopolymer–homopolymer blends, with one exception, mix irregularly. Regularity in backbone structure seems to ordain irregularity in mixing. Values of X_E for all the homopolymer blends studied are listed in Table 8.12.

Based on PRISM theory, Singh and Schweizer have shown that changes in local packing with mixing can lead to either positive or negative values of X_E[79]. A suggestion that local packing and the irregularity pattern are somehow related also arose from the simulation-derived results of Maranas et al.[80] for the pair-distribution function

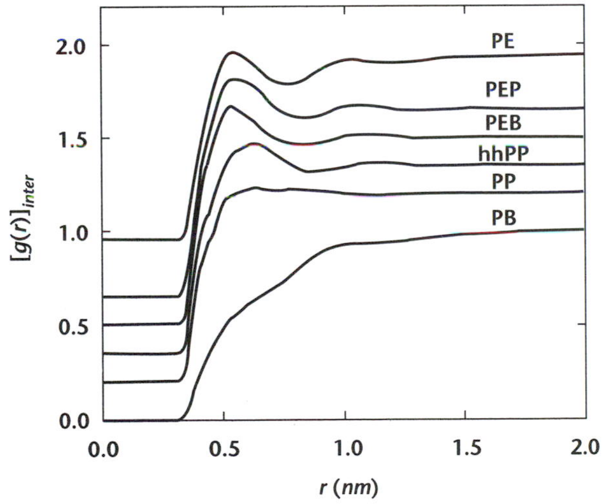

FIGURE 8.19 Intermolecular pair distribution function for selected polyolefins. Data obtained by molecular dynamics simulations as described by Maranas and coworkers[71,143].

TABLE 8.13 Mixing behavior of homopolymer pairs and their liquid structure difference[80]

Blend	$[g(r)]_{inter}$ Shapes	Effect
PP/PB	similar	destabilizing ($X_E > 0$)
PEB/hhPP	similar	destabilizing ($X_E > 0$)
PEP/hhPP	similar	destabilizing ($X_E > 0$)
PEP/PEB	intermediate	regular ($X_E \sim 0$)
PP/hhPP	different	stabilizing ($X_E < 0$)
PP/PEP	different	stabilizing ($X_E < 0$)
PB/PEP	different	stabilizing ($X_E < 0$)
PE/PEP	different	stabilizing ($X_E < 0$)
PE/PB	different	stabilizing ($X_E < 0$)
PIB/hhPP	different	stabilizing ($X_E < 0$)

$g(r)$. As shown in Figure 8.19, the local order reflected in the intermolecular part, $[g(r)]_{inter}$, varies significantly among the various species of olefin homopolymers. The values of X_E for their blends were found to correlate with the difference in the component $[g(r)]_{inter}$ forms. As recorded in Table 8.13, the mildly positive X_E values correspond to similarity in $[g(r)]_{inter}$ form, then pass over to large negative values for components having large differences in $[g(r)]_{inter}$.

The Maranas et al. explanation offered for this behavior grew from a supposition about how the local packing of chains change with mixing. Thus, polymer chains must satisfy some distribution of local conformations, dictated primarily by the intramolecular energetics of the species. The energetics supply the set of local geometries that must not only pack together but also satisfy the strong space-filling demand of the dense liquid state. If two homopolymers with similar local structures are mixed, the choice of local packing geometries is little changed, the difference in cohesive energy dominates, and X_E is essentially zero or weakly positive. If the components have different local structures, however, the mixing increases the range of local geometries available to each species, a situation that might permit more efficient space filling and introduces the possibility of a large net attraction (that is, negative values of X_E). The same line of reasoning suggests that statistical copolymers, being already more structurally diverse than homopolymers, have relatively less to gain from the expanded range of packing choices that would come with mixing. Thus, in the absence of other considerations, one might expect copolymer–copolymer pairs to mix regularly, as observed. A homopolymer–copolymer blend might go either way.

The above commentary is obviously very speculative. Some method is needed to obtain data about how the conflicts of local chain packing in the melt are resolved. As discussed in Section 2.2.1, wide-angle x-ray or neutron scattering can provide information about packing through the pair-distribution function $g(r)$. Londono et al.[81]

have examined the scattering by several polyolefins from this point of view and have indeed found differences among the various species. Comparisons made recently between the experimentally determined pair distributions and those obtained with united-atom simulations show remarkable agreement[82]. Unfortunately, the pair distribution for polymers is dominated by the intramolecular part $[g(r)]_{intra}$, and the details of $[g(r)]_{inter}$, the part that contains the packing information, are largely obscured.

Simulations to obtain $[g(r)]_{inter}$ are an attractive course to follow, but something beyond united atom models along with careful tuning and verification with experimental data will almost certainly be required. Two routes to experimental verification are probably feasible: A PRISM-based calculation of local structure, with results incorporated into a rotational isomeric state model for extrapolation to long chains, could supply an estimate of $[g(r)]_{\mathrm{intra}}$. Subtraction from the measured $g(r)$ would then provide an estimate of $[g(r)]_{inter}$ that is in part experimentally based. A second and fully experimental route to $[g(r)]_{inter}$ is available through wide-angle neutron scattering with appropriate deuterium labeling. Exploration and simulation verification of local packing in both pure melts and single-phase blends could be undertaken with the labeled-chain method. These various approaches require skills of several sorts and perhaps several years to accomplish.

8.4 Hydrocarbon Polymer Blends

Phase behavior in polyolefin blends is evidently governed by two rather different sorts of interactions. Dispersive interactions produce systematic and reasonably well understood net repulsions between species. Local packing effects lead to net attractions of uncertain origin and unpredictable magnitude. Free volume mismatches, another potential source of net repulsion, are easily shown to be small for the polyolefin systems already discussed. This section deals more generally with the mixing behavior of hydrocarbon polymers. Although still composed of only carbon and hydrogen and still relatively nonpolar, the molecules may now contain cyclic and unsaturated substituents. Data on mixing behavior is less complete for species with such substituents, so the first consideration is what role solubility parameters and free volume might play. Estimates of those contributions can be made using data on the pure component PVT properties.

Values of the FOV-based P^* and T^*, thermal expansion coefficient α, and PVT-based solubility parameter δ_{PVT} for sixteen hydrocarbon polymer species at 167°C are listed in Table 8.14. The first is polystyrene, a species containing aromatic unsaturation in its phenyl side group. The second is polycyclohexylethylene (PCHE), a species derived from polystyrene by hydrogenation of the phenyl ring. The cyclic substituent, one per monomeric unit, distinguishes PCHE topologically from the

TABLE 8.14 FOV and PVT-based solubility parameters for selected hydrocarbon polymer species at 167°C

Species	$\alpha(K^{-1}) \times 10^4$	$P^*(MPa)$	$T^*(K)$	$\delta_{PVT}(MPa^{1/2})$
PS	5.84	479	7647	17.77
PCHE*	5.99	448	8121	16.99
PIB	6.08	469	7693	17.38
PE	7.80	490	6827	17.67
PEP	7.33	458	6888	16.74
PP	7.52	403	6738	15.87
PB	7.01	409	7032	15.86
PBD-8	7.24	508	6827	18.18
PBD-24	7.19	480	6873	17.91
PBD-40	6.94	456	6914	17.50
PBD-50	6.99	424	6994	17.33
PBD-87	6.66	395	7040	16.40
PI-8	6.69	417	7067	17.03
PI-14	6.88	455	7065	17.74
PI-41	6.90	448	7056	17.68
PI-56	6.99	473	6977	17.80

*Values at 200°C.

saturated but open-structured polyolefins. The next five are polyolefins. The last nine are polydienes, species that contain nonaromatic unsaturation. The first five polydienes are polybutadienes with various 1,2 (vinyl) contents; the last four are polyisoprenes with various 3,4 contents. As discussed briefly in Chapter 1, polydienes have many isomeric forms. All those listed were made by anionic polymerization (organolithium-derived[25] and are thus of the mixed microstructure, atactic variety).

The FOV parameters for the first seven species in the table were obtained by fitting PVT data over the pressure range $10 < P(MPa) < 200$ and temperature range $150 < T(°C) < 250$[12]. (For PCHE, the range $220 < T(°C) < 290$ was used, owing to its high glass temperature, $T_g = 138°C$.) The thermal expansion coefficient α and isothermal compressibility β at 167°C and low pressure were obtained by fitting more limited blocks of data, as described in Section 8.3.3. The solubility parameter was obtained using Eqs. 8.16 and 8.18, with the assumption that the ratio of cohesive energy density to internal pressure is of order unity and the same for all polymer species:

$$\Pi_{CED} = B\Pi_{IP} \quad (B \text{ independent of species}) \tag{8.57}$$

The FOV parameters for the polydienes are those reported by Yi and Zoller[83]; the solubility parameter at 167°C was obtained from α and β at low pressure, as calculated with the Tait equation coefficients provided. Characteristic pressure P^*, which contains information on cohesive energy density, though less precisely than δ_{PVT}, is included mainly for completeness.

The phase behavior of block copolymers offers an additional source of interaction data for some pairs in this group. Diblock copolymers, chains of different species joined to one another by an end, undergo an order–disorder transition that is analogous to phase separation in blends[7]. The transition temperature T_{ODT} depends on the interaction parameter for the species and their respective chain lengths. For symmetric diblocks, the block molar volumes are equal, $V_1 = V_2$, and related to the interaction parameter at the transition by[84]:

$$\frac{\chi_{ODT}}{V_{ref}} = \frac{10.5}{V_1 + V_2} \tag{8.58}$$

The numerical coefficient increases with increasing diblock asymmetry in a roughly predictable way, reaching ~ 45 for $V_1 = V_2/9$ or $V_1 = 9V_2$, for example. The transition can be detected rheologically or via scattering methods[85]. For purposes here, the value lies in permitting the determination of larger values of χ than does phase separation in blends. Thus, for symmetric blends the coefficient in Eq. 8.58 is 4, so a factor of at least 2.5 is gained, or even 10 with highly asymmetric diblocks.

Interaction predictions and observations for twenty-two pairings of the species listed in Table 8.14 are shown in Table 8.15. The observed interaction parameters at approximately 167°C, if available or reasonably estimated, are given in the last column. Also listed are interaction strengths calculated from the pure component PVT properties in Table 8.14. Contributions from cohesive energy density mismatch, $X_{CED} = (\delta_1 - \delta_2)^2$ from Eq. 3.96, are estimated using values of δ_{PVT} through Eqs. 8.1 and 8.2. For comparison purposes, two estimates of the contribution from free-volume difference X_{FV} are made, one based on the T^* difference [Eq. 8.4] and the other on the difference in α [Eq. 8.6]. An interaction parameter was calculated from these results using:

$$\chi = \frac{V_{ref}}{RT}(X_{CED} + X_{FV}) \tag{8.59}$$

in which the T^*-based value of X_{FV} was used with V_{ref} as the geometric average volume per mer.*

*The values of X_{FV} from ΔT^* and from $\Delta\alpha$ obviously differ, but not in a systematic or particularly harmful way. When X_{FV} is small relative to X_{CED}, the values of X_{FV} from the two sources can differ widely—ΔT^* and $\Delta\alpha$ are small and their errors dominate—but the effect on the calculated value of χ is small. When X_{FV} is significant, however, the disagreement is smaller. Both X_{FV} and X_{CED} are themselves only estimates, so great accuracy is not to be expected.

TABLE 8.15 Estimates of cohesive and free volume contributions to the interactions for blends of selected hydrocarbon polymers at 167°C

Blend	Components	X_{sp} (MPa)	X_{fv}(MPa) from $\Delta T*$	X_{fv}(MPa) from $\Delta\alpha$	$\chi \times 10^3$(calc)	$\chi \times 10^3$(obs)
1	PCHE/PS[(a)]	0.61	0.17	0.01	15	>75
2	PCHE/PB[(a)]	1.28	1.6	0.53	90	17
3	PCHE/PEP[(a)]	0.06	1.25	0.68	36	42
4	PCHE/PE[(a)]	0.46	1.4	1.03	32	63
5	PS/PEP	1.06	0.50	0.70	37	97
6	PS/PE	0.01	0.60	1.1	13	>75
7	PS/PBD-8	0.17	0.59	0.62	16	79
8	PS/PI-8	0.55	0.29	0.25	20	59
9	PEP/PI-8	0.08	0.03	0.11	2.3	24
10	PB/PEP	1.12	0.02	0.03	22	9.5
11	PE/PEP	0.87	0.0	0.05	17	3.6
12	PE/PB	3.3	0.04	0.16	58	23
13	PE/PP	3.2	0.01	0.02	56	>33
14	PB/PP	0.0	0.08	0.07	1.5	0.9
15	PI-8/PBD-8	1.32	0.055	0.09	26	~0
16	PI-8/PBD-87	0.40	0.0	0.0	7.4	<0
17	PI-41/PBD-40	0.032	0.02	0.0	1.0	2.3
18	PBD-8/PBD-50	0.72	0.03	0.02	13	1.3
19	PBD-8/PBD-87[(b)]	3.2	0.04	0.10	55	>4.8
20	PBD-24/PBD-50[(b)]	0.34	0.01	0.01	6.0	<1
21	PBD-40/PBD-87[(b)]	1.2	0.02	0.02	21	>2.5
22	PBD-50/PBD-87[(b)]	0.87	0.0	0.03	15	>2.6

(a) At 200°C.
(b) At 25°C.

The comparisons of observed and calculated values of χ in Table 8.15 are discussed in two groups. The first group is a collection of *inter-family* blends: Blends 1 through 9 pair species that contain different types of substituents—cyclic, noncyclic, aromatic, unsaturated. The rest are *intra-family* blends: Blends 10 through 14 pair polyolefins, and blends 15 through 22 pair polydienes.

8.4.1 Inter-Family Blends.

The first blend system listed in Table 8.15 is one of many pairings of PCHE with other hydrocarbon polymers that have been screened for miscibility by Lin[86]. All blends with PCHE remained two-phase up to 200°C for chain lengths as short as ~25 mers. Neither solubility parameter difference nor free volume difference is large, and yet the observed interaction for the PCHE/PS system is at least 15 times more repulsive than calculated. Blends 2 through 4 are the only PCHE-hydrocarbon polymer pairings with observed values of χ, inferred from the order–disorder transition temperature of diblock copolymers[85,87]. Agreement is adequate for the PCHE/PEP system, but the repulsion is somewhat stronger than calculated for PCHE/PE and much weaker than calculated for PCHE/PB. The Cochran-Bates diblock work is additionally notable for its test of internal consistency for the PCHE, PE, PEP, PB interactions with a solubility parameter scheme alone. These researchers demonstrated rather good numerical agreement with relative assignments that led to the ordering $\delta_{PCHE} < \delta_{PB} < \delta_{PEP} < \delta_{PE}$. This is the same PB, PEP, PE ordering as found by blend studies[12], although the numerical assignments differ. The main difference between the Cochran-Bates results and the estimates in Table 8.15 is traceable to the value of δ_{PVT} for PCHE as reported in Table 8.14, and this may reflect a departure from Eq. 8.57 relating Π_{IP} and Π_{CED}.

The values of χ for blends 5 through 8 were obtained from the T_{ODT} of diblock copolymers, although cloud-point data for blends of short chains provided corroboration in some cases[58,85,88]. In all four systems, the observed χ is significantly larger than calculated, whether the polystyrene pairing is with polyolefins, blends 5 and 6, or with polydienes, 7 and 8. Blend 9, PEP/PI-8[89], is the only example found of a single-phase polyolefin–polydiene pairing. The observed χ from both SANS and cloud-point data for a blend is larger by factor of 10 than the calculation, apparently another example of anomalously large repulsion.

8.4.2 Intra-Family Blends

In contrast with the inter-family blends, the polyolefin pairings, blends 10 through 14 in Table 8.15, suggest a pattern of anomalous extra attraction, as discussed in Section 8.3.5. Interaction data for the PB/PEP, PE/PEP, and PE/PB pairs were provided by SANS measurements on the corresponding block copolymers and low molecular weight blends[65,90]. These data demonstrate the anomalous attraction discussed earlier: The calculated values of χ are too large by factors of from 2 to 5. The interaction inequality for the PE/PP pairing is based on solubility observations for the corresponding block copolymer[91,92] and leaves open the possibility of agreement with the calculated χ. The calculation for PB/PP agrees fairly well with SANS data on the blend at 167°C[56].

Polydienes have peculiar mixing properties, which was apparently known at least back to the early 1970s[93]. Roland noted in 1987 that blends of cis 1,4 polyisoprene and atactic high vinyl polybutadiene (86% 1,2; PBD86) are miscible at high molecular weights ($M \sim 300{,}000$) over a wide range of temperatures[94]. For similar molecular weights, blends of the polyisoprene with atactic low-vinyl PBD and blends of low- and high-vinyl PBD are both, however, two-phase at room temperature. A variety of supporting evidence for miscibility was gathered[95], including an observation that, in contrast to the atactic variety, syndiotactic high-vinyl PBD is immiscible with cis 1,4 polyisoprene. Later SANS studies on blends of PI-7 and PBD-97 (both prepared anionically) revealed a negative interaction parameter with an LCT for long chain components around 160°C[96].

For blends of low 3,4 PI with PBD, Kawahara et al.[97] showed a crossover at room temperature from two-phase to single-phase near 33% vinyl and an LCT near 60°C for a blend of that composition. Han and coworkers have used SANS to evaluate the interactions in PI/PBD blends[98–100] and in PBD/PBD blends[101,102]. Schwann and Willner[103] have also studied PBD/PBD blends using SANS, and Krishnamoorti et al.[104] have conducted studies of miscibility at 25°C on PBD/PBD blends. The polydiene–polydiene blend informations given in the last column of Table 8.15 for blends 15 through 22, was gleaned from these various studies. Those results suggest an algebraic decrease in net interaction between low 3,4 PI and PBD as the PBD vinyl content increases. The data for blends 15 and 16 are consistent with that picture, as is the Heffner and Mirau inference from NMR of specific association between PI methyl and PBD vinyl groups[13]. The values of χ are inferred from the Thudium-Han[100] data, but are also consistent with the Hasegawa et al.[99] and Tomlin-Roland[96] results for similar blends. Both are LCT systems. Indeed, all PI/PBD blends tested are LCT systems except blend 15, which is just barely UCT (χ is nearly independent of temperature over the range tested[100]).

It is difficult to extract unambiguous information about the interactions in PBD/PBD blends. The value of χ for blend 17 is consistent with results obtained for similar blends by Jannai et al,[102] Krishnamoorti et al.,[104] and Schwann and Willner.[103] The bounds on χ indicated for blends 19 through 22 are primarily from the Krishnamoorti et al.[104] study, conducted however at a much lower temperature; these bounds are not inconsistent with values found in other studies at higher temperatures. It is clear that, as in the PI/PBD blends, negative χ and LCT behavior occur for PBD/PBD blends in the compositional midrange. Also like the PI/PBD blends, the observed interactions for PBD/PBD blends are much less repulsive than the calculations of χ would suggest.

Why are blends of polydienes so complex when compared with polyolefin blends, and why, after so many studies, is so little known with certainty? At least three complications arise, one being of course specific association in the polydiene blends. Another is the complexity of polydiene microstructure, giving many more possible local interactions to consider. Deriving model polyolefins by saturating polydiene double bonds eliminates the cis-trans variations and results in less structural variability. The third complication has to do with deuterium labeling. Structurally matched pairs of model polyolefins, labeled and unlabeled, are obtained by saturating separate batches with D_2 and H_2. Label switching can thus be used to correct for the isotopic interactions. For the polydienes, each labeled or unlabeled sample must be synthesized separately, structural matching is difficult, and label switching becomes expensive. In all SANS studies of PI/PBD blends to date, only the PBD has been deuterated, so the effect of labeling cannot be assessed. Schwann and Willner[103] performed a few experiments with label-switched PBD pairs, but the molecular weights were too small for the useful evaluations of polymeric interaction densities.

8.4.3 Commentary

The information on inter-family pairings, blends 1 through 9, suggest a pattern—with some exceptions—of large extra repulsion. This is a curious counterpart to the large extra attraction found in many intra-family polyolefin blends.

It is conceivable that the origin of both anomalous attraction and anomalous repulsion resides in the complexities of local accommodation to shape differences in the presence of the strong cohesive forces that enforce volume filling in dense liquids. Thus, extending the earlier argument of Maranas et al.[80] (Section 8.3.5), within a family, the local shapes of species differ, but the basic structural elements are the same. The "sticks" and "bends" that define the species skeleton locally—the covalent bond lengths and angles—are the same within a family like the polyolefins; this may make possible the net attraction of species owing to a gain in packing choices with mixing[80]. If, on the other hand, the species have different bond lengths and angles, and different local structural elements, then accommodation may exact an energy penalty. In polyolefins, the carbon-carbon bond lengths are all 0.153 *nm* within some close tolerance, and the bond geometry is nearly tetrahedral. In polydienes, some bond lengths drop to 0.133 *nm* and some bond geometries are planar. Species with aliphatic cyclic units, such as PCHE, or with aromatic cyclic structures, such as PS, have their own sets of quantum mechanically enforced requirements.

Thus, in addition to the cohesive energy and free volume mismatches—taken into account only crudely here—intra- and inter-family local packing mismatches may be important as well. Unfortunately, this line of reasoning, whether correct or not, provides at the current state of the field no more than an unverifiable rationalization of experimental observations that have already been made. Another approach, which also has to do with currently unobservable features, can perhaps be framed broadly enough for molecular simulations to play a role. The central importance of the intermolecular pair distribution function has already been discussed (Section 8.3.5). Cohesive energy density, intimately dependent on $[g(r)]_{inter}$, is one of the key parameters in mixing behavior. Some evidence suggests that molecular dynamics—perhaps even at a level of detail no more intricate than united atom representations—can provide useful information on the relationship between Π_{CED} and the internal pressure Π_{IP}, a macroscopically measurable property[71]. In effect, the object would be to evaluate the parameter B in Eq. 8.57 for various polymer species of interest.

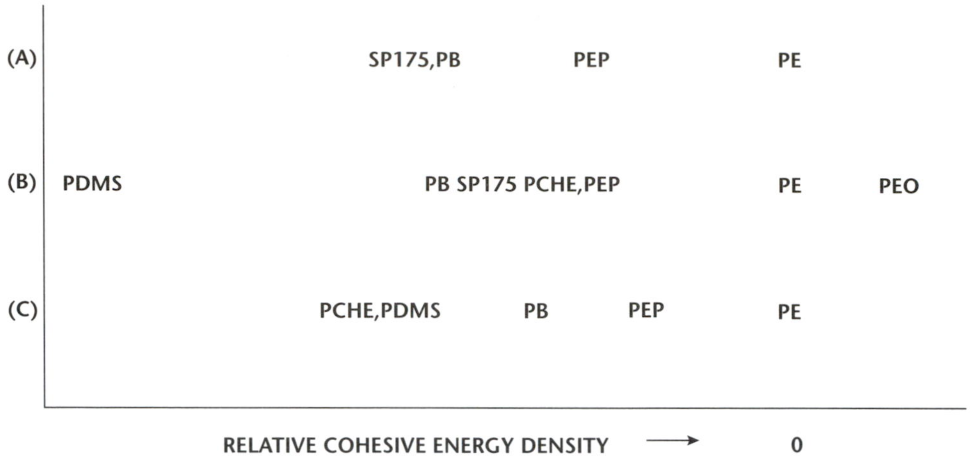

FIGURE 8.20 Relative cohesive energy density ordering of selected polymer species as obtained by three methods: (A) solubility parameter assignment from SANS-based interaction data for linked sequences of blends[56,57], (B) internal pressure from pure component PVT measurements[12,56], and (C) order–disorder transition determinations of diblock copolymers with common components[87].

That this approach might be useful is suggested by the estimates of relative cohesive energy density that have already been provided by experimental data. Figure 8.20 compares the energetic positioning semiquantitatively at about 180°C relative to polyethylene as benchmark (PE is vertically aligned on all three scales). One source is the self-consistent assignment of relative solubility parameters from polyolefin blends in Table 8.5. Another is the internal pressure collection as obtained from PVT data and the choice $B = 1$ in Eq. 8.57. The values of Π_{IP} for the two nonhydrocarbon polymers included in the comparisons, polydimethylsiloxane and polyethylene oxide, were calculated using thermal expansion coefficients and compressibilities from Orwoll[105]. A third source is the set of self-consistent assignments made on the basis of block copolymer studies[87,106].

Note that both PEP and PB from blends and PVT in Figure 8.20 are vertically aligned (reflecting agreement within the errors for those two in Figure 8.14), whereas the values for SPI75, which differ beyond the errors from blends and PVT[56], are offset. The block copolymer-based values for PEP and PB are displaced from those on the other two scales, providing examples of anomalous attraction (Section 8.3.5) or, alternatively, an effect of coil stretching in the copolymer blocks[90]. The PVT- and copolymer-based results are quite different for PCHE, but a large free volume mismatch exists for all the PCHE-containing systems [see Eq. 8.6]; allowing for that would move the PVT-based value down. Free volume effects are negligible for the relevant PDMS and PEO systems, yet the two PDMS values differ enormously. The PVT-based assignment for PEO places it above PE, but unfortunately a self-consistent diblock-based assignment was not possible.

Each of the three methods of estimating cohesive energies has obvious limitations. The development of a reliable prediction of Π_{CED}/Π_{IP} from local chemical structure would offer the benefit of relating one of the most important factors in governing blend interactions to macroscopic observables.

8.5 Blends of Copolymers

Many of the blend components so far considered are copolymers. Thus, aside from some minority substituents, PEP and PEB are equivalent structurally to alternating copolymers of ethylene with propylene and butene-1. Structurally, the EB series are statistical copolymers of ethylene and butene-1, and the metallocene-derived copolymers of ethylene and α-olefins (see Figure 8.12) are also statistical, but probably with sequencing statistics different from the EB series. Indeed, atactic polypropylene and polystyrene are statistical copolymers of syndiotactic and isotactic moieties. Even those species that are called syndiotactic or isotactic are in reality also statistical copolymers, merely having a preponderance of one or the other isomeric form. Viewed in this detailed way, very few "structurally pure" synthetic polymers actually exist. Only polymers derived from highly symmetric monomers—linear polyethylene, polydimethylsiloxane, and (leaving aside head-to-head incorporation) polyisobutylene, for example—are literally true homopolymers. Such considerations raise two rather distinct questions about the interactions that govern phase behavior: For a fixed overall mer ratio, does the sequencing of mers along the chain matter? And, How are the intermolecular interactions of a copolymer related to those of the individual homopolymers?

8.5.1 Sequencing Effects

From the qualitative variations in phase behavior between block copolymers and statistical copolymers, sequencing is obviously important at the extremes of in-chain segregation, so the question is whether it matters for more local differences, such as those between alternating and statistical copolymers. A theory for sequencing contributions at the triad level, considering nearest in-chain neighbor effects, has been developed by Balazs and coworkers[107–110]. Unfortunately, no data are available on systems that are sufficiently well defined to give the theory a proper test. Some data, however, address the question of whether nearest neighbor effects are important, but these are too few to permit a general conclusion.

The polyolefins offer a few examples suggesting that sequence distribution is relatively unimportant. Thus, the alternating PEB has nearly the same nominal ethylene to 1-butene ratio as the statistical EB66 and, from Table 8.5, the solubility parameters of PEB and EB66 are practically the same. Also, PEP is equivalent to EP 60, an ethylene–propylene copolymer with 60 wt % propylene. From Table 8.5, for the alternating PEP and an estimate for the statistical metallocene-derived EPs based on Figure 8.12, the solubility parameters at 167°C agree quite well. Indeed, the excellent general agreement between the model and metallocene-derived EBs in Figure 8.12 suggests that whatever sequencing differences exist between the two groups of statistical EB copolymers have little effect on their interactions.

The first test for sequencing effects, and a clear demonstration of their importance, was performed by Balazs et al. in 1985[108]. These researchers showed that blends of polyvinyl chloride—effectively an alternating copolymer of CH_2 and CHCl—and chlorinated polyethylene—effectively a statistical copolymer—were immiscible when their average compositions were matched. The carefully designed studies of polystyrene–polymethylmethacrylate interactions by Galvin, Winey, and coworkers[111,112] tell a similar story. Using electron microscopy image analysis, they established phase diagrams at 150°C for two ternary blends: PS and PMMA with an alternating PS/PMMA copolymer in one, the same PS and PMMA with a statistical PS/PMMA copolymer in the other. The copolymers were closely matched in average composition, average molecular weight, and molecular weight distribution. The phase diagrams are clearly different, a result that indicates—given the careful controls and analytical procedures applied—a significant contribution from local sequence distribution. The alternating copolymer led to enhanced miscibility in this case[112], a result seen, although less clearly, in other systems in which sequencing might perhaps contribute. There appears to be nothing definitive beyond the Balazs et al.[110] and Winey-Galvin studies[111,112] and the polyolefins examples quoted above. The influence of local sequencing on blend interactions in general remains an open question.

8.5.2 Mean Composition Effects

In the absence of sequencing effects, the interaction parameter for copolymer blends depends on the various homopolymer interaction parameters and the component compositions. It is treated in the lattice version of the simple mixture model (Section 3.4): The mers mix randomly, each interaction energy depends only on the identity of adjacent mers, and the total interaction energy is the sum of pairs, each weighted by the corresponding homopolymer pair interaction energy. The problem was considered first by Scott in 1952[113], and then later by several others[114–117].

The development by ten Brinke, Karasz, and MacKnight[116] for binary blends of a copolymer AB with A and B mers and a copolymer CD with C and D mers leads in a straightforward way to a formula for the interaction parameter of an AB/CD blend:

$$\chi = \varphi_A\varphi_C\chi_{AC} + (1-\varphi_A)\varphi_C\chi_{BC} + \varphi_A(1-\varphi_C)\chi_{AD} + (1-\varphi_A)(1-\varphi_C)\chi_{BD} - \varphi_A(1-\varphi_A)\chi_{AB} - \varphi_C(1-\varphi_C)\chi_{CD} \tag{8.60}$$

The volume fractions of A in AB and C in CD are φ_A and φ_C respectively, and the six χ_{ij} are homopolymer interaction coefficients, all defined with respect to the same reference volume, for example $(V_A V_B V_C V_D)^{1/4}$, where the four V_i are mer molar volumes. Two special cases are of common interest. For blends of homopolymer A with copolymer CD,

$$\chi = \varphi_C \chi_{AC} + (1 - \varphi_C)\chi_{AD} - \varphi_C(1 - \varphi_C)\chi_{CD} \tag{8.61}$$

where the χ_{ij} are all defined with respect to the same reference volume, for example $(V_A V_C V_D)^{1/3}$ in this case. For blends of two AB copolymers having A mer volume fractions φ_1 and φ_2,

$$\chi = (\varphi_1 - \varphi_2)^2 \chi_{AB} \tag{8.62}$$

when χ for the copolymer blend and χ_{AB} for the homopolymer blend have the same reference volume. All three equations can be written for any choices of reference volumes by replacing χ and the various χ_{ij} throughout by the ratios χ/V_{ref} and $\chi_{ij}/(V_{ref})_{ij}$, respecting the additivity of energy density, as carefully explained by Paul and Barlow[117]. Thus, for Eq. 8.62,

$$\chi = V_{ref}(\varphi_1 - \varphi_2)^2 \chi_{AB}/(V_{ref})_{AB} \tag{8.63}$$

Note that Eqs. 8.60 and 8.61 indicate that all χ_{ij} for a system may be positive, yet depending on their magnitudes and the component compositions, the interaction coefficient of the blend may be negative[115]. Accordingly, a blend of two copolymers may be miscible even though the component homopolymers are all immiscible with one another. The explanation of this surprising possibility can be understood most readily from Eq. 8.61. Thus, if the constituents C and D of the copolymer are strongly repulsive (χ_{CD} large and positive), the large negative contribution that term makes to χ may more than offset the positive contributions from the other two terms. Stated another way, a reduction in the highly repulsive C-D contacts in the copolymer by dilution with an A homopolymer may more than balance the mildly repulsive A-C and A-D contacts that are generated. Examples of systems that demonstrate the "copolymer effect" are given elsewhere[116,118].

The quantitative aspects of copolymer equations such as Eqs. 8.60 through 8.63 have been tested in detail with only a few systems. The first test was conducted by Kampour and Bendler[119] with homopolymers and statistical copolymers of styrene and parabromostyrene, synthesized by brominating some fraction x, $0 < x < 1$, of the phenyl ring para positions on nearly monodisperse polystyrene samples, $4000 < M < 860{,}000$. The reaction is clean, and the result is a nearly monodisperse statistical copolymer with mole fraction x bromostyryl mers (-SBr-) and the remainder unreacted styryl mers (-S-). Blends of these copolymers form two liquid phases with phase diagrams of the UCT type and a critical temperature that increases with increasing component chain lengths and increasing difference in component compositions, $x_1 - x_2$. The refractive index contrast was large enough to make the presence of two phases readily detectable visually. Equal weights of the two components were dissolved in a common solvent, and films were cast by evaporation at room temperature. The films were classified as cloudy (strongly two-phase), clear (single phase), and opalescent (judged to be near the critical region). Unfortunately, the temperature corresponding to the development and "freezing-in" of these features is not well defined. However, as rightly pointed out by the authors,[119] the phase boundary locations depend so strongly on component composition and molecular weight that strict control over the conditions of phase separation is perhaps not of great importance.

The near-critical blends were grouped into two sets. Set A were blends of each parent polystyrene with whichever of its derived copolymers gave the opalescent signature. Set B were blends of the various copolymers derived from the same polystyrene parent, $M = 97{,}000$, which gave the opalescent signal. With volume fractions approximated by the mole fractions, the data were analyzed with the simple mixture expressions for critical conditions to show rather good, although not precise, agreement with Eq. 8.62. A few years later, Koningsveld and MacKnight closed the apparent gap between theory and experiment with a more detailed analysis[120].

The Koningsveld-MacKnight analysis replaces mole fractions by volume fractions, as the simple mixture model requires, and takes into account the change in molar volume with bromination. From Eq. 3.81, the interaction parameter at the critical point can be written:

$$\chi_c = \frac{V_{ref}}{2V_1}\left[1 + \left(\frac{V_1}{V_2}\right)^{1/2}\right]^2 \tag{8.64}$$

Thus, with Eq. 8.63,

$$|\varphi_1 - \varphi_2|_c = \left[\frac{(V_{ref})_{S/BrS}}{2\chi_{S/BrS}}\right]^{1/2} \frac{1}{V_1^{1/2}}\left[1 + \left(\frac{V_1}{V_2}\right)^{1/2}\right] \tag{8.65}$$

in which *S/BrS* refers to blends of styrene and parabromostyrene homopolymers. For the Kambour-Bendler set A, let 1 refer to the parent PS. Accordingly, $|\varphi_1 - \varphi_2|_c = \varphi_c$; $V_1 = M/\rho(0)$, where M is the parent PS molecular weight and $\rho(0)$ the mass density of styrene homopolymer; and $V_1/V_2 = \rho(x_c)/\rho(0)$, the component mass density ratio. Thus for set A, agreement with the copolymer equation requires:

$$\varphi_c \propto \Omega M^{-1/2} \tag{8.66}$$

where $\Omega = [1 + [\rho(x_c)/\rho(0)]^{1/2}]/2$. For set B, $V_1 = 97{,}000/\rho(x_{1c})$ and $V_1/V_2 = \rho(x_{2c})/\rho(x_{1c})$, so from Eq. 8.65, agreement with the copolymer equation requires the ratio

$$\Gamma = \frac{2|\varphi_1 - \varphi_2|_c}{[[\rho(x_{1c})/\rho(0)]^{1/2} + [\rho(x_{2c})/\rho(0)]^{1/2}]} \tag{8.67}$$

to be independent of the average component composition, $\langle\varphi\rangle = (\varphi_1 + \varphi_2)/2$.

An assumption of mer volume additivity and the known mass densities and mer weights of the two homopolymers—1.04 *gm/cm*^{3} and 104 for PS, 1.57 *gm/cm*^{3} and 183 for PBrS—permit estimates of copolymer mass densities and conversions from the reported mole fractions to volume fractions. The results permit the calculation of Ω and Γ and thus tests of the direct proportionality predicted by Eq. 8.66 and the constancy of Γ as defined in Eq. 8.67. The results are shown in Figures 8.21a and 8.21b. The critical composition proportionality for set A is maintained over a molecular weight range of more than a factor of 200. For set B, the values of Γ relative to the average are essentially within the estimated error bars reported by Kambour and Bendler. The validity of Eq. 8.63 for wide ranges of chain length and copolymer composition is thus well established for the styrene–parabromostyrene system. Subsequent to this work, the phase behavior of PS/BPS blends has been studied by other methods[19,121,122]. The polystyrene–polyparabromostyrene system has now become a model for investigating the influence of copolymer blending on various properties[123–125].

Only one other examination of Eq. 8.63 at the styrene–parabromostyrene level of detail has been made, and its conclusion is quite different. Ethylene-butene-1 (EB) copolymers, model materials derived by hydrogenating polybutadienes, were discussed briefly in Section 8.3.3. These are statistical copolymers of the linear and branched C_4 units enchained by butadiene polymerization, 1,4-mers and 1.2-mers whose relative amounts can be varied over wide ranges by polymerization modifiers[73]. Unlike the PS/BPS system, the EB mer volumes defined in this way differ by a negligible amount (<3 %), so mole fraction x of a co-mer is negligibly different from its volume fraction φ. Also, values of χ defined with respect to the mer volumes are available for blends of the copolymers over the entire composition range. Thus, Eq. 8.62 differs only negligibly from Eq. 8.63 and can be tested directly. Interaction parameters for the homopolymer pair, as calculated from Eq. 8.62, $\chi = \chi_{PB/PE}/(\Delta\varphi)^2$, at various mean pair compositions, $\langle\varphi\rangle = (\varphi_1 + \varphi_2)/2$, can be compared. The strong variation with mean composition in Figure 8.22—a difference factor of five between "ethylene rich" and "butene rich" pairings—makes it evident that the copolymer equation does not apply in this system. The same general conclusion could be anticipated from the nonlinear dependence of solubility parameters on branched C_4 content, as seen in Figure 8.11. The applicability of the solubility parameter scheme and inconsistency with the copolymer equation can evidently occur within the same

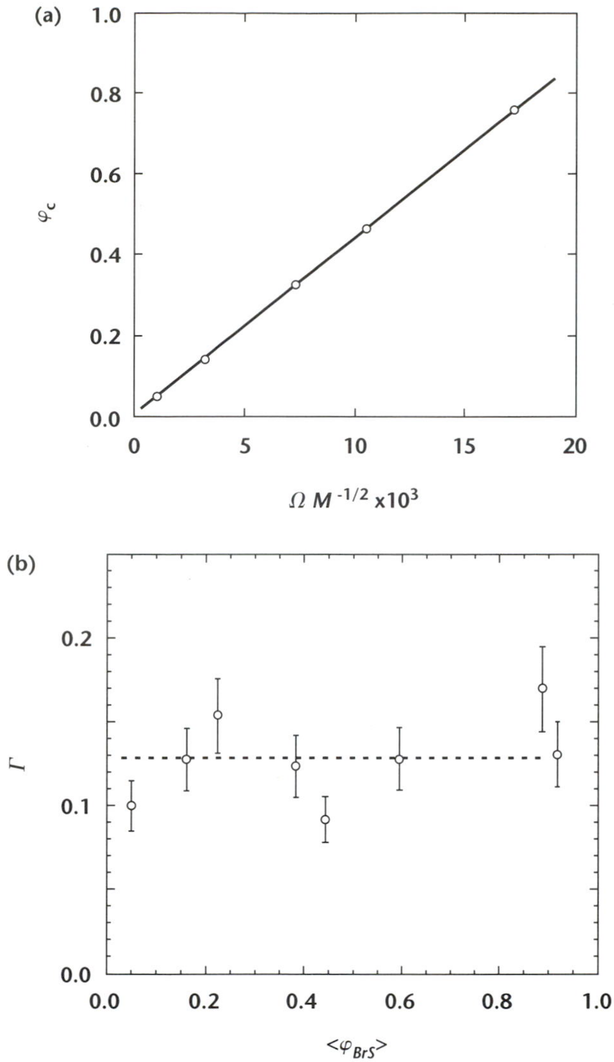

FIGURE 8.21 Tests of the copolymer equation for the polystyrene–poly (parabromostyrene) system. Data from Kambour and Bendler[119]. (a) Set A test. Comparison of data with the predicted relationship between critical copolymer composition and parent molecular weight for blends of S/BrS copolymer and the parent PS. (b) Set B test. Comparison of data with prediction for composition difference at criticality in blends of copolymers derived from the same PS parent. The ratio $\Gamma = |\varphi_1 - \varphi_2|_c / \langle(\rho(x_c)/\rho(0))^{1/2}\rangle$ is independent of the average of the copolymer compositions if the copolymer equation is obeyed [see Eq. 8.62].

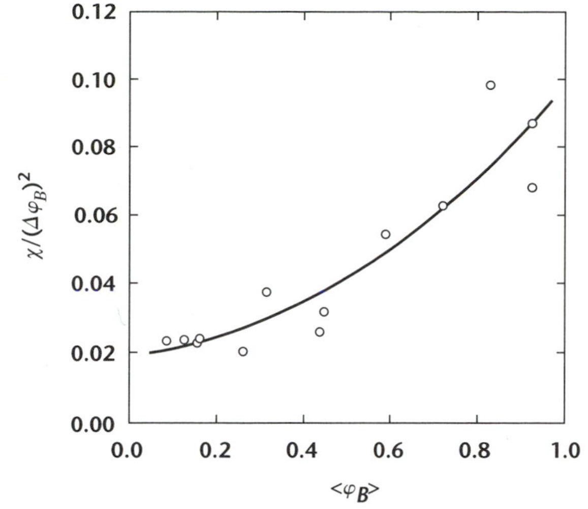

FIGURE 8.22 Test of the copolymer equation for ethylene–butene (linear C_4/branched C_4) system. Data from Graessley et al.[73]. The ratio $\chi/(\Delta\varphi)^2$ is independent of the average of the copolymer compositions if the copolymer equation is obeyed.

family of materials. Unfortunately, at this stage in the field, it has only been shown that the copolymer equation works within some copolymer families but not in others.

8.6 Blends with Specific Associations

The polystyrene–polyvinylmethylether (PS/PVME) system is a well-studied example of specific association influence on the component interactions and phase behavior in polymer blends. Other chemically dissimilar pairs are miscible over some range of temperature. Polyethylene oxide and polymethylmethacrylate are miscible over a very wide range[126]. The PEO/PMMA blends have a small negative interaction parameter ($|\chi| \lesssim 3 \times 10^{-3}$) that hardly varies at all with temperature[127], reminiscent of the polydiene blends (Section 8.4.2). Polystyrene–poly(1,4-xylenyl ether) (PS/PXE) blends are miscible and exhibit LCT behavior at elevated temperatures. Well below T_c, their interaction coefficient is negative, much larger than for PEO/PMMA blends ($|\chi| > 20 \times 10^{-3}$), and varies more strongly with temperature as well [128]. In those respects, the PS/PXE system resembles the PS/PVME system chosen for examination here.

Polystyrene and polyvinylmethylether are miscible at room temperature and remain so up to about 100°C. At higher temperatures, PS/PVME blends undergo LCT phase separation, with critical temperatures ranging upward with decreasing

TABLE 8.16 PVT Parameters at 120°C for Polystyrene and Polyvinylmethylether[105,134]

Species	$P^*(MPa)$	$T^*(K)$	$\alpha(°C^{-1}) \times 10^4$	$\beta(MPa^{-1}) \times 10^4$	δ_{PVT} $(MPa^{1/2})$
PS	506	7948	5.79	6.0	19.47
PVME	483	7051	7.06	8.2	17.37

component molecular weights. For $M_{PVME} = 51{,}000$, T_c is about 100°C for $M_{PS} > 10^5$, about 150°C for $M_{PS} \sim 20{,}000$, and above 200°C for $M_{PS} \sim 10,000$[129,130]. This behavior resembles the high temperature LCT phenomenon in polymer solutions (Figure 7.22, for example) brought on by a repulsive interaction that increases with increasing temperature. The FTIR[131,132] and NMR[133] spectra of the blend change with temperature and phase separation in a manner consistent with a decreasing contribution from an attractive interaction involving the C-O bonds in PVME and the aromatic electrons in PS.

The pure component PVT properties suggest a significant repulsive interaction between PS and PVME, from cohesive energy and free volume mismatches. Shiomi et al.[134] report pure component FOV parameters at $T \sim 100$°C and Orwoll[105] reports the compressibilities and expansion coefficients that are shown in Table 8.16. Values of δ_{PVT} were calculated using Eq. 8.1, and $\chi = 0.11$ was then obtained using Eqs. 8.2, 8.4, and 8.59 for the average mer volume $V_{ref} = (v_S v_{VME})^{1/2} = 74\ cm^3$. Blends of polymers of any significant size at all are immiscible for $\chi = 0.11$, so without some offsetting effect, such as a specific attraction, PS/PVME blends would lie deeply in the two-phase region.

The substitution of perdeuterated PS increases the cloud-point temperatures about 40°C[135], a direction that, from the discussion in Section 8.2.1, is consistent with $\delta_{PS} > \delta_{PVME}$. This expectation is also supported by the values of δ_{PVT} for the two components (see Table 8.16). The upward shift is also consistent with the algebraically higher values of $\chi(T)$ for blends of PVME with fully hydrogenous PS[62], as determined by the ternary SANS method (see Section 8.3.3). The glass transition temperature for PS/PVME blends is insensitive to deuteration, and it depends on blend composition[135] as shown in Figure 8.23. An interpretation of SANS data obtained on systems below T_g is therefore subject to some ambiguity owing to possible nonequilibrium effects.

The interactions in blends of PVME and perdeutero-PS have been investigated extensively via SANS measurements as interpreted using the de Gennes equation. (Eq. 8.30). Values of χ/V_{ref} were determined as a function of blend composition near 25°C for blends of narrow distribution PS ($M_w = 252{,}000$) and polydisperse PVME ($M_w = 99{,}000$)[136]. Another group used fractionated PVME to determine the effects of molecular weight, composition, and temperature on the interactions in perdeutero-PS/PVME blends[137] with three PS/PVME pairs of narrow distribution

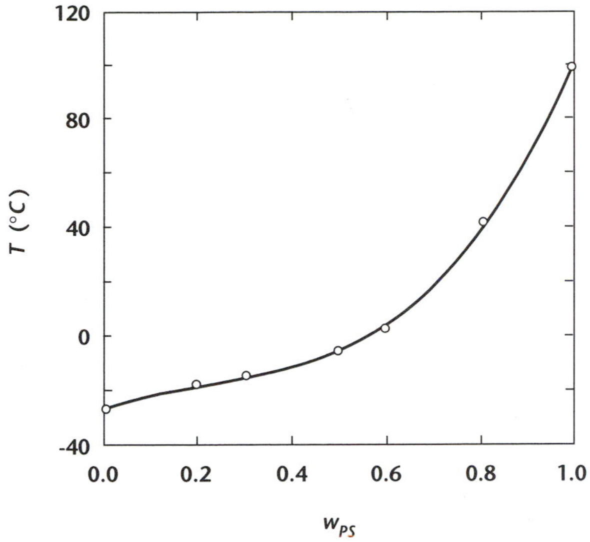

FIGURE 8.23 Glass transition temperature as a function of blend composition for the PS/PVME system. Data from Yang et al.[135]

components. The dependence of χ/V_{ref} on blend composition was determined at several temperatures for each of the three pairs—series L, M, H with M_{PS} = 230,000, 402,000, and 593,000 paired respectively with M_{PVME} = 389,000, 210,000, and 1,100,000.

The values of χ at 130°C, defined with respect to $V_{ref} = (\upsilon_{PS}\upsilon_{PVME})^{1/2} = 74\ cm^3$, are shown as a function of blend composition in Figure 8.24a[137]. The values are negative, indicating a net attraction of the components. No evidence of chain length dependence appears for χ, but a relatively strong composition dependence is evident, with attraction growing with increasing PS content. The variation with ϕ_{PS} is monotonic and essentially linear over the range investigated, quite different from the unexplained upturn in the composition wings commonly seen for $\chi(\phi)$ in isotopic and polyolefin blends (Figures 8.4 and 8.17a). The composition dependence at other temperatures is shown in Figure 8.24b[136,137]. The variation remains linear, with perhaps some steepening as the temperature is reduced.

An observable volume change occurs with mixing in this system[134,135,138], so the composition dependence could conceivably be a finite compressibility effect[139–143]. The volume change is negative, although the magnitudes differ significantly among investigators. A compressible model that accounts for most of the composition dependence also predicts a negative volume change with mixing, with a magnitude within reported values for PS/PVME blends[141].

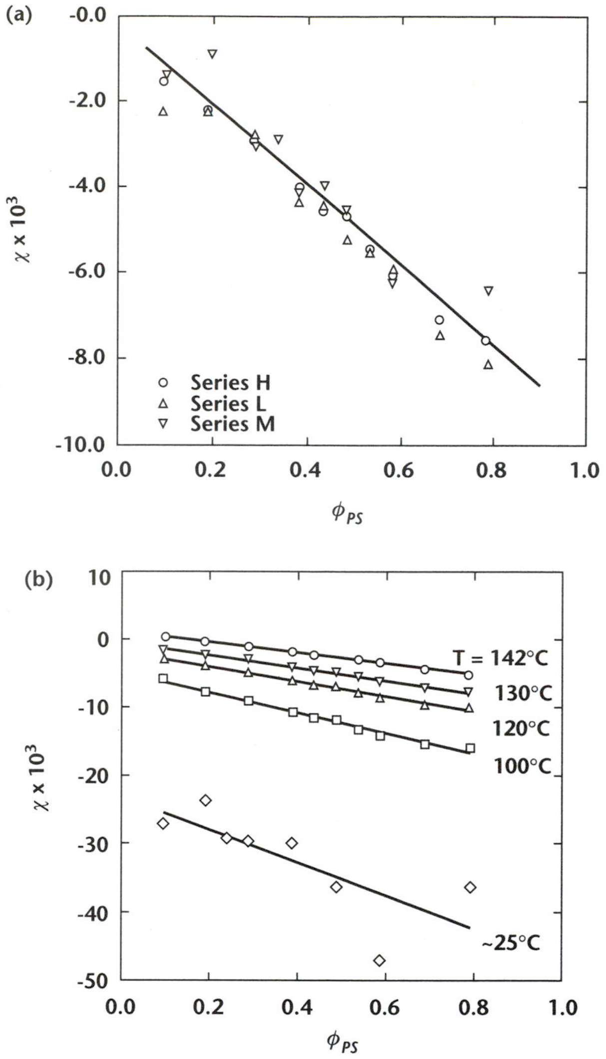

FIGURE 8.24 Interaction parameter as a function of composition for PS(d)/PVME blends. (a) Values at 130°C for three pairs of component molecular weights. Data from Han et al.[137]. (b) Values at several temperatures. Data from Shibayama et al.[136] and Han et al.[137]

REFERENCES

1. Scott R.L. 1949. *J. Chem. Phys.* 17:279.
2. Paul D.R. and Bucknall C.B. (eds.). 2000. *Polymer Blends*, (Vol. 1 and 2). New York: Wiley-Interscience.
3. Helfand E. and A.M. Sapse. 1975. *J. Chem. Phys.* 62:1327.
4. Creton C., E.J. Kramer, H.R. Brown, and C.-Y. Hui. 2002. *Adv. Polym. Sci.* 156:53.
5. Sauer B.B. and G.T. Dee. 2002. *Macromolecules* 35:7024.
6. Datta S. and Lohse D.J. 1996. *Polymeric Compatibilizers*, Munich: Hanser Publishers.
7. Hamley I.M. 1999. *Block Copolymers*, Oxford: Oxford University Press.
8. Noolandi J. 1996. In *Polymeric Materials Encyclopedia* (Vol. 10), Salamone J.C. (ed.). Boca Raton, FL: CRC Press, p. 8207.
9. Sheehan C.J. and Bisio A.L. 1966. *Rub. Chem. Tech.* 39:149.
10. Barton A.F.M. 1983. *Handbook of Solubility Parameters and other Cohesion Parameters*, Boca Raton, FL: CRC Press.
11. Coleman M.M., Graf J.F., and Painter P.C. 1991. *Specific Interactions and Miscibility of Polymer Blends*, Lancaster, PA: Technomic.
12. Krishnamoorti R., W.W. Graessley, G.T. Dee, D.J. Walsh, L.J. Fetters, and D.J. Lohse. 1996. *Macromolecules* 29:367.
13. Heffner S.A. and P.A. Mirau. 1994. *Macromolecules* 27:7283.
14. Chung G.-C., J.A. Kornfield, and S.D. Smith. 1994. *Macromolecules* 27:964.
15. Roland C.M. 1988. *J. Polym. Sci., Part B: Polym. Phys.* 26:839.
16. Schmidt-Rohr K. and Spiess H.W. 1994. *Multidimensional Solid-State NMR and Polymers*, New York: Academic Press.
17. Wang X. and J.L. White. 2002. *Macromolecules* 35:3795.
18. Müller M., K. Binder, and L. Schäfer. 2000. *Macromolecules* 33:4568.
19. Koch T. and G.R. Strobl. 1990. *J. Polym. Sci., Part B: Polym. Phys.* 28:343.
20. Rabeony M., H. Shao, K.S. Liang, E. Siakali-Kioulafa, and N. Hadjichristidis. 1997. *Macromolecules* 30:7332.
21. Strazielle C. and H. Benoit. 1975. *Macromolecules* 8:203.
22. Atkin E.L., L.A. Kleintjens, R. Koningsveld, and L.J. Fetters. 1982. *Polym. Bull.* 8:347.
23. Yang H., G. Hadziioannou, and R.S. Stein. 1983. *J. Polym. Sci., Part B: Polym. Phys.* 21:159.
24. Lapp A., C. Picot, and H. Benoit. 1985. *Macromolecules* 18:2437.
25. Morton M. and L.J. Fetters. 1975. *Rub. Chem. Tech.* 48:359.
26. Hsieh H.L. and Quirk R.P. 1996. *Anionic Polymerization, Principles and Practical Applications*, New York: Marcel Dekker.
27. Moberly C.W. 1967. *Encyclopedia of Polymer Science & Technology* (Vol. 7) New York: J. Wiley, p. 557.
28. Rachapudy H., G.G. Smith, V.R. Raju, and W.W. Graessley. 1979. *J. Polym. Sci., Polym. Phys. Ed.* 17:1211.
29. Gehlsen M.D., P.A. Weimann, F.S. Bates, S. Harville, J.W. Mays, and G.D. Wignall. 1995. *J. Polym. Sci., Part B: Polym. Phys.* 33:1527.
30. Reichart G.C., R.A. Register, W.W. Graessley, R. Krishnamoorti, and D.J. Lohse. 1995. *Macromolecules* 28:8862.
31. Bates F.S., Chu C., Fredrickson G.H., and Hahn S.F. 2000. *PCT Int. Appl.* World.
32. Buckingham A.D. and H.G.E. Hentschel. 1980. *J. Polym. Sci., Part B: Polym. Phys.* 18:853.

33. Bates F.S., L.J. Fetters, and G.D. Wignall. 1988. *Macromolecules* 21:1086.
34. Bartell L.S. and R.R. Roskos. 1966. *J. Chem. Phys.* 44:457.
35. Bell R.P. 1942. *Trans. Faraday Soc.* 38:422.
36. Bates F.S. and G.D. Wignall. 1986. *Macromolecules* 19:932.
37. Bates F.S., G.D. Wignall, and W.C. Koehler. 1985. *Phys. Rev. Lett.* 55:2425.
38. Bates F.S. and G.D. Wignall. 1986. *Phys. Rev. Lett.* 57:1429.
39. Bates F.S., G.D. Wignall, and S.B. Dierker. 1986. *Macromolecules* 19:1938.
40. Gehlsen M.D., J.H. Rosedale, F.S. Bates, G.D. Wignall, L. Hansen, and K. Almdal. 1992. *Phys. Rev. Lett.* 68:2452.
41. Londono J.D., A.H. Narten, G.D. Wignall, K.G. Honnell, E.T. Hsieh, T.W. Johnson, and F.S. Bates. 1994. *Macromolecules* 27:2864.
42. Yang H., R.S. Stein, C.C. Han, B.J. Bauer, and E.J. Kramer. 1986. *Polym. Comm.* 27:132.
43. Schwahn D., K. Hahn, J. Streib, and T. Springer. 1989. *J. Chem. Phys.* 93:8383.
44. Krishnamoorti R., W.W. Graessley, N.P. Balsara, and D.J. Lohse. 1994. *J. Chem. Phys.* 100:3894.
45. Balsara N.P., A.A. Lefebvre, J.H. Lee, C.C. Lin, and B. Hammouda. 1998. *AIChE J.* 44:2515.
46. Lohse D.J. 1986. *Polym. Eng. Sci.* 26:1500.
47. Nicholson, J.C., T.M. Finerman, and B. Crist. 1990. *Polymer* 31:2287.
48. Scheffold F., E. Eiser, A. Budkowski, U. Steiner, J. Klein, and L. J. Fetters. 1996. *J. Chem. Phys.* 104:8786.
49. Klein J., T. Kerle, F. Zink, and E. Eiser. 2000. *Macromolecules* 33:1298.
50. Balsara N.P., L.J. Fetters, N. Hadjichristidis, D.J. Lohse, C.C. Han, W.W. Graessley, and R. Krishnamoorti. 1992. *Macromolecules* 25:6137.
51. Rhee J. and B. Crist. 1993. *J. Chem. Phys.* 98:4174.
52. Graessley W.W., R. Krishnamoorti, N.P. Balsara, L.J. Fetters, D.J. Lohse, D.N. Schulz, and J.A. Sissano. 1993. *Macromolecules* 26:1137.
53. Budkowski A., J. Klein, E. Eiser, U. Steiner, and L.J. Fetters. 1993. *Macromolecules* 26:3858.
54. Krishnamoorti R., W.W. Graessley, L.J. Fetters, R.T. Garner, and D.J. Lohse. 1995. *Macromolecules* 28:1252.
55. Krishnamoorti R. 1994. Doctoral dissertation, Princeton University.
56. Reichart G.C., W.W. Graessley, R.A. Register, R. Krishnamoorti, and D.J. Lohse. 1997. *Macromolecules* 30:3036.
57. Graessley W.W., R. Krishnamoorti, G.C. Reichart, N.P. Balsara, L.J. Fetters, and D.J. Lohse. 1995. *Macromolecules* 28:1260.
58. Lin J.-L. and R.-J. Roe. 1987. *Macromolecules* 20:2168.
59. Graessley W.W., R. Krishnamoorti, N.P. Balsara, L.J. Fetters, D.J. Lohse, D.N. Schulz, and J.A. Sissano. 1994. *Macromolecules* 27:2574.
60. Krishnamoorti R., W.W. Graessley, N.P. Balsara, and D.J. Lohse. 1994. *Macromolecules* 27:3073.
61. Reichart G.C., W.W. Graessley, R.A. Register, and D.J. Lohse. 1998. *Macromolecules* 31:7886.
62. Hammouda B., R.M. Briber, and B.J. Bauer. 1992. *Polymer* 33:1785.
63. Hammouda B. 1993. *Adv. Polym. Sci.* 106:87.
64. Reichart G.C., W.W. Graessley, R.A. Register, R. Krishnamoorti, and D.J. Lohse. 1997. *Macromolecules* 30:3363.

65. Bates F.S., M.F. Schulz, J.H. Rosedale, and K. Almdal. 1992. *Macromolecules* 25:5547.
66. Bates F.S. and G.H. Fredrickson. 1994. *Macromolecules* 27:1065.
67. Fredrickson G.H., A.J. Liu, and F.S. Bates. 1994. *Macromolecules* 27:2503.
68. Schweizer K.S. and C. Singh. 1995. *Macromolecules* 28:2063.
69. Zoller P., P. Bolli, V. Pahud, and H. Ackermann. 1976. *Rev. Sci. Instrum.* 47:948.
70. Maranas J.K., M. Mondello, G.S. Grest, S.K. Kumar, P.G. Debenedetti, and W.W. Graessley. 1998. *Macromolecules* 31:6991.
71. Indrakanti A., J.K. Maranas, and S.K. Kumar. 2000. *Macromolecules* 33:8865.
72. Jorgensen W.L., J.D. Madura, and C.J. Swenson. 1984. *J. Am. Chem. Soc.* 106:6638.
73. Graessley W.W., R. Krishnamoorti, N.P. Balsara, R.J. Butera, L.J. Fetters, D.J. Lohse, D.N. Schulz, and J.A. Sissano. 1994. *Macromolecules* 27:3896.
74. Lefebvre A.A., J.H. Lee, N.P. Balsara, and B. Hammouda. 2000. *Macromolecules* 33:7977.
75. Rabeony M., D.J. Lohse, R.T. Garner, S.J. Han, W.W. Graessley, and K.B. Migler. 1998. *Macromolecules* 31:6511.
76. Han S.J., D.J. Lohse, P.D. Condo, and L.H. Sperling. 1999. *J. Polym. Sci., Pt. B, Polym. Phys.* 37:2835.
77. Crist B. 1998. *Macromolecules* 31:5853.
78. Melenkevitz J., B. Crist, and S.K. Kumar. 2000. *Macromolecules* 33:6869.
79. Singh C. and K.S. Schweizer. 1997. *Macromolecules* 30:1490.
80. Maranas J.K., S.K. Kumar, P.G. Debenedetti, W.W. Graessley, M. Mondello, and G.S. Grest. 1998. *Macromolecules* 31:6998.
81. Londono J.D., A. Habenschuss, J.G. Curro, and J.J. Rajasekaran. 1996. *J. Polym. Sci., Part B: Polym. Phys.* 34:3055.
82. Londono J.D., J.K. Maranas, M. Mondello, A. Habenschuss, G.S. Grest, P.G. Debenedetti, W.W. Graessley, and S.K. Kumar. 1998. *J. Polym. Sci., Part B: Polym. Phys.* 36:3001.
83. Yi Y.X. and P. Zoller. 1993. *J. Polym. Sci., Part B: Polym. Phys.* 31:779.
84. Matsen M.W. and F.S. Bates. 1997. *J. Polym. Sci. B: Polym. Phys.* 35:945.
85. Adams J.L., D.J. Quiram, W.W. Graessley, R.A. Register, and G.R. Marchand. 1998. *Macromolecules* 31:201.
86. Lin J.C. 1997. Internal report, Princeton University.
87. Cochran E.W. and F.S. Bates. 2002. *Macromolecules* 35:7368.
88. Rounds N.A. 1970. Doctoral dissertation, Akron University.
89. Bates F.S., J.H. Rosedale, P. Stepanek, T.P. Lodge, P. Wiltzius, G.H. Fredrickson, and R.P. Hjelm, Jr. 1990. *Phys. Rev. Lett.* 65:1893.
90. Maurer W.W., F.S. Bates, T.P. Lodge, K. Almdal, K. Mortensen, and G.H. Fredrickson. 1998. *J. Chem. Phys.* 108:2989.
91. Sakurai K., W.J. MacKnight, D.J. Lohse, D.N. Schulz, J.A. Sissano, W. Wedler, and H.H. Winter. 1996. *Polymer* 37:5159.
92. Sakurai K., W.J. MacKnight, D.J. Lohse, D.N. Schulz, J.A. Sissano, J.S. Lin, and M. Agamalyan. 1996. *Polymer* 37:4443.
93. Akiyama S. and Kawahara S. 1996. In *Polymeric Materials Encyclopedia.* Vol. 1, Salamone J. C. (ed.). Boca Raton, FL: CRC Press, p. 699.
94. Roland C.M. 1987. *Macromolecules* 20:2557.
95. Trask C.A. and C.M. Roland. 1989. *Macromolecules* 22:256.
96. Tomlins D.W. and C.M. Roland. 1992. *Macromolecules* 25:2994.

97. Kawahara S., S. Akiyama, and A. Ueda. 1989. *Polym. J.* 21:221.
98. Sakurai S., H. Jinnai, H. Hasegawa, T. Hashimoto, and C.C. Han. 1991. *Macromolecules* 24:4839.
99. Hasegawa H., S. Sakurai, M. Takenaka, T. Hashimoto, and C.C. Han. 1991. *Macromolecules* 24:1813.
100. Thudium R.N. and C.C. Han. 1996. *Macromolecules* 29:2143.
101. Sakurai S., H. Hasegawa, T. Hashimoto, I.G. Hargis, S.L. Aggarwal, and C.C. Han. 1990. *Macromolecules* 23:451.
102. Jinnai H., H. Hasegawa, T. Hashimoto, and C.C. Han. 1992. *Macromolecules* 25:6078.
103. Schwahn D. and L. Willner. 2002. *Macromolecules* 35:239.
104. Krishnamoorti R., W.W. Graessley, L.J. Fetters, R.T. Garner, and D.J. Lohse. 1998. *Macromolecules* 31:2312.
105. Orwoll R.A. 1996. In *Physical Properties of Polymers Handbook*, Mark J.E. (ed.). Woodbury, N.Y.: AIP Press, pp. 81–89.
106. Almdal K., M.A. Hillmyer, and F.S. Bates. 2002. *Macromolecules* 35:7685.
107. Balazs A.C., I.C. Sanchez, I.R. Epstein, F.E. Karasz, and W.J. MacKnight. 1985. *Macromolecules* 18:2188.
108. Balazs A.C., F.E. Karasz, W.J. MacKnight, H. Ueda, and I.C. Sanchez. 1985. *Macromolecules* 18:2784.
109. Van Hunsel J., A.C. Balazs, R. Koningsveld, and W.J. MacKnight. 1988. *Macromolecules* 21:1528.
110. Balazs A.C. and M.T. DeMeuse. 1989. *Macromolecules* 22:4260.
111. Galvin M.E. 1991. *Macromolecules* 24:6354.
112. Winey K.I., M.L. Berba, and M.E. Galvin. 1996. *Macromolecules* 29:2868.
113. Scott R.L. 1952. *J. Polym. Sci.* 9:423.
114. Krause S., A.L. Smith, and M. Duden. 1965. *J. Chem. Phys.* 43:2144.
115. Kambour R.P., J.T. Bendler, and R.C. Bopp. 1983. *Macromolecules* 16:1827.
116. ten Brinke G., F.E. Karasz, and W.J. MacKnight. 1983. *Macromolecules* 16:1827.
117. Paul D.R. and J.W. Barlow. 1984. *Polymer* 25:487.
118. Merfeld G.D. and Paul D.R. 2000. *Polymer Blends* (Vol. 1). Paul D.R. and Bucknall C.B. (eds.). New York: John Wiley & Sons, Inc., pp. 55–88.
119. Kambour R.P. and J.T. Bendler. 1986. *Macromolecules* 19:2679.
120. Koningsveld R. and W.J. MacKnight. 1989. *Makromol. Chem.* 190:419.
121. Strobl G.R., J.T. Bendler, R.P. Kambour, and A.R. Shultz. 1986. *Macromolecules* 19:2683.
122. Strobl G.R. and G. Urban. 1988. *Coll. Poly. Sci.* 266:398.
123. Bruder F., R. Brenn, B. Stühn, and G.R. Strobl. 1989. *Macromolecules* 22:4434.
124. Bruder F. and R. Brenn. 1991. *Macromolecules* 24:5552.
125. Gorga R.E., E.L. Jablonski, P. Thiyagarajan, S. Seifert, and B. Narasimhan. 2002. *J. Polym. Sci.: Pt. B: Polym. Phys.* 40:255..
126. Colby R.H. 1989. *Polymer* 30:1275.
127. Ito H., T.P. Russell, and G.D. Wignall. 1987. *Macromolecules* 20:2213.
128. Composto R.J., E.J. Kramer, and D.M. White. 1988. *Macromolecules* 21:2580.
129. Nishi T. and T.K. Kwai. 1975. *Polymer* 16:285.
130. Walsh D.J., G.T. Dee, J.L. Halary, J.M. Ubiche, M. Millequant, J. Lesec, and L. Monnerie. 1989. *Macromolecules* 22:3395.
131. Lu F.J., E. Benedetti, and S.L. Hsu. 1983. *Macromolecules* 16:525.
132. Garcia D. 1984. *J. Polym. Sci.: Polym. Phys. Ed.* 22:1773.

133. Mirau P.A. and F.A. Bovey. 1990. *Macromolecules* 23:4548.
134. Shoimi T., F. Hamada, T. Nasako, K. Yoneda, K. Imai, and A. Nakajima. 1990. *Macromolecules* 23:229.
135. Yang H., M. Shibayama, R.S. Stein, N. Shimizu, and T. Hashimoto. 1986. *Macromolecules* 19:1667.
136. Shibayama M., H. Yang, R.S. Stein, and C.C. Han. 1985. *Macromolecules* 18:2179.
137. Han C.C., B.J. Bauer, J.C. Clark, Y. Muroga, Y. Matsushita, M. Okada, Q. Tran-cong, T. Chang, and I.C. Sanchez. 1988. *Polymer* 29:2002.
138. Kwei T.K., T. Nishi, and R.F. Roberts. 1974. *Macromolecules* 7:667.
139. Dudowicz J. and K.F. Freed. 1992. *J. Chem. Phys.* 96:9147.
140. Bidkar U. and I.C. Sanchez. 1995. *Macromolecules* 28:3963.
141. Taylor-Maranas J.K., P.G. Debenedetti, W.W. Graessley, and S.K. Kumar. 1997. *Macromolecules* 30:6943.
142. Bates F.S., M. Muthukumar, G.D. Wignall, and L.J. Fetters. 1988. *J. Chem. Phys.* 89:535.
143. Maranas J.K. 2002. Private communication.

CHAPTER 9

Network Structure and Elasticity

This chapter begins with a discussion of network structure as inferred from considerations based on the equal reactivity principle and the random linking assumption. The structure of networks formed by linking in the undiluted state or a concentrated polymer solution is emphasized. Two varieties are considered: crosslinked networks and endlinked networks. One issue is network connectivity, the two extremes being treelike with few local loops and latticelike with many local loops. The dimensions of network strands is another issue, considering whether network formation changes the equilibrium coil sizes. Methods are then described for calibrating and testing the network forming chemistry, primarily from molecular weight changes in the pre-gelation region and gel fraction changes in the post-gelation region. The estimation of elasticity-related network properties from random linking assumptions and linking chemistry calibrations—elastically active strand and junction concentrations, entanglement trapping factors—is then explained. The small-deformation elastic modulus according three network models based on Gaussian strand assumptions—affine, phantom, and entangled—are then considered, and tests of their validity based on small-deformation modulus measurements are described.

The elastic properties of polymer networks have long been the subject of both practical and scientific interest[1,2]. The earliest studies were made with natural rubber, a polyisoprene of highly regular microstructure (>95% cis 1,4 head-to-tail placement), obtained from the sap of *Hevea braziliensis* trees. Discovered in antiquity and used first in a game, the waterproof coatings, electrical insulation, and pneumatic tires made of Hevea rubber are still important articles of commerce. Much more pliable than other materials known in earlier times, rubber has a unique capacity to withstand stretching to several times its original length, then to spring back spontaneously when released.

In the fullness of time, other less obvious features of rubber elasticity were also noted. In 1805, Gough recorded two observations about the *thermoelastic* characteristics of natural rubber[3]:

- Rubber emits heat upon stretching and absorbs heat upon retraction.
- Rubber stretched by a constant force shortens when heated and extends when cooled.

Gough's discoveries, confirmed some 50 years later by Joule[4], became known as the Gough-Joule effects. The benefits of *vulcanization*—curing or crosslinking in modern terminology—had also been discovered by the mid-1800s, and the first quantitative study of thermoelasticity was conducted by Joule with vulcanized natural rubber. Eventually, other natural products, and later a multitude of synthetic substances, were found to display properties similar to those of natural rubber, including the Gough-Joule effects. An increased understanding of molecular structure brought the knowledge that rubberlike elasticity is a general phenomenon, originating from the action of flexible-chain molecules per se, and that vulcanization can be brought about any chemical reaction connecting those molecules into a system-spanning molecular network.

Crosslinking is used in the manufacture of rubber tires and many other elastomeric items of commerce[5]. Long chain polymers with moderately low glass transition temperatures, $T_g \sim -60°C$, are the basic ingredients needed for classification as *elastomers*. Covalent bonds between molecules are commonly introduced by chemical reactions at elevated temperature between unsaturated polymeric units (typically olefinic double bonds) and peroxides, elemental sulfur, and other substances by a diversity of chemistries. Still other substances are frequently present—carbon black, silica, diluents, antioxidants—not to influence crosslinking, but to improve processing behavior, strength and endurance, and to lower costs. Polybutadienes and polyisoprenes of various chemical microstructures are widely used in elastomeric formulations. The concentration of polymeric double bonds is somewhat reduced in the statistical styrene–butadiene copolymers used to make SBR rubber. Double bonds originate from the diene units only. Double bonds are still less plentiful in the isobutylene–isoprene copolymers used to make butyl rubber, and in the ethylene–propylene–diene terpolymers used to make EPDM rubber. Although only about 1% of the mers contain double bonds in these cases, enough crosslinking takes place to form effective networks if the starting molecules—the primary chains—are long enough.

Polymeric double bonds are not a necessity for network formation. Peroxides or other free-radical sources, including high-energy radiation, can crosslink fully saturated species such as polyethylene, polydimethyl siloxane, and polystyrene[6,7]. *Multilinking*, in which more than two chains are joined by the same linking event, can also occur[8]. Crosslinks can also be introduced as part of the polymerization process. Styrene–divinylbenzene copolymerizations are used to synthesize ion-exchange resins and the separation media for size-exclusion chromatography[9]. Polyurethane foams and polyester networks in fiberglass composites are other examples. *Endlinking,* the joining of functional groups on polymeric chain ends by multifunctional linking agents, is also widely used to form networks[10–12].

This chapter and Chapter 10 are devoted to the molecular aspects of equilibrium elasticity in *rubberlike networks,* also called *elastomeric networks*. Networks of this type are commonly but not exclusively those made in the undiluted state by chemically linking a small fraction of the monomeric units on long chains. The discussion is focused entirely on unfilled network systems and on the fundamental understanding of elastomeric network properties.

This chapter treats the formation and architectural features of such networks—the growth in molecular size and complexity as links are added; the *gel point,* at which an indefinitely proliferated network first appears; and the post-gelation region, in which the structures that govern elasticity are formed. It also describes the molecular origins of elasticity in networks of Gaussian strands—the classical and entanglement models of elasticity—emphasizing the factors that govern the initial or small-deformation modulus. Chapter 10 deals more generally with deformation properties, such as stress-strain behavior and the effects of diluents and temperature, including the Gough-Joule effects. It also deals with the microscopic aspects of networks, primarily as revealed by scattering methods, and with theories of behavior arising from the entanglement interaction.

Certain fundamental aspects—finite strand length effects, contributions of stress-induced crystallization—are omitted, in keeping with a primary interest throughout the book in the properties of random coil systems. The continuum formulation of rubberlike elasticity, the methods for measuring mechanical properties, and the viscoelastic properties of elastomeric networks are treated in Volume 2.

9.1 Equal Reactivity and Random Linking

Flory and Stockmayer laid the scientific foundation for the relationship between network structure and reaction mechanisms in the early 1940s, admirably summarized in Flory's 1953 book[13]. Random crosslinking is considered first. Long polymer chains at high concentrations ($\phi > \phi^{\ddagger} \sim 0.1$), the *primary chains* before links are added, have random coil conformations and low self-concentrations (Section 7.1.2). Some pairs of neighboring monomeric units belong to the same primary chain, either because they are next to one another along the chain, or because the chain has looped back to place them in a position to couple at the moment of link formation[14]. Links that connect such units produce *local rings* and are sterile: They do not extend the network structure. The fraction of such intramolecular links is governed by the same species-specific interactions that determine the statistical segment length and the intramolecular part of the pair distribution function (Chapters 4 and 8). The intra-linked fraction is typically small and independent of chain length for long chains. Its main effect is a harmless scaling down of all apparent intermolecular reaction rates by the same factor. Thus, for a collection of long *primary chains* at high concentrations, *random intermolecular linking* is reasonable to assume if, from the equal-reactivity standpoint[13], all neighboring pairs have the same intrinsic likelihood of link participation.

Departures from the predictions of random linking can come about for many reasons. Obviously, the randomness assumption fails if for any reason the groups involved have different reactivities. Gelation may still occur, but its quantitative description now requires more information and a more elaborate analysis[15]. The assumption that local ring formation has negligible effect may also be violated. Ring formation in the pre-gel region is always significant for thermosetting systems, such as the epoxy resins, in which the starting materials are small molecules[16]. Such systems also pass through the semidilute region in the early stages of reaction, making percolation theory[17] more appropriate for analyzing their structural evolution[18].

The dilution of the chains in crosslinking systems also promotes the formation of rings and, when carried far enough, can reduce interchain linking sufficiently to suppress macroscopic network formation altogether. High dilution ($\phi < \phi^*$) is used purposely in the synthesis of microgels[19,20]. Ring formation can also be important in endlinked networks, where the reactive groups themselves are dilute. Various other instances of nonrandom linking and dilution-related ring formation are discussed in the literature[12,21,22]. Another variety of network ring is considered in the following sections, the large-scale macrocycles that are inevitably formed in networks and play a crucial role in their elasticity.

The first effect of intermolecular linking is to create long-chain branching and increase molecular weight. The properties of the reacting system itself then undergo a qualitative change when an average of about one crosslink per primary chain has been added. What had been a liquid, a material that has no preferred shape, becomes a solid. A solid material is elastic: It has a *rest state,* a shape taken up spontaneously when no external forces act upon it. The network formed from an equilibrated body of polymeric liquid has, as its rest state, the shape of the body when the crosslinks were introduced. Microscopically, the transformation from liquid to solid occurs when a connected molecular structure that extends across and penetrates throughout the entire body is formed. The structure evolves further as still more crosslinks are added. Some crosslinks simply join previously separate molecules to the structure. Others connect the structure to itself, forming the macrocycles that define a three-dimensional template of its as-formed rest state.

Randomness in the pairing of units for linking to form networks is assumed. Randomness seems a plausible assumption for many network-forming systems but is never certain. The possibility of unanticipated differences in reactivity and of competing reactions such as *chain scission* must always be considered and accounted for on a case-by-case basis. The need to control reactant concentrations precisely, avoiding or compensating for side reactions or differences in reactivity, is particularly important for endlinked networks. No practical way exists to take a network apart and confirm whether the course of the reaction went as expected—to do a *post-mortem* on the network, so to speak. Serious investigators must assume the worst and perform as many cross-checks as possible. As described below, a determination of extractables, the *sol fraction,* exemplifies such tests.

Because networks do not form solutions, the usual methods for determining or confirming macromolecular structure, described in Chapter 5, cannot be directly applied. Although sometimes claimed otherwise, the concentrations of reacted groups are too low even for detection in most cases, much less for useful quantification. Such elementary questions as "What concentration of crosslinks has been achieved?" and "What fraction are elastically effective?" cannot be answered with assurance by even the most advanced spectroscopic methods. Understanding network properties in terms of molecular structure thus places a heavy burden on inference from theory, from simulation[23–25], and from peripheral measurements of various kinds and reliabilities. Prudence accordingly dictates more than the usual circumspection in interpreting the data offered and the supporting claims about network structure.

9.1.1 Network Connectivity

Consider the long-range structural connections that are established in a concentrated system of primary chains when links are introduced between randomly selected pairs of neighboring mers. The purpose is to estimate the size of the macrocycles that govern network elasticity, beginning from the introduction to random crosslinking in Section 1.4.2. All mers have the same reactivity, and α is the crosslink density, the fraction of mers that participate in crosslinks. Monodisperse primary chains, $(P_n)_o = r_o$, are considered for simplicity, and the volume that each occupies is $r_o \upsilon_m$, with υ_m being the volume per mer. The chains are long ($r_o \gg 1$), and α is small ($\alpha \ll 1$). Accordingly, all primary chains retain approximately the same reactivity throughout the linking process (saturation of reactivity has negligible effect), whereas the average number of crosslinked units per primary chain (the crosslink index $\gamma = \alpha r_o$) may range up to 10 or more without compromising the randomness assumption.

A general definition for the self-concentration of any volume-pervading object, or the ratio of occupied volume to pervaded volume, is given by Eq. 6.11:

$$\phi_{self} = \frac{V_{occ}}{V_{per}} \tag{9.1}$$

For the primary chains,

$$\phi_{self} = \frac{r_o \upsilon_m}{\upsilon_{per}} \tag{9.2}$$

With Eq. 6.7 defining the pervaded volume,

$$\phi_{self} = \frac{3V_p}{4\pi R_g^3 N_a} \tag{9.3}$$

where $V_p = M/\rho$ is the molar volume of the primary chains. At the high concentrations of polymer being considered, the chain dimensions are unperturbed: $R_g = (R_g)_o$. For typical values—$r_o = 10^4$, $v_m = 0.1\,nm^3$, $(R_g)_o = 20\,nm$—the primary chain self-concentration is about 0.03.

The crosslink index $\gamma = 1$ corresponds to the gel point for monodisperse primary chains [see Eq. 1.22]. The local rings from crosslinks that join mers on the same primary chain are ignored. Other rings can be formed, however, as a result of redundant links between primary chains. Thus, two-segment rings are formed by links between already connected pairs of chains, three-segment rings by links between already connected triplets, and so on through a hierarchy of macrocyclic structures. These macrocycles provide closed paths of covalent bonds and form the network circuits that pass through and around the various parts of the network. The average length of such circuits—the average number of chains participating in them—is a structural parameter that characterizes the *network connectivity*. It varies from one network to another and can affect their elastic properties. An estimate of this number can be obtained.

Consider first the connections that lead away from some randomly selected primary chain after crosslinking has taken place. With loop formation ignored for the moment, the chain is connected to about γ other primary chains, and each of those is linked to about γ others, and so on through successive generations of connections. The average number of primary chains in the *ith* generation, counting from any starting chain, is γ^i. The number in the entire sequence of generations through the *ith* is thus $1 + \gamma + \gamma^2 + \cdots + \gamma^i$, or $(\gamma^{i+1} - 1)/(\gamma - 1)$. Thus, the occupied volume of this collection of primary chains grows exponentially with i:

$$(v_{occ})_i = \frac{\gamma^{i+1} - 1}{\gamma - 1} r_o v_m \tag{9.4}$$

The volume pervaded by the collection also increases with increasing i, but more slowly than the occupied volume. Thus, a link joining two chains whose centers of mass are separated by some position vector $\mathbf{s}'$ is a step in distance $|\mathbf{s}'| = (\mathbf{s}' \cdot \mathbf{s}')^{1/2}$ between successive generations. The connected sequence through the generations is an i-step random walk, with step length equal to the average, $\langle \mathbf{s}' \cdot \mathbf{s}' \rangle^{1/2}$, and the average space swept out by such walks is a measure of the pervaded volume of this connected array of primary chains:

$$(v_{per})_i = \frac{4\pi}{3} (\langle \mathbf{s}' \cdot \mathbf{s}' \rangle i)^{3/2} \tag{9.5}$$

To estimate $\langle \mathbf{s}' \cdot \mathbf{s}' \rangle$, consider two primary chains separated by distance $\mathbf{s}'$, and approximate their segment density distributions using Gaussian functions:

$$\begin{aligned} \rho_1 &= A \exp\left[-\beta \mathbf{s} \cdot \mathbf{s}\right] \\ \rho_2 &= A \exp\left[-\beta \left(\mathbf{s} - \mathbf{s}'\right) \cdot \left(\mathbf{s} - \mathbf{s}'\right)\right] \end{aligned} \tag{9.6}$$

To become linked, the two chains must have mers at the same location, so the probability of linking is proportional to $\Gamma(\mathbf{s}') = \int \rho_1(\mathbf{s})\rho_2(\mathbf{s} - \mathbf{s}')d\mathbf{s}$. The fraction of crosslinks that join chains separated by $\mathbf{s}'$ in $d\mathbf{s}'$ is therefore $f(\mathbf{s}') = \Gamma(\mathbf{s}')d\mathbf{s}' / \int \Gamma(\mathbf{s}')d\mathbf{s}'$. Accordingly, the mean square distance between linked chains is $\int \mathbf{s}' \cdot \mathbf{s}' f(\mathbf{s}')d\mathbf{s}'$. The result is:

$$\langle \mathbf{s}' \cdot \mathbf{s}' \rangle = 2(R_g^2)_o \tag{9.7}$$

Then, from Eq. 9.5, the average volume pervaded by an *ith* generation array is a power law in i:

$$(\upsilon_i)_{per} = (2i)^{3/2}\, \upsilon_{per} \tag{9.8}$$

Finally, using Eqs. 9.1, 9.4, and 9.8, the self-concentration of an *ith* generation array is

$$(\phi_{self})_i = \frac{(\gamma^{i+1} - 1)}{(\gamma - 1)(2i)^{3/2}} (\phi_{self})_o \tag{9.9}$$

where $(\phi_{self})_o$ is now the self-concentration of the primary chains.

Although the self-concentration of primary chains is small, the multiplying factor for $\gamma > 1$ grows rapidly with i. At some generation $i = N$, whose value depends on γ and $(\phi_{self})_o$, the self-concentration of the array approaches unity. Higher generation crosslinks only serve to close loops. The onset of this condition is gradual, but the point is that loop closure dominates eventually. Thus, setting $(\phi_{self})_i = 1$ leads to an expression for the mean macrocycle size N, the average number of primary chains on the shortest paths around the various macrocycles:

$$\frac{\gamma^N - 1}{\gamma - 1} = \frac{(2N)^{3/2}}{(\phi_{self})_o} \tag{9.10}$$

Using the typical $(\phi_{self})_o \sim 0.03$, the calculated macrocycle sizes for $\gamma = 2, 4$, and 10 are about 12, 6, and 4 respectively. When both $\varepsilon = \gamma - 1$ and $(\phi_{self})_o$ are small,

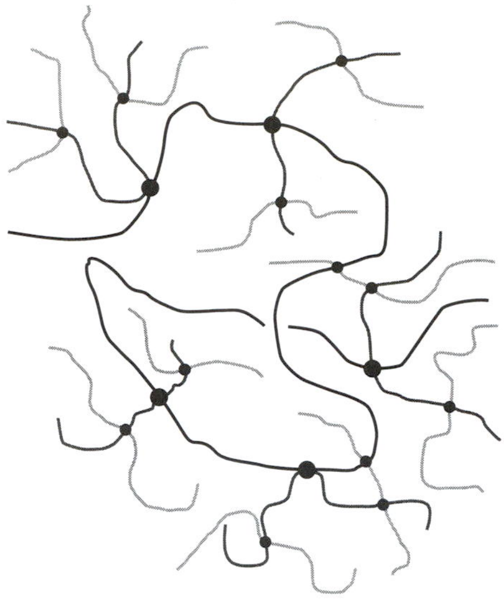

FIGURE 9.1 A representative primary chain and the first two generations of its successive links to others by random crosslinking. The gradation of gray indicates the connections hierarchy: black for the primary chain, mid-gray for directly linked chains, light gray for next connected chains.

Eq. 9.10 can be written:

$$N = \frac{(\phi_{self})_o^2}{8}\left[\frac{\exp(N\varepsilon) - 1}{N\varepsilon}\right]^2 \tag{9.11}$$

from which it is easily shown that N varies approximately as ε^{-1}. Accordingly, macrocycle circumference is infinite at the gel point and decreases with increasing primary chain self-concentration. The same considerations can be applied to endlinked networks; typical primary chains for that case have smaller sizes, $(R_g)_o \sim 3.5\,nm$, and larger self-concentrations, $(\phi_{self})_o \sim 0.1$. Macrocycle sizes from endlinking are likely smaller than those from crosslinking.

As depicted in Figures 9.1 and 9.2, mean loop size and network connectivity patterns are related properties. Figure 9.1 shows a representative primary chain in a melt of long chains that has been crosslinked well beyond the gel point ($\gamma \gg 1$). The representative chain is linked to several other primary chains, and those have links to still others, and so forth. Remoteness in connection with the representative chain is indicated in Figure 9.1 by lightness in the gray shade. From the discussion, relatively few macrocycle closures are expected through the first few generations of connected chains. The chain paths that lead from one link to other links, and from those to still others, should accordingly exhibit *locally treelike connectivity,* as shown in Figure 9.2. Remoteness in the connectivity of strands from the representative junction is again denoted by gray scale lightness. The other extreme from treelike connectivity—a local loop-dominated, *lattice connectivity*—is unrealistic for

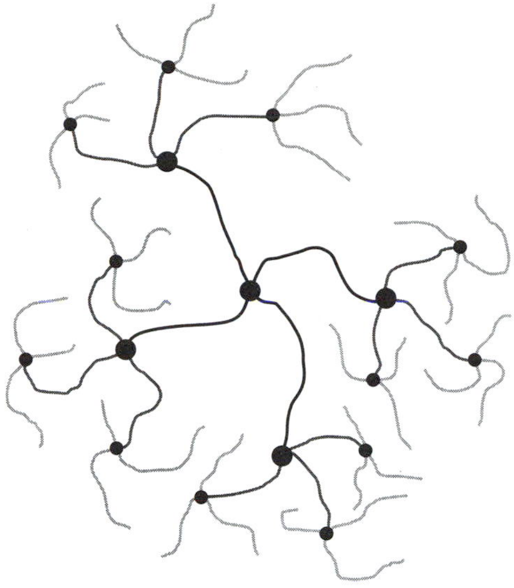

FIGURE 9.2 A representative junction and the first two generations of successive junctions for a network with treelike connectivity. The gradation of gray indicates the connections hierarchy: black for connected directly strands, mid-gray for once removed strands, light gray for twice removed strands.

randomly crosslinked concentrated systems of long chains. The lattice pattern is appropriate for thermosetting networks. It would also be a reasonable starting point for discussing the architecture of microgels or networks formed at high degrees of dilution.

One consequence of locally treelike connectivity is that two mers on different primary chains might be spatial neighbors, yet distant as measured along the shortest network path connecting them. Architecturally, the networks are arrays of locally treelike structures that fill space by overlapping extensively, but are connected only remotely through macrocycles containing segments of many primary chains. Another connectivity-related consideration applies mainly to endlinked networks, *reactant starvation* in the last stages of linking. Unreacted functional groups become increasingly isolated at high conversions. Their mean separations become large, while their mobilities become increasingly restricted owing to the tethering effects of reacted groups on their primary chains. The network is flexible, so unreacted functions can eventually find one another and react, but the result can be highly distorted strands and a heterogeneous pattern of network connectivity.

Computer simulations of network formation by crosslinking and endlinking[24–27] are beginning to provide valuable insights about network structure, including the frequency of local loop formation and functional group isolation. The results of Duering et al.[27] are especially noteworthy in confirming the random statistics of structural development during endlinking. The Gilra et al. study[25] clearly demonstrated the validity of the optimized network technique, compensating for side reaction effects on endlinked network formation (see Section 9.7.1).

9.1.2 Strand Dimensions

Ring formation creates a driving force for the contraction of the chain dimensions (Section 4.3.2). Thus, depending on the relative rates of link formation and size equilibration, the molecular self-concentrations may increase, at least in principle, and thereby accelerate ring closure. Our main interest here is elastomeric polymer networks, and these are generally formed under conditions that strongly promote intermolecular linking; that is, high concentrations and long primary chains. Thus, as discussed earlier, ring formation should be a relatively rare event in the pre-gelation region. Macrocycles are always profusely formed beyond the gel point, however, so a driving force for contraction is certainly present. However, Eq. 9.10 suggests that the network macrocycles are large, consisting of segments from many primary chains, until well beyond the gel point. From Eq. 4.57 and the related discussion, the contraction of individual segments at equilibrium would therefore be relatively slight. Macrocycle interconnection, when combined with the constraint that chains cannot cross through one another, would also tend to suppress independent contraction. Cooperative contraction of the entire network, at least for crosslinking in the melt, would be effectively ruled out by volume exclusion*.

Thus, on the various grounds already described, significant coil contraction in as-formed elastomeric networks, without syneresis or addition or removal of diluents, seems highly unlikely. That possibility was a major concern to many early investigators. It remained a lingering uncertainty until the application of small-angle neutron scattering (SANS) to networks in the mid-1970s[30]. The results of that work, summarized in 1982[31], left little doubt that the dimensions of chains in polymer networks in the as-formed state are indistinguishable from those of free chains in the same environment.

9.1.3 Linking Statistics

We now consider some further consequences of random crosslinking in a melt of primary chains in which each mer has the same linking probability α, and multiple crosslink participation by mers is negligible. The primary chain distribution is arbitrary because we deal mostly with subgroups of chains having the same number of mers. As before, the chains are long, $r \gg 1$; the crosslink density is small, $\alpha \ll 1$; crosslinking is random; and local loop formation is negligible.

---

*Networks formed in solution can undergo *syneresis*, a macroscopic contraction that "squeezes out" diluent[28,29]. That process is driven by macrocycle formation, generates a network pressure (see Section 9.5.3), and would certainly be accompanied by coil contraction. Coil dimensions must surely change when networks are swollen from their as-formed state by the absorption of diluents and when diluent is removed from networks formed in solution, but such naturally expected behavior is not the question here.

First consider $F_i(r, \alpha)$, the fraction of the r-mer primary chains that have exactly i crosslinked units when the fraction of mers participating in crosslinks is α. This is a classical probability problem[32], with α the outcome probability for success (crosslink participation) in an individual trial, and $F_i(r, \alpha)$ the probability of that outcome occurring in exactly i of r independent trials, regardless of order. The result is a *binomial distribution:*

$$F_i(r, \alpha) = \frac{r!}{(r-i)!\,i!}\alpha^i (1-\alpha)^{r-i} \tag{9.12}$$

The average number of crosslinked units per r-mer primary chain is thus:

$$\bar{n} = \sum_{i=0}^{r} i\,F_i(r, \alpha) = r\alpha \tag{9.13}$$

With $r \gg 1$ and $\alpha \ll 1$, the term $(1-\alpha)^r$ can be harmlessly replaced by $\exp(-\alpha r)$. For a sufficiently small linked mer fraction, $i \ll r$, the approximations $r!/(r-i)! = r^i$ and $(1-\alpha)^i = 1$ can be used. Eq. 9.12 simplifies to a *Poisson distribution:*

$$F_i(r, \alpha) = \frac{1}{i!}(\alpha r)^i \exp(-\alpha r) \tag{9.14}$$

Consider now some consequences of Eq. 9.14 for various values of $\bar{n} = r\alpha$. When $\bar{n} = 1$, the gel point for monodisperse primary chains, a significant fraction of them have no links at all: $F_0(r, \alpha) = e^{-1} = 0.368$, and about 25%, $1 - 2e^{-1}$, have two or more links. These fractions change at $\bar{n} = 2$, to about 0.135 for no links and about 0.59 for two links or more. Even at $\bar{n} = 4$, well beyond the gel point of a monodisperse system, nearly 2% of the primary chains still have no crosslinked units. For polydisperse primary chains, the average linking frequency per chain at fixed α is directly proportional to chain length. Thus, for example, the nonlinked fractions are larger for the shorter chains. In effect, random crosslinking of mers preferentially selects the longer chains for incorporation into the network.

The addition of crosslinks introduces *junctions* that partition the primary chains into *segments;* a linear primary chain with i crosslinked units participates in i junctions and consists of $i+1$ segments. As $\bar{n}$ becomes large, the segment length distribution becomes independent of the primary chain distribution and, as long as the condition $\alpha \ll 1$ is still met, tends toward the *exponential distribution:*

$$p(\alpha, s) = \alpha \exp(-\alpha s) \tag{9.15}$$

in which s is segment length, and $\bar{s} = 1/\alpha$. This distribution, confirmed for dense collections of chains by computer simulations[26], was derived in Section 1.4.2 for the case of chain length distribution in linear condensation polymers and for random chain scission. The principle used there applies here too: Namely, that the probability of an uninterrupted run of equally likely outcomes decreases exponentially with run length. In this case, the segment lengths vary in networks made by random crosslinking, although in a known and well-defined way.

These considerations of segment length distribution and chain scission are special features of crosslinked networks [see Eq. 1.24 and accompanying discussion]. In endlinking, chain scission is infrequent because linking conditions are mild. The segment length distribution in endlinked networks is simply the primary chain distribution, which, depending on the synthetic route, can be tailored to yield a variety of distributions and network properties[33].

9.2 Structurally Related Observables

Understanding the architectural features of polymer networks is necessary for several reasons. Mechanical properties such as the modulus of elasticity depend directly on crosslink density, but only a small fraction of monomeric units in elastomeric networks are reacted, typically less than 1%. As noted earlier, the linked unit concentrations are so low that it is seldom possible to directly measure crosslink density with sufficient accuracy using conventional analytical techniques. The use of indirect methods, such as applying random linking theory to infer crosslink density from observables, such as the gel point and gel curve, is typically the only alternative. The properties of crosslinked networks also depend on other architectural features, such as the amount and size of dangling network structures and topologically trapped chain entanglements. The pattern of local connectivity (Section 9.1.1), which is important in developing theories about properties, is also known only by inference. To quantify these features, which themselves are essentially unobservable, the theory of random linking offers virtually the only recourse. Simulation methods are growing in both power and scope and, wherever appropriate, the contributions they have made to current understanding are included.

These comments, made in the context of randomly crosslinked and multilinked networks, most certainly apply to endlinked networks as well. Endlinked networks offer, however, one significant advantage over crosslinked networks for fundamental studies. In a perfect endlinked network, the concentrations of network junctions and strands can be calculated from the molecular weights and functionalities of the starting materials alone. Perfection is unfortunately a hazardous assumption. Perfect endlinked networks are formed under conditions of exact stoichiometry, with preformed chains having functional groups on each end that link to completion with negligible side reactions. Independent verification or corrections based on such consistency tests as are available—pre-gelation molecular weight growth, gel point detection, post-gelation extractables content—is obviously essential.

9.2.1 Pre-Gelation Region

A brief introduction to the evolution of molecular weight distribution with random crosslinking was given in Section 1.4.2. The random linking of units on long primary chains with low self-concentrations, but having otherwise arbitrary architecture and distribution, are considered. Let $F(r)$ be the fraction of r-mers in the primary chains and α the fraction of mers that participate in crosslinks. The equation describing how the distribution changes with α, Eq. 1.20, was converted to a set of equations for the distribution moments, $Q_i(\alpha), i = 0, 1, 2, \ldots$, Eq. 1.21, which are solvable sequentially. Expressions are thus obtained for as many of the average polymerization indices as desired $P_n = Q_1/Q_o, \ldots$[17]:

$$\begin{aligned} P_n &= \frac{P_n^{\text{o}}}{1-\alpha P_n^{\text{o}}/2} \\ P_w &= \frac{P_w^{\text{o}}}{1-\alpha P_w^{\text{o}}} \\ P_z &= \frac{P_z^{\text{o}}}{(1-\alpha P_w^{\text{o}})^2} \\ P_{z+\xi} &\propto \frac{1}{(1-\alpha P_w^{\text{o}})^2} \qquad \xi = 1, 2, \ldots \end{aligned} \tag{9.16}$$

in which P_n^{o}, P_w^{o}, and the other averages are the primary chain values.

Several features of Eq. 9.16 are noteworthy. Regardless of the primary chain distribution, P_w and all higher averages diverge at the same crosslink density. The divergence marks the gel point location and coincides with the first appearance of a network structure. The crosslink density at the gel point depends only on the weight-average polymerization index of the primary chains:

$$\alpha_g = 1/P_w^{\text{o}} \tag{9.17}$$

The extent of crosslinking is sometimes expressed in reduced form as the crosslink index:

$$\gamma = \frac{\alpha}{\alpha_g} = \alpha P_w^{\text{o}} \tag{9.18}$$

The molecular weight distribution evidently broadens enormously on approaching the gel point: Even the dispersity ratios P_w/P_n and P_z/P_w diverge at the gel point. The change in the number-average polymerization index, however, is quite modest:

$$\frac{P_n^{\mathrm{g}} - P_n^{\mathrm{o}}}{P_n^{\mathrm{o}}} = \frac{1}{2\left(P_w^{\mathrm{o}}/P_n^{\mathrm{o}}\right) - 1} \tag{9.19}$$

Thus, $(P_n^{\mathrm{g}}/P_n^{\mathrm{o}}) < 2$ for all primary chain distributions.

Similar characteristics in the *pre-gel* region are found for endlinked systems. Consider, for example, a melt of long primary chains with a reactive group A on each end. This group then couples randomly to form A-B bonds with B groups on an f-functional nonpolymeric linking agent. For a system with equal numbers of A and B groups (matched stoichiometry), the fractional extent of reaction of either A or B functions is the conversion p, $0 < p < 1$, and the conversion at the gel point is[13]:

$$p_{\mathrm{g}} = \left(\frac{1}{f-1}\right)^{1/2} \tag{9.20}$$

The chain-length expressions for crosslinking, Eq. 9.16, also apply to endlinking systems, when α is replaced everywhere by the quantity $p/[p_{\mathrm{g}} P_w^{\mathrm{o}}]$.

The dependence of average chain length on crosslink density changes when chain scission is also taking place. The change in number-average is obtained directly by subtracting one molecule for each crosslink and adding one for each scission:

$$\frac{1}{P_n} = \frac{1}{P_n^{\mathrm{o}}} + \beta - \frac{\alpha}{2}$$

or (9.21)

$$\frac{M_n^{\mathrm{o}}}{M_n} = 1 - \left(\frac{\alpha}{2} - \beta\right)\frac{M_n^{\mathrm{o}}}{m_{\mathrm{o}}}$$

where m_{o} is the mer molecular weight and β is the scission probability, or the fraction of backbone bonds broken. The other averages change in a manner that depends on the primary distribution. For an exponential primary distribution, however, the form of the distribution does not change [see Eq. 1.24], and the following is easily shown:

$$\frac{1}{P_w} = \frac{1}{P_w^{\mathrm{o}}} + \frac{\beta}{2} - \alpha$$

or (9.22)

$$\frac{M_w^{\mathrm{o}}}{M_w} = 1 - \left(\alpha - \frac{\beta}{2}\right)\frac{M_w^{\mathrm{o}}}{m_{\mathrm{o}}}$$

Such equations are sometimes useful for calibrating the crosslinking reactions used to form networks. A series of samples is prepared with different doses of crosslinking agent—sulfur, peroxide, radiation—and molecular weights are determined for those in the pre-gelation range. The crosslink and chain scission densities are assumed directly proportional to the crosslinking agent dose:

$$\begin{aligned} \alpha &= k_X D \\ \beta &= k_S D \end{aligned} \tag{9.23}$$

The object is not only to determine the values of k_X and k_S, which permit extrapolations to estimate network structure parameters, but also to test consistency with the various assumptions used, such as randomness of linking and the unimportance of competing reactions.

Molecular weight data for polystyrene crosslinked by gamma radiation[34] are shown in Figures 9.3a and 9.3b. The primary chain distribution is approximately exponential for all the samples, so Eqs. 9.21 and 9.22 should apply. The data are plotted in reduced fashion, M_w^o/M_w versus $M_w^o D$ and similarly for M_n, so a single straight line should fit in each case. That is certainly true within the errors for the weight-average values: The number-average values hardly change at all over the same range. The results demonstrate consistency with the randomness assumption, and the M_w versus D data provide a chain-length independent value for $k_X - k_S/2$. The scatter in M_n versus D is too large to obtain $k_X/2 - k_S$ reliably. Those results appear also to have been affected by a coupling reaction at the chain ends for some of the samples[34]. In any case, the post-gelation behavior described next permits the separation of k_X and k_S, and the line drawn in Figure 9.3a was obtained with the resulting values.

9.2.2 The Gel Curve

Beyond the gel point, the structure consists of two physically distinct and separable portions, the *gel fraction* and the *sol fraction*. The gel is the network itself, and the *sol* is that part of the initial polymer that is not attached to the network. The sol and gel fractions are separable by solvent extraction. The extraction of the soluble portion leaves behind the network portion, swollen but not dissolved by the solvent. It is important to have a large excess of solvent, solvent changes, and the monitoring of the extract to assure completeness of solubles removal. The gel is a macroscopic object, as illustrated in Figure 9.4. It can be plucked from the excess solution, dried, and finally weighed to determine the gel fraction, or the fractional weight of the starting polymer that is in the network. Although the gel fraction can be determined relatively easily far beyond the gel point, networks near the gel point are fragile and care in manipulation is necessary.

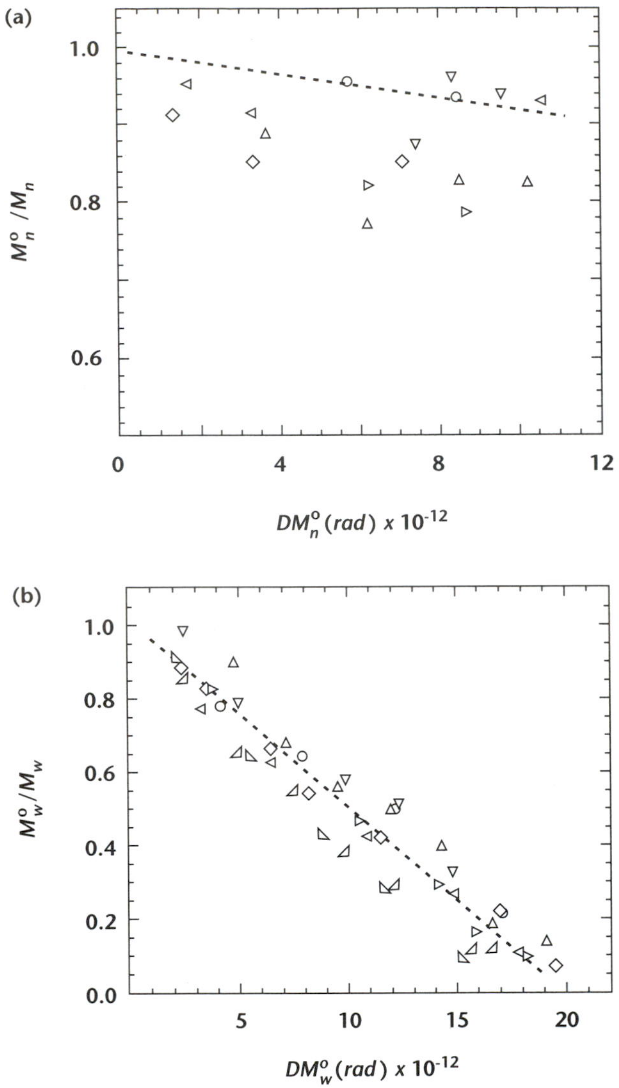

FIGURE 9.3 Molecular weight changes in the pre-gel region for polystyrene samples with nearly exponential primary molecular weight distributions when crosslinked with various doses of gamma radiation[34]: (a) Values of M_n obtained by membrane osmometry are plotted according to Eqs. 9.21 and 9.23. Six primary chain samples with $3.12 \times 10^4 \leq M_n^o \leq 4.56 \times 10^5$ are represented by the various symbols. The dashed line in the prediction from post-gel point data (Fig. 9.5). (b) Values of M_w from static light scattering are plotted according to Eqs. 9.22 and 9.23. Nine primary chain samples with $5.62 \times 10^4 \leq M_w^o \leq 3.37 \times 10^6$ are represented by the various symbols. The dashed line in the prediction from post-gel point data (Fig. 9.5).

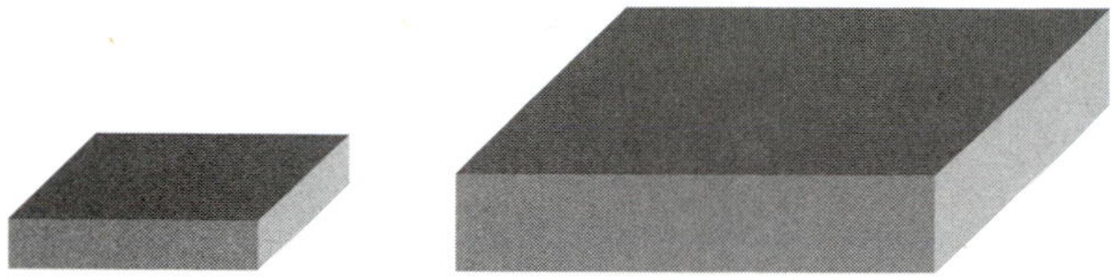

FIGURE 9.4 A network swollen with solvent in the course of sol extraction and determination of the equilibrium swell ratio.

The gel fraction w is zero up to the gel point. A swollen but nonsoluble fraction appears beyond the gel point and increases as more crosslinks are added. If link formation is the only reaction taking place, the gel fraction eventually approaches unity with increasing linking. The sol fraction $s = 1 - w$ correspondingly decreases with linking density and finally approaches zero. If scission accompanies crosslinking, the sol fraction levels off at some non-zero value. For random linking, the gel fraction is related directly to the crosslink density and the primary chain distribution, a very useful feature for calibration purposes, as discussed next.

Consider as before a population of long primary chains with arbitrary distribution and architecture undergoing random crosslinking, but now in the post-gelation region. A simplified probability argument is used here to relate the initial weight distribution W_r and the gel fraction $w(\alpha)$. It applies for the long chain case of interest here; a more elaborate calculation is required otherwise[35]. The probability that a mer selected at random does not participate in a crosslink is $1 - \alpha$. The probability that it does participate, but that the crosslink does not connect it to the gel, is the product $\alpha(1 - w)$. The sum, $1 - \alpha + \alpha(1 - w) = 1 - \alpha w$ (not crosslinked or crosslinked but not to the gel), is therefore the probability that a randomly selected mer in the system is not directly linked to the gel.

Now apply that result to an entire chain. The probability that no monomeric unit in a randomly selected r-mer primary chain is linked to the gel is $(1 - \alpha w)^r$. The weight fraction of sol in the system is the sum of weight fractions of primary chains that have no links to the gel:

$$1 - w = \sum_{r=1}^{\infty} W_r \, (1 - \alpha w)^r \tag{9.24}$$

Because $\alpha \ll 1, r \gg 1$ in nearly all cases, and w is never larger than one, the approximation $(1 - \alpha w)^r = \exp(-\alpha w r)$ can be used. Replacing summation by integration and α by γ / P_w° [from Eq. 9.18] leads to the following implicit relationship between primary chain distribution and the gel curve:

$$1 - w = \int_{0}^{\infty} W(r)\exp[-wr\gamma/P_w^{o}]\,dr \qquad (9.25)$$

From Eq. 9.25, the gel curve formula for monodisperse primary chains is:

$$1 - w = \exp(-\gamma w) \qquad (9.26)$$

or:

$$\gamma = -\frac{\ln s}{1 - s} \qquad (9.27)$$

For primary chains with an exponential distribution [Eq. 1.17], Eq. 9.25 leads to:

$$1 - w = \frac{1}{(1 + \gamma w/2)^2} \qquad (9.28)$$

or, with some rearrangement, to:

$$s + s^{1/2} = 2/\gamma \qquad (9.29)$$

Equation (9.29) forms the basis of the *Charlesby-Pinner plot*[36], a widely use procedure for analyzing gel curve data. For exponential primary distributions, it can accommodate random scission as well as crosslinking. Thus, using Eqs. 9.18 and 9.22, Eq. 9.29 becomes:

$$s + s^{1/2} = \beta/\alpha + 2/\alpha P_w^{o}$$

or: (9.30)

$$s + s^{1/2} = \beta/\alpha + 2m_o/\alpha M_w^{o}$$

Charlesby-Pinner plots of gelation data for polystyrene samples with exponential distributions crosslinked by gamma radiation[34] are shown in Figure 9.5. As in Figure 9.3, reduced variables are used, and the abscissa is $1/DM_w^{o}$ in this case. Accordingly, if Eq. 9.23 is valid regardless of chain length, a single straight line should fit all the data. That is evidently true for the data in Figure 9.5, the slope and intercept provide values for the proportionality constants $k_X = 6.37 \times 10^{-12}\ rad^{-1}$ and $k_S/k_X = 0.35$.

Other methods for calibration, such as determining the gel points for a series of samples with widely different molecular weights, are also used. Flory describes

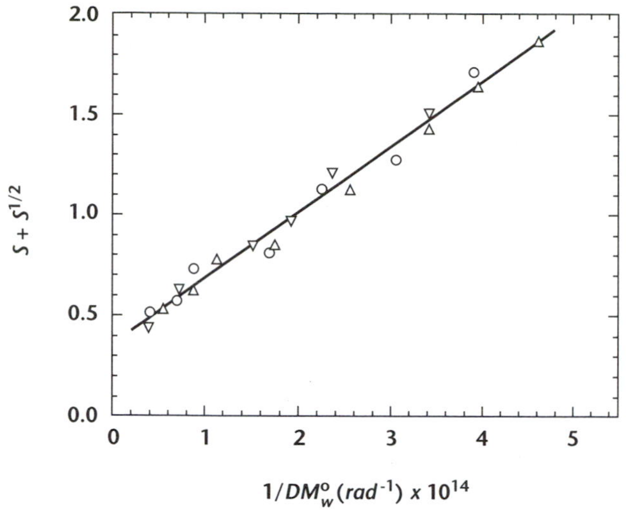

FIGURE 9.5 Sol fractions in the post-gel region for three polystyrene samples with exponential primary molecular weight distributions when crosslinked with various doses of gamma radiation[34]. Values of s obtained by solvent extraction plotted in the Charlesby-Pinner manner [Eqs. 9.23 and (9.29)] for $M_w^o = 9.35 \times 10^5 (\nabla)$, $1.82 \times 10^6 (\Delta)$, and $3.37 \times 10^6 (\circ)$.

other examples of crosslink density calibration[13]. Procedures for predicting gel curves from molecular weight distribution are available for both crosslinked and multilinked networks with chain scission, even without restriction to long primary chains[35]. Gel curves can also be predicted by the Miller-Macosko method for many types of stoichiometric and nonstoichiometric endlinked and crosslinked networks[11,37].

9.3 Elasticity-Related Properties

The equilibrium stress-strain behavior of polymer networks depends on a number of properties of the network architecture. The concentration of *elastically active network junctions* and the concentration of *elastically active network strands* depend on architecture in the same ways for all polymer species. Equilibrium elasticity also depends on the contributions of *trapped entanglements,* originating from the mutual "uncrossability" of chain backbones and strongly dependent on polymer species as well as architecture. The viscoelastic behavior of networks depends on all the structural features that govern equilibrium elasticity as well as on the size, architecture, and entanglement of the *dangling structures*[38]. All these various network features are unobservable. Testing theories of elasticity, however, requires that independent estimates be made. The only recourse is inference from observables—gelation-related properties with predictions arising from the randomness assumption.

9.3.1 Active Junctions and Strands

Consider the junctions introduced by crosslinking a melt of long primary chains. Each crosslink is a junction and has four chain paths that emanate from it. Each junction thus belongs to one of five structural categories, depending on the other connections along these four paths. A chain path leads to the network if its mers remain part of the gel, even if its connection to the junction being considered were to be severed. Let P_i be the probability that exactly i paths ($i = 0, 1, 2, 3, 4$) leading from a junction are independently connected to the network. The possible combinations are shown schematically in Table 9.1.

The elastic activity of each junction is determined by a criterion proposed in 1960 by Scanlan[39] and contemporaneously by Case[40]. Following upon and refining extensive earlier work of others[41], Scanlan and Case suggested that a junction contributes to the shape memory of a network, and thus to its equilibrium elasticity, if it anchors three or more independent chain paths to the network. Accordingly, the number density of elastically active junctions in randomly crosslinked systems is:

$$\mu = \nu_c(P_3 + P_4) \tag{9.31}$$

where ν_c is the number density of crosslinks in the system. Scanlan and Case also developed a criterion for the elastic activity of strands: Each elastically active strand is a chain path that connects two elastically active junctions. For crosslinking, the number of elastically active strands is thus one half the number of elastically active chain ends emanating from junctions. Hence,

$$\nu = \frac{\nu_c}{2}(3P_3 + 4P_4) \tag{9.32}$$

Expanded to multilinked networks generally, and including also endlinked networks,

$$\begin{aligned} \mu &= \nu_c \sum_{i=3}^{\infty} P_i \\ \nu &= \frac{\nu_c}{2} \sum_{i=3}^{\infty} i\, P_i \end{aligned} \tag{9.33}$$

Now consider the network in relation to its crosslinked mers. Each such mer belongs to one of the primary chains, and two paths along its chain lead away from the mer. With the randomness assumption, the P_i can then be expressed in terms of

TABLE 9.1 Post-gelation crosslink categories

P_i	Representative Junction	Equilibrium Elastic Activity
P_0	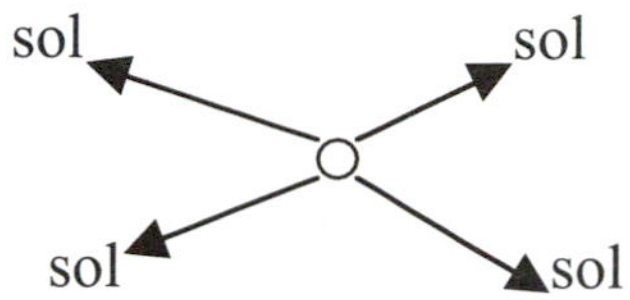	Sol, **inactive**
P_1	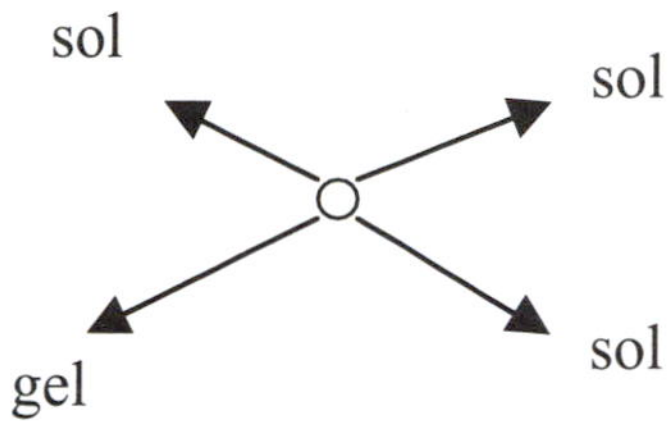	Dangling structure, **inactive**
P_2	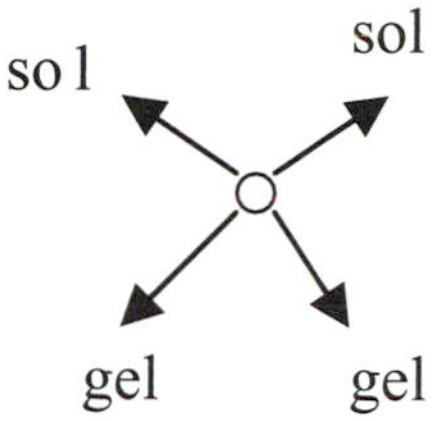 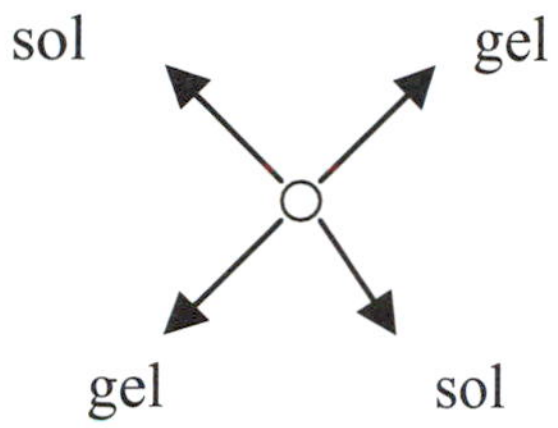	Dangling structure tie;network strand part, **neutral**
P_3	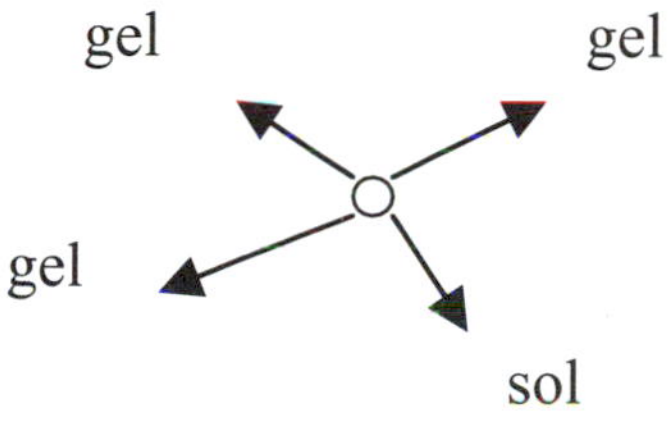	Network junction, **active**
P_4	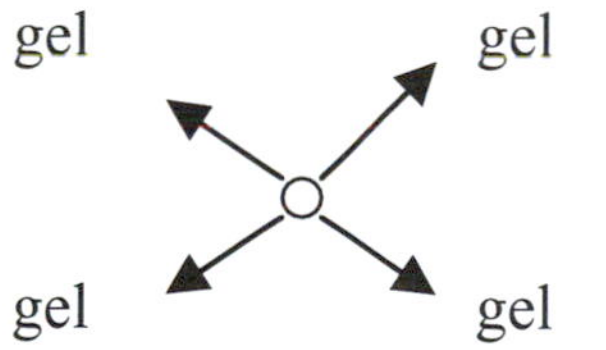	Network junction, **active**

three *linking probabilities* for those two paths p_i, $i = 0, 1, 2$, corresponding to 0, 1, or 2 paths leading from the mer to the gel. Thus, a crosslinked unit on the $r'th$ mer of an r-mer primary chain partitions it into two parts, one with r' mers, the other with $r - r'$ mers. From the gel curve calculation (Section 9.2.2), the probability of no connection to the gel in a run of r' mers is $\exp(-\alpha wr')$, so the probability of finding at least one connection is $1 - \exp(-\alpha wr')$. Let $A(u) = \exp(-\alpha wu)$ and $B(u) = 1 - A(u)$.

The three linking probabilities can be expressed as summations—or integrals, because the chains are very long—in terms of these two auxiliary functions:

$$
\begin{aligned}
p_o &= \frac{1}{r}\int_0^r A(r')A(r-r')\,dr' = \exp(-\alpha wr) \\
p_1 &= \frac{2}{r}\int_0^r A(r')B(r-r')\,dr' = 2\left[\frac{1-\exp(-\alpha wr)}{\alpha wr} - \exp(-\alpha wr)\right] \\
p_2 &= \frac{1}{r}\int_0^r B(r')B(r-r')\,dr' = 1+\exp(-\alpha wr) - 2\left[\frac{1-\exp(-\alpha wr)}{\alpha wr}\right]
\end{aligned}
\tag{9.34}
$$

For polydisperse primary chains, the probability that a crosslink joins an r-mer primary chain and an r'-mer primary chain is the product of their weight fractions, the product of the fraction of all mers that each contains is $W(r)W(r')$. Thus, for example,

$$
P_o = \int_0^\infty \int_0^\infty p_o(r)p_o(s)W(r)W(s)drds \tag{9.35}
$$

which factors into the product of two identical integrals:

$$
P_o = \left[\int_0^\infty W(r)p_o(r)dr\right]^2 = \bar{p}_o^2 \tag{9.36}
$$

The other P_i are obtained in like manner, with the results:

$$
\begin{aligned}
P_o &= \bar{p}_o^2 \\
P_1 &= 2\bar{p}_o\bar{p}_1 \\
P_2 &= 2\bar{p}_o\bar{p}_2 + \bar{p}_1^2 \\
P_3 &= 2\bar{p}_1\bar{p}_2 \\
P_4 &= \bar{p}_2^2
\end{aligned}
\tag{9.37}
$$

in which the *mean linking probabilities* are:

$$
\bar{p}_i = \int_0^\infty W(r)p_i(r)\,dr \tag{9.38}
$$

The calculation of μ and ν for monodisperse primary chains is straightforward. Thus, $\bar{p}_i = p_i$ for this case, so substitution into Eq. 9.37, then into Eqs. 9.31 and 9.32, and finally replacement of $\exp(-\alpha wr)$ by the sol fraction s [Eq. 9.26] leads to:

$$\mu = \nu_c \left[1 + s + \frac{2(1-s)}{\ln s}\right]\left[1 - 3s - \frac{2(1-s)}{\ln s}\right]$$
$$\nu = 2\nu_c \left[1 + s + \frac{2(1-s)}{\ln s}\right]\left[1 - 2s - \frac{(1-s)}{\ln s}\right] \quad (9.39)$$

Results for most-probable primary chain distributions, including effects of scission, are given in Table 9.2. Expressions for ν and μ in multilinked and endlinked networks can be derived by procedures described elsewhere[11,35,37]. Results for a representative endlinked network system, long monodisperse B_2 chains linked by A_4 units with equal stoichiometry, obtained by the recursive procedures of Miller and Macosko[11,37], are also given in Table 9.2. Also included in the table for future use are the expressions for $\nu - \mu$, the *cycle rank*[42], the number density of independent macrocycles in the network.

9.3.2 Topological and Dynamics-Related Properties

Certain features of the spatial relationships among chain molecules in the liquid state are rendered permanent by network formation. Some have to do with relative location—linked mers can never diffuse apart, linked chains must always share the same small region of space, and so forth—and are for the most part obvious. Others, having to do with the mutual uncrossibility of chain backbones and with networks being ensembles of interpenetrating macrocycles, are subtler. Entanglement effects, arising from mutual uncrossability, have a strong influence on the dynamic behavior of polymer melts and concentrated solutions. Their effects are seen most clearly in the *plateau modulus,* a property of the liquid that much evidence suggests originates from the *uncrossability interaction*[41] and retards large-scale chain motions and conformational rearrangement rates. Such characteristics of the liquid can become *topological;* that is, nonbonded but nonetheless permanent features of the structure, upon network formation. A retardation in dynamics turns into suppression as the interacting chains become permanent parts of different network macrocycles, partitioned into mutually exclusive classes, such as the entwined and not-entwined states illustrated in Figure 9.6. This aspect of network structure has to do with *conserved spatial relationships* arising from backbone uncrossability rather than, as the others have been, with *conserved connectivity relationships*. Interestingly, trapped entanglements appear to contribute beneficially to various technological properties of rubber[43]. The entanglement interaction is discussed extensively in Volume 2.

TABLE 9.2 Structural parameters for representative crosslinked and endlinked networks

Property	Random Crosslinking and Scission[(a)] (Exponential Primary Chain Distribution)	Endlinking[(b)] (Monodisperse Primary Chain Distribution)
Gel Curve	$s + s^{1/2} = \beta/\alpha + 2/[(n_w)_o \alpha]$	$s^{1/2} = \left[(1/p^2 - 3/4)^{1/2} - 3/2 + p\right]/p$
Active Junction Concentration μ	$\nu_c(1 - s^{1/2})^3(1 + 3s^{1/2})$	$\nu_c(1 - s^{1/2})^3 p^3[4 - 3p(1 - s^{1/2})]$
Active Strand Concentration ν	$2\nu_c(1 - s^{1/2})^3(1 + 2s^{1/2})$	$2\nu_c(1 - s^{1/2})^3 p^3[3 - 2p(1 - s^{1/2})]$
Cycle Rank $\nu - \mu$	$\nu_c(1 - s^{1/2})^3(1 + s^{1/2})$	$\nu_c(1 - s^{1/2})^3 p^3[2 - p(1 - s^{1/2})]$
Active Network Fraction f_a	$(1 - s^{1/2})^2$	$(1 - s^{1/2})^2$
Dangling Fraction f_d	$2s^{1/2}(1 - s^{1/2})$	$2s^{1/2}(1 - s^{1/2})$
Entanglement Trapping Factor T_e	$(1 - s^{1/2})^4$	$(1 - s^{1/2})^4$

(a) α = fraction of mers that participate in crosslinks; β = fraction of backbone bonds that are broken; ν_c = number density of crosslinks.
(b) $f = 4$; equal stoichiometry; p = fractional conversion of functional groups; ν_c = number density of linking agent.

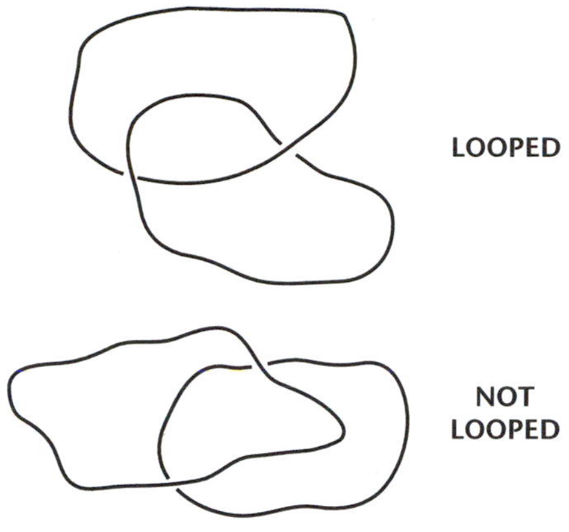

FIGURE 9.6 Topologically looped and non-looped macrocycle pairs.

Topological problems of this sort are extremely difficult to treat rigorously[41,44–46], so some sensibly simplified way for handling the topological entrapment aspect is needed. The entanglement contribution to the equilibrium modulus of networks was assumed to be expressible as the product of two terms. One term is the plateau modulus of the polymeric liquid G_N^o, a property that depends on the polymer species. The other term is a trapping factor, T_e, a property related to structural features of the network alone. Following a long history of development[41], a physically plausible and practically useable approximation for the trapping factor was proposed by Langley in 1968[47]. He suggested T_e should be the fraction of all pairwise relationships between network strands that were made permanent when the network was formed. He set that equal to the fraction of all mer pairs for which each of the four chain paths leading away from the two mers are connected independently to the network.

The Langley trapping criterion is shown in Figure 9.7. Averaged over all pairs, T_e translates to a connectivity parameter that counts as permanent only those relationships between pairs of mers on strands that are elastically active—that is, that are already parts of the network macrocycles. The trapping factor and some other properties of interest are related directly to the mean linking probabilities—$\bar{p}_o$, $\bar{p}_1$, and $\bar{p}_2$—used in Eq. 9.37 to evaluate junction connectivity. Thus, the mers fall into three categories, depending on the number of chain paths leading independently to the gel. Those with two such connections belong to the gel as parts of elastically active strands; those with one connection belong to the gel, but as parts of elastically inactive *dangling structures;* and those with no connections belong to the sol. A moment's reflection leads to the realization that the fraction of mers from long primary chains in the *active fraction* f_a, the dangling fraction f_d, and the sol fraction $f_s = s$ categories are respectively $\bar{p}_2$, $\bar{p}_1$, and $\bar{p}_o$, as given by Eqs. 9.34 and (9.38). Further, the

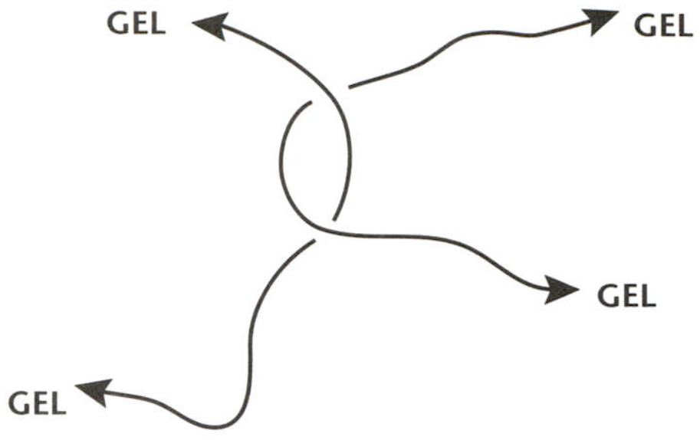

FIGURE 9.7 The Langley condition for topologically trapping entanglements.

entanglement trapping factor T_e, as defined by Langley, is $\bar{p}_2^2$. Summarizing,

$$\begin{aligned} f_s &= \bar{p}_0 \\ f_d &= \bar{p}_1 \\ f_a &= \bar{p}_2 \\ T_e &= \bar{p}_2^2 \end{aligned} \tag{9.40}$$

For the example of random crosslinking of monodisperse primary chains, the observable quantity, the sol fraction $f_s = s$, is related to crosslink density by Eq. 9.27, and:

$$\begin{aligned} f_d &= -2\left[s + \frac{(1-s)}{\ln s}\right] \\ f_a &= 1 + s + \frac{2(1-s)}{\ln s} \\ T_e &= \left[1 + s + \frac{2(1-s)}{\ln s}\right]^2 \end{aligned} \tag{9.41}$$

The behavior of the network parameters—sol fraction s; active strand and junction concentrations, ν and μ; and entanglement trapping factor T_e—are shown for exponential primary chain distributions as functions of the crosslinking index in Figure 9.8. The gel fraction rises the most rapidly and the entanglement trapping factor much more slowly. The active strand and junction concentrations grow in an intermediate fashion. Expressions for f_d, f_a, and T_e for the random crosslinking and scission of primary chains with exponential distribution are included in Table 9.2. The corresponding expressions for an endlinking example, $A_4 - B_2$ networks with equal stoichiometry and long monodisperse B_2 chains, are also given there. The expressions for random multilinking, including random crosslinking and some

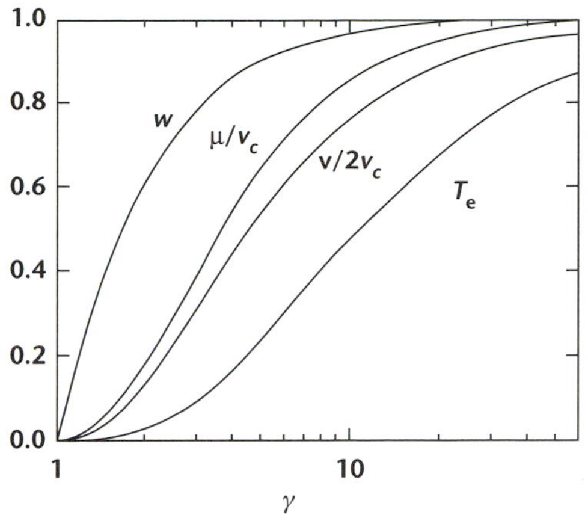

FIGURE 9.8 Gel fraction, elastically active strand, and junction concentrations, and entanglement trapping factor as functions of crosslink index from random crosslinking of primary chains with exponential molecular weight distribution.

proportionate amount of chain scission as well, are given in Pearson-Graessley[35]; the expressions for various cases of random crosslinking and endlinking are also given in the Miller-Macosko papers[11,37], and in the references they contain.

9.3.3 Architectural Characterization

Aside from general features such as locally treelike versus locally latticelike connectivity, a precise determination of network architecture is not required to investigate most aspects of network behavior. However, in some cases it is essential. Examples of the latter include assessing the contribution of chain entanglement to network elasticity (Sections 9.7.1 and 9.7.2) and exploring the microscopics of elastic deformation by scattering methods (Section 10.4.1). Especially serious for crosslinking are chain scission and chain end defects; especially serious for endlinking are incomplete linking and side reactions of the linking moieties.

The only handles on the relevant architectural structure are indirect—even the best spectroscopic methods are helpless when such quantities as the concentration of elastically active strands and the trapped entanglement fraction are required. Sol fraction determination—measurement of the gel curve, location of the gel point, and evaluation of primary chain length effects for evidence of nonrandom linking—is the main source of information. Typically, gels are mechanically fragile until well beyond the gel point. Handling networks with sol fractions greater than 20 wt% or so is delicate work.

An accurate determination of the sol fraction, the weight fraction of polymer not attached to the network, is straightforward but needs careful attention to detail. The exhaustive solvent extraction of weighed network aliquots, followed by drying and reweighing, is the usual procedure. Sometimes the starting materials contain nonvolatile soluble impurities—polydimethylsiloxane commonly contains ~3 wt% of oligomeric PDMS rings, for example. Sometimes crosslinking agents or parts of them are combined into the network and hence are unextractable. Sometimes they leave unattached, nonvolatile, and hence extractable residues, and sometimes they form volatile by-products that never show up in either original or extracted weights. These affect the interpretation of extraction data, so that the careful calibration of the particular system, frequent crosschecks, and a redundancy of testing are essential. The exercise is pointless otherwise. Crosslinking with high-energy radiation is attractive because it doesn't require polymeric double bonds and usually leaves no nonvolatile residues. High energy radiation is rather indiscriminate, however, and introduces enough chain scission in some polymer species to make it unsuitable.

Mildness of linking chemistry to minimize concomitant chain scission and the use of long primary chains to minimize chain end effects are natural requirements for networks formed by crosslinking. Crosslinking polydienes near room temperature with sulfur monochloride is a possible example[48]. Another is the use of mesitylenedinitrileoxide[49–52], which at moderate temperatures can crosslink many polymeric species. Careful selection of linking reaction, analysis of reactant structural purity and variations of stoichiometry—the effects of linking ingredient ratio on sol fraction, network swell ratio, or modulus—are natural requirements for networks formed by endlinking. Merrill and coworkers[53], Macosko and Benjamin[54], and Cohen and coworkers[55] provide some examples of this careful characterization with exhaustive attention to detail required for proper endlinked network studies.

9.4 General Considerations on Network Elasticity

The modern understanding of network elasticity began in the early 1930s. Debate about the existence and nature of macromolecules was closing, and ideas about rubber elasticity based on the accepted molecular structure and flexibility were beginning to appear. In 1932, chain segments with freely extendible random conformations were already considered the basis of reversible extension and retraction by Busse[56], and associated with the Gough-Joule effects by Meyer et al.[57]. In 1934, Guth and Mark[58] announced the force-extension law for random coils, $\boldsymbol{F} = (3kT/\langle R^2\rangle_o)\boldsymbol{r}$ [Eq. 4.85] and suggested that the modulus of a rubber is proportional to the concentration of random coil elements. Kuhn published the first molecular theory of rubber elasticity in 1936[59]. Based on the random coil model, it provided the first absolute prediction of mechanical modulus from network structure. Comparison with the 1935 data of Meyer and Ferri[60] left little room for doubting that the mechanism responsible for rubber elasticity is random coil deformation.

Before proceeding into detailed considerations, it is instructive to see what sort of theory can be constructed merely with the results available in 1936. Consider

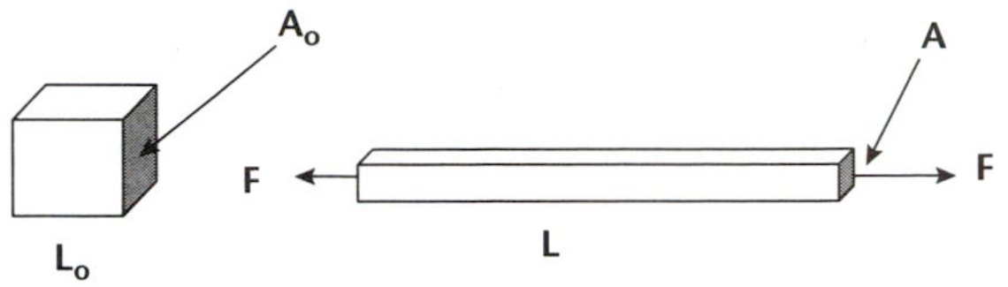

FIGURE 9.9 Uniaxial extensional deformation.

a test sample of rubber shown in Figure 9.9. Its rest length and cross-sectional area are L_o and A_o, and it is stretched in the x-direction to length L by a tensile force F. The sample volume L_oA_o does not change significantly with stretching, so the area is reduced to$A = A_oL_o/L$. The *tensile stress* σ is F/A, the *tensile strain* ε is $(L - L_o)/L_o$, and the *tensile modulus* is the stress-strain ratio for small strains, $E = \sigma/\varepsilon$. The object is to express the equilibrium force–length relationship in terms of sample geometry and the relevant molecular structure of the sample.

Consider a rubber sample with number density ν of network strands and in particular those that straddle an arbitrary plane through the sample normal to the stretch direction, as shown in Figure 9.10. The strands are random coils, and they can supply a restoring force by acting as linear springs, $f = -(kT/\langle x^2\rangle_o)x$. Assume the junctions move in proportion to the stretch such that x_e, the x-component of end separation at rest, increases by the amount $x_e(L - L_o)/L_o$. Use the average, $\langle x^2\rangle_o^{1/2}$, to approximate x_e for each chain, obtaining thereby an average increment of force per chain, $\langle f\rangle = (kT/\langle x^2\rangle_o)\langle x^2\rangle_o^{1/2}((L - L_o)/L_o)$. The net number of plane-straddling strands is approximately $\nu A_o\langle x^2\rangle_o^{1/2}$, the number within an end-to-end distance of the plane. Accordingly, the total restoring force is about $\nu A_o\langle x^2\rangle_o^{1/2}\langle f\rangle$, so the stress-stretch relationship is approximately

$$F = \frac{A_o\nu kT(L - L_o)}{L_o} \tag{9.42}$$

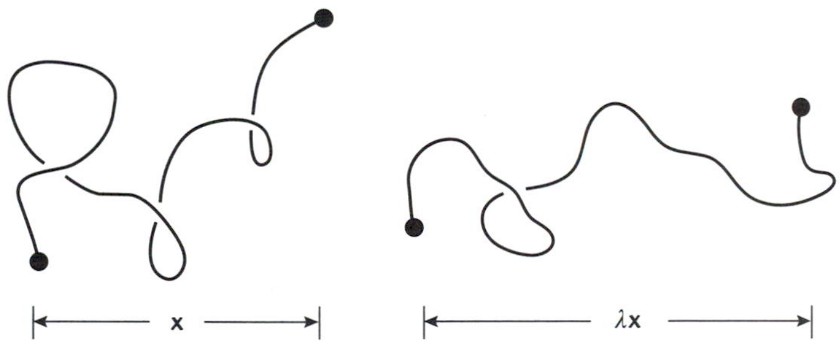

FIGURE 9.10 A representative network strand before and after extension.

Equation (9.42) supplies an approximate expression for the stress-strain behavior of a Gaussian network. Note that it depends only on one structural parameter of the material, the number density of network strands, a quantity that can in principle be estimated independently. Only crude estimates of ν were available in 1936. Nevertheless, they sufficed to show the predicted tensile modulus for typical rubber materials ($E = \nu kT$ from Eq. 9.42) to be in the observed range—smaller by a factor of 10^3 to 10^4 than the modulus of conventional solids.

Although useful for estimating magnitudes, Eq. 9.42 is unsatisfactory for reasons that go beyond even the obvious oversimplification in its derivation. It is not generally applicable for large stretch ratios or for other types of deformation. Even for tensile tests at small deformations, the predicted modulus is seldom quantitatively correct. In later work, beginning in 1942 and continuing through the 1970s with Wall[61], Treloar[62], and James and Guth[63], the derivations become more refined. The strain measures that result from them have at least the virtues of generality and internal consistency. However, unless features beyond noninteracting strands are included, the predicted stress-strain forms and modulus magnitudes agree only marginally better with experiments than Eq. 9.42. In fact, the transition to a fully adequate understanding of real networks is even now far from complete.

Only *Gaussian networks,* consisting of flexible strands whose distribution of conformations can be described adequately by the Gaussian approximation (see Section 4.5.2), are considered here. Three models are considered in this chapter; the *affine network,* the *phantom network,* and the *entangled network*. The affine network is the simplest, and its analysis was essentially complete by 1960. The phantom network, the most natural to use as a reference, is far more difficult to develop. Its form was finally settled only in the early 1980s. The term "phantom" in fact properly applies to both the affine and phantom models, because both assume that each strand is free to take up conformations independently of the conformations of the other strands and that each has the ghostlike capacity of passing freely through the backbone contours of other strands. This obviously unrealistic assumption is made for the most understandable of reasons: The calculation of elastic properties cannot proceed at all if backbone uncrossability and all the complexity that entails is insisted upon. The distinction between the two models is in the handling of the network junctions. The junctions are free to fluctuate spatially in the phantom network. Junction fluctuations are suppressed in the affine network, and the junctions are compelled to move affinely, in direct proportion to the macroscopic displacements of the network. It would perhaps have been more appropriate to name them according to their inventors, Kuhn instead of affine and James-Guth instead of phantom, but usage has already made the choice.

The entangled network includes the effect of chain uncrossability, but in a still tentative and empirical manner. In contrast with the first two, the entangled network is not, strictly speaking, a model at all but rather a framework for organizing the data obtained on real networks. As such, it is still a work in progress. Several molecularly motivated theories of the effects of uncrossability, seeking to mimic them in a sound fashion, are discussed in Section 10.5. The intervening discussions are divided into two parts, one on the *initial modulus,* the factor that scales magnitude in the stress-strain relationship; the other on the *reduced stress,* the form of the stress-strain relationship.

Consider any material body acted upon mechanically by pressure forces alone. From the first and second laws of thermodynamics, the increments of internal energy, entropy, and volume along any path of its equilibrium states are related by the *Gibbs equation:*

$$dU = TdS - PdV \tag{9.43}$$

Consider now an elastic body, an object that has a unique *rest state*—a shape to which the body returns when all externally applied forces are removed. All other shapes are *deformed states.* The interest here is the particular case of *isotropic solids,* those materials whose properties in the rest state are independent of orientation; that is, the properties are the same in all directions.

Including deformed states in the thermodynamics of isotropic solids requires an additional term expressing how deformation changes the internal energy. For the particular case of uniaxial extension, shown in Figure 9.9, the new variables are the tensile force F and the length L. Consider an increment of work on the solid FdL along some path of constant entropy and volume. From the first law, the internal energy increment is equal to the work, $dU = FdL$, and hence $F = (\partial U/\partial L)_{S,V}$. Accordingly, for any path of equilibrium states, including states of tensile deformation,

$$dU = TdS - PdV + FdL \tag{9.44}$$

This relationship can just as well be expressed in terms of the more convenient terms of Helmholtz and Gibbs free energies. Thus, with their definitions, $A = U - TS$ and $G = A + PV = H - TS$, Eq. 9.44 leads to:

$$dA = -SdT - PdV + FdL \tag{9.45}$$

$$dG = -SdT + VdP + FdL \tag{9.46}$$

Tensile force is related to tensile deformation. Thus, with $\Delta A = A(T, V, L) - A(T, V, L_o)$ and $\Delta G = G(T, P, L) - G(T, P, L_o)$, from Eqs. 9.45 and 9.46,

$$F = (\partial A/\partial L)_{V,T} = (\partial \Delta A/\partial L)_{V,T} \tag{9.47}$$

$$F = (\partial G/\partial L)_{P,T} = (\partial \Delta G/\partial L)_{P,T} \tag{9.48}$$

The potential energy stored in the solid by an elastic deformation is commonly expressed as an *elastic energy density* W, so the elastic energy of deformation is VW. The general relationship in mechanics between potential energy and force, when applied to a tensile deformation leads to $F = (\partial(VW)/\partial L)_{V,T}$ or $F = (\partial(VW)/\partial L)_{P,T}$ along constant V, T or P, T paths. These expressions and Eq. 9.47 then connect stored energy density with the free energy of deformation:

$$V(\partial W/\partial L)_{V,T} = (\partial A/\partial L)_{V,T} = (\partial \Delta A/\partial L)_{V,T}$$
$$(\partial(VW)/\partial L)_{P,T} = (\partial G/\partial L)_{P,T} = (\partial \Delta G/\partial L)_{P,T} \tag{9.49}$$

When no uniaxial force is applied, pressure is related to the volume dependence of free energy. Thus, for example, from Eq. 9.45:

$$P = -(\partial A/\partial V)_T = -(\partial VW/\partial V)_T \tag{9.50}$$

The development for uniaxial extension has thus far been general. However, the elastic modulus of rubber, $E \lesssim 10^{-2} GPa$, is much smaller in magnitude than the *bulk modulus* or reciprocal of isothermal compressibility: $1/\beta \sim 1GPa$, for the dense liquid state. As shown in Volume 2, it is permissible to ignore compressibility effects in deformation when $\beta E \ll 1$. (Note, however, the exceptional case of thermoelastic analysis in Section 10.3.) Thus, for some modest range of stresses and pressures, the volume effectively depends only on temperature. For isothermal tests, $V = V_o$ and $AL = A_o L_o$, where A is the cross-section area, and the subscripts denote the rest state. From Eq. 9.49,

$$\left(\frac{\partial W}{\partial L}\right)_T = \left[\frac{\partial(\Delta A/V)}{\partial L}\right]_T = \left[\frac{\partial(\Delta G/V)}{\partial L}\right]_T \tag{9.51}$$

Accordingly, from either Eq. 9.47 or Eq. 9.48,

$$F = V\left(\frac{\partial W}{\partial L}\right)_T \tag{9.52}$$

The *tensile stress* σ is F/A, the tensile force/cross-section area; and the *stretch ratio* λ is L/L_o, the stretched length/rest length, as illustrated in Figure 9.9. Thus, using Eq. 9.52,

$$\sigma = \lambda\left(\frac{\partial W}{\partial \lambda}\right)_T \tag{9.53}$$

The *tensile modulus* is the stress-strain ratio in the small strain limit:

$$E_o = \lim_{\lambda \to 1}\left(\frac{\sigma}{\lambda - 1}\right) \tag{9.54}$$

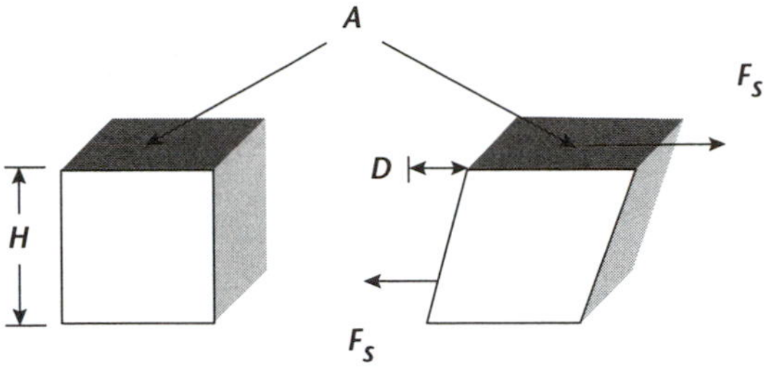

FIGURE 9.11 Simple shear deformation.

Another class of deformation is *simple shear,* illustrated in Figure 9.11. A force F acts on a surface of area A parallel to the surface, causing a parallel displacement D of that surface relative to a parallel fixed surface at distance H. The ratio of force to area, the *shear stress* σ_s, is F/A, shear force/area in this case; and the *shear strain* γ is D/H, displacement/separation. The *shear modulus* is the stress-strain ratio in the small strain limit:

$$G_o = \lim_{\gamma \to 0} \left(\frac{\sigma_s}{\gamma} \right) \tag{9.55}$$

Both shear modulus and tensile modulus can be measured for solids. They are related properties for isotropic incompressible solids:

$$E_o = 3G_o \qquad (\beta E \ll 1) \tag{9.56}$$

This relationship and others appropriate for network materials, along with the methods for determining mechanical behavior in simple shear and uniaxial extension or compression, are treated in Volume 2.

9.5 The Affine Network

The idea to regard network junctions as being fixed to material points and moving affinely with macroscopic deformations was first introduced by Kuhn[64], seemingly as a mere expedient for simplification. In 1942–43, that idea was adopted as a physical assumption by Wall[61] and Treloar[62] in their more precise analyses of the Kuhn model. In 1943, Flory and Rehner[65] and James and Guth[63] investigated the effect of relaxing the fixed junction constraint by setting some free to fluctuate. Both found that the *mean positions* of free junctions in networks of freely crossing Gaussian strands move affinely with deformation. Thus, for junction i and an extension in the x-direction,

$$\bar{x}_i = \lambda(\bar{x}_i)_o \tag{9.57}$$

James and Guth discovered another property of free junctions in freely crossing Gaussian networks[63]—that the span of the junction fluctuations about this mean is independent of deformation:

$$\overline{(\Delta x_i)^2} = \overline{(\Delta x_i)_o^2} \tag{9.58}$$

in which $\Delta(x_i)_o = x_i(t) - (\bar{x}_i)_o$ and $\Delta x_i = x_i(t) - \lambda(\bar{x}_i)_o$ are fluctuations from the mean. Apparently, neither pair of investigators recognized the implications of Eq. 9.58, because both Flory-Rehner and James-Guth concluded from Eq. 9.57 that the assumption of affine junction displacement was harmless.

Wall and Treloar based their respective analyses of the affine model on a free energy calculation, from which they derived the stress-strain behavior. Following Kuhn, each attributed the change in Helmholtz free energy with deformation—the origin of the restoring force—to the change in configurational entropy of the strands:

$$\Delta A = -T\Delta S_{conf} \tag{9.59}$$

Their analyses proceeded along the following lines.

Consider a network occupying some volume V, formed from undiluted polymer at some temperature T, and tested at those conditions. The network consists of νV and μV elastically active strands and junctions; some fraction F_n of the strands have n backbone bonds. Any junctions and strands in dangling structures or the sol fraction make no contribution to equilibrium elasticity and are therefore ignored here. Assume the distribution of end-to-end vectors for n-strands in the unstrained network to be Gaussian, with the same mean-square end-to-end distance as free n-chains at the conditions of network formation. Thus, from Eq. 4.76,

$$P_n(x, y, z)dxdydz = (\beta_n/\pi)^{3/2}\exp[-\beta_n(x^2 + y^2 + z^2)]dxdydz \tag{9.60}$$

in which x, y, z are the end-to-end vector components, and $\beta_n = 3\langle R_o^2\rangle_n/2$.

Consider now a uniform deformation of the network, described by the *macroscopic stretch ratios* $\lambda_x, \lambda_y, \lambda_z$. As shown schematically in Figure 9.12, strands with end-to-end components x, y, z at rest have coordinates $x' = \lambda_x x,\ y' = \lambda_y y,\ z' = \lambda_z z$ in the deformed state. From Eq. 4.93, the change in free energy per n-strand of free chains undergoing the same uniform deformation is:

$$\Delta A_n = \frac{kT}{2}\left(\lambda_x^2 + \lambda_y^2 + \lambda_z^2 - 3\right) - kT\ln(\lambda_x\lambda_y\lambda_z) \tag{9.61}$$

Note the absence of an n-dependence in this expression. It means that, insofar as Eq. 9.61 applies to networks, the free energy change with deformation is directly proportional to the total number of elastically active strands in the network. The distribution of strand lengths has no effect at all.

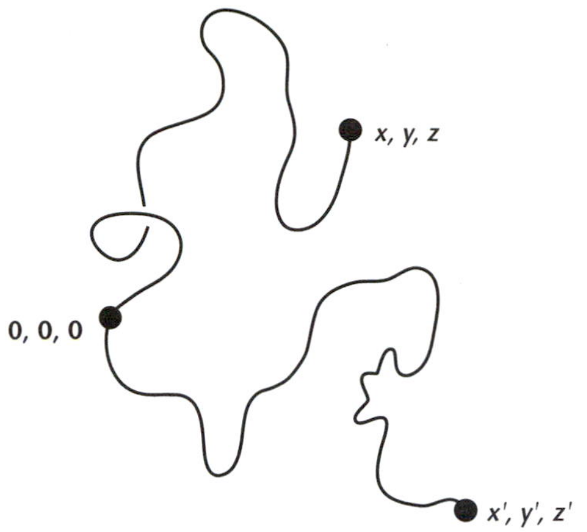

FIGURE 9.12 Affine displacement of strand end-to-end distance for an arbitrary deformation.

9.5.1 The Logarithmic Contribution

The existence and contribution of the logarithmic term in network elasticity was a matter of dispute for some time and has never been properly settled. For constant volume deformations, the product $\lambda_x \lambda_y \lambda_z$ is unity, so for such cases the contribution of the logarithmic term is identically zero, and the question is moot. It must be considered, however, whenever the network undergoes a volume change. Treloar[1] took the seemingly reasonable position that the affine assumption was meant literally—that a junction at some location $\boldsymbol{r}$ moves with deformation precisely to another location $\boldsymbol{r}'$, with the result that only strand stretching need be considered. Stated differently, the term $\ln(dx\,dy\,dz)$ inside the brackets in Eq. 4.89 does not deform to $\ln(\lambda_x dx\,\lambda_y dy\lambda_z dz)$ and give the $\ln(\lambda_x \lambda_y \lambda_z)$ term in Eq. 4.93 for the strands of affine networks. Instead, $dx\,dy\,dz$ in this case is only the usual volume element in integrating over ensembles and, for affine networks, it does not change with deformation. Accordingly, the logarithmic term in Eq. 9.61 is zero for affine networks and, in the Treloar view, the free energy of deformation for as-formed affine networks of Gaussian strands is:

$$\frac{\Delta A}{VkT} = \nu\left(\frac{\lambda_x^2 + \lambda_y^2 + \lambda_z^2 - 3}{2}\right) \tag{9.62}$$

Flory returned to the matter later and arrived at a different result[13,66]. His argument is most easily understood in terms of isotropic deformations, of which the swelling of networks by solvents (see Section 10.1.2) is an important example. Consider as before a network formed at some volume V and having νV and μV elastically active strands and junctions. Now expand it uniformly and isotropically by a factor Q to a volume QV. If the change in free energy per active strand were the same as that for free chains with the same expansion, then per strand from Eq. 4.93,

$$\frac{\Delta A}{kT} = \frac{3\left(Q^{2/3} - 1\right)}{2} - \ln Q \tag{9.63}$$

The free-chain result [Eq. 9.63] expresses the decrease in configurational entropy from the stretching of random coils as partially offset by an increase in the translational entropy of chain ends owing to the expanded space that stretching provides.

The configurational contribution is the same for network strands, but the translational offset is smaller because all strand end positions cannot be selected independently. Suppose, for example, some strand ends take up relative positions as if they were free-chain ends. Equation (9.63) would then apply for those "free-chain" strands. However, other strands attached to the free-strand junctions cannot then choose their end positions independently: They must have the same positions as those already selected by the free-strand ends. Equation (9.63) without the logarithmic term would apply to these strands. One end of each strand simply localizes the strand without thermodynamic consequence in the affine model. For the free-chain strands, the other end locates one of the junctions. Thus, a total of μV free-chain strands are required to set all junction positions in the network, each of them thereby contributing a full free-chain logarithmic term $-\ln q$. The remaining $(\nu - \mu)V$ strands contribute no logarithmic term. Accordingly, the expansion free energy for the network is given by:

$$\frac{\Delta A}{VkT} = \nu\frac{3(Q^{2/3} - 1)}{2} - \mu \ln Q \tag{9.64}$$

The corresponding Treloar expression is:

$$\frac{\Delta A}{VkT} = \nu\frac{3(Q^{2/3} - 1)}{2} \tag{9.65}$$

The derivation of Eq. 9.64 given above differs from the Flory method[13], although the result is the same. The Flory analysis is based on an argument having to do with the creation of networks and positioning of junctions in both the swollen and unswollen states, which preserves the affine picture but, to this author at least, seems rather unconvincing. The derivation above involves a relaxation of the strict affine condition, by pinning down one end of each strand while permitting the other end to fluctuate. In any case, both procedures lead to the Flory expression for the free energy of deformation for as-formed affine networks of Gaussian strands[13]:

$$\frac{\Delta A}{VkT} = \nu\left(\frac{\lambda_x^2 + \lambda_y^2 + \lambda_z^2 - 3}{2}\right) - \mu \ln\left(\lambda_x \lambda_y \lambda_z\right) \tag{9.66}$$

9.5.2 Effects of Temperature and Volume Change

The molecular architecture of a network is established when it is formed. The conditions during formation—temperature, state of dilution, and the like—provide a network *reference state*. Elastic properties at other conditions are commonly of interest, and the affine model is readily extended to such cases. Thus, T_{ref} and V_{ref} are the temperature and volume of the network at formation, and T and V are the conditions of interest. The characteristic ratios $(C_\infty)_{ref}$ and C_∞ are the corresponding properties for free chains of the polymer species. Three factors can then influence network elasticity at the conditions of interest. The end-to-end components of all active strands change with volume, the number density of strands changes with volume, and the size of free chains changes with temperature:

$$\begin{aligned} x/x_{ref} &= y/y_{ref} = z/z_{ref} = (V/V_{ref})^{1/3} \\ \nu/\nu_{ref} &= V_{ref}/V \\ \langle R_n^2\rangle_{ref}/\langle R_n^2\rangle &= (C_\infty)_{ref}/C_\infty \end{aligned} \tag{9.67}$$

When these expressions are used in Eq. 9.60, and Eq. 9.66 is applied, the free energy of deformation for n-strands at the conditions of interest is obtained:

$$\Delta A_n = \nu V k T F_n \int\int_{-\infty}^{\infty}\int [\Phi_n]_{V,T}\,[P_n(x,y,z)]_{V_{ref},T_{ref}}\,dx_{ref}\,dy_{ref}\,dz_{ref} \tag{9.68}$$

where

$$[\Phi_n]_{V,T} = \beta_{n,T}\,(V/V_{ref})^{2/3}\left[x_{ref}^2(\lambda_x^2-1) + y_{ref}^2(\lambda_y^2-1) + z_{ref}^2(\lambda_z^2-1)\right] - (\mu/\nu)\ln(\lambda_x\lambda_y\lambda_z)$$

and

$$[P_n(x,y,z)]_{T_{ref},V_{ref}} = \left(\beta_{n,T_{ref}}/\pi\right)^{3/2}\exp\left[-\beta_{n,T_{ref}}\left(x_{ref}^2 + y_{ref}^2 + z_{ref}^2\right)\right] \tag{9.69}$$

Integration and summation then give the free energy of deformation for Gaussian networks according to the affine model with the Flory log term:

$$\frac{\Delta A_{aff}}{VkT} = \nu_{ref}\frac{(C_\infty)_{ref}}{C_\infty}\left(\frac{V_{ref}}{V}\right)^{1/3}\left(\frac{\lambda_x^2+\lambda_y^2+\lambda_z^2-3}{2}\right) - \mu_{ref}\ln\lambda_x\lambda_y\lambda_z \tag{9.70}$$

For future reference, note that the factor $(V_{ref}/V)^{1/3}$ in Eq. 9.70 is the product of two terms: $(V/V_{ref})^{2/3}V_{ref}/V$. The first comes from the conversion of x, y, z chain coordinates to $\alpha_x x_{ref}, \alpha_y y_{ref}, \alpha_z z_{ref}$ on performing the integrations [see Eq. 9.69], thus reflecting end-to-end extension or contractions from x_{ref}, y_{ref}, z_{ref} to x, y, z when $V \neq V_{ref}$. The second term converts from ν_{ref} to ν, reflecting an increase or decrease of active strand concentration when $V \neq V_{ref}$.

9.5.3 Network Pressure

Consider a network that is formed and then tested under the conditions of formation, V_{ref} thus being the rest volume V_o for the tests. Use the same subscript to designate the other reference terms in Eq. 9.70 and assume $(C_\infty)_{ref}/C_\infty = 1$ for simplicity. Equation (9.66), the Flory modification of the affine model, then reduces to:

$$\frac{\Delta A_{aff}}{V_o kT} = \nu_o \left(\frac{\lambda_x^2 + \lambda_y^2 + \lambda_z^2 - 3}{2} \right) - \mu_o \ln \lambda_x \lambda_y \lambda_z \qquad (9.71)$$

Consider now an isotropic deformation to some new volume V: $\lambda_x = \lambda_y = \lambda_z = (V/V_o)^{1/3}$, and,

$$\frac{\Delta A_{aff}}{V_o kT} = \frac{3\nu_o}{2} \left(\frac{V}{V_o} - 1 \right)^{2/3} - \mu_o \ln \frac{V}{V_o} \qquad (9.72)$$

From Eq. 9.50, the network pressure contribution for the Flory modification is:

$$\frac{P}{kT} = -\nu_o \left(\frac{V_o}{V} \right)^{1/3} + \mu_o \frac{V_o}{V} \qquad (9.73)$$

In its reference state, a Flory network is under tension ($P < 0$):

$$P = -(\nu_o - \mu_o) kT \qquad (V = V_o) \qquad (9.74)$$

The tension grows smaller if the network is permitted to shrink. Thus, $V(P)$ goes to some limiting value as P goes finally to zero: $V(0) = V_o(\mu_o/\nu_o)^{3/2}$. For tetrahedral networks ($\nu = 2\mu$), $V(0) = V_o/2^{3/2}$. The tension at $V = V_o$ is larger for a Treloar network, [Eq. 9.62],

$$\frac{P}{kT} = -\nu_o \left(\frac{V_o}{V} \right)^{1/3} \qquad (9.75)$$

$$P = -\nu_o kT \qquad (V = V_o) \qquad (9.76)$$

and the network tension goes to zero only in the $V = 0$ limit.

The Treloar formulas are obtained from the Flory formulas simply by omitting the logarithmic term from Eqs. 9.70 through 9.72. Both versions of the affine model are carried along for comparison purposes later. In both, the network-generated tension produces a pressure on the volume-excluding elements (the James-Guth space-filling liquid component, see Section 9.6) in real networks. If that tension is large enough, *syneresis* can occur: Solvents or other unattached species are expelled from the network to form a separate phase (Section 10.2).

9.6 The Phantom Network

The affine network suffers from two artificialities: It ignores the mutual uncrossability of network strands, and it arbitrarily suppresses, or treats as a special case, the spatial fluctuations of network junctions. The phantom network removes the second of these by permitting all but a few junctions to fluctuate freely. It would therefore seem to be the logical starting point for molecular network theory. The phantom network, however, is far more difficult to analyze and also has its own ambiguities. For example, networks contain many macrocyclic rings—they would not be elastic solids otherwise—and the equilibrium dimensions of chain segments in loops are always smaller than free chain dimensions [Eq. 4.57]. As discussed in Section 9.1.2, however, any phantom network model for rubber networks formed from melts and concentrated solutions should not have contracted strand dimensions. The affine network skirts the issue through its artificiality of fixing elastically active junctions as they form to material locations. There is no such way out for the phantom network, and something must be added to prevent collapse. Simply to impose a deformation on phantom networks presents another difficulty. How to take hold of a phantom network without localizing, constraining, or otherwise suppressing the fluctuations is not obvious.

By 1943, James and Guth had already recognized the collapse and deformation problems and devised an approach to circumvent them[63]. They envisioned the real network, pervading and filling up some definite space, to be a kind of two-component system. One component is a freely fluctuating mathematical network with the same architecture as the real network but no occupied volume; it is held in place and shape by a relatively few points fixed to the boundary. The other component is a cohering but otherwise inert fluid whose only role is to fill up the space occupied by the network. James and Guth showed that a phantom network without some force to preserve the volume would shrink to microscopic dimensions, a result consistent with the absence of a logarithmic term in the free energy (see Section 9.5.3). The fixed points in their picture play the role of the metaphorical inert fluid in that they prevent the collapse and also supply a means for imposing deformations. They derived Eqs. 9.57 and 9.58 for phantom networks using Gaussian strands. They found that the stress-strain form was the same as that from the affine model form and suggested the possibility that the affine and phantom models were equivalent, as already noted. They raised objections to the structural aspects of the affine model, particularly to its universal proportionality of modulus magnitude to strand concentration. They also expressed what could be interpreted as a concern about crosslinking rate effects on the network architecture. (They also considered the non-Gaussian behavior of highly extended networks, a topic not discussed here in any detail.)

In 1947, James explained the derivations in the 1943 paper in more detail[67]. Boggs presented a nice elaboration of the James-Guth two-component model in 1952[68]. James and Guth returned to the question of modulus magnitude and its relationship to the evolution of structure during network formation in another 1947 paper[69]. Unfortunately, their discussion mixed together two aspects of network structure in a confusing way. One aspect is the evaluation of elastically active junction and strand concentrations, an unsolved problem at the time but now (as explained in Section 9.3) fairly well in hand for the locally treelike networks of interest. The other aspect is the competition between the rate of addition of crosslinks and the rate of structural rearrangement in response to crosslink formation. Such rearrangements could in principle lead to coil contraction and thus alter the strain response of the strands. Indeed, for latticelike networks such as those illustrated in James[67], contraction during linking could well occur because the loops being formed are small. For reasons explained in Sections 9.1.2, however, contraction during linking in the melt state is unlikely for treelike networks. Also, latticelike structure is physically unrealistic except for networks formed from highly diluted states. In any case, based on later experiments, coil contraction from network formation is not observed (Section 9.1.2).

Near the end of their 1947 paper, James and Guth offered an estimate for the modulus magnitude[69]. They concluded, by rather obscure reasoning based on a putative rate-determined coil contraction, that ΔA would be smaller than the affine prediction by approximately a factor of two. That inference can in no way be construed as a firmly based deduction. It happens to be a correct result for tetrafunctional phantom networks, but that agreement is clearly fortuitous.

Later articles by James and Guth[70–74] were mostly concerned with lesser matters, such as the various shortcoming of the affine model. An important special point was made in 1953[71] and later articles that the free energy for phantom networks contains no logarithmic term. Beyond that, little was added to what must be regarded as their quite monumental contribution of 1943.

The modulus of phantom networks remained unsettled until Duiser and Staverman showed in 1965 that junction fluctuations have a significant effect on network elasticity[75]. They evaluated the deformation free energy for an ensemble of elementary network elements, essentially the Flory-Rehner tetrahedron model[65] illustrated in Figure 9.13. Each element consists of four Gaussian strands, joined at one end to a common junction and at the other to a fixed point. The common junction is free to fluctuate, but the fixed points are permanently attached to material points

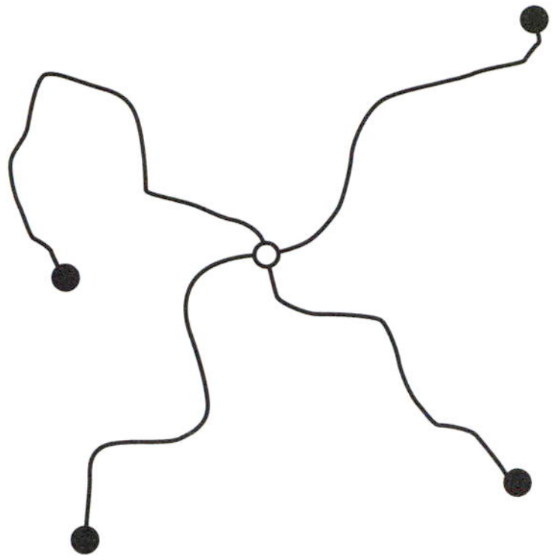

FIGURE 9.13 The Duiser-Staverman tetrahedral network element with fluctuating central junction (○) connected to four fixed junctions (●).

that are distributed spatially according to the number of element configurations and move affinely with macroscopic deformations. In solving for the Rouse dynamics of the ensemble (see Volume 2), they found three normal modes of motion per tetrahedron with infinite relaxation time, each such mode contributing kT to the free energy of deformation. A network with νV strands corresponds to $\nu V/4$ tetrahedra or $3\nu V/4$ elastic modes, less than the νV modes expected from the affine model. Accordingly, the correct free energy of deformation for an ensemble of Flory-Rehner tetrahedra is:

$$\Delta A = \frac{3\nu VkT}{4}\left(\frac{\lambda_x^2 + \lambda_y^2 + \lambda_z^2 - 3}{2}\right) \tag{9.77}$$

Duiser and Staverman went on to confirm Eq. 9.77 by the more conventional method of counting configurations. They then inferred, based on some incompletely explained principle, that the appropriate expression for a macroscopic phantom network would be:

$$\Delta A = (\nu - \mu)VkT\left(\frac{\lambda_x^2 + \lambda_y^2 + \lambda_z^2 - 3}{2}\right) \tag{9.78}$$

The possibility of a logarithmic term was not discussed.

The idea of inferring phantom network behavior from ensembles of microscopic subsystems was later extended beyond the tetrahedron model[76,77]. The affine and tetrahedral models are low-order members of a hierarchy of *locally treelike micronetworks* (see Figure 9.14). The junctions most remote in connectivity from the center are represented by fixed points that anchor the micronetwork and move affinely with deformation. All micronetworks in the ensemble are structurally identical, but the relative locations of their fixed points are distributed according to the

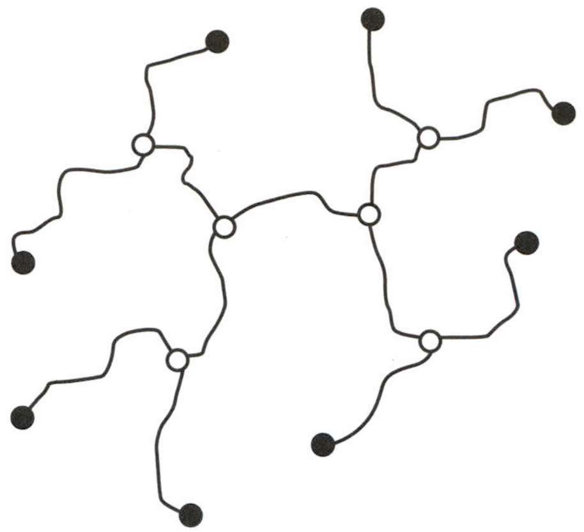

FIGURE 9.14 A second-order trifunctional micronetwork with fluctuating interior junctions (○) and fixed exterior junctions (●).

number of configurations that result[76]. All interior junctions are free to fluctuate, and the Gaussian approximation for the end-to-end distributions of the strands is assumed. The free energy of deformation is obtained in the manner of Eqs. 4.88 through (4.93), as done for the affine network. The result for ensembles of treelike micronetworks with N_F fixed points and N_S strands is:

$$\Delta A = \frac{N_F - 1}{N_S} \nu V k T \left(\frac{\lambda_x^2 + \lambda_y^2 + \lambda_z^2 - 3}{2} \right) \tag{9.79}$$

This result is independent of strand length distribution, junction functionality distribution, and indeed all aspects of micronetwork structure except its treelike (ringless) connectivity[77]. Note, however, that two categories of strands exist: those with one fluctuating end and those with two. Except for the relatively few fixed ends, required to prevent collapse by the incomplete nature of the model itself, both ends of all strands in phantom networks are free to fluctuate.

Accordingly, it is necessary to decompose this total for the micronetwork model into contributions from each category, attributing the phantom network behavior to strands of the second category only. For the example of networks with the same functionality f for all junctions, one can easily demonstrate the identity:

$$\frac{N_F - 1}{N_S} = \frac{f-1}{f}\frac{N_F}{N_S} + \frac{f-2}{f}\left(1 - \frac{N_F}{N_S}\right) \tag{9.80}$$

in which $N_F = f(f-1)^j$ and $N_S = f[(f-1)^j - 1]/(f-2)$, and j is the micronetwork order (see Figure 9.14). Thus, for any order j of micronetwork structure, the term $(N_F - 1)/N_S$ in Eq. 9.79 is the sum of contributions from strands with one end free, equal to N_F, and strands with both ends free, equal to $N_S - N_F$. The latter contribution is the phantom network part; so by inference, the free energy of networks without coil shrinkage and all junctions free to fluctuate is:

$$\Delta A = \frac{f-2}{f} \nu VkT \left(\frac{\lambda_x^2 + \lambda_y^2 + \lambda_z^2 - 3}{2} \right) \tag{9.81}$$

The same argument can be carried through, with only slightly more trouble, for micronetworks with an arbitrary distribution of junction functionalities. The results, like those in Eq. 9.81, are consistent with the Duiser-Staverman[75] formula for the phantom network. It is easily shown, through the fixed-point approach initiated by James and Guth[63], that the various modifications of the affine model, allowing for conditions other than those at the reference state, [Eq. 9.70] carry over unchanged to the phantom model:

$$\frac{\Delta A_{ph}}{VkT} = (\nu_{ref} - \mu_{ref}) \frac{(C_\infty)_{ref}}{C_\infty} \left(\frac{V_{ref}}{V} \right)^{1/3} \left(\frac{\lambda_x^2 + \lambda_y^2 + \lambda_z^2 - 3}{2} \right) \tag{9.82}$$

A logarithmic contribution for the phantom network was introduced erroneously in the second micronetworks paper[77]. The James-Guth finding of no logarithmic term for phantom networks, obtained also by others, is the correct result. Equation (9.82) describes the locally treelike phantom network of interest here.

By the method leading to Eq. 9.73, the elastic pressure for phantom networks is:

$$\frac{P}{kT} = -(\nu_o - \mu_o) \left(\frac{V_o}{V} \right)^{1/3} \tag{9.83}$$

The volume dependence of pressure for a phantom network differs from that for the Treloar version of an affine network only in the proportionality factor. The Flory version behaves differently from both.

By other methods, Edwards[78] in 1971 and Deam and Edwards[79] in 1976 obtained $(\Delta A)_{ph}/(\Delta A)_{aff} = 1/2$ for tetrafunctional crosslinked networks ($f = 4$), a result consistent with the Duiser-Staverman formula in Eq. 9.78. They also found no logarithmic term in the free energy. In 1976[80], Flory confirmed Eq. 9.82 and the absence of a logarithmic term, extending the method he had used to obtain a logarithmic contribution in his version of the affine model[66]. He also pointed out, following the lead of Eichinger[42], that the term $(\nu - \mu)V$ is the cycle rank of the network. That identification is especially satisfying, because the macrocycles alone confer shape memory to a network and thereby establish its mechanical identity as an elastic solid. The locally treelike phantom network received an elegant development in a 1982 review article by Staverman[81] and more recently in the Boyd-Phillips textbook[82].

9.7 The Entangled Network

The behavior of real networks nearly always departs from the affine and phantom predictions. For mechanical properties, two sorts of departures occur: those for the modulus magnitude, considered here, and those for the shape of the stress-strain relationship, discussed in Chapter 10.

Consider the extension of a rubber sample as shown in Figure 9.9. The compressibility effects are typically negligible, as explained earlier, so volume is conserved. Thus, for uniaxial extension, the two lateral dimensions must contract; the contraction ratio for each is $(L_o/L)^{1/2}$, so $\lambda_x = \lambda$ and $\lambda_y = \lambda_z = \lambda^{-1/2}$. With free energy of deformation ΔA replaced by WV [Eq. 9.49], the elastic energy density of either an affine network or a phantom network relative to its rest state can be expressed as:

$$W = \frac{\Gamma}{2}\left[\lambda^2 + \frac{2}{\lambda} - 3\right] \tag{9.84}$$

where the parameter Γ is a material constant, $\nu_{ref} kT(V_{ref}/V)^{1/3}(C_\infty)_{ref}/C_\infty$ for the affine model and $(\nu_{ref} - \mu_{ref})/\nu_{ref}$ multiplied by that quantity for the phantom model. The application of Eq. 9.53 then leads to:

$$\sigma = \Gamma\left(\lambda^2 - 1/\lambda\right) \tag{9.85}$$

For both models the tensile stress is thus proportional to the *strain measure* $\lambda^2 - 1/\lambda$, and the ratio $\sigma/(\lambda^2 - 1/\lambda)$ is independent of the strain. Recall that tensile stress σ is F/A, that the extension ratio λ is L/L_o, and that $L/L_o = A_o/A$ from the incompressibility approximation. Accordingly,

$$\frac{\sigma}{\lambda^2 - 1/\lambda} = \frac{F}{A_o(\lambda - 1/\lambda^2)} \tag{9.86}$$

and Eq. 9.85 is sometimes written with this harmless substitution.

Equation (9.85) applies to *uniaxial compression* as well as extension. In compression, $\lambda_x = \lambda$ is less than unity, so $\lambda^2 - 1/\lambda$ becomes negative. However, the sign of the stress also reverses on going from tension to compression, and the equation applies to phantom networks over the entire range. The tensile modulus is obtained by applying Eq. 9.54:

$$E_o = \Gamma \lim_{\lambda \to 1}\left(\frac{\lambda^2 - 1/\lambda}{\lambda - 1}\right) = 3\Gamma \tag{9.87}$$

When compressibility effects are negligible, the initial shear and tensile moduli are related through $E = 3G_o$ [Eq. 9.56]. Accordingly, the parameter Γ is the shear modulus for both models, and both predict the relationship:

$$\sigma/\left(\lambda^2 - 1/\lambda\right) = G_o \tag{9.88}$$

9.7.1 Modulus Magnitude

The comparisons of observed and predicted moduli are of course only possible when the chemistry of network formation is appropriately calibrated and confirmed, as discussed in Sections 9.1 through 9.3. Only when proper calibration is established can relevant structural parameters, such as the elastically active strand and junction concentrations, be estimated. The early results of this kind of absolute comparison, obtained when only the affine model was available, were commonly expressed as a *front factor*, or the ratio g of observed modulus to the modulus predicted using the affine network formula. In this scheme, the front factor for a phantom network would be:

$$g_{ph} = \frac{\nu - \mu}{\nu} \tag{9.89}$$

with $g = 1/2$ corresponding to the phantom network prediction for tetrafunctional junctions ($f = 4; \mu = \nu/2$).

Before neutron scattering began to provide information about the dimensions of polymer chains and network strands in concentrated systems, coil size uncertainties clouded the subject as well. Reviews of the pre-1975 literature[28,41] noted numerous instances of studies done with sufficient care taken in structural characterization to provide useful estimates of active strand concentration ν, and hence of the affine model modulus. In some cases, $g \sim 0.5$ was found, indicating consistency with the phantom network prediction. Others found $g \sim 1$, indicating consistency with the affine model. From some results, however, $g > 1$ was obtained, and thus moduli even larger than affine network prediction[13,83,84]. Formulas for estimating active strand concentration and entanglement effects became more refined throughout this period.

By the mid-1970s, neutron scattering had cleared up the uncertainties about chain dimensions in melts and networks. Corrections of the modulus for entanglement contributions were also being applied routinely at this time, with g now redefined as the front factor after such corrections were made. Through these later "entanglement-corrected" procedures, $g \sim 0.5$ was obtained for a series of solution-formed polystyrene networks by Allen, Walsh et al.[85–87], and $g \sim 0.6$ for a series of melt-formed polydimethylsiloxane networks by Langley and Polmanteer[88].

The current formulation for incorporating entanglement effects into the analysis evolved from the Langley approach[47,88,89]. The shear modulus, measured or obtained from tensile measurements using Eq. 9.56, is expressed as the sum of two contributions—those from the network architecture (the structural term) and those

TABLE 9.3 Plateau modulus and entanglement molecular weight for selected polymer species[90,91]

Polymer	T(°C)	G_N^o (MPa)	M_e
polyethylene	140	2.60	1,050
polyethylene oxide	140	1.80	2,000
1,4 polybutadiene	140	1.25	2,250
1,2 polybutadiene	25	0.57	3,900
polypropylene	140	0.47	5,750
1,4 polyisoprene	140	0.42	6,750
polyvinylmethylether	30	0.41	6,400
polyvinyl acetate	75	0.36	8,600
polyisobutylene	140	0.32	9,100
polymethylmethacrylate	140	0.30	12,500
polystyrene	140	0.20	16,600
polydimethylsiloxane	140	0.20	15,500
poly(1-butene)	140	0.20	13,800
Polycyclohexylethylene	160	0.068	49,000

from trapped entanglements (the topological term). At the conditions of network formation,

$$G_o = (\nu - h\mu)kT + T_e G_e \tag{9.90}$$

in which T_e is the Langley trapping factor (see Section 9.3.2). The modulus term G_e is the maximum entanglement contribution, a temperature-insensitive property of the polymer species and concentration during network formation. Although used as a fitting parameter for network data, it is conceptually the plateau modulus of the system at the moment of network formation:

$$G_e = G_N^o \tag{9.91}$$

The topological term $T_e G_e$ reflects the contribution of entanglement constraints on the network strands. The values of plateau modulus and entanglement molecular weight, $M_e = \rho RT/G_N^o$, are listed for some representative polymer species[90,91] in Table 9.3. The plateau modulus, an important parameter in polymeric liquid dynamics, is discussed extensively in Volume 2.

The structural term $(\nu - h\mu)\, kT$ contains the *fluctuation span coefficient h*, reflecting the influence of entanglement constraints on the junction fluctuations. The possibility of entanglement-induced suppression of junction fluctuations was suggested in 1975 by Ronca and Allegra[92], and h was designed to accommodate that effect[93]. The value of this coefficient, used as a second fitting parameter for network data, is expected to lie between $h = 0$ (suppressed junction fluctuations: the affine network form) to $h = 1$ (freely fluctuating junctions: the phantom network form). Thus, the front factor g, which had much earlier reflected uncertainties in both coil size and network imperfection effects became $1 - h\mu/\nu$, and the various representations of the topological contributions became $T_e G_N^o$.

Two related issues on network elasticity were being actively debated in the late 1970s. One had to do with the evaluation of network structure, mainly the elastically active strand and junction concentrations. The second issue concerned the contribution of entanglements. Counting the links in crosslinked networks gave particular difficulty, because spectroscopic methods were impractical—concentrations are low, absolute calibration is difficult, and assigning elastic activity spectroscopically in any case is impossible. Indirect counting methods, such as gel curve measurements, rely on the randomness of linking and intricate data analyses to deduce the structural parameters of interest, and they depend mainly on internal consistency for verification.

Endlinked networks are inherently better for quantitative studies of elasticity because the concentration of potential linking sites can be estimated independently, through the study of the precursors. They do, however, carry the danger of unwarranted belief in "perfect" linking chemistry. Methods of endlinking had greatly improved by the mid-1970s, especially for the formation of polydimethylsiloxane networks, and several laboratories undertook studies of their elastic properties[53,94–98]. Macosko and coworkers in particular, but others as well, used measurements of sol fractions and variations of stoichiometry to demonstrate that even with favorable linking chemistry, endlinking reactions are seldom perfect and that even small degrees of imperfection can produce significant effects on network elasticity. A massive analysis of data from the several laboratories demonstrated rather conclusively that a proper handling of such data indicates a substantial elevation of the initial modulus beyond the prediction of affine or phantom models alone[99].

To appreciate the difficulties of dealing with network elasticity quantitatively, even for the relatively clean polydimethylsiloxane endlinking system, consider the informal inter-laboratory comparison of initial moduli gathered from the literature in Table 9.4. Tetrafunctional linking agents were used, three strand lengths are represented, and no corrections have been applied. The results do not constitute a statistically valid comparison—different crosslinking systems are represented, results from the early 1970s to the present time are included, and reactant stoichiometry has been varied in some cases to optimize the modulus. Nonetheless, the spread of values is remarkably large.

Linking chemistry and the approaches used to minimize its imperfection have advanced since the early 1980s. Results published in 1992 by Cohen and coworkers[55] are shown in Table 9.5. These results are based on optimized networks, the ratio of ingredients having been varied to maximize the modulus or minimize the swell ratio (see Section 10.2). A confirmation of the validity of this procedure by the Gilra et al. simulations[25] has already been mentioned. The experimental modulus G_o for a series

TABLE 9.4 Modulus comparison for endlinked polydimethylsiloxane networks from tetrafunctional linking agents

	G_o (*MPa*)		
Laboratory	$M_s \sim 10{,}000$	$M_s \sim 20{,}000$	$M_s \sim 30{,}000$
A	0.37	0.27	0.21
B	0.31	0.23	0.15
C	0.30	0.20	0.16
D	0.29	0.27	0.30
E	0.29	0.24	—
F	0.29	0.18	0.15
G	0.22	0.13	0.10
H	0.20	0.10	—

of tetrafunctional endlinked networks of polydimethylsiloxane at 30°C are compared with the modulus predicted by the affine model $\nu_s RT$, where the strand number density ν_s is the ratio of mass density to strand precursor molecular weight ρ/M_{B_2}. The affine prediction works only for fairly short strands and becomes significantly smaller than the observed value as the strand length increases. For long strands, G_o becomes relatively constant at a value close to the plateau modulus of PDMS, $G_N^o = 0.20\ MPa$, as suggested by Eq. 9.91.

9.7.2 Experimental Tests

Using a properly calibrated linking chemistry, based on a systematic study of sol fraction versus reactant stoichiometry or crosslinking agent concentration for example, the network parameters ν, μ, and T_e required by Eq. 9.90 can be estimated (Section 9.3). The maximum entanglement contribution G_e is expected to be approximately G_N^o, the plateau modulus of the polymeric liquid prior to linking, as mentioned earlier. No way has been found to predict the span of junction fluctuations from network structure, so the value of h remains wholly empirical. However, values near unity seem plausible for lightly entangled networks, $(\nu - h\mu)kT \gtrsim T_e G_e$, and near zero for heavily entangled ones, $(\nu - h\mu)kT \lesssim T_e G_e$.

Equation (9.90) has been tested extensively and carefully since the mid-1970s on both endlinked and crosslinked networks. Several polymer species having widely differing values of G_N^o were used, and the networks were prepared and tested in both undiluted and concentrated solution reference states. During this time, the chemistry as well as pitfalls of endlinking also became better understood and thus increasingly subject to calibration and control. Gelation studies supplied calibration or the confirmation of linking densities. The major post-1975 results are summarized in Table 9.6. The ratio of experimental modulus to affine model prediction, $G_o/\nu kT$, describes the scale of values to be understood. Thus, $G_o/\nu kT > 1$ indicates a modulus

TABLE 9.5 Experimental shear modulus for tetrafunctional endlinked PDMS networks and the affine model predictions for perfect and observed linking chemistries[55]

$M_s^{(a)} \times 10^{-3}$	$\rho/M_s^{(b)}$ $(mol\ cm^{-3}) \times 10^4$	$\nu_s^{(c)}$ $(mol\ cm^{-3}) \times 10^4$	G_o (MPa)	$G_o M_s/\rho RT^{(d)}$	$G_o/\nu_s RT^{(e)}$
2.46	3.94	2.42	0.763	0.77	1.25
6.16	1.57	1.00	0.463	1.18	1.85
8.35	1.16	0.645	0.375	1.29	2.32
8.35	1.16	0.677	0.388	1.34	2.30
10.3	0.94	0.509	0.393	1.67	3.08
10.9	0.89	0.438	0.375	1.69	3.43
16.8	0.58	0.240	0.187	1.29	3.12
18.5	0.525	0.315	0.327	2.50	4.17
18.5	0.525	0.249	0.257	1.96	4.13
19.8	0.49	0.260	0.269	2.20	4.15
19.8	0.49	0.229	0.274	2.24	4.79
20.1	0.48	0.240	0.207	1.72	3.44
28.2	0.345	0.199	0.207	2.41	4.18
53.5	0.181	0.103	0.200	4.41	7.75
58.0	0.167	0.0687	0.187	4.47	10.87

(a) M_s is precursor strand molecular weight. (b) ρ/M_s is the network strand density for perfect linking chemistry. (c) ν_s is the elastically active strand density as corrected with sol fraction information. (d) and (e) are ratios of measured modulus to affine model prediction for uncorrected and corrected elastically active strand densities.

that exceeds the maximum attainable value of the structural term in Eq. 9.90, a number obtained from the data with a minimum of correction and analysis. The range covered by each series is listed in the table.

The Langley trapping factor T_e characterizes the degree of network perfection. For endlinking, it varies with the stoichiometric ratio of reactant groups. For crosslinking, it varies with the length of the primary chains. In general, T_e increases with distance beyond the gel point, and its upper limit is set by various types of side reactions. The sol fraction provides an approximate guide through the formula $T_e = \left(1 - s^{1/2}\right)^4$ in Table 9.2. Thus, a 1% sol fraction corresponds to $T_e = 0.66$ and 3% sol to $T_e = 0.36$. The trapped entanglement contribution to the modulus is very sensitive to network perfection. The range of T_e for each series is shown in Table 9.6.

The experimental uncertainties are too large to permit assigning individual values of h to each network, so all networks within a series are typically assumed to have the same average fluctuation-span parameter $\bar{h}$. The data for most series in Table 9.6 were analyzed by means of a *Langley plot,* a rearranged form of Eq. 9.90:

$$G_o/T_e = G_e + A\nu RT/T_e \tag{9.92}$$

in which $A = 1 - h\mu/\nu$. A plot of G_o/T_e versus $\nu RT/T_e$ for a series of networks should be a straight line if the quantity $h\mu/\nu$ does not vary too much over the range. The average functionality $< f >= 2\nu/\mu$ usually varies little within a series, and an average value of h can be obtained from the slope:

$$\bar{h} = \overline{h\mu/\nu}/\overline{\mu/\nu} = (1 - A) < f > /2 \tag{9.93}$$

Thus, $\bar{h}$ and G_e are fitted quantities, obtained by analyzing the modulus data using Eq. 9.92.

The plateau modulus for an undiluted polymer liquid is a property of the species, commonly lying in the range, $0.1 < G_N^o(MPa) < 3$. It is insensitive to large-scale molecular architecture and the temperature, but it varies significantly with polymer concentration (see Volume 2). Interestingly, regardless of polymer and diluent species, the dependence on polymer volume fraction ϕ is essentially universal:

$$G_N^o \propto \phi^a \; (2.1 \lesssim a \lesssim 2.3) \tag{9.94}$$

Values of the plateau modulus are known for many polymer species[90,91,100]. In addition, a reasonable estimate can be made for any species with a known packing length, including statistical copolymers, by means of the *Lin-Fetters equation*[90,101]:

$$G_N^o = 0.48\frac{kT}{l_p^3} \tag{9.95}$$

Packing lengths for some common species are listed in Table 4.2.

TABLE 9.6 Summary of structural and entanglement contributions to elasticity for well-characterized networks

Network Type	Formation State	$G_o/\nu kT$	T_e	$\langle f \rangle = 2\nu/\mu$	$\bar{h}^{(a)}$	$G_e\,(MPa)$	$G_N^o\,(MPa)$	References
				Polydimethylsiloxane				
endlinked, $f = 3$	undiluted	0.80–4.8	0.15–1.00	3.0	0.91	0.23	0.20	99,102
endlinked, $f = 3$	$0.23 < \phi < 0.88$	0.40–1.16	0.03–1.00	3.0	1.09	$0.24\phi^{2.3}$	$0.20\phi^{2.3}$	102
endlinked, $f = 4$	undiluted	0.83–3.50	0.27–0.87	3.45	0.77	0.22	0.20	99
endlinked, $f = 4$	undiluted	1.3–11	0.48–0.82	3.25	0.14	0.29	0.20	55
endlinked, $f = 5$	undiluted	—	—	—	—	0.29	0.20	103
endlinked, $f > 10$	undiluted	1.6–5.4	0.40–0.88	>10	(b)	0.24	0.20	99
crosslinked	undiluted	0.89–1.62	0.06–0.85	3.4	0.69	0.25	0.20	99
				Polybutadiene (8 % Vinyl)				
multilinked	undiluted	2.8–7.4	0.62–0.83	4.3	~0	1.20	1.15	93
crosslinked	undiluted	1.0–1.7	0.43–0.86	3.6	0.66	0.96	0.96(c)	104
multilinked	$\phi = 0.50$	2.1–3.3	0.59–0.75	4.1	~0	0.26	0.24(c)	93
				Polybutadiene (43 % Vinyl)				
crosslinked	undiluted	0.57–1.55	0.22–0.96	3.9	0.91	0.50	0.70(c)	104
				Polybutadiene (88 % Vinyl)				
crosslinked	undiluted	~4	—	—	—	0.70	0.66(c)	105,106
				Ethylene-Propylene Copolymer (60 mole % Ethylene)				
crosslinked	undiluted	2.4–4.4	0.28–0.40	3.45	~0	1.74	1.7(c)	108

(a) $\bar{h} \equiv \langle hf \rangle / \langle f \rangle$, (b) Results insensitive to $\bar{h}$, (c) Plateau modulus as determined by the authors; other values taken from ref. 90.

Polydimethylsiloxane (PDMS) networks formed by endlinking have been studied extensively and are well represented in Table 9.6. The paper of Gottlieb et al.[99] is a comprehensive review in its own right, covering twelve studies of PDMS networks at several laboratories up to 1981. The more recent results for PDMS networks[55,102,103] are included as well. Entanglement trapping factors and the concentrations of elastically active strands and junctions were estimated from stoichiometry and sol fraction measurements using the Miller-Macosko equations[11].

The remainder of Table 9.6 deals with networks formed by crosslinking well characterized and usually nearly monodisperse primary chains. Network architecture—values of ν, μ, and T_e—were estimated from gel curve measurements and primary chain polydispersities using the Pearson equations[35]. The species represented are polybutadiene with vinyl contents of 8%[93,104], 43%[104], and 88%[105,106]—PBD8, PBD43, PBD88—polydimethylsiloxane[88,107] and statistical ethylene–propylene copolymers containing ~60 mol% ethylene—EP60[108]. Gelation data for the Dossin and Pearson networks are shown in Figures 9.15a and 9.15b. In both systems, sol fraction is essentially a function of the radiation dose–primary chain length product DM_w^o alone, consistent with the similar polydispersity within each series and indicating that crosslinking and scission probabilities are independent of chain length (Section 9.2.2). Expressed in terms of Eq. 9.23, a detailed analysis yields $k_X = 2.36 \times 10^{-10}\ rad^{-1}$, $k_S/k_X = 0.036$, and a slight amount of multilinking for PBD8[93]. For EP60, $k_X = 0.196 \times 10^{-10}\ rad^{-1}$ and $k_S/k_X = 0.24$[108]. Values of ν, μ, and T_e follow directly from this information on α and β.

The represented species span a wide range of plateau moduli, $0.20 \lesssim G_N^o\ (MPa) \lesssim 1.7$, which encompasses the values for nearly all polymer species, certainly for all common elastomers. Langley plots for three series of undiluted networks, catalytically endlinked PDMS, electron beam multilinked PB8, and electron beam crosslinked EP60, are shown in Figure 9.16. The PB8 studies also include electron beam multilinking in a diluent ($\phi = 0.5$) and crosslinking by catalytic coupling. The values of $\bar{h}$ differ among the three series: $\bar{h} = 0.66$ for the catalytically crosslinked networks[104]; and $\bar{h} \sim 0$ for the networks crosslinked by electron beam[93], suggesting the possibility of differences in mechanical behavior owing to differences in linking mechanism. Conversely, as shown in Figure 9.17, the results of catalytic crosslinking and crosslinking by gamma radiation agree fairly well in the case of PDMS networks. Also, differences in $\bar{h}$ are small among the series shown in Figure 9.16, despite distinctly different linking methods, and large differences appear in $\bar{h}$ among endlinked PDMS network series, despite similar linking chemistry. Overall, there may be a weak trend toward smaller $\bar{h}$ with increasing $G/\nu kT$, the intuitively expected result, but certainly no clear proof exists.

The results for PB88 were obtained by the *two-network method* of Ferry and coworkers[105,106], based on the following procedure. An entangled polymer melt is deformed from a rest state (extension ratios $\lambda_x^o, \lambda_y^o, \lambda_z^o$) and crosslinks are added

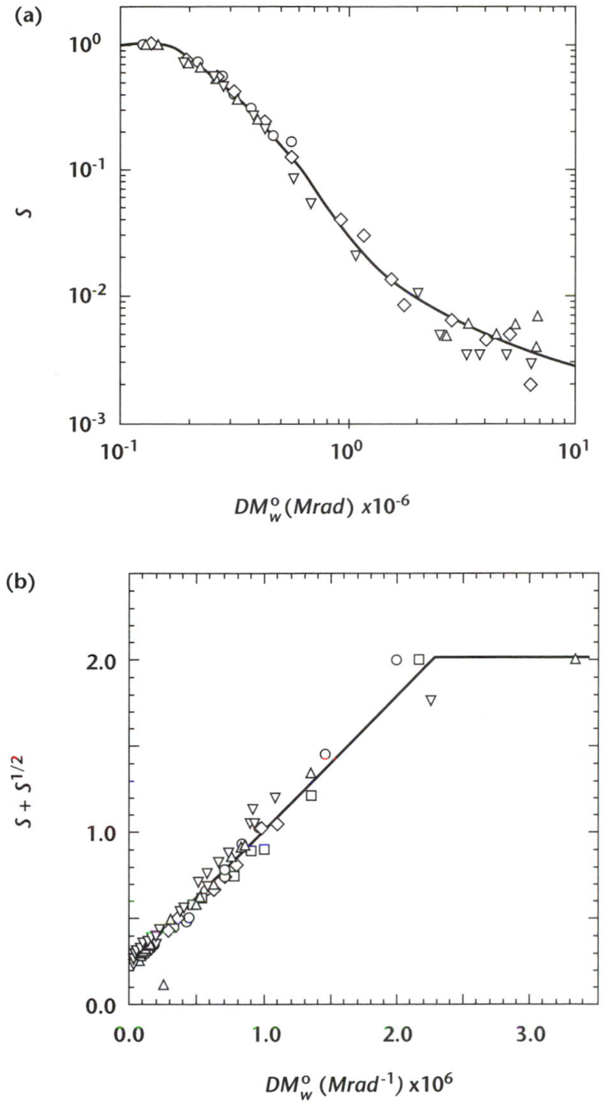

FIGURE 9.15 Experimental gel curves for well-characterized polymers linked by various doses of electron beam radiation, fitted to random linking theory to obtain linking and scission frequencies. (a) Data of Dossin for 1,4 polybutadienes with $22{,}500 \leq M_w^o \leq 210{,}000$ and nearly monodisperse primary molecular weight distributions[93]. (b) Data of Pearson for ethylene–propylene copolymer fractions with $90{,}000 \leq M_w^o \leq 310{,}000$ and narrow distributions, plotted in the Charlesby-Pinner manner[108].

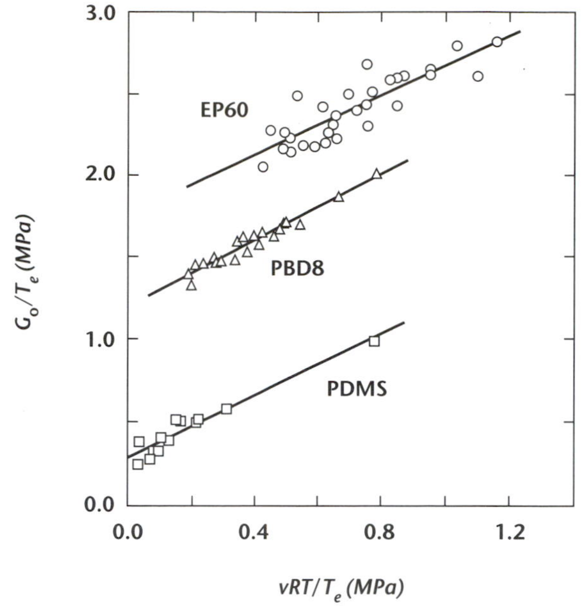

FIGURE 9.16 Langley plots of initial modulus data for well-characterized networks. Data of Pearson for ethylene–propylene copolymer networks[108] (○) and Dossin for 1,4-polybutadiene networks[93] (Δ), both linked by various electron-beam radiation doses, and data of Patel et al. for endlinked ($f = 4$) polydimethylsiloxane networks[55](□).

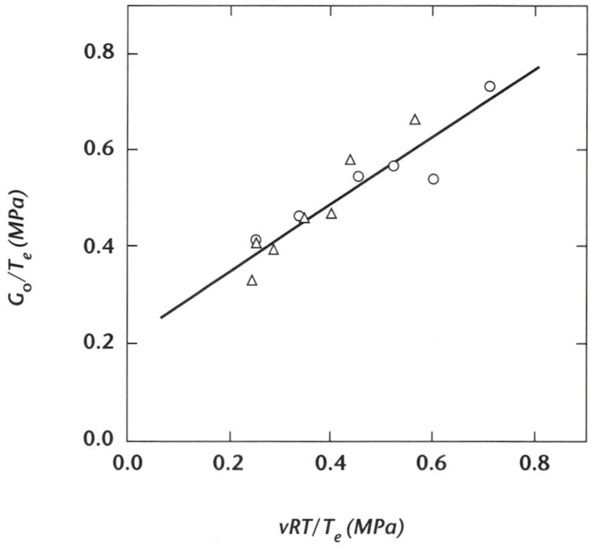

FIGURE 9.17 Langley plot of modulus data for crosslinked polydimethylsiloxane networks. Results obtained by chemical linking from Gottlieb et el.[107] (Δ) and by various gamma radiation doses from Langley and Polmanteer[88](○).

quickly, before significant relaxation occurs. The resulting network is then released and allowed to equilibrate (extension ratios $\lambda_x^s, \lambda_y^s, \lambda_z^s$ with respect to the original state). A formula based on the affine network model relates the original modulus of a crosslinked network, the modulus after some second set of crosslinks is added in the deformed state, and the various λ^o and λ^s values[109]. In the Ferry method, entangled chains comprise the original "network." Final modulus data is obtained for a range of imposed strains, then extrapolated to the small imposed-strain limit to obtain G_c, the modulus of the chemical network. The two-network formula is then applied to obtain the modulus of the original network, which is the entanglement contribution G_e in this case. The results for various types of deformation gave the results for PB88 in Table 9.6. For that system, the value listed for $G_o/\nu kT$ is the ratio $(G_c + G_e)/G_c$.

Further evidence regarding topological contributions to network elasticity have been obtained by other methods. Kramer and coworkers[110,111] developed an elegant variant of the two-network method that demonstrates, in an essentially model-free manner, the reality of entanglement trapping. Thus, instead of releasing the stress after the crosslinks are added and permitting the network to find its rest state, the system is allowed to equilibrate while the imposed deformation $\lambda_x^o, \lambda_y^o, \lambda_z^o$ is maintained. The crosslinks were introduced at that state so, according to the affine network theory[109], the network formed by those crosslinks cannot contribute to the equilibrium stress. The stress should thus decay away to zero if the network strands can equilibrate by passing through one another, the latter being one of the affine network's assumptions. A significant residual stress at equilibrium, however, implies that the topological relationships of chains are not only trapped permanently, but also that the trapped topology contributes to equilibrium elasticity.

The results for a PB88 sample[112] are shown in Figure 9.18. A series of nine experiments is represented, each with a different time lapse between the imposition of deformation and the addition of crosslinks. The filled points show stress at the moment of crosslinking divided by the imposed strain. These points describe a viscoelastic property of the uncrosslinked PB88 liquid called the *stress relaxation modulus*[38]. The open points show the stress-deformation ratios after full equilibration—$\lambda_x^o, \lambda_y^o, \lambda_z^o$ being maintained throughout—thus representing the stress supported by topological entrapment alone. The abscissa is the time lapse between deformation to crosslinking. For practical reasons, the experiments were conducted at various temperatures. The shift factors a_T and T_o/T adjust all values to a common reference temperature T_o[38]. Note that significant relaxation occurs for the short time lapses, but the final stress-deformation ratio is essentially independent of time lapse, thus indicating a constant entrapment modulus. That value is very near $G_e = 0.70\ MPa$, the entrapment modulus obtained for the PB88 species by the Ferry two-network system. Reasonably, it is only for long time lapses before crosslinking, corresponding to significant large-scale configurational rearrangement, that the full potential entrapment contribution is not attained.

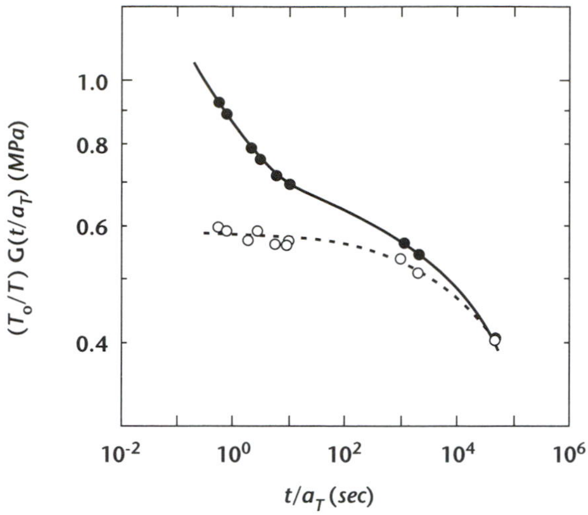

FIGURE 9.18 Results on entanglement trapping by the Kramer variant of the two-network method[112]. The filled circles are shear stress at a fixed strain versus time for a high molecular weight sample of 1,2-polybutadiene. The open circles are shear stress at the same fixed strain following strain for various time intervals, then quenching to the glass, crosslinking by various doses of electron beam radiation, and finally reheating to permit equilibration.

Network elasticity has also been examined computationally, by deforming computer-generated networks formed from melts of chains and recording their mechanical response. Strand uncrossability and conformational fluctuations are important in networks, and molecular dynamics simulations, if the computational time permits, are attractive in such cases. Exploring equilibrium stress-strain properties always requires the relaxation of any time-dependent contributions: Network conformations must readjust to each deformation. The heterogeneous nature of randomly crosslinked networks leads to very long equilibration times, so long in fact that modulus determination for realistic networks of long strands is computationally awkward, even for the fastest computers. Some progress has been made with endlinked networks, however, in obtaining rapidly equilibrating forms while still employing physically realistic linking conditions. The first computer-based results for the initial modulus of such endlinked networks, inferred from an analysis of the undeformed network dynamics, were reported in 1994[27]. The authors later concluded that this indirect method gave values of G_0 that were only approximate. They proceeded to determine the stress-strain properties directly, by imposing uniaxial stretches and compressions through changes in the simulation box dimensions[113]. The results for the modulus of simulated endlinked networks[114], including revised data for $N = 100$ networks[115], analyzed with the Langley equation, are shown in Figure 9.19.

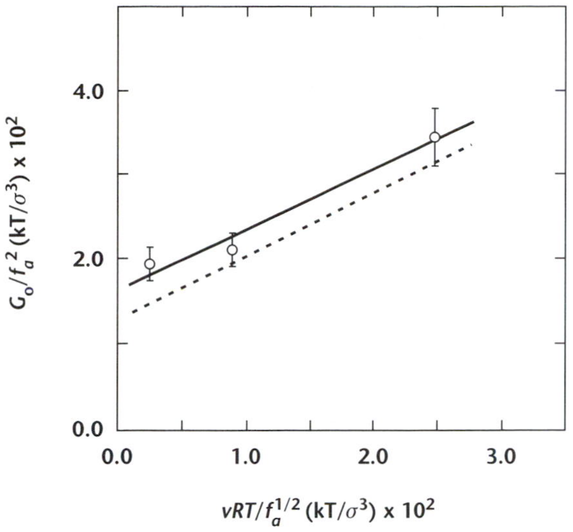

FIGURE 9.19 Modified Langley plot of initial modulus data from off-lattice Monte Carlo simulations of endlinked network formation and extensional deformation from Grest et al.[114].

The network strands are effectively modeled, as described using Eq. 6.60 and the related discussion[116], as uncrossable strings of N spheres with diameter σ that interact by hard core exclusion. The number density of spheres for all networks is fixed at $0.85/\sigma^3$, well into the dense liquid range, so that the number density of strands is $0.85/\sigma^3 N$. An initial modulus was determined for three strand lengths $N = 35$, 100 and 350: $G_o = 0.0318$, 0.0177, and 0.159 in units of kT/σ^3 with respective elastically active strand fractions, $f_a = 0.96$, 0.918, and 0.908. From the formulas in Table 9.2 and for such relatively defect-free networks as these ($s < 0.005$ from f_a and the Table 9.2 formula), νRT in Eq. 9.92 is approximately $0.85 f_a^{3/2} kT/\sigma^3 N$, and the appropriate Langley form for analyzing the data is:

$$\frac{G_o}{f_a^2} = G_e + A\frac{0.85}{f_a^{1/2} N} \tag{9.96}$$

The modulus uncertainties are $\sim 10\%$, shown as error bars in Figure 9.19. Within those uncertainties, the fit seems acceptable. From the slope ($A = 0.72$) and Eq. 9.93, $\bar{h} \sim 0.56$ for this set of computer-generated networks, within the reported values for the chemical networks in Table 9.6. From the intercept, $G_e = 0.0163$ in kT/σ^3 units, corresponding to $N_e \sim 38$ from Eq. 9.96, close to the value $N_e \sim 35$ obtained from diffusion coefficient measurements for linear chains with the same simulation model and density[117]. The three networks thus span the range from slightly entangled to highly entangled, and the results pass the rough equivalent of the G_e, G_N^o equivalency tests in Table 9.6. The dashed line is the result without corrections ($f_a = 1$); the numbers change only slightly.

9.7.3 Comments

Evidence gathered from many sources demonstrates rather clearly that the initial modulus of elastomeric polymer networks depends significantly on the topological interactions of both strands and junctions and that the results are in general accord with the Langley formulation, Eq. 9.90. The numerical agreement between G_e and G_N^o is generally good and especially convincing when one considers the strict requirements of accuracy in so many aspects of the experiments. The results for PB43 show the largest discrepancy, but in that case, the difference is still smaller than the rather large error bars on the value of G_e. The validity of Eq. 9.90 seems secure, both as an organizing principle for experimental data and (with G_e replaced by the plateau modulus G_N^o) as a means for assessing the relative importance of entanglement interactions in particular situations.

REFERENCES

1. Treloar L.R.G. 1975. *The Physics of Rubber Elasticity*, 3rd ed., Oxford: Oxford University Press.
2. Morawetz H. 2000. *Rub. Chem. Tech.* 73:405.
3. Gough J. 1805. *Mem. Lit. Phil. Soc. Manchester* 1:288.
4. Joule J.P. 1859. *Phil. Trans. Roy. Soc.* 149:91.
5. Morton M. (ed.). 1987. *Rubber Technology*, (Vol. 1). New York: Van Nostrand Reinhold.
6. Baldwin F.P. and G. Ver Strate. 1972. *Rub. Chem. Tech.* 45:709.
7. Dole M. (ed.). 1973. *The Radiation Chemistry of Macromolecules*, (Vol. 1). New York: Academic Press.
8. Pearson D.S., B.J. Skutnik, and G.G.A. Böhm. 1974. *J. Polym. Sci.: Polym. Phys. Ed.* 12:925.
9. Lloyd W.G. and T. Alfrey. 1962. *J. Polym. Sci.* 62:301.
10. Macosko C.W. and D.R. Miller. 1976. *Macromolecules* 9:199.
11. Miller D.R. and C.W. Macosko. 1976. *Macromolecules* 9:206.
12. Dutton S., H. Rolfes, and R.F.T. Stepto. 1994. *Polymer* 35:4521.
13. Flory P.J. 1953. *Principles of Polymer Chemistry*, Ithaca, NY: Cornell University Press.
14. Jacobson H. and W.H. Stockmayer. 1950. *J. Chem. Phys.* 18:1600.
15. Miller D.R. and C.W. Macosko. 1980. *Macromolecules* 13:1063.
16. Gillham J.K. and J.B. Enns. 1994. *Trends Polym. Sci.* 2:4061.
17. Stauffer D. and Aharony A. 1992. *Introduction to Percolation Theory*, 2nd ed., Philadelphia: Taylor & Francis.
18. Lusignan C.P., T.H. Mourey, J.C. Wilson, and R.H. Colby. 1995. *Phys. Rev. E.* 52:6271.
19. Allen G., J. Burgess, S.F. Edwards, and D.J. Walsh. 1973. *Proc. R. Soc. London* A 334:477.
20. Antonietti M. and C. Rosenauer. 1991. *Macromolecules* 24:3434.

21. Dušek K. 1986. *Adv. Polym. Sci.* 78:1.
22. Macosko C.W. and D.R. Miller. 1991. *Makrol. Chem.* 192:377.
23. Lee K.-J. and B.E. Eichinger. 1990. *Polymer* 31:414.
24. Stepto R.F.T. 1992. In *Comprehensive Polymer Science*, Vol. 10, Aggarwal S.L. and Russo S., (eds.). Oxford, UK: Pergamon Press, p. 199.
25. Gilra N., C. Cohen, and A.Z. Panagiotopoulos. 2000. *J. Chem. Phys.* 112:6910.
26. Grest G.S. and K. Kremer. 1990. *Macromolecules* 23:4994.
27. Duering E.R., K. Kremer, and G.S. Grest. 1994. *J. Chem. Phys.* 101:8169.
28. Dušek K. and W. Prins. 1969. *Adv. Polym. Sci.* 6:1.
29. Mark J.E. 1970. *J. Am. Chem. Soc.* 92:7257.
30. Benoit H., D. Decker, R. Duplessix, C. Picot, P. Rempp, J.C. Cotton, B. Farnoux, G. Jannink, and R. Ober. 1976. *J. Polym. Sci.: Polym. Phys. Ed.* 14:2119.
31. Candau S., J. Bastide, and M. Delsanti. 1982. *Adv. Polym. Sci.* 44:27.
32. Feller W. 1971. *An Introduction to Probability Theory and its Applications*, New York: J. Wiley & Sons.
33. Mark J.E. 1994. *Acc. Chem. Res.* 27:271.
34. Alberino L.M. and W.W. Graessley. 1968. *J. Phys. Chem.* 72:4229.
35. Pearson D.S. and W.W. Graessley. 1978. *Macromolecules* 11:528.
36. Charlesby A. and S.H. Pinner. 1959. *Proc. R. Soc. (London)* A 249:367.
37. Miller D.R. and C.W. Macosko. 1988. *J. Polym. Sci., Part B: Polym. Phys.* 26:1.
38. Ferry J.D. 1980. *Viscoelastic Properties of Polymers,* 3rd ed., New York: J. Wiley & Sons.
39. Scanlan J. 1960. *J. Polym. Sci.* 43:501.
40. Case L.C. 1960. *J. Polymer Sci.* 45:397.
41. Graessley W.W. 1974. *Adv. Polym. Sci.* 16:1.
42. Eichinger B.E. 1972. *Macromolecules* 5:496.
43. Ver Strate G. and Lohse D.J. 1994. In *Science and Technology of Rubber,* 2nd ed., New York: Academic Press, p. 95.
44. Edwards S.F. 1967. *Proc. Phys. Soc., London* 91:513.
45. Graessley W.W. and D.S. Pearson. 1977. *J. Chem. Phys.* 66:3363.
46. Iwata K. 1985. *J. Chem. Phys.* 83:1969.
47. Langley N.R. 1968. *Macromolecules* 1:348.
48. Halasa A. 2002. Personal communication.
49. Boiko V.V., N.D. Malaya, and L.M. Klimenko. 1993. *Int. Polym. Sci. Tech.* 20:T51.
50. Vlasyuk M.G., V.V. Boiko, and L.G. Prokof'era. 1996. *Int. Polym. Sci. Tech.* 23:T44.
51. Breton F.J. and Parker D.K. 2001. U.S. Patent, 6,252,009.
52. Parker D.K. 2002. Personal communication.
53. Meyers K.O., M.L. Bye, and E.W. Merrill. 1980. *Macromolecules* 13:1045.
54. Macosko C.W. and G.S. Benjamin. 1981. *Pure & Appl.Chem.* 53:1505.
55. Patel K.P., S. Malone, C. Cohen, J.R. Gillmor, and R.H. Colby. 1992. *Macromolecules* 25:5241.
56. Busse W.F. 1932. *J. Phys. Chem.* 36:2862.
57. Meyer K.H., G. von Suslich, and E. Valkó. 1932. *Kolloid Z.* 59:208.
58. Mark H. and E. Guth. 1934. *Monatsh. Chem.* 65:93.
59. Kuhn W. 1936. *Kolloid Z.* 76:358.
60. Meyer K.H. and C. Ferri. 1935. *Helv. Chim. Acta* 18:570.
61. Wall F.T. 1942. *J. Chem. Phys.* 10:485.
62. Treloar L.R.G. 1943. *Trans. Faraday Soc.* 39:36.

63. James H.M. and E. Guth. 1943. *J. Chem. Phys.* 11:455.
64. Kuhn W. 1936. *Kolloid Z.* 76:258.
65. Flory P.J. and J. Rehner. 1943. *J. Chem. Phys.* 11:512.
66. Flory P.J. 1950. *J. Chem. Phys.* 18:108.
67. James H.M. 1947. *J. Chem. Phys.* 15:651.
68. Boggs F.W. 1952. *J. Chem. Phys.* 20:1952.
69. James H.M. and E. Guth. 1947. *J. Chem. Phys.* 15:669.
70. James H.M. and E. Guth. 1949. *J. Polymer Sci.* 4:153.
71. James H.M. and E. Guth. 1953. *J. Chem. Phys.* 21:1039.
72. Guth E. 1966. *J. Polymer Sci., Pt. C* 12:89.
73. Guth E. 1979. *J. Appl. Polym. Sci.: Appl. Polym. Symp.* 35:1.
74. Guth E. and H.F. Mark. 1991. *J. Polym. Sci., Part B: Polym. Phys.* 29:627.
75. Duiser J.A. and Staverman A.J. 1965. In *Progress in Non-Crystalline Solids,* Prins J.A., (ed.). Amsterdam: North-Holland Publ., p. 176.
76. Graessley W.W. 1975. *Macromolecules* 8:186.
77. Graessley W.W. 1975. *Macromolecules* 8:865.
78. Edwards S.F. 1971. In *Polymer Networks, Structure and Mechanical Properties,* Chompff A.J. and Newman S., (eds.). New York: Plenum Press, pp. 83–110.
79. Deam R.T. and S.F. Edwards. 1976. *Philos. Trans. R. Soc. London, Ser. A* 280:317.
80. Flory P.J. 1976. *Proc. R. Soc. London* A 351:351.
81. Staverman A.J. 1982. *Adv. Polym. Sci.* 44:73.
82. Boyd R.H. and Phillips P.J. 1993. *The Science of Polymer Molecules,* Cambridge: Cambridge University Press.
83. Moore C.G. and W.F. Watson. 1956. *J. Polym. Sci.* 19:237.
84. Mullins L. 1959. *J. Appl. Polym. Sci.* 2:1.
85. Allen G., P.A. Holmes, and D.J. Walsh. 1974. *Discuss. Faraday C Soc.* 57:19.
86. Walsh D.J., G. Allen, and G. Ballard. 1974. *Polymer* 15:366.
87. Allen G., P.L. Egerton, and D.J. Walsh. 1976. *Polymer* 17:65.
88. Langley N.R. and K.E. Polmanteer. 1974. *J. Polym. Sci.: Polym. Phys. Ed.* 12:1023.
89. Langley N.R. and J.D. Ferry. 1968. *Macromolecules* 1:363.
90. Fetters L.J., D.J. Lohse, D. Richter, T.A. Witten, and A. Zirkel. 1994. *Macromolecules* 27:4639.
91. Fetters L.J., D.J. Lohse, and W.W. Graessley. 1999. *J. Polym. Sci.: Pt B: Polym. Phys.* 37:1023.
92. Ronca G. and G. Allegra. 1975. *J. Chem. Phys.* 63:4990.
93. Dossin L.M. and W.W. Graessley. 1979. *Macromolecules* 12:123.
94. Mark J.E., R.R. Rahalkar, and J.L. Sullivan. 1979. *Macromolecules* 71:1794.
95. Llorente M.A. and J.E. Mark. 1980. *Macromolecules* 13:681.
96. Valles E.M. and C.W. Macosko. 1979. *Macromolecules* 12:673.
97. Granick S., S. Pedersen, G.W. Nelb, J.D. Ferry, and C.W. Macosko. 1981. *J. Polym. Sci., Polym. Phys. Ed.* 19:1745.
98. Meyers K.O. 1980. Doctoral dissertation, MIT.
99. Gottlieb M., C.W. Macosko, G.S. Benjamin, K.O. Meyers, and E.W. Merrill. 1981. *Macromolecules* 14:1039.
100. Fetters L.J., D.J. Lohse, S.T. Milner, and W.W. Graessley. 1999. *Macromolecules* 32:6847.
101. Lin Y.H. 1987. *Macromolecules* 20:3080.
102. Valles E.M., E.J. Rost, and C.W. Macosko. 1984. *Rub. Chem. Tech.* 57:55.

103. Oppermann W. and N. Rennar. 1987. *Prog. Polym. Coll. Sci.* 75:49.
104. Aranguren M.I. and C.W. Macosko. 1988. *Macromolecules* 21:2484.
105. Hvidt S., O. Kramer, W. Batsberg, and J.D. Ferry. 1980. *Macromolecules* 13:933.
106. Twardowski T. and O. Kramer. 1991. *Macromolecules* 24:5769.
107. Gottlieb M., C.W. Macosko, and T.C. Lepsch. 1981. *J. Polym. Sci., Polym. Phys. Ed.* 19:1603.
108. Pearson D.S. and W.W. Graessley. 1980. *Macromolecules* 13:1001.
109. Flory P.J. 1960. *Trans. Faraday Soc.* 56:722.
110. Batsberg W. and O. Kramer. 1981. *J. Chem. Phys.* 74:6507.
111. Batsberg W., Hvidt S., Kramer O., and Fetters L.J. 1988. In *Biol. Synth. Polym. Networks,* Kramer O., (ed.). Amsterdam: Elsevier, pp. 509–516.
112. Kramer O. 1997. Private communication.
113. Everaers R. and K. Kremer. 1996. *Phys. Rev. E.* 53:R37.
114. Grest G.S., M. Putz, R. Everaers, and K. Kremer. 2000. *J. Non-Cryst. Solids* 274:139.
115. Grest G.S. 2002. Private communication.
116. Graessley W.W., R.C. Hayward, and G.S. Grest. 1999. *Macromolecules* 32:3510.
117. Kremer K. and G.S. Grest. 1990. *J. Chem. Phys.* 92:5057.

CHAPTER 10

Network Properties

Selected properties and theories for networks without fillers are examined in this chapter. Stress-strain behavior in moderate uniaxial extensions from the state of network formation differs in form from the affine and phantom network predictions, but those data can be organized reasonably well through the Mooney-Rivlin formulation, with the coefficients C_1 and C_2 characterizing the behavior. The fraction $C_2/(C_1 + C_2)$ is shown to correlate with the fraction of the small-deformation modulus contributed by trapped entanglements. The coefficients also change in different but seemingly general ways when a network is swollen with diluents, or when deswollen in the case of networks formed in solution. These results comprise a well-defined database for testing entanglement theories of real network behavior, a still open subject. As described next, equilibrium swelling behavior also presents unsettled questions. In contrast, thermoelastic behavior, the following topic, appears to be a settled matter, presenting clear and compelling evidence for the Gaussian strand model in network elasticity. The chapter finishes with an examination of current theories of entangled network elasticity as judged by predictions of response to moderate strains in both uniaxial extension and compression.

The mechanical response of elastomeric networks to small deformations, expressed primarily in terms of the initial shear modulus G_o, was discussed in Chapter 9. Here, the focus expands to include large deformations, but still only a few deformation types. Most mechanical tests of networks are conducted in uniaxial extension. Studies that include uniaxial compression are less common, but the two together, applied to the same network, provide a more complete and interesting picture of its properties. Volume change, brought about by adding or withdrawing diluents after network formation, is the third deformation type of concern here. Large deformation phenomena—strain-induced crystallization, finite extensibility effects and failure—are not included. As in Chapter 9, only unfilled networks are considered.

The emphasis in this chapter is behavior at moderate deformations, in which the dimensions of network strands change, but the Gaussian approximation remains valid.

The goal is to understand the structure–property relations of networks at the molecular level under the simplest possible circumstances. Section 10.1 deals with the systematics of response to uniaxial extension; Section 10.2 with network swelling, the balance between network expansion, and diluent–network thermodynamics; and Section 10.3 with thermoelasticity, the temperature dependence of elastic response. Section 10.4 surveys the microscopics of network deformation and summarizes the systematics of stress-strain behavior to be accounted for theoretically. The current state of molecular theory for network elasticity, mainly extensions to finite deformations of the entangled network picture described in Section 9.7, is summarized in Section 10.5.

10.1 Stress-Strain Behavior

Relatively few studies have been reported of both uniaxial extension and compression behavior for a single sample over wide ranges of strain. One of the most extensive is also one of the oldest, conducted by Treloar in the early 1940s and discussed in his book[1]. The sample is crosslinked natural rubber, prepared under conditions that minimize strain-induced crystallization. Tensile force divided by the cross-sectional area at rest, F/A_o, commonly called the *engineering stress*. This is shown as a function of stretch ratio over the full range of study in Figure 10.1. Note that the force is negative for $\lambda < 1$. Compressive force is negative and tensile force is positive in the terminology of mechanics.

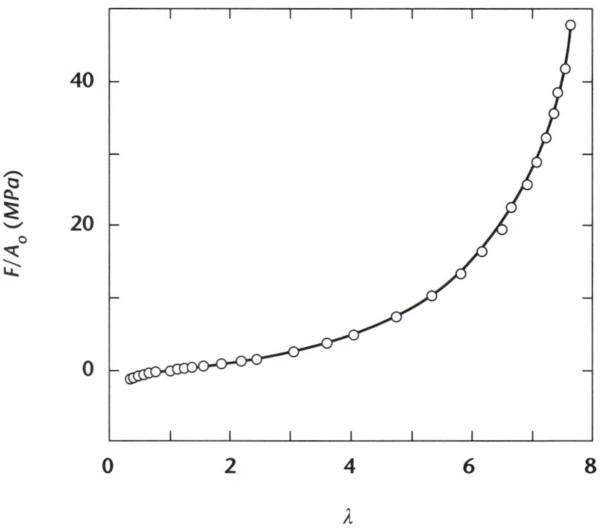

FIGURE 10.1 Nominal extensional stress as a function of stretch ratio for a natural rubber network. Data from Treloar[1].

Relaxation to an equilibrium stress after stretch or compression is applied can be slow. As explained by Ferry, many factors influence the equilibration time in networks[2]. (The time dependent behavior of elastomeric networks will be considered in Volume 2.) Equilibration time may be short enough to offer no problem, but it also may extend over many hours, days, years, or even longer, such that the equilibrium elasticity of many systems simply cannot be studied in any practical sense. A widely practiced preliminary procedure—"exercising" the test specimen by extending it briefly to the maximum strain intended for the study—significantly reduces the equilibration time for subsequent tests at smaller strains. This strain history dependence, called the *Mullins effect*[3], is not well understood. Chapter 16 of Ferry's book[2] is a definitive source of information on network relaxation. Achieving equilibrium is, of course, a major concern in quantitative studies, such as those described here and in Chapter 9. Time-dependent effects were carefully assessed and minimized for the data in Figure 10.1.

Strain-induced crystallization, which causes strong upturns in stress at high extensions, commonly occurs in networks of those species with highly regular chemical microstructure, such as natural rubber. Local chain irregularities, such as those introduced by copolymerization, can prevent crystallization. Noncrystallizing networks tend to break at lower extensions, either during stretching or, more commonly, while equilibrating. Although rare, examples are known of upturns at high strains in equilibrated noncrystallizing systems, and these are likely caused by the finite extensibility of the coils[4]. Despite the precautions taken in network preparation, the strong upturn beyond $\lambda \sim 4$ in Figure 10.1 is almost certainly the result of strain-induced crystallization[1]. The same data over a more restricted range, $0.3 < \lambda < 4$, in which strain-induced crystallization is not yet evident, are shown in Figure 10.2.

Stress-strain behavior in uniaxial extension and compression can be expressed as a *reduced stress,* $\sigma/(\lambda^2 - 1/\lambda)$, sometimes called the *reduced force*[5]. [To avoid confusion in reading other sources, it must be realized that $\sigma(\lambda^2 - 1/\lambda)$ is equal to $F/A_o(\lambda - 1/\lambda^2)$, where F/A_o is the engineering stress.] Reduced stress is independent of stretch ratio for affine and phantom networks. Constancy of the ratio $\sigma/(\lambda^2 - 1/\lambda)$ over a moderate range of λ is occasionally found for real networks. Strain independence of $\sigma/(\lambda^2 - 1/\lambda)$ in uniaxial extension and compression is a property of the *neo-Hookean solid,* one of the continuum models used in finite elasticity theory[6,7]. Both affine and phantom networks are *neo-Hookean solids,* defined by the following relationship between elastic energy density, shear modulus, and the stretch ratios:

$$W = (G_o/2)\left(\lambda_x^2 + \lambda_y^2 + \lambda_z^2 - 3\right) \tag{10.1}$$

The definitions and properties of some continuum models are discussed in Volume 2.

Variation of $\sigma/(\lambda^2 - 1/\lambda)$ with λ, as shown by the Treloar data in Figure 10.3, is the more common response for real networks. For any isotropic, effectively

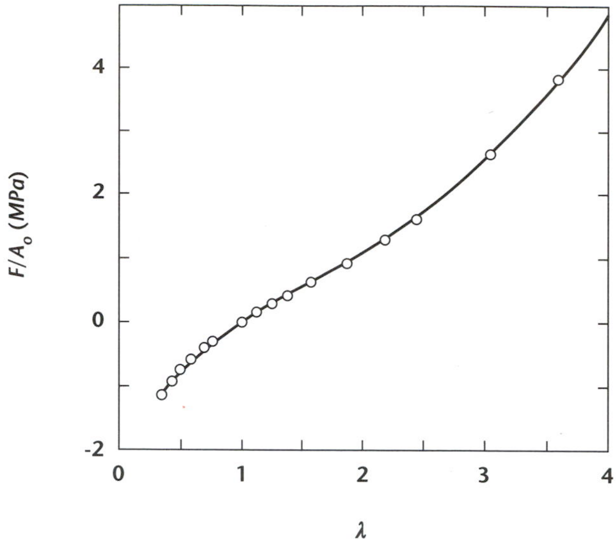

FIGURE 10.2 The data in Figure 10.1 in the low stretch region ($\lambda < 4$), where the effects of stress-induced crystallization are negligible.

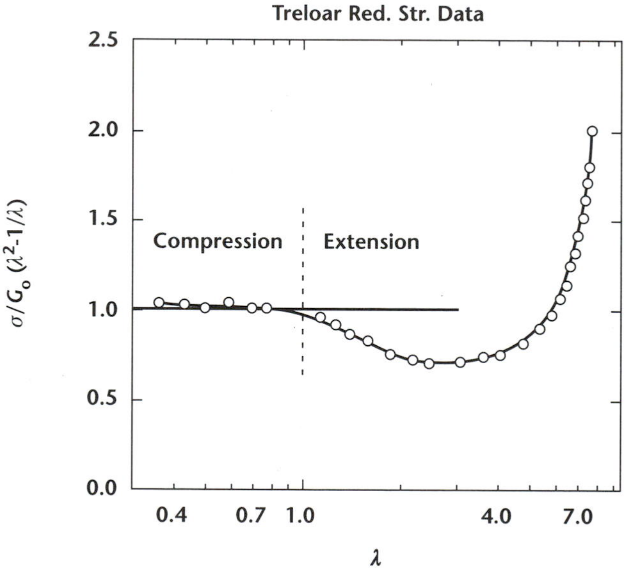

FIGURE 10.3 Reduced stress for the data in Figure 10.1, ($G_o = 0.39$ *MPa*), showing neo-Hookean behavior for compression and systematic reduction in the reduced stress for mild extensions ($1 < \lambda \lesssim 3$).

incompressible solid ($\beta E_o \ll 1$), the initial shear modulus is given by the small strain limit:

$$G_o = \lim_{\lambda \to 1} \frac{\sigma}{\lambda^2 - 1/\lambda} \tag{10.2}$$

In Figure 10.3, the reduced stress data has been normalized by the initial modulus, $G_o = 0.39$ *MPa,* thereby forcing the stress-strain curve that results to pass through unity at $\lambda = 1$. When plotted in this manner, the data for all neo-Hookean networks collapse to a single horizontal line, $\sigma/(\lambda^2 - 1/\lambda)G_o = 1$. When data for different networks are compared in this way, distinctions are based on stress-strain functional form alone. Reduced stress normalized by initial modulus, a combination used frequently in this chapter, is the *dimensionless stress:*

$$S(\lambda) = \frac{\sigma(\lambda)}{(\lambda^2 - 1/\lambda)G_o} \tag{10.3}$$

Although judged by the results of an unfortunately limited range of species and careful studies[8–14], the main features of stress-strain behavior seen in Figure 10.3 appear to be fairly general. The dimensionless stress on the compression side decreases relatively slowly with increasing strain, or as in the example here, even agrees in a rough way with the neo-Hookean form. It then commonly passes through a shallow maximum on approaching the strain-free state $\lambda = 1$, then goes over to a relatively rapid descent with increasing strain on the extension side. With further increase in extension, the values in Figure 10.3 pass through a minimum and rise again rather abruptly, owing to crystallization (as discussed earlier).

The behavior shown in Figure 10.3 is representative of elastomeric networks generally. For moderate strains, these networks behave in a nearly neo-Hookean manner in compression but "soften" conspicuously relative to that form in extension. Departures from the neo-Hookean form appear prominently in networks with significant entanglement contributions and have been variously attributed to the ability of entanglements to respond to deformation in a softer way—to slide along the constraining chain contours, so to speak—than do chemical crosslinks[15]. The amount of experimental investigation in the various regions of uniaxial extension–compression behavior varies widely. Uniaxial extension is relatively easy to study, and the major portion of experimental work is limited to that region. Uniaxial compression is more difficult experimentally, especially if the desired data cover a wide range of strains and include extension for the same network. In this section, behavior in uniaxial extension alone is considered first, then the effect of diluents, and finally the correlation with entanglement contributions. Theories for uniaxial extension–compression curves are discussed in Section 10.5.

10.1.1 The Mooney-Rivlin Form

The inadequacy of the neo-Hookean form is especially noticeable in uniaxial extension, which from the beginning has also been the most commonly used mechanical test of elastomeric network properties. A useful expression for organizing and quantifying such data was introduced in 1940 by Mooney[16]. In 1948, Rivlin demonstrated its grounding as a special case in the continuum theory of isotropic solids elasticity[6] (see Volume 2). Properties of the incompressible *Mooney-Rivlin solid* are derived from the following relationship between elastic energy density and the stretch ratios:

$$W = C_1\left(\lambda_x^2 + \lambda_y^2 + \lambda_z^2 - 3\right) + C_2\left(\lambda_z^{-2} + \lambda_y^{-2} + \lambda_z^{-2} - 3\right) \tag{10.4}$$

The coefficients C_1 and C_2 are empirical parameters. The first term has the neo-Hookean form [Eq. 10.1]; the second expresses an extensional softening relative to that form.

Although consistent with continuum elasticity, Eq. 10.4 has no intrinsic molecular content; the values of C_1 and C_2 must be determined for each network experimentally. The evaluation of tensile force for uniaxial extension ($\lambda_x = \lambda; \lambda_y = \lambda_z = \lambda^{-1/2}$) as before, with $F = V(\partial W/\partial L)_{V,T}$, leads to the *Mooney-Rivlin equation:*

$$\frac{\sigma}{\lambda^2 - 1/\lambda} = 2C_1 + 2C_2/\lambda \qquad (10.5)^*$$

The small-strain shear modulus G_o for a Mooney-Rivlin solid is evidently $2C_1 + 2C_2$. The *Mooney-Rivlin plot,* $\sigma/(\lambda^2 - 1/\lambda)$ versus $1/\lambda$ is a straight line with intercept $2C_1$ and slope $2C_2$. The Treloar uniaxial compression–extension data are plotted in Mooney-Rivlin fashion in Figure 10.4. Similar data for an endlinked polydimethylsiloxane network, although over a less extended range, are shown in Figure 10.5a[17,18]. Data for two networks obtained by molecular dynamics simulations of end-linked networks are shown in Figure 10.5b[19]. The departure of strain dependence in compression from the Mooney-Rivlin form is a general feature of polymeric network elasticity.

At one time, it was thought that departures from the neo-Hookean form were artifacts, that $C_2 \neq 0$ was a sign of nonequilibrium data. By the mid 1960s, however, it had been established that, although large but transient values of C_2 could indeed be observed during testing, they did not in fact approach zero with full equilibration[20,21]. Over modest ranges of uniaxial extension, $1 \lesssim \lambda \lesssim 2$, the equilibrium behavior of most polymer networks can be represented reasonably well by the Mooney-Rivlin form. As seen in Figure 10.4 and Figure 10.5, however, this realm of consistency with experiment does not continue across the rest state ($\lambda = 1$). Equation (10.5) describes only

---

*Some authors prefer to absorb the factors of 2 into the definitions of C_1 and C_2.

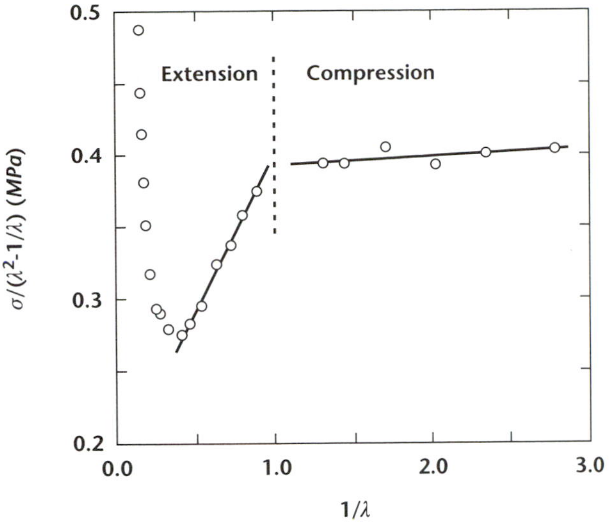

FIGURE 10.4 Mooney-Rivlin plot for the data in Figure 10.1, showing linear behavior for mild extensions ($1 > 1/\lambda \gtrsim 0.4$), but nearly constant reduced stress in compression.

a limited range of uniaxial extension. It does, however, confer the significant advantage of being a neutral framework for expressing the physical behavior of networks in a relatively simple and widely used test. The ratio of fitted coefficients, $\Lambda = C_2/C_1$, is a convenient numerical index of departure from the neo-Hookean form. For the Treloar natural rubber crosslinked network, $2C_1 \sim 0.18\ MPa$, $2C_2 \sim 0.21\ MPa$ in the extension regime ($\lambda^{-1} < 1$), and thus $\Lambda \sim 1.2$. For the Rennar-Oppermann endlinked polydimethylsiloxane network, $2C_1 \sim 0.17\ MPa$, $2C_2 \sim 0.10\ MPa$, and thus $\Lambda \sim 0.6$.

The extensive investigations of stress-strain curve shape for polymer networks prior to 1975 were based on uniaxial extension tests and expressed in terms of the Mooney-Rivlin ratio $\Lambda = C_2/C_1$. The British Rubber Producers Research Association housed the leading group engaged on elastomeric network fundamentals prior to 1970. Their work, as valuable today as then, is described in their own papers as well as surveyed in reviews on networks in 1965[20], 1969[21], and 1975[22]. The leading academic group, Allen, Price, and coworkers at Manchester, also reviewed these data in summary papers[23,24]. Along with some more recent results, these works are drawn on extensively to outline here what appear to be the general features of uniaxial extension behavior in the range $1 < \lambda \lesssim 3$ for noncrystallizing elastomeric networks.

For networks formed and tested in the absence of diluents, three significant general facts stand out clearly. First, in uniaxial extension, negative departures from the neo-Hookean form, $F \propto \lambda - 1/\lambda^2$, are the norm. Second, the coefficients C_1

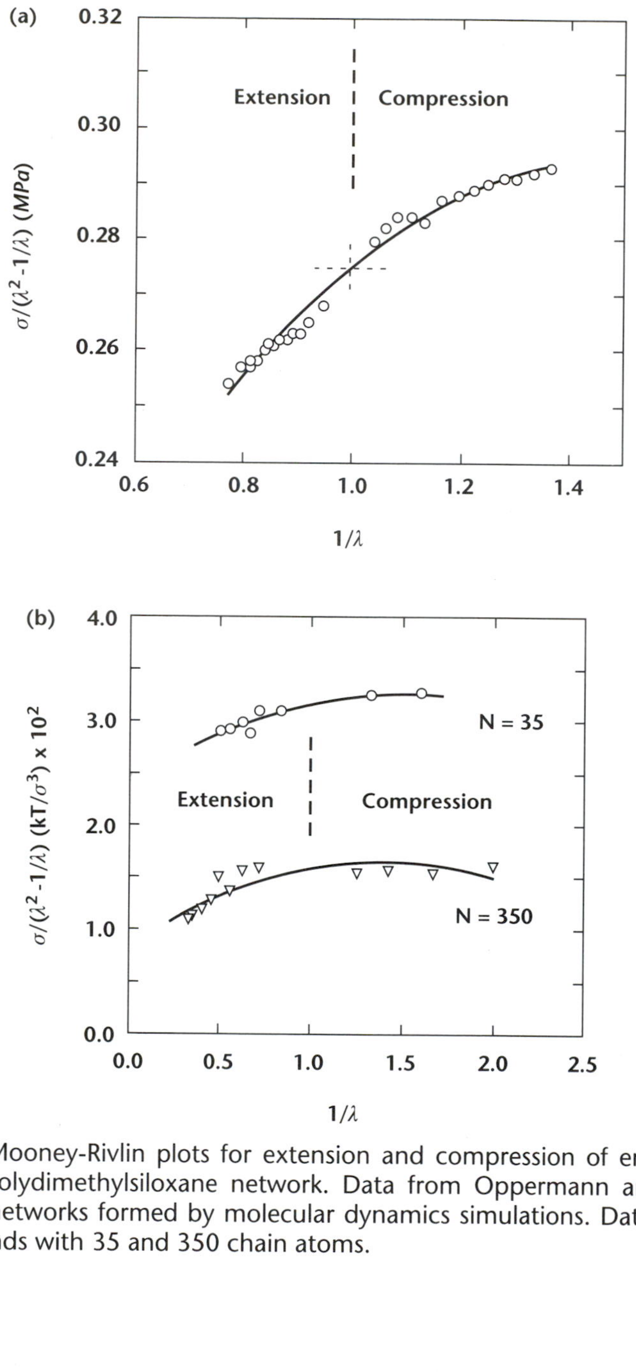

FIGURE 10.5 Mooney-Rivlin plots for extension and compression of endlinked networks. (a) A polydimethylsiloxane network. Data from Oppermann and Rennar[17]. (b) Endlinked networks formed by molecular dynamics simulations. Data from Grest et al.[19] for strands with 35 and 350 chain atoms.

and C_2 used to represent those departures have essentially the same temperature dependence. Their ratio $\Lambda = C_2/C_1$, and thus stress-strain curve shape, is for all practical purposes independent of temperature[22,25]. Third, values of Λ near unity are common for both crosslinked and endlinked networks. Some early data had suggested Λ was significantly smaller for endlinked networks, but extensive work by Mark and coworkers [14,26] have shown this to be untrue.

The effects of diluent on C_1 and C_2 are described in the following section. The correlation of C_2 with entanglement contributions is described in Section 10.1.3.

10.1.2 Swelling and Supercoiling Effects

Solvents for the precursor polymer are readily absorbed into elastomeric networks. Total volume is essentially unchanged by the absorption, so diluent addition expands the network from its dry volume V_{net} to $V_{net} + V_{dil}$, a process called *swelling*. Conversely, diluents present during network formation can be removed, thus shrinking the network from $V_{net} + V_{dil}$ to its dry volume V_{net}, a process called *supercoiling*. In general, network expansion and contraction alter properties by changing the dimensions and concentrations of the network strands. Swelling and supercoiling affect the stress-strain behavior of both affine and phantom networks in the same way, by changing the shear modulus magnitude. Thus, from Eq. 10.1 for uniaxial extension, with polymer volume fraction $\phi = V_{net}/(V_{net} + V_{dil})$ and either Eq. 9.70 or 9.82,

$$G_o(\phi_{ref}, \phi) = G_o(\phi_{ref}, \phi_{ref})(\phi/\phi_{ref})^{1/3} \qquad (10.6)$$

in which $G_o(\phi_{ref}, \phi_{ref})$ is the shear modulus for networks formed and tested at ϕ_{ref}, while $G_o(\phi_{ref}, \phi)$ is that for the same network formed at ϕ_{ref} but tested at ϕ. Accordingly, the modulus for phantom and affine networks decreases with swelling, $\phi < \phi_{ref}$, and increases with supercoiling, $\phi > \phi_{ref}$.

Suppose now that the presence of a diluent does not affect the chemistry of linking, that networks with the same architecture can be formed at several diluent concentrations. A series of crosslinked networks formed from the same primary chain precursors and having the same average number of crosslinked units per primary chain could satisfy the criterion. A series of endlinked networks formed from the same difunctional strand precursors and coupled to the same extent of reaction could also serve. Under these circumstances, the modulus of an affine or phantom network formed at polymeric volume fraction $\phi = \phi_{ref}$ and then tested there would be related to the modulus of one formed and tested in the undiluted state by:

$$G_o(\phi_{ref}, \phi_{ref}) = \phi_{ref}\, G_o(1, 1) \qquad (10.7)$$

Thus, assuming that all networks were formed in the concentrated solution region, $\phi_{ref} > \phi^{\ddagger}$, the mean coil dimensions would be unperturbed and thus the same for all members in their reference states. The only difference among the networks would therefore be the densities of strands, junctions, and the like, all scaling as ϕ_{ref} and causing the modulus to scale likewise.

The following rules apply to affine and phantom networks for some special cases:

$$G_o(1, \phi) = G_o(1, 1)\phi^{1/3} \quad \text{(Formed dry, tested at } \phi) \tag{10.8}$$

$$G_o(\phi_{ref}, 1) = G_o(\phi_{ref}, \phi_{ref})\phi_{ref}^{-1/3} \quad \text{(Formed at } \phi_{ref}\text{, tested dry)} \tag{10.9}$$

$$G_o(\phi_{ref}, 1) = G_o(1, 1)\phi_{ref}^{2/3} \quad \text{(Formed at } \phi = \phi_{ref} \text{ or dry, tested dry)} \tag{10.10}$$

Equations 10.8 and 10.9 refer to the same network tested in two states. Equation 10.10 refers to two networks, each conforming to the conditions required by Eq. 10.7 and tested in the same state, one formed at ϕ_{ref} and tested dry, the other formed dry and tested dry.

The effect on the stress-strain relationship of adding various amounts of diluent, n-decane in this case, to a natural rubber network formed by peroxide crosslinking in the melt[25] is shown in Figure 10.6. The variation in the Mooney-Rivlin coefficients with swelling is shown in Figure 10.7a. The coefficient C_1 is proportional to $\phi^{1/3}$ in the range $0.2 < \phi < 1$: The product $C_1\phi^{-1/3}$ is essentially constant, in accord with the phantom-affine network modulus prediction in Eq. 10.8. The coefficient

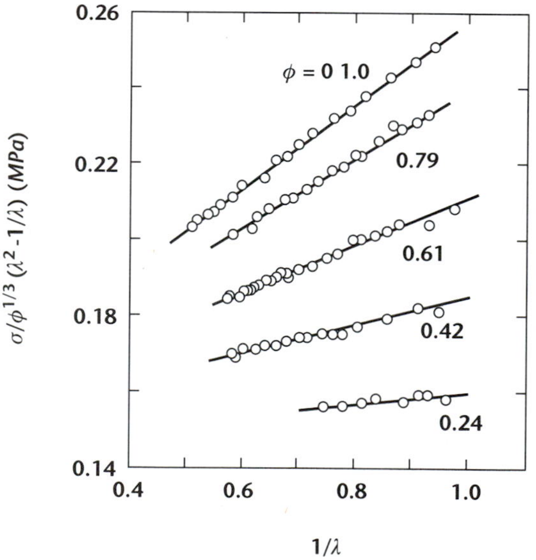

FIGURE 10.6 Mooney-Rivlin plots for a natural rubber network, dry and swollen to various extents. Data from Allen et al.[25].

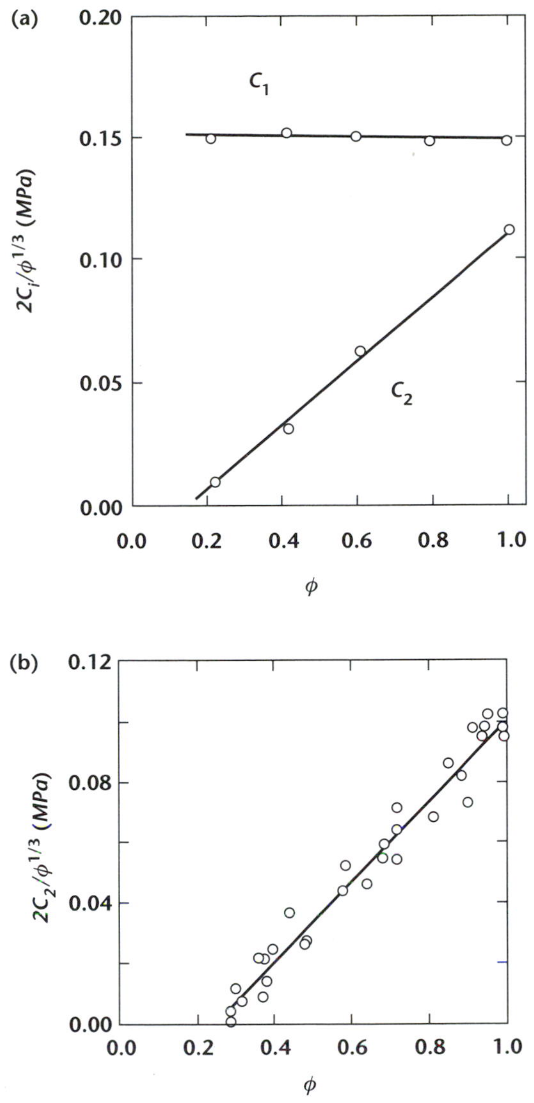

FIGURE 10.7 Dependence of the Mooney-Rivlin coefficients on volume fraction of crosslinked natural rubber networks. (a) Data for a peroxide-cured network swollen to various extents with n-decane. Data from Allen et al.[25]. (b) Data for several sulfur-cured natural rubber networks swollen to various extents with a variety of diluents. Data from Gumbrell et al.[27].

TABLE 10.1 Mooney-Rivlin coefficients for unfilled natural rubber networks from accelerated sulfur vulcanization[27]

Vulcanizate	Sulfur (pph)	Cure Time (min)	$2C_1$ (*MPa*)	$2C_2$ (*MPa*)
A	3.0	45	0.115	0.098
B	3.0	10	0.098	0.098
C	3.0	30	0.140	0.101
D	3.0	20	0.169	0.103
E	3.25	60	0.171	0.103
F	4.0	10	0.249	0.102
G	4.0	12	0.302	0.100

Compounded with 5 pph (parts per hundred of rubber) zinc oxide, 1 ppm stearic acid, 1 ppm antioxidant, and 0.5–1.5 ppm of various accelerant mixtures; cured at 141.5°C.

C_2 changes more rapidly: The product $C_2\phi^{-1/3}$ decreases linearly with the polymer concentration, going toward zero at $\phi_o \sim 0.13$. This extraordinary behavior of C_2 is not a special case. The effect of diluent on another natural rubber network in the same study was the same except $\phi_o \sim 0.25$ was obtained in that case. Other networks, several of natural rubber and an entire series based on styrene–butadiene copolymers in a comprehensive earlier study by Gumbrell, Mullins, and Rivlin[27] gave similar results and $0.20 < \phi_o < 0.25$. The values of C_1 in that study varied slightly less strongly than $\phi^{1/3}$, but still in essential agreement with Eq. 10.8. The behavior of $C_2\phi^{-1/3}$ for the series of sulfur-cured natural rubber networks described in Table 10.1, swollen by a variety of diluents, is shown in Figure 10.7b. Variations of C_1 and C_2 with dilution are independent of the diluent species, ranging in polarity from decane to nitrobenzene. The effects of dilution are evidently physical[27]. They are well described for these many networks, all natural and synthetic rubbers, by the following empirical formulas:

$$C_1(\phi) = C_1(1)\phi^{1/3} \tag{10.11}$$

$$\begin{aligned} C_2(\phi) &= C_2(1)\phi^{1/3}\left(\frac{\phi - \phi_o}{1 - \phi_o}\right) && 1 \gtrsim \phi \gtrsim \phi_o \\ C_2(\phi) &= 0 && \phi_o \gtrsim \phi \end{aligned} \tag{10.12}$$

where $\phi_o \sim 0.2$ is typical.

Some effects of supercoiling on the stress-strain behavior of natural rubber networks[28] are shown in Table 10.2. Aliquots of a natural rubber sample were dissolved in various diluents at concentrations $0.15 \le \phi \le 0.40$, crosslinked with peroxide, then tested with the diluent extracted. Except for network 9, the ratio of peroxide to polymer was the same for all, offering the possibility of forming architecturally

TABLE 10.2 Values of the Mooney-Rivlin coefficients for solution-cured natural rubber measured on the dry state[28]

Network	Diluent	ϕ_{ref}	DCP* (*pph*)	$2C_1$ (*MPa*)	$2C_2$ (*MPa*)	$2C_1\phi_{ref}^{-2/3}$ (*MPa*)
1	decalin	0.15	3.0	0.050	0.000	0.177
2	decalin	0.18	3.0	0.071	0.025	0.221
3	decalin	0.22	3.0	0.100	0.017	0.274
4	decalin	0.31	3.0	0.137	0.025	0.299
5	n-decane	0.20	3.0	0.108	0.000	0.316
6	n-decane	0.28	3.0	0.123	0.005	0.286
7	n-decane	0.28	3.0	0.119	0.007	0.278
8	n-decane	0.40	3.0	0.174	0.010	0.320
9	o-dichlorobenzene	0.25	4.0	0.177	0.000	0.445

*Parts of dicumyl peroxide per 100 parts by weight of rubber; cured at 140°C for 2 hrs.

similar networks as required for Eqs. 10.7 and 10.10 to apply. Judged by the last column in the table, and aside from networks 1, 2, and 9, the values of C_1 decrease with the concentration of polymer during network formation in rough accord with the phantom-affine modulus prediction, Eq. 10.10. Relative to the values of C_1, those of C_2 have decreased much more rapidly with diluent addition during crosslinking. The results in Figure 10.8 are representative. All the networks formed over this range of diluent concentrations and tested dry are effectively neo-Hookean in uniaxial extension. In contrast, significant values of C_2 ($\Lambda \sim 0.35$) were reported for natural rubber

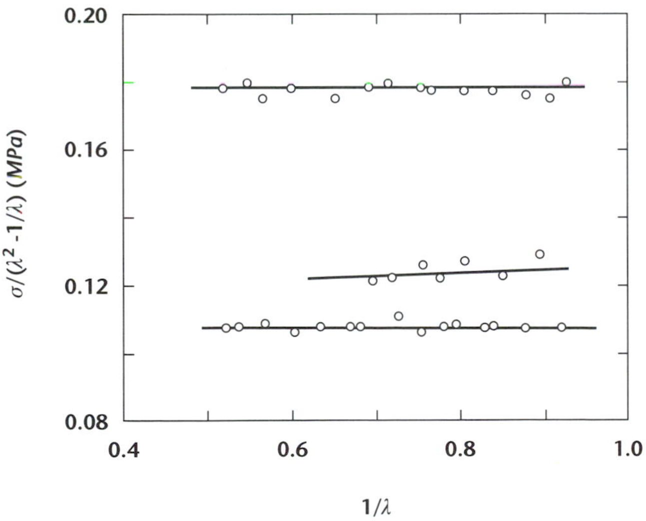

FIGURE 10.8 Mooney-Rivlin plots for supercoiled natural rubber networks, formed in solution and tested dry. Data from Price et al.[28].

networks formed from similar precursors and the same ratio of peroxide to rubber, but cured and tested without diluent present. Later work demonstrated similar behavior for supercoiled natural rubber networks in compression[29] and for supercoiled polybutadiene networks[30]. An extensive study of supercoiled polydimethylsiloxane networks[31] demonstrated similar but more gradual reductions in C_2 with increasing amounts of diluent present during crosslinking.

In summary of this early work, the behavior of C_1 in both swollen and supercoiled networks is fairly well described by the phantom-affine network model. For $\phi \gtrsim 0.2$, the variation of C_1 with diluent concentration for individual networks is about as predicted by Eq. 10.8, $C_1 \propto \phi^{1/3}$. The values of C_1 for networks with similar architecture formed from solution and tested dry are related about as predicted by Eq. 10.10, $C_1 \propto \phi_{ref}^{2/3}$. The behavior of C_2 with swelling and supercoiling, however, is considerably different. The values change with dilution more rapidly than C_1 in both cases. Empirically, the values of C_2 go roughly as $\phi^{4/3}$ with swelling according to Eq. 10.12. How rapid the variation with supercoiling has not been established.

An extensive examination of swelling and supercoiling effects on the initial modulus for endlinked polydimethyl siloxane networks has been reported recently by Sivasailam and Cohen[32]. Solutions of difunctional PDMS chains in nonreactive PDMS oligomer ($M_n = 3900$) were prepared to give $0.4 \lesssim \phi_{ref} \lesssim 1.0$. They were than mixed and catalytically reacted with a tetrafunctional linking agent to form the networks. (In separate experiments, the A_4 to B_2 ratio was varied to identify compositions of maximum modulus and minimum swell at equilibrium, as employed to prepare the Patel et al. networks[33], whose initial moduli were examined in Section 9.7). Equilibrium swelling ratios in toluene and the PDMS oligomer were determined for the resulting "optimized" networks, and small-deformation moduli were determined at three states: the reference state $\phi = \phi_{ref}$, the equilibrium swollen state in PDMS oligomer $\phi = \phi_{eq}$, and the dry state $\phi = 1$.

The results of that study, summarized in Table 10.3a, provide an opportunity to test some of the observations on crosslinked networks in the context of endlinked networks. From Eqs. 9.90 and 9.91 with $h = 0$, and from Eq. 9.94 with $a = 2.3$, the shear modulus at the reference state is the sum of crosslink and entanglement contributions:

$$G_{ref} = \nu_{ref} kT + T_e G_N^o \phi_{ref}^{2.3} \tag{10.13}$$

where G_N^o is the plateau modulus of the undiluted liquid polymer and T_e is the entanglement trapping factor. Suppose now the crosslink contribution $\nu_{ref} kT$, like the Mooney-Rivlin C_1 term, changes with concentration in the phantom-affine manner, according to Eq. 10.6. Suppose further that all networks made with the

TABLE 10.3a Observations and predictions of shear modulus for swollen and supercoiled endlinked polydimethylsiloxane networks in PDMS oligomer[32]

			G_{ref} (MPa)		$G_{swollen}$ (MPa)		G_{dry} (MPa)	
M_{B_2}	ϕ_{ref}	ϕ_{eq}	obs	calc	obs	calc	obs	calc
101,700	1.00	0.38	0.111	0.111	0.059	0.037	0.111	0.111
	0.89	0.37	0.095	0.087	0.062	0.032	0.105	0.087
	0.79	0.34	0.072	0.067	0.045	0.027	0.082	0.068
	0.74	0.32	0.062	0.059	0.036	0.023	0.071	0.060
	0.67	0.23	0.050	0.048	0.016	0.016	0.057	0.049
71,500	1.00	0.48	0.176	0.191	0.129	0.084	0.175	0.191
	0.89	0.44	0.143	0.149	0.110	0.070	0.148	0.150
	0.78	0.37	0.098	0.114	0.061	0.052	0.105	0.116
	0.72		0.089	0.097				
	0.60		0.071	0.067				
	0.50	0.30	0.048	0.047	0.036	0.029	0.077	0.051
	0.40	0.23	0.027	0.031			0.037	0.035
30,200	1.00	0.51	0.210	0.229	0.153	0.121	0.210	0.229
	0.89	0.47	0.168	0.183	0.130	0.101	0.172	0.185
	0.78	0.45	0.150	0.143	0.109	0.087	0.159	0.148
	0.70		0.106	0.118			0.135	0.123
	0.59	0.37	0.085	0.087	0.067	0.059	0.098	0.095
	0.50	0.33	0.061	0.066	0.048	0.048	0.072	0.074
	0.40		0.039	0.046				
9,900	1.00	0.62	0.343	0.363	0.280	0.258	0.342	0.363
	0.89	0.59	0.277	0.303	0.240	0.228	0.286	0.310
	0.80	0.56	0.251	0.258	0.212	0.203	0.263	0.270
	0.69	0.52	0.193	0.208	0.161	0.175	0.208	0.226
	0.57	0.47	0.147	0.159	0.133	0.141	0.172	0.182
	0.51		0.123	0.137				

same B_2 precursor are similar in structure, such that Eq. 10.7 applies to their crosslink contributions. Suppose finally that the entanglement contribution $T_e G_N^o \phi_{ref}^{2.3}$ varies with concentration like the C_2 term, according to Eq. 10.12 for swelling and, lacking clear information, remains unchanged with supercoiling. The following expressions then describe the reference state, swollen, and dry network moduli:

$$G_{ref} = G_o(\phi_{ref}, \phi_{ref}) = G(1,1)\phi_{ref} + T_e G_N^o \phi_{ref}^{2.3} \tag{10.14}$$

$$G_{swollen} = G_o(\phi_{ref}, \phi) = G(1,1)\phi_{ref}^{2/3}\phi^{1/3} + T_e G_N^o \phi_{ref}\phi^{1.3} \tag{10.15}$$

$$G_{dry} = G_o(\phi_{ref}, 1) = G(1,1)\phi_{ref}^{2/3} + T_e G_N^o \phi_{ref}^{2.3} \tag{10.16}$$

TABLE 10.3b Observations and predictions of shear modulus for swollen and supercoiled endlinked polydimethylsiloxane networks in toluene[36,37]

		G_{ref} (MPa)			G_{eq} (MPa)	
M_{B_2}	ϕ_{ref}	obs	calc	ϕ_{eq}	obs	calc
29,400	1.0	0.113	0.1183	0.187	0.033	0.029
	0.852	0.093	0.0873	0.155	0.027	0.023
	0.709	0.063	0.0624	0.126	0.0173	0.0177
	0.544	0.030	0.0394	0.0933	0.0073	0.0125
	0.411	0.019	0.0250	0.0778	0.0047	0.0093
	0.281	0.0087	0.0142	0.0549	0.0023	0.0061
	0.179	0.0030	0.0077	0.0381	0.0009	0.0039
4400	1.0	0.680	0.6132	0.275	0.251	0.323
	0.777	0.423	0.4410	0.216	0.149	0.246
	0.654	0.330	0.3558	0.208	0.137	0.215
	0.601	0.297	0.3211	0.194	0.119	0.197
	0.584	0.301	0.3102	0.195	0.113	0.194
	0.504	0.241	0.2606	0.180	0.100	0.170
	0.381	0.150	0.1892	0.140	0.053	0.128
	0.298	0.061	0.1442	0.109	0.029	0.099

where $G(1, 1)$ is νkT, the crosslink contribution for the member of each series that was formed and tested in the dry state.

For PDMS, $G_N^\circ = 0.20\ MPa$. If the linking chemistry for the four network series in Table 10.3a were perfect, $T_e = 1$ and $\nu kT = \rho RT/M_{B_2}$. The sol fractions for the member formed and tested dry are known, however—$s = 0.026$ for $M_{B_2} = 101{,}700$, $s \sim 0.0025$ for the others—so corrections can be estimated[34]. These corrections are $\nu/\nu_{perfect} = 0.52$ and $T_e = 0.49$ for $s = 0.026$, and $\nu/\nu_{perfect} = 0.825$ and $T_e = 0.813$ for $s = 0.0025$. Using these data and Eqs. 10.14 through 10.16, values for all the various observed moduli listed in Table 10.3a can be calculated from ϕ_{ref} and $\phi = \phi_{eq}$ as given with no adjustable parameters. No systematic divergence of observed and predicted moduli is evident; the overall agreement is good. A recently initiated study of these phenomena by molecular simulation methods offers some promise for a more microscopic understanding of swollen network behavior[35].

The analysis described above was also performed on other solution-formed PDMS endlinked networks[36,37], but this led to somewhat different results. Moduli in the reference and swollen states were determined with toluene as diluent for two series, $M_{B_2} = 29{,}400$ and 4400. Network structure was less thoroughly characterized than in the Sivasailam-Cohen study: Nothing was reported on sol fractions except that they were all less than 8% in one study and all less than 10% in the other. Comparison of G_0 for the undiluted reference state, with results reported for tetrafunctional linking agents by other laboratories at similar values of M_{B_2}[34], led to estimates of $\nu/\nu_{perfect} \sim T_e \sim 0.4$ for $M_{B_2} = 29{,}400$ and $\nu/\nu_{perfect} \sim T_e \sim 0.83$ for $M_{B_2} = 4000$. Comparisons of observed and calculated results are shown in Table 10.3b. Although agreeing fairly well at high concentrations, the observations and predictions systematically diverge with decreasing values of both ϕ_{ref} and ϕ_{eq}. One interpretation is that the structures within a series are not similar, so that Eq. 10.7 is not applicable.

Although far from perfect, the agreement in Table 10.3a between observed and predicted moduli in the extensive and evidently careful Sivasailam-Cohen study is good enough to support the notion that swelling and supercoiling effects on initial modulus can be understood as a sum of two rather distinct contributions. One of these is a crosslink contribution, behaving in phantom-affine fashion, and the other is an entanglement contribution, behaving in a manner that parallels the behavior of the Mooney-Rivlin term C_2.

The idea that departures from neo-Hookean behavior in elastomeric networks are related to chain entanglements is not a new one. It dates back to the 1953 Gumbrell-Mullins-Rivlin paper[27] and perhaps even earlier. That idea provides a reasonably consistent explanation of other observations, such as the seemingly unexpected behavior of the data in Table 10.1: C_1 varies throughout the series with reaction conditions, ones that presumably lead to differences in crosslink density, whereas C_2 is relatively constant. Thus, in general, the plateau modulus G_N^o is independent of chain length in melts of long chains and represents a latent entanglement contribution in the networks formed from such chains. Crosslinks trap some fraction of that potential contribution into the structure governing equilibrium elasticity. As indicated by the variations with crosslinking index γ in Figure 9.8, that trapped fraction—and hence presumably C_2—does not vary greatly with crosslink density when values of γ go beyond several crosslinked units per primary chain. In the same range of γ, the strand density—and hence presumably C_1—becomes roughly proportional to the crosslink density. Similar arguments can be applied to endlinked networks.

The following section, describing an extensive analysis of C_1,C_2 data by Ferry and Kan,[38] lends additional support to the interpretation outlined here.

10.1.3 The Ferry-Kan Formulation

The possibility that C_1, which quantifies the neo-Hookean contribution to stress-strain behavior in axial extension, is associated with crosslinks, and that C_2, which quantifies the departures, is associated with entanglements has been frequently suggested. Indeed, a good deal of circumstantial evidence, as described above, supports that idea. Even beyond those observations, large-strain stress relaxation measurements on polymer melts, highly entangled but not crosslinked, display a significant C_2 contribution at intermediate times[39], suggesting the entanglement underlies the C_2 contribution in networks as well. Finally, the additivity of crosslink and entanglement

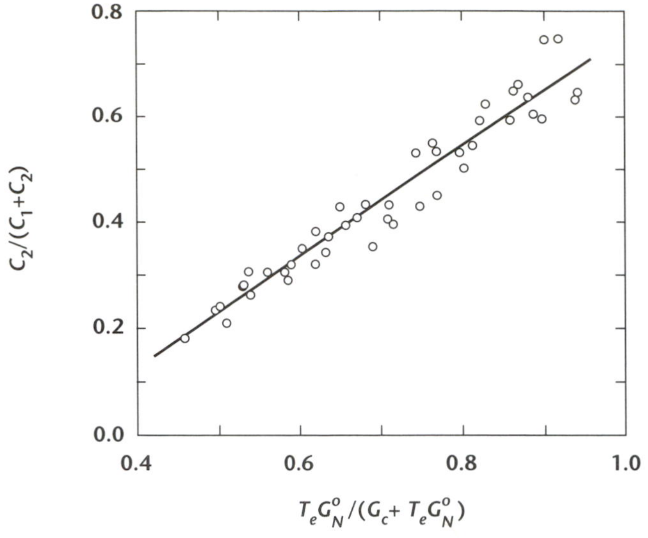

FIGURE 10.9 Relationship between Mooney-Rivlin coefficients and entanglement contributions to the initial modulus, based on data from many crosslinked network systems. Data from Ferry and Kan[38].

contributions in Eq. 9.90, which organizes initial modulus data rather well, parallels the additivity of the Mooney-Rivlin form, Eq. 10.5.

All this suggests the possibility that stress-strain curve shape depends in some general manner on the relative proportion of crosslink and entanglement contributions to the initial modulus. Ferry and Kan[38] examined that possibility by searching for a relationship between $2C_2/(2C_1+2C_2)$, a measure of the stress-strain curve shape for a network, and $T_eG_e/(G_c+T_eG_e)$, the fractional contribution of trapped entanglements to its initial modulus. The crosslink contribution G_c, $(\nu - h\mu)kT$ in Eq. 9.90, was taken simply to be νkT. Available values of plateau modulus G_N^o gave G_e; estimates of crosslink densities and trapping factors were based on Langley[40]; and C_2 was assumed to arise from trapped entanglements alone. The data are for crosslinked networks of five polymer species—natural rubber or 1,4-polyisoprene[41,42], 1,4-polybutadiene[43], cis-polybutadiene[44], polydimethylsiloxane[31], and 1,2-polybutadiene—made and tested in the undiluted state. These values represent wide ranges of both plateau modulus and trapping factor. The resulting relationship is shown in Figure 10.9. A straight line with a slope of unity was fitted to the values, leading to the *Ferry-Kan formula:*

$$\frac{2C_2}{2C_1+2C_2} = \frac{T_eG_N^o}{G_c+T_eG_N^o} - A \qquad (10.17)$$

with $A = 0.275$. Because $G_o = 2C_1 + 2C_2 = G_c + T_e G_N^o$, Eq. 10.17 can also be written:

$$2C_2 = T_e G_N^o - AG_o \tag{10.18}$$

leading to expressions for the two coefficients:

$$\begin{aligned} 2C_1 &= (1 + A)G_c + AT_e G_N^o \\ 2C_2 &= (1 - A)T_e G_N^o - AG_c \end{aligned} \tag{10.19}$$

Note that if A were zero, C_1 would come entirely from the chemical crosslinks and C_2 entirely from trapped entanglements. This is the simple and intuitive expectation, but the data over the rather wide ranges of the variables tested seem to indicate otherwise. The implication is that $C_1$and C_2 each depend on both crosslinks and trapped entanglements, the mix varying with the relative magnitudes of G_c and $T_e G_N^o$.

Ferry and Kan went on to examine data for PDMS networks that had been crosslinked in solution but tested in the dry state by Johnson and Mark[31]. They made due allowance, along the lines of Eq. 10.16, for the reduced entanglement contribution for networks formed in solution and for the modulus changes associated with diluent removal. They found reasonable agreement with Eq. 10.17, concluding that the decrease in C_2/C_1, a common feature of supercoiled networks, is mainly the result of a reduction in available entanglements for networks crosslinked in solution. Ferry and Kan were unable to incorporate the data on swollen networks into the scheme, however.

The values of C_1 and C_2 for crosslinked networks of highly entangled 1,4-polybutadiene[45] and ethylene–propylene copolymer[46] suggest slightly different expressions for the Mooney-Rivlin coefficients:

$$\begin{aligned} 2C_1 &= (\nu - h\mu)kT + 0.5T_e G_N^o \\ 2C_2 &= 0.5T_e G_N^o \end{aligned} \tag{10.20}$$

Although agreeing in magnitude with calculations based on the Ferry-Kan equation, the values of $2C_2/(2C_1 + 2C_2)$ exhibit considerable scatter, and the observed values drift increasingly below the predictions of Eq. 10.17 with increasing values of $T_e G_e/(G_c + T_e G_e)$. Meyers et al.[47] found similar agreement in magnitude between their extensive compilation of data for endlinked polydimethylsiloxane networks and the Eq. 10.17 predictions.

Thus, although the Ferry-Kan equation does not render precise agreement with all available data, it predicts magnitudes very well. The mere existence of even an approximate relationship of the Ferry-Kan type—across many species and network types—supports the notion of a general connection between the stress-strain form, as characterized by C_1 and C_2, and the trapped entanglement contribution to G_o. In the limit of entanglement domination, $T_e G_e \gg G_c$, the Eqs. 10.19 predict $C_2/C_1 \sim 2.5$, providing what is perhaps a useful number for comparing with the predictions of molecular theories for entangled network stress-strain behavior. Ferry and Kan noted that this ratio lies in the range of values obtained experimentally in the two-network studies of trapped entanglements[38,48–50].

10.2 Swelling Equilibrium

A network in contact with an unlimited amount of a liquid imbibes and expands up to some limit of diluent concentration that depends on both network structure and the liquid species. If the network is structurally uniform, its expansion might be expected to be approximately uniform as well, and affine displacement of the average locations of network junctions is a reasonable expectation. The network strands are accordingly stretched on average by swelling, and an elastic restoring pressure is developed. Diluent is driven into the gel phase by its chemical potential gradient, and equilibrium is reached when the osmotic pressure in the gel is equal to the elastic stress. This idea was put forward by Frenkel in 1938[51], although he indicates in a note added in proof that it had been independently suggested even earlier by Haller in 1931[52]. It remained for Flory and Rehner, in 1943, to express the idea in a clear and testable form[53]. They wrote the free energy of the network as the sum of two contributions, one from mixing polymer segments and diluent molecules as given by the simple mixture theory (Section 7.2.1) and the other from stretching the network strands. The sum was then minimized with respect to diluent concentration to provide a relationship at equilibrium between diluent concentration and the network structure. (This is essentially the procedure applied a few years later by Flory in his treatment of individual coil expansion in good solvents[54].)

The Flory-Rehner expression can be obtained in a slightly different, although entirely equivalent way, by following the Frenkel idea of setting the sum of osmotic pressure π_{op} and elastic pressure π_{el} to zero. Thus, as in the treatment of excluded volume expansion in Section 6.1.1,

$$\pi_{op} + \pi_{el} = 0 \tag{10.21}$$

The osmotic pressure is taken to be that for a solution of a polymer with infinite molecular weight at the concentration of the swollen network. From Eq. 7.35, the simple mixture expression is:

$$\pi_{op} = -\frac{RT}{V_d}[\ln(1-\phi) + \phi + \chi\phi^2] \tag{10.22}$$

where $\phi = V_o/V$ is the polymer volume fraction, V_o and V are the dry and swollen network volumes, V_d is the molar volume of the diluent, χ is the polymer–diluent interaction parameter, and $r = V_p/V_d = \infty$ is used. The elastic pressure is taken from the Flory version of affine network [Eq. 9.73]:

$$\pi_{el} = -G_o\left(\phi^{1/3} - \frac{2}{\langle f\rangle}\phi\right) \tag{10.23}$$

where G_o is the affine shear modulus in the dry state, and $\langle f\rangle$ is $2\nu/\mu$, the average junction functionality in the network. Elastic pressure for both the Treloar affine network [Eq. 9.75] and for the phantom network [Eq. 9.83] can be expressed as:

$$\pi_{el} = -G_o\phi^{1/3} \tag{10.24}$$

The difference between the Flory version and the other two is the "log term" in the elastic free energy [see Eqs. 9.64, 9.65, and 9.82]. All three versions can be accommodated by writing:

$$\pi_{el} = -G_o\left(\phi^{1/3} - Z\frac{2}{\langle f\rangle}\phi\right) \tag{10.25}$$

in which $Z = 1$ for the Flory version and $Z = 0$ for the phantom and Treloar versions. The substitution of Eqs. 10.22 and 10.25 into the equilibrium condition Eq. 10.21 leads to the *Flory-Rehner equation*[55,56]:

$$\ln\left(\frac{Q-1}{Q}\right) + \frac{1}{Q} + \chi\frac{1}{Q^2} = -\frac{V_d G_o}{RT}\left[\left(\frac{1}{Q}\right)^{1/3} - Z\frac{2}{\langle f\rangle}\frac{1}{Q}\right] \tag{10.26}$$

in which, following common practice, the polymer volume fraction at equilibrium is expressed in terms of the *swell ratio* $Q = 1/\phi_{eq}$.

The unambiguous confirmation of the existence and magnitude of the "log term" contribution, embodied in Eq. 10.26 as $2/\langle f\rangle Q$, is difficult experimentally and remained the subject of controversy for many years. The log term contribution comes into play only with volume change, so it has no influence on stress-strain measurements as normally conducted. A number of approaches were devised to test for its presence and to quantify it[21,57,58]. Most require additional assumptions about network structure and elasticity and about the solution thermodynamics as well, thus tending to add uncertainty. The preponderance of evidence does seem to support the existence of some extra contribution, and one that is not inconsistent with calculations based on the Flory formula [Eq. 9.64]. The precise form and magnitude of the extra contribution, and its relationship to the other elements of network structure, remain open to question.

Under certain conditions, equilibrium swelling measurements offer the possibility of examining the nature of the log term rather directly. The logarithmic contribution reduces the pressure generated by network formation. The network pressure resists swelling, so including the log term contribution, $-2/\langle f \rangle Q$, raises the solubility limit of diluents. Its effect is most important relative to the $1/Q^{1/3}$ term near $Q = 1$, so concentrations of sparingly soluble species are those most strongly affected. Further, it is easy to show using Eq. 10.26 that polymeric diluents having large molar volumes are inherently only sparingly soluble. Using long unattached chains of the same species as the network diluent, and using a network whose strands are long, has the additional advantage of making $\chi = 0$, although apparently even that seemingly sensible assumption is not certain[59]. In any case, for $\chi = 0$ and $Z = 1$, Eq. 10.26 simplifies to:

$$\ln\left(\frac{Q-1}{Q}\right) + \frac{1}{Q} = \frac{V_d G_o}{RT}\left(\frac{2}{\langle f \rangle Q} - \frac{1}{Q^{1/3}}\right) \tag{10.27}$$

Solubility thus depends only on average junction functionality $\langle f \rangle$ and the readily determinable parameters G_o and V_d.

In light of this discussion, the data on solubility of unattached polydimethylsiloxane (PDMS) chains in PDMS networks, obtained some time ago by Tobias and Gent[60,61] for another purpose, is of interest. They determined the solubility of unattached PDMS chains of various molecular weights in three nominally tetrafunctional PDMS networks made by endlinking. The A_4 to B_2 ratio was varied to obtain the optimum stoichiometry (minimum Q in benzene). The properties of the optimized networks used in the study are given in Table 10.4. The values of G_o, estimated from separately calibrated swelling measurements in benzene and from stress-strain and swelling data on networks made similarly[61], are given there as well.

For comparison purposes, the values of $G_o M_{B_2}/\rho RT$, the ratio of observed modulus to the modulus calculated for an affine network and perfect linking chemistry, are also included. Molecular weights of the diluent chains and the swelling ratios measured at 70°C are listed in Table 10.5. Gent and Tobias suggested $\langle f \rangle = 3.5$ for the networks; comparisons with other data on PDMS networks[34] suggest $\langle f \rangle = 3.2$.

Figure 10.10a compares the observed swell ratios with values calculated using Eq. 10.27 for $\langle f \rangle = 3, 4$ and ∞, the choice $\langle f \rangle = \infty$ corresponding to the absence

TABLE 10.4 Properties of endlinked polydimethylsiloxane networks[60,61]

Network	M_s	$G_o(MPa)$obs.	$G_o M_s/\rho RT$calc.
A	11,500	0.302	1.39
B	22,500	0.201	1.81
C	36,000	0.157	2.28

TABLE 10.5 Equilibrium swelling results for polydimethylsiloxane diluents[60]

Network	M_d	Q	$G_o M_d/\rho RT$
A	4,700	1.477	0.535
A	11,500	1.271	1.31
A	22,500	1.149	2.56
A	24,900	1.151	2.84
A	36,000	1.109	4.10
A	38,900	1.111	4.43
B	4,700	1.610	0.356
B	11,500	1.375	0.872
B	22,500	1.220	1.71
B	24,900	1.219	1.89
B	36,000	1.162	2.73
B	38,900	1.165	2.95
C	4,700	1.773	0.278
C	11,500	1.572	0.681
C	22,500	1.283	1.33
C	24,900	1.287	1.48
C	36,000	1.211	2.13
C	38,900	1.213	2.30

of the log term, the Treloar-phantom prediction. The dashed line in the figure corresponds to $Q_{calc} = Q_{obs}$, that is, agreement between calculated and observed swell ratios. The Treloar-phantom prediction is seriously in error for low swelling ratios, $Q \sim 1$, whereas the predictions for $\langle f \rangle = 3$ and $\langle f \rangle = 4$ agree reasonably well with the data. At higher Q, however, the $\langle f \rangle = 3$ and $\langle f \rangle = 4$ calculations predict too much swelling, and in this range the Treloar-phantom prediction seems to give a somewhat better description.

The details of behavior are seen more clearly in Figure 10.10b. The quantity $2/\langle f \rangle$ is treated as an unknown in this case and calculated using Eq. 10.27 for each network–diluent pair. The resulting values are shown as a function of Q. Most data for the three networks fall roughly along a single curve and, below $Q \sim 1.3$, most lie in the range corresponding to $3 \lesssim f \lesssim 4$. However, the apparent log term contribution decreases systematically with increasing Q throughout the entire range, reaching zero near $Q \sim 1.5$. It is not at all clear how to account for such bizarre behavior. Scattering studies of unattached chains in networks [62,63] are discussed briefly in Section 10.4.1.

Unfortunately, the strangeness of polymeric solubility adds one more element of complexity to the equilibrium swelling of networks. The Flory-Rehner derivation and the Flory modification assume affine (neo-Hookean) network behavior. From the modulus ratio, the last column in Table 10.4, all three networks have significant entanglement contributions, so their stress-strain properties probably differ from

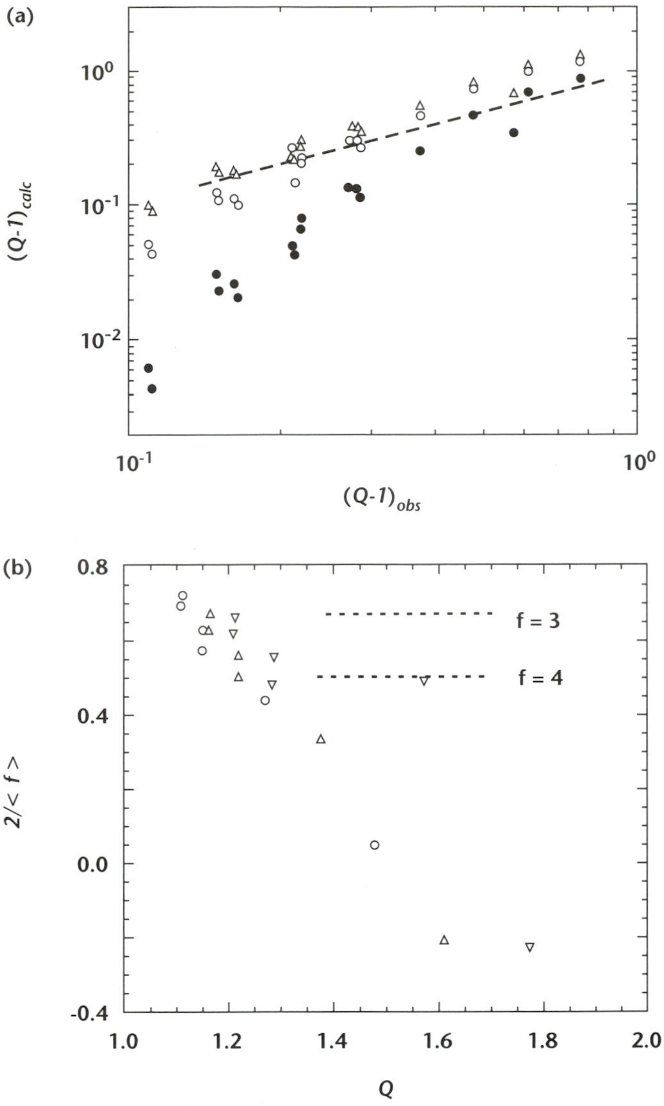

FIGURE 10.10 Test of the Flory-Rehner formula based on the equilibrium swelling of well-characterized endlinked polydimethylsiloxane networks by unattached polydimethylsiloxane chains Data from Gent and Tobias[60], and from Tobias[61]. (a) Comparison of observed swell ratio and swell ratio calculated using Eq. 10.27 for $f = 3(\Delta)$, $f = 4(\circ)$, and $f = \infty(\bullet)$, the dashed line corresponding to exact agreement with the formula. (b) Network functionality calculated from observed swell ratios with Eq. 10.27, the various symbols indicating the different networks used in the tests.

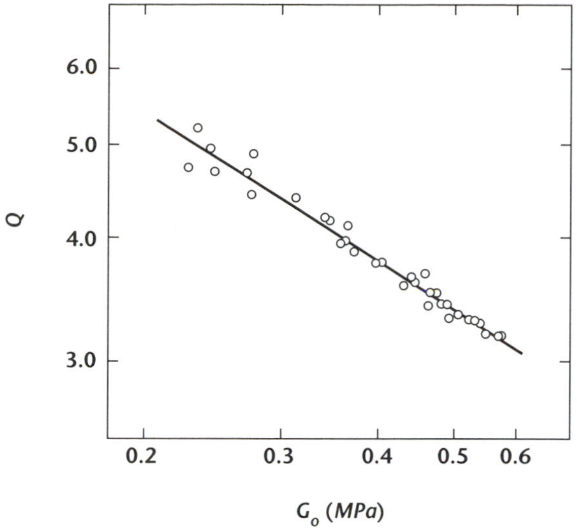

FIGURE 10.11 Equilibrium swell ratio in n-decane as a function of initial modulus in the dry state for various crosslinked natural rubber networks. Data from Mullins[41].

the neo-Hookean form*. In addition, McKenna and coworkers make the disquieting suggestion, based on extensive data, that network formation itself introduces new polymer–diluent interactions[59], resulting in $\chi > 0$ even for diluent chains and network strands of the same species. Finally, the painstaking work of Eichinger and his students[64–67] led them to conclude that free energy additivity, an assumption that underlies swelling theory, is simply invalid. Placed beside these imponderables, the existence of the log term contribution seems a very minor matter.

Despite these uncertainties, a strong incentive exists to employ swelling measurements because they offer an experimentally simple way to quantify network elasticity. At the very least, equilibrium swelling provides a means for correlating modulus data in network systems of similar species. For example, the relationship between the modulus of natural rubber networks and swell ratio in n-decane, shown in Figure 10.11, was established by careful measurements of both properties[41]. The correlation was then used extensively as a quick, convenient, and probably fairly accurate way to estimate the modulus of other natural rubber networks from swell ratio, by far the more easily measured property[68]. Kraus[69,70] has shown that with calibration, the swelling data on fully formulated rubber can provide useful information on the crosslink density, apparently unobscured by filler–polymer interactions. Bristow

*Forcing agreement with a constant value of $2/\langle f \rangle$ upon the Gent-Tobias data would require a strain-hardening modulus, quite the opposite of the expected strain softening.

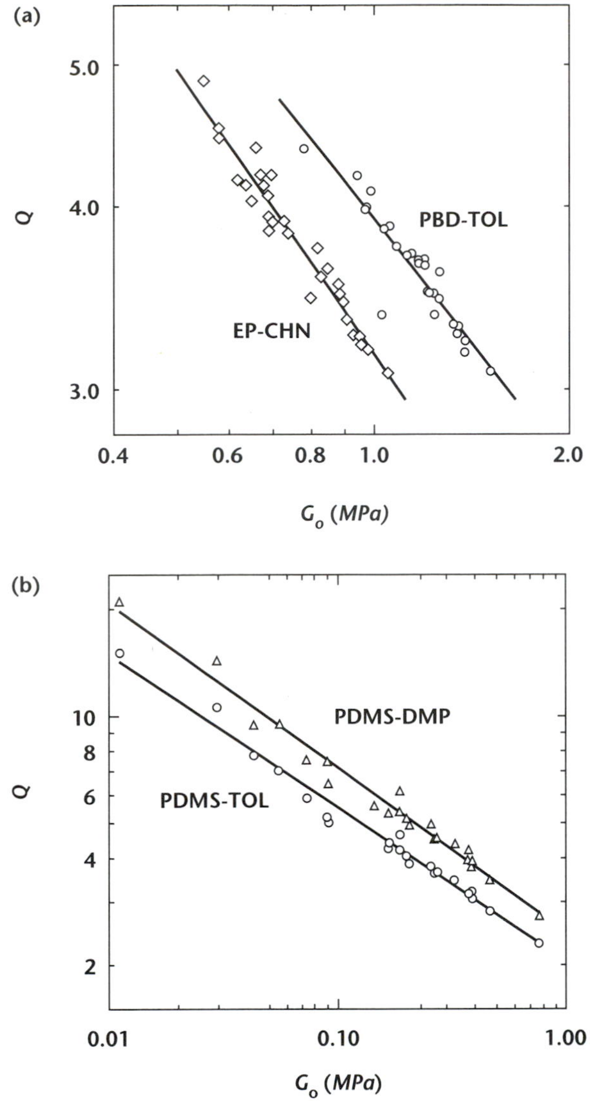

FIGURE 10.12 Equilibrium swell ratio as a function of initial modulus in the dry state for selected polymer network–diluent systems. (a) Crosslinked networks of 1,4 polybutadiene swollen in toluene from Dossin[73] and ethylene–propylene copolymer swollen in cyclohexane from Pearson[74]. (b) Endlinked polydimethylsiloxane networks swollen in toluene from Patel et al.[33] and dimethylpentane from Malone et al.[75].

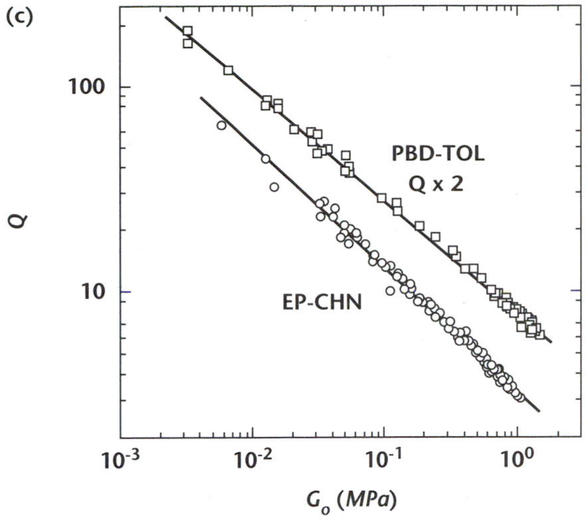

FIGURE 10.12 (*Continued*) (c) Crosslinked networks of 1,4 polybutadiene swollen in toluene and ethylene–propylene copolymer swollen in cyclohexane, extended to include data sets values with initial modulus estimated with gelation and modulus structure correlations[73,74].

and Watson[71] used swelling data for a range of diluents to assign solubility parameters to polymers. Sheehan and Bisio[72] assigned polymer–solvent interaction parameters from swelling data. Equilibrium swelling data can have considerable value when used in an appropriate manner. Only some experimental examples, observations, and comparisons with theory for a few well-defined systems are given here.

The relationship between initial modulus in the dry (reference) state and swell ratio is shown in Figure 10.12a for two crosslinked systems: 1,4 polybutadiene in toluene (PBD–TOL)[73] and ethylene–propylene copolymers in cyclohexane (EP–CHN)[74]. The results for endlinked polydimethylsiloxane in toluene (PDMS–TOL)[33] and dimethylpentane (PDMS–DMP)[75] are shown in Figure 10.12b. Only measured modulus values were used in all these cases. A power law describes the relationship for each of these systems fairly well. The results, along with that for the data in Figure 10.11, with modulus expressed in the dimensionless form $G_o V_d/RT$, are given in Table 10.6.

In the Flory-Rehner formulation, power laws relating swell ratio and network modulus $Q \propto (G_o)^p$ arise quite naturally in the semidilute regime. Thus, expanding Eq. 10.26 in powers of $1/Q$ and retaining the dominant term leads to:

$$Q = (1/2 - \chi)^{3/5} \left(\frac{V_d G_o}{RT} \right)^{-3/5} \qquad (\chi < 1/2) \qquad (10.28)$$

TABLE 10.6 Power law fits to the swelling data for selected network systems

System	Figure	Fit	Range
NR–DEC	10.11	$Q = 0.63(G_o V_d/RT)^{-0.52}$	$3 \lesssim Q \lesssim 5.5$
PBD–TOL	10.12a	$Q = 0.57(G_o V_d/RT)^{-0.62}$	$3.0 \lesssim Q \lesssim 4.2$
EP–CHN	10.12a	$Q = 0.43(G_o V_d/RT)^{-0.64}$	$3 \lesssim Q \lesssim 5$
PDMS–TOL	10.12b	$Q = 0.52(G_o V_d/RT)^{-0.43}$	$2 \lesssim Q \lesssim 20$
PDMS–DMP	10.12b	$Q = 0.65(G_o V_d/RT)^{-0.47}$	$2 \lesssim Q \lesssim 20$
PBD–TOL extended	10.12c	$Q = 0.70(G_o V_d/RT)^{-0.55}$	$3 \lesssim Q \lesssim 100$
EP–CHN extended	10.12c	$Q = 0.51(G_o V_d/RT)^{-0.60}$	$3 \lesssim Q \lesssim 60$
Eq. 10.26, Z=0 $\chi = 0.40$	10.13	$Q = 0.74(G_o V_d/RT)^{-0.45}$	$2.8 \lesssim Q \lesssim 5.5$
Eq. 10.26, Z=0 $\chi = 0.45$	10.13	$Q = 0.75(G_o V_d/RT)^{-0.40}$	$2.8 \lesssim Q \lesssim 5.5$
Eq. 10.26, Z=0 $\chi = 0.50$	10.13	$Q = 0.83(G_o V_d/RT)^{-0.34}$	$2.8 \lesssim Q \lesssim 5.5$

The power-law exponent $p = -0.60$ is a consequence of applying Flory-Huggins theory (Eq. 7.44): $\pi_{op} \propto \phi^2$ for good solvents in the semidilute range. As described in Section 7.2, however, $\pi_{op} \propto \phi^{2.25}$ is a better approximation, and the use of that expression in Eq. 10.21 leads to a somewhat weaker power law in good solvents[76]:

$$Q \propto (G_o)^{-0.52} \qquad (\chi < 1/2) \tag{10.29}$$

At the theta condition, $\pi_{op} \propto \phi^3$ for semidilute solutions, leading to:

$$Q = \frac{1}{3^{3/8}} \left(\frac{V_d G_o}{RT} \right)^{-3/8} \qquad (\chi = 1/2) \tag{10.30}$$

an even weaker power-law dependence.

Based on earlier considerations (Section 7.1.2), these various semidilute formulas are probably applicable, only for $\phi_{eq} \lesssim 0.15$ or $Q \gtrsim 6$. In fact, however, the three systems with exponents near the predicted good solvent values—NR–DEC, PBD–TOL, EP–CHN—cover data in the concentrated region $Q \lesssim 6$. The two PDMS systems extend well into the semidilute region but their exponents are somewhat lower than the prediction. Figure 10.12c shows swelling results for extended PBD–TOL and EP–CHN data sets, supplemented by data for networks whose dry state moduli could not be measured directly because of equilibration time constraints[73,74]. Instead, the dry state moduli were calculated from the results of gelation studies and modulus–structure correlations, Eq. 9.90 and Table 9.6, which were established separately. Power laws, given in Table 10.6, nicely describe these data, which now

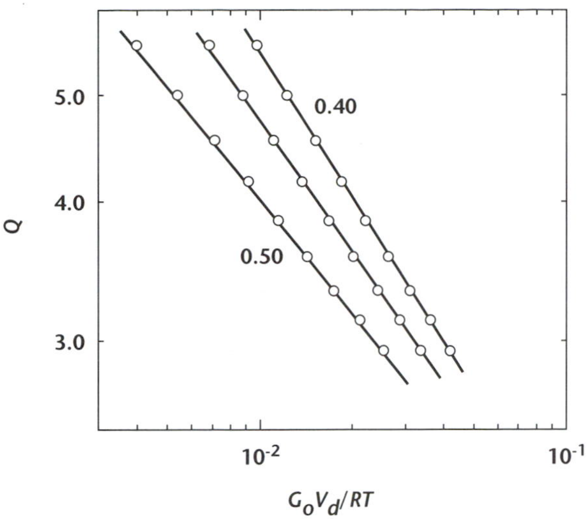

FIGURE 10.13 Relationship between equilibrium swell ratio in the concentrated regime and initial modulus in the dry state according to the Flory-Rehner expression, Eq. 10.26 Treloar version ($Z = 0$), for typical values of χ.

extend deeply into the semidilute region. The exponents are slightly reduced but still in reasonable agreement with the semidilute predictions.

Consider now the predictions of the Flory-Rehner equation in the concentrated region covered by the NR–DEC, PBD–TOL, and EP–CHN systems in Figures 10.11 and 10.12a, $2.8 \lesssim Q \lesssim 5.5$. Empirically, power laws describe fairly well the $G_o - Q$ relationship predicted by Eq. 10.26 in that region. The results for the Treloar version, $Z = 0$, are shown in Figure 10.13 for three interaction parameter covering the likely span, $\chi = 0.40, 0.45$, and 0.50. The slight curvatures are well below the experimental detection level; the power law fits are given in Table 10.6. The Flory version behaves similarly, although with slightly different exponents. The exponents from either version vary somewhat with χ but are always clearly smaller than the limiting exponents in Eqs. 10.28 and 10.29 for systems. Of course, the interaction parameter itself depends on concentration (see Section 7.4.2), so other power law exponents, or even a departure from power-law behavior altogether, are entirely possible.

The influence of interaction imponderables can be examined in another way, by recasting the swelling data into a method for determining a concentration-dependent χ. Figure 10.14 shows values of $\chi(\phi)$ calculated from G_o and Q data using Eq. 10.26, again with the Treloar version. Three systems are represented—PDMS–TOL, PBD–TOL, EP–CHN—and only experimental moduli are used. Values of χ are available from osmometry for the PDMS–TOL system[77]: $\chi = 0.445 + 0.297\phi$. The agreement

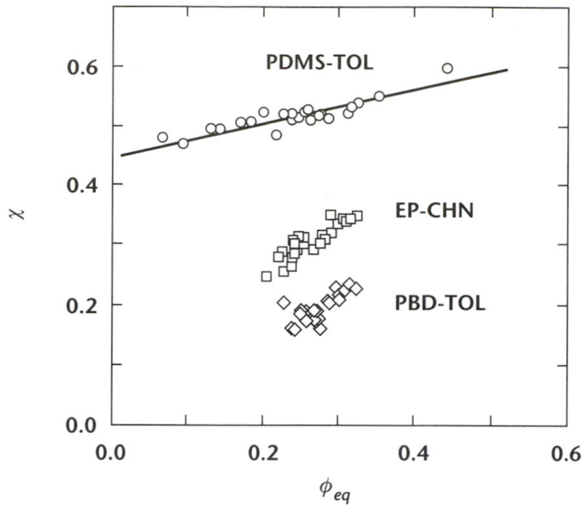

FIGURE 10.14 Interaction parameter calculated from swelling data and dry network modulus with the Flory-Rehner expression, Eq. 10.26 Treloar version ($Z = 0$) for three systems: polydimethylsiloxane–toluene, polybutadiene–toluene, and ethylene propylene copolymer-cyclohexane (see Figures 10.12a and 10.12b).

between these values and results from swelling, as seen in Figure 10.14 and found earlier[33], is excellent. Independent values of χ are not available for the other two systems. The trend with concentration, χ rising gently with increasing ϕ, is consistent with behavior in good solvent systems generally (see Section 7.4.2). However, owing to the seemingly universal free volume effects in polymer solutions (as discussed in Section 7.4.2), the magnitude of χ in both systems is simply too small to be correct. Noting the rapid relative decrease of the Mooney-Rivlin coefficient C_2 with dilution [see Eq. 10.12], one might consider using $2C_1$ in place of G_o for the modulus. That would decrease all moduli for the two systems by factors of about two, with the result that the calculated values of χ rise significantly. Using the Flory version of Eq. 10.26 boosts χ a bit more, into the range of $\chi = 0.4$, a reasonable value for good solvent systems. Unfortunately, applying these same adjustments to the PDMS–TOL calculation spoils the agreement with independently obtained values of χ.

All this complexity of swelling equilibrium seems especially incongruous, considering the apparent simplicity of stress-strain response in swollen networks [Eqs. 10.14 through 10.16 and Table 10.3a]. At the present state of knowledge, however, converting the equilibrium swell ratio into a reliable quantitative vehicle for network characterization is obviously impossible. Other investigators of the swelling–modulus relationship[1,56] have reached similar conclusions about the Flory-Rehner formula—qualitative consistency between observations and predictions, but variable enough in individual cases to rule out a general reliance upon it. Molecular simulations, now increasingly active in the area of swollen networks[35,78], offer the main hope for improved molecular understanding. Regardless of the outcome of that work, the Flory-Rehner formula provides a useful foundation for developing empirical correlations between the elastic and swelling properties within families of chemically similar networks. The log term contribution remains a mystery.

10.3 Thermoelasticity

The relationship between equilibrium stress-strain properties and temperature can be treated as a topic of macroscopic thermodynamics alone, to be investigated without regard to molecular considerations. This was, in fact, the approach taken by Joule in his mid-nineteenth century studies of rubber elasticity[79], mentioned briefly in the first pages of Chapter 9. In 1976, Price[23] presented an elegant modern account of that early work, describing its important role in the then new science of thermodynamics. Joule demonstrated experimentally the essentially entropic nature of the restoring force in natural rubber networks, showing it to be fundamentally different from its energetic nature in other materials. He also demonstrated the correctness of the relationship between the caloric and mechanical manifestations of elasticity demanded by the second law.

Macroscopic thermodynamics played an important role in a much later round of studies, beginning in the late 1930s, which eventually revealed the existence of an energetic species-dependent component in the restoring force in elastomeric networks. A third stage began in the late 1950s, dealing with the molecular aspects of thermoelasticity and particularly with the relationship between the energetic contribution and the rotational potentials of the network chain backbones.

Equations 9.45 and 9.46 provide the basis for resolving the restoring force into its energetic and entropic components. The most commonly used experiments are isothermal measurements of restoring force as a function of length in uniaxial extension, conducted at several temperatures and at ambient pressures $P_a \sim 0.1\ MPa$ to provide the data set $F(L, T)$. An early example of such data was published in 1942 by Anthony, Caston, and Guth[80].

The network was sulfur-cured natural rubber, formulated without additives and under conditions chosen to avoid strain-induced crystallization. Based on extensive preliminary tests, experimental protocols were also devised to minimize the effect of such troublesome nonequilibrium complications as hysteresis and permanent set. The data were first examined as plots of F versus T for various fixed lengths. For small extensions, the thermal expansion caused the rest length to increase and the force to decrease with increasing temperature. To eliminate that contribution, the data were recast as F versus T at various fixed stretch ratios. Thus, restoring force divided by the rest cross section at 25°C, forming a nominal stress $F/A_o\ (25)$, was

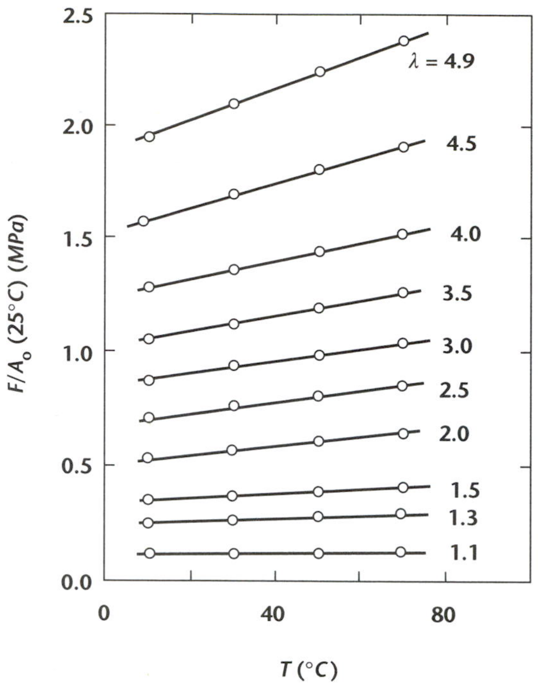

FIGURE 10.15 Restoring force as a function of temperature at constant stretch ratio for a natural rubber network. Data from Anthony et al.[80].

plotted as a function of temperature for various ratios of length to the rest length $\lambda = L/L_o(T)$, as shown in Figure 10.15. Straight lines fit the data fairly well, giving slopes and intercepts that depend on the stretch ratio:

$$F/A_o(25) = a(\lambda) + b(\lambda)T \tag{10.31}$$

From Eq. 9.46 and the definition $G = H - TS$,

$$F = \frac{1}{L_o(T)}\left[\left(\frac{\partial H}{\partial \lambda}\right)_{P,T} - T\left(\frac{\partial S}{\partial \lambda}\right)_{P,T}\right] \tag{10.32}$$

Accordingly, the enthalpic component of the force is $(\partial H/\partial\lambda)_{P,T}/L_o(T)$ and the enthalpic fraction of total force is:

$$\left[\frac{F_h(\lambda)}{F(\lambda)}\right]_{P,T} = \frac{a(\lambda)}{a(\lambda) + b(\lambda)T} \tag{10.33}$$

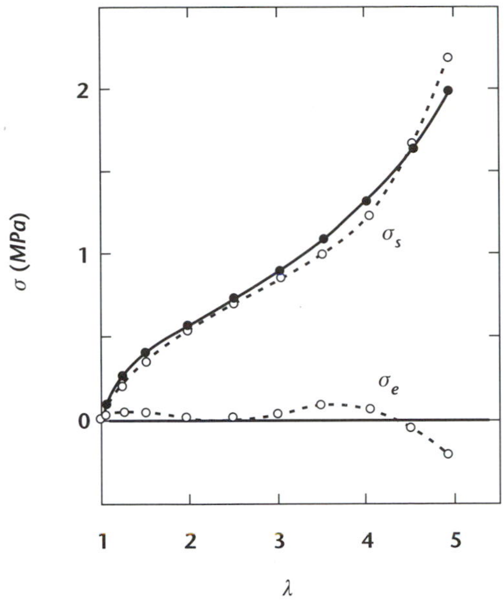

FIGURE 10.16 Enthalpic and entropic components of restoring stress for natural rubber as deduced from the data in Figure 10.15.

The results can be used to construct the enthalpic and entropic components of the stress-strain curve shown in Figure 10.16. The values of $(F_h/F)_{P,T}$ scatter around zero, and the authors concluded that, within the errors, the restoring force for a natural rubber network is entirely entropic.

Nothing is fundamentally wrong with the above analysis of data, but an important distinction must be recognized. As discussed in Section 9.6, a rubber network can be usefully regarded as having two coexisting aspects in its response: that of the network and that of a volume-excluding liquid[81,82]. Only the network energetics is of interest here. When a tensile force is applied to deform the rubber, it also causes a volume expansion or dilatation, and even a small volume change can significantly change the energy of a liquid. Accordingly, the restoring force for tensile deformations performed at constant pressure contains both a network-deformation part and a liquid-dilatation part. Obtaining the network contribution alone requires either experiments performed under constant volume conditions or, in constant pressure experiments, corrections to be applied for the liquidlike response. The important distinction is thus between $(\partial H/\partial \lambda)_{P,T}$ or any other determination where volume changes can occur, and a constant volume quantity such as $(\partial U/\partial L)_{V,T}$, which reflects the network deformation contribution alone.

The equations for analyzing constant volume measurements of $F(L, T)$ can be derived from Eq. 9.45 and the $A = U - TS$ definition:

$$F = \left(\frac{\partial U}{\partial L}\right)_{V,T} - T\left(\frac{\partial S}{\partial L}\right)_{V,T}$$

$$\left(\frac{\partial S}{\partial L}\right)_{V,T} = -\left(\frac{\partial F}{\partial T}\right)_{V,T} \tag{10.34}$$

$$\left(\frac{\partial U}{\partial L}\right)_{V,T} = F - T\left(\frac{\partial F}{\partial T}\right)_{V,T} = \left[\frac{\partial(F/T)}{\partial(1/T)}\right]_{V,L}$$

Accordingly, the energetic fraction of the restoring force,

$$f_e \equiv \left(\frac{F_e}{F}\right)_{V,T} = \frac{1}{F}\left(\frac{\partial U}{\partial L}\right)_{V,T} \tag{10.35}$$

is thus related to the force data by:

$$f_e = 1 - \frac{T}{F}\left(\frac{\partial F}{\partial T}\right)_{V,T} = -\left[\frac{\partial \ln(F/T)}{\partial \ln T}\right]_{V,L} \tag{10.36}$$

Accurately measuring the restoring force at various temperatures and extensions with the volume held constant is an extraordinarily difficult undertaking. One group successfully managed such experiments[25] and was thereby able to obtain f_e directly by applying Eqs. 10.34. Most studies, like those of Anthony et al.[80], used constant pressure data, but then corrected for the volumetric contribution. A change of variable from P to V leads to the following correction formula:

$$\left(\frac{\partial U}{\partial L}\right)_{V,T} = \left(\frac{\partial H}{\partial L}\right)_{P,T} - \Delta_{P,L} \tag{10.37}$$

where:

$$\Delta_{P,L} = \left(\frac{\partial H}{\partial V}\right)_{L,T}\left(\frac{\partial V}{\partial L}\right)_{P,T} \tag{10.38}$$

A correction can be written for the entropic part of the force as well, $-T(\partial S/\partial L)_{P,T}$, but that is simply equal and opposite to the energetic correction. From Eqs. 9.45 and 9.46, $(\partial G/\partial L)_{P,T} = (\partial A/\partial L)_{V,T}$; and substituting the definitions of A, G, and $\Delta_{P,L}$ into this expression leads to $T[(\partial S/\partial L)_{P,T} - (\partial S/\partial L)_{V,T}] = -\Delta_{P,L}$.

By inspection, the correction term $\Delta_{P,L}$ is always a positive quantity for organic liquids. Thus, $(\partial U/\partial V)_{L,T}(\partial V/\partial L)_{P,T}$ is the product of the internal pressure (always positive for organic liquids, see Chapter 2) and a dilatation coefficient, which on physical grounds must also be positive. Thus, the $\Delta_{P,L}$ correction invariably reduces (algebraically) the apparent energetic part and invariably increases the apparent entropic part by an equal amount.

Various methods are available for estimating $P \rightarrow V$ corrections, all requiring an expression for the force–length relationship. For $F(L, T)$ data in uniaxial extension, Flory in 1961 developed an expression for the dilatational coefficient and the $P \rightarrow V$ correction based on the phantom network form[83], $F \propto (\lambda - 1/\lambda)^2$:

$$\left(\frac{\partial V}{\partial L}\right)_{P,T} = \frac{\beta f}{\lambda^3 - 1} \tag{10.39}$$

$$\Delta_{P,L} = \frac{\alpha T}{\lambda^3 - 1} F \tag{10.40}$$

Corrections for other types of constant pressure measurements were worked out similarly[1]. The only one of immediate concern here is uniaxial extension analyzed in terms of stretch ratio rather than length. For that case, also based on the phantom-network form[83],

$$\Delta_{P,\lambda} = -\frac{\alpha T}{3} F \tag{10.41}$$

Thus, the Anthony-Caston-Guth null result for natural rubber networks at 25°C corresponds to $f_e \sim (6.6 \times 10^{-4})(298)/3 = +0.066$, based on Eq. 10.41.

Equations 10.40 and 10.41 provided the sole means of estimating f_e until the Manchester group built a uniaxial extension apparatus for measuring $(\partial U/\partial L)_{V,T}$ directly[23,25,84]. They gathered stress-strain data at constant volume for a variety of elastomeric networks, both undiluted and swollen with solvents. Values of f_e were obtained for a range of extensions in the networks, which differed in stress-strain behavior by varying amounts from the phantom network form. Within the errors, the values of f_e obtained for natural rubber networks are independent of stretch ratio λ, of diluent type and concentration, and of stress-strain form as measured by C_2/C_1. The average of those results for natural rubber is $f_e = +0.124$. However, the average from constant pressure measurements and the application of Eq. 10.40 is significantly larger: $f_e = +0.183$.

Allen, Price, and coworkers also independently measured sample dilatations and $P \rightarrow V$ corrections. Their results showed both Eq. 10.39 and Eq. 10.40 to be seriously in error. Indeed, as others have noted[85,86], a fundamental understanding of the tension–dilatation relationship in rubber is lacking, and remains so to this day.

Price[23] went on to note that a good account of the dilatational observations was given by a formula that had in fact proposed much earlier by Gee[87]:

$$\left(\frac{\partial V}{\partial L}\right)_{P,T} = \frac{\beta L}{3}\left(\frac{\partial F}{\partial L}\right)_{P,T} \tag{10.42}$$

Working out the consequences of the Price observation with the phantom network form, leads to a new correction formula[88]:

$$\Delta_{P,L} = \frac{\alpha T}{3}\left(\frac{2+\lambda^3}{\lambda^3-1}\right)F \tag{10.43}$$

This formula, although based on the empirical Eq. 10.42, works remarkably well. The calculated values closely track the experimental values of $(\partial U/\partial V)_{L,T}(\partial V/\partial L)_{P,T}$ with no sign at all of systematic departures. These values also lead to values of f_e that are in excellent agreement with those from the constant volume experiments. The average, $f_e = +0.117$, is practically identical to the one based on constant volume measurements, cited above. An alternative procedure, one based on torsional measurements at constant pressure, which causes much smaller dilatational effects[89], gives $f_e = +0.126$. This value is in excellent agreement with both the constant volume results and the constant pressure extensional results based on Eq. 10.43.

Values of f_e for a variety of polymer species[90] are listed in Table 10.7. Some are from constant volume measurements, but most are based on ambient pressure measurements of $F(T, L)$ as corrected using Eq. 10.43. The values have been shown to be insensitive to temperature, deformation ratio, the stress-strain form, test geometry, network architecture, network formation method, and diluent concentration[1,5,90].

The molecular origin of f_e was a matter of debate for some time[20,21]. Specific interactions, such as changes in local packing with deformation, are ruled out as a source by the various insensitivities noted above, particularly the insensitivity of f_e to the nature and amounts of network diluent. It seemed then that the explanation for f_e must be found within the network itself. In fact, an explanation of that sort was available already from molecular network theory, and it does not conflict with the experimental observations[91–93]. Thus, from Eq. 9.82, restoring force for the uniaxial extension of a phantom network at constant pressure is the following function of temperature and length:

$$F(L,T) = A_{ref}(\nu_{ref} - \mu_{ref})kT\left(\frac{V_{ref}}{V_o(T)}\right)^{1/3}\frac{(C_\infty)_{ref}}{C_\infty(T)}\left[\frac{L}{L_o(T)} - \frac{L_o^2(T)}{L^2}\right] \tag{10.44}$$

TABLE 10.7 Thermoelastic and chain dimension temperature coefficients for various polymer species

Polymer Species	f_e	$(d \ln C_\infty/dT)(°C^{-1}) \times 10^3$
cis 1,4 polyisoprene	0.10	0.33
trans 1,4 polyisoprene	−0.16	−0.47
cis 1,4 polybutadiene	0.05	0.19
trans 1,4 polybutadiene	−0.32	−0.86
polyethylene oxide	0.01	0.03
polydimethyl siloxane	0.12	0.39
polyethylene	−0.49	−1.25
polyisobutylene	−0.13	−0.39
ethylene-propylene copolymer	−0.50	−1.5
poly(1-butene) atactic	0.15	0.30
poly(1-butene) isotactic	−0.03	−0.11
poly(1-pentene) atactic	0.13	0.33
poly(1-pentene) isotactic	0.06	0.14

(a) Selected values from Mark, ref. 90, corrected according to Graessley and Fetters, ref. 88.

in which L_o and V_o are rest state values at the temperature of measurement. The energetic fraction of the restoring force can be obtained by applying Eq. 10.35. Both L_o and V_o are fixed in constant volume experiments, so the only parts of Eq. 10.44 that vary with temperature are $C_\infty(T)$ and T. Thus,

$$\ln(F/T) = \text{const} - \ln C_\infty(T) \tag{10.45}$$

The same expression is obtained with the affine model as well. For both models,

$$f_e = \frac{d \ln C_\infty}{d \ln T} = T\frac{d \ln C_\infty}{dT} \tag{10.46}$$

Equation 10.46 provided the basis for a determination of the temperature coefficient of chain dimensions, $\kappa = d \ln C_\infty/dT$, from thermoelastic measurements on networks. Results so obtained, along with those from R_g versus T data in melts (see Section 7.1.1) and from $[\eta]$ versus T in good solvents[94], are listed in Table 10.8. Speculative at first, because the effect is a delicate one, the Volkinstein-Flory interpretation of f_e has thus proved to be quite reliable. The confirmation of Eq. 10.46 is important for a more general reason, however. It was derived from Eq. 10.44, thus showing that, although inadequate in many other ways for real networks, the phantom model nevertheless predicts very well indeed the observed temperature dependence of the modulus. Some use of this point is made later.

TABLE 10.8 Comparison of $\kappa = d \ln C_\infty/dT$ obtained by various methods[(a)]

	$d \ln C_\infty/dT$ (°C^{-1}) $\times 10^3$		
Polymer Species	**From f_e (melt state)**	**From SANS (melt state)**	**From $d[\eta]/dT$ (athermal solvent)**
polyethylene	−1.2	−1.2	−1.2
poly(ethylene-propylene)	−1.5	−1.1	
a-polybutene	+0.30	+0.40	
a-polypentene	+0.33		+0.52
polyisobutylene	−0.25		−0.28
~8% vinyl polybutadiene	+0.16		~0
high cis-polyisoprene[(b)]	+0.41	+0.40	
polydimethylsiloxane	+0.78		+0.71
a-polystyrene	+0.17	~0	
a-polymethylmethacrylate	−0.10	+0.10	
polyoxyethylene	+0.03	−0.30	+0.20

(a) Values from Table 4 of ref. 88.

Finally, the demonstration of non-zero values of f_e has had an unfortunate side effect. Calling f_e the energetic fraction of the elastic force, although literally correct in some austere thermodynamic sense, has left the impression that some fundamental defect exists in the Gaussian coil basis for rubber elasticity—that only the entropic part, $1 - f_e$, comes from coil stretching and the rest from some nonnetwork, species-specific source. The confirmation of Eq. 10.46 means quite the opposite: Within the errors, *all* the elastic force originates from the coil deformations. It supports the Volkinstein analysis, described in Section 4.5.2, that the Gaussian coil model for network elasticity requires precisely the temperature dependence found, a restoring force at constant volume proportional to absolute temperature with a correction from the temperature dependence of chain dimensions for the species. It is ironic that what is really a triumph of the Gaussian network idea, the confirmation Eq. 10.46, has led to such misunderstandings.

10.4 Observations on Networks

The section deals with some nonmechanical aspects on elastomeric networks. Orientation correlations of network strands, readily observable by a variety of techniques and widely found in polymeric liquids generally, are considered briefly. The application of small-angle neutron scattering, aimed at determination of strand dimensions in both deformed and undeformed networks, is also reviewed. The section ends with a summary of experimental observations on network properties, preparatory to considerations of theories of entangled networks in the following section.

10.4.1 Microscopic Features

This section summarizes what has been learned from the microscopic examination of relaxed and deformed networks using NMR and related methods and by small-angle neutron scattering (SANS) methods.

Orientational correlations. Many investigations have focused on networks at the atomic and mer levels. Some techniques provide information on local interactions, such as those producing correlations in the relative orientation of the constituent molecules. Orientational correlation at the atomic level in polymeric liquids is a well-established phenomenon, detected initially by depolarized light scattering and magnetic birefringence[95]. More recent methods include deuterium NMR[96–99] and infrared dichroism[100–102]. It appears, however, that local orientational correlation in itself does not significantly affect either large-scale molecular dimensions in polymeric liquids or their macroscopic properties.

Gent demonstrated in 1969 the insensitivity of polymeric network properties to orientational correlations in an elegantly simple way[103], with measurements of the stress-optical coefficient C for polyisoprene networks swollen with a variety of diluents. He found that C, which governs the birefringence induced by stress and reflects the optical anisotropy of the chain units, depends strongly on the geometrical asymmetry of the diluent molecules, thus indicating local alignment with the network strands. He also found, however, that the stress itself depends only on network concentration and strain. According to the Gent findings, the stress is independent of all the species-dependent characteristics of the diluent, including its molecular asymmetry. Accordingly, the strength of orientational correlation has no effect on the network response. The analysis of infrared dichroism signals during the relaxation of stress in mixtures of long and short chains leads to the same conclusion[100]. Local orientation per se is thus considered no further here.

Neutron scattering. Only those SANS studies for which information related to network quality is reported, such as sol fraction s, are considered here. As discussed earlier, even a few percent of unattached network precursors implies a large proportion of dangling structures, which contributes nothing to equilibrium elasticity but confuses the SANS interpretation. In addition, only networks formed in a melt or solution state and characterized reasonably near that state are considered. Networks far away from the reference concentration are already highly strained or supercoiled by the volume change and lack a basis for comparison.

The earliest systematic SANS study of networks was conducted with a set of polydimethylsiloxane (PDMS) networks by Beltzung and coworkers[104–106]. Mixtures of labeled and unlabeled PDMS precursors, matched in length, were used to form endlinked networks having various strand lengths and branch point functionalities at reference concentrations that ranged from $\phi = 0.6$ to $\phi = 1$. Although the networks contained 3 to 4 wt % of extractables, the majority was traced to inert contaminants in the precursors, leaving $s \lesssim 0.01$ based on active precursor alone. Values of R_g for both network strands and their precursors were determined from SANS data on the dry state[104]. No difference in size was found between precursor and strand, thus establishing (for this case at least) that chain dimensions do not change significantly with network formation. Both precursor and strand were shown to have unperturbed dimensions: The ratio $R_g/M^{1/2} = 0.025\,nm$ is the same for both and in good agreement with theta solvent value of $0.0265\,nm$.

On a related question, Liu et al.[62] examined the size of unattached chains in a series of unstrained networks of varying crosslink density. Dilute solutions of deuterated polystyrene (PS) in styrene–divinylbenzene (DVB) mixtures were polymerized to form PS networks with various crosslink densities, controlled by the DVB–S ratio, and containing unattached labeled PS chains. Zimm plots were constructed with the SANS data to obtain R_g and A_2 for the unattached species as a function of the network modulus; the latter was estimated from the DVB–S ratio as well as by equilibrium swelling measurements in toluene. Initially, R_g decreased slowly from its unperturbed value with increasing network modulus while A_2 became increasingly negative. As the modulus rose further, the limit of solubility was eventually reached, and the unattached chains aggregated. This sequence of events is nicely consistent with the Gent-Tobias interpretation of polymeric solubility in networks[60], although the Liu et al. data[62] appear to be the first to have shown R_g and A_2 variations on approaching phase separation for such systems. However, Gilra et al.[107] found the size of unattached PDMS chains in undiluted endlinked PDMS networks to be independent of strand length and equal to the unperturbed size.

Beltzung et al.[105] investigated the size of network strands in the series of endlinked PDMS networks above[104], but now swollen to equilibrium in cyclohexane. They found $R_{g,s}(nm) = 0.017M_s^{0.58}$, a result almost identical with the size–length relationship they obtained for free PDMS chains in the same solvent over the same range of lengths, $3100 \leq M \leq 25{,}000$. Judged by the results reviewed in Chapter 6, their observation of fully developed good solvent behavior for such short free chains is somewhat surprising. Despite this mild ambiguity, however, their results clearly demonstrate that expanding the network well beyond the reference concentration, in this case to swelling equilibrium, can significantly increase $R_{g,s}$. Bastide et al.[108] found a different behavior for $R_{g,s}$ in an endlinked PS network ($M_s = 26{,}000$), formed with ~5% labeled strands in solution at $\phi = 0.1$. For the dry network, they found $R_{g,s} = 4.4\,nm$, essentially the unperturbed size for the free chain, which increased to about $5.2\,nm$ upon reswelling to $\phi \sim 0.25$, and thereafter increased slowly with further expansion to reach only about $5.6\,nm$ at swelling equilibrium, $\phi \sim 0.055$. From the following discussion, the results of Beltzung et al.[105] and Bastide et al.[108] are not as contradictory as might first appear.

Swelling and supercoiling are forms of network deformation, and molecular theories generally suppose the end-to-end strand dimensions of the active strands to vary proportionately:

$$\frac{\langle R^2\rangle}{\langle R^2\rangle_{ref}} = \left(\frac{\phi}{\phi_{ref}}\right)^{-2/3} \tag{10.47}$$

However, scattering experiments on endlinked networks can only provide the strand radius of gyration R_g, and there is no reason to think that R_g and $\langle R^2\rangle^{1/2}$ would vary with ϕ in the same way. For phantom networks, according to Eq. 4.95,

$$R_g^2 = \frac{(R_g^2)_{ref}}{2}\left(1 + \frac{\langle R^2\rangle}{\langle R^2\rangle_{ref}}\right) \tag{10.48}$$

so even in the hypothetical event of network collapse, corresponding to $\phi = \infty$, all junctions are at the same point. So, $\langle R^2\rangle^{1/2}/\langle R^2\rangle^{1/2}_{ref} = 0$, but $R_g/(R_g)_{ref}$ has only decreased to about 0.7. Accordingly, R_g might change little with supercoiling, which is most of the range explored by Bastide et al. Behavior might well be different in the swollen region, $\phi > \phi_{ref}$, but for the extreme case of swelling equilibrium, the swelling ratio is much larger in the Beltzung et al.[106] study, $\phi_{ref}/\phi_{eq} = 5$ to 10 compared with < 2 for Bastide et al.

Some efforts have been made to determine the microscopic stretch ratios by SANS. For endlinked networks, the most successful has been the 1984 Beltzung et al.[106] study of uniaxial stretching with the PDMS networks used in their 1982 and 1983 studies. They determined the quantities $(R_g^2)_{\parallel}$ and $(R_g^2)_{\perp}$, which are the mean-square components of radius of gyration for the network strands parallel and perpendicular to the stretch direction, as functions of the stretch ratio λ. Predictions for f-functional endlinked networks are available for three network models. The *fully affine network*, one having each part of all strands displaced affinely[108a], is a possible representation for highly entangled networks. The scattering behavior for the Kuhn model, having only the junctions displaced affinely, and for the James-Guth phantom network were derived by Pearson[109]:

$$\begin{aligned} (R_g^2)_{\parallel} &= \lambda^2 \frac{(R_g^2)_{\lambda=1}}{3} \\ (R_g^2)_{\perp} &= \frac{1}{\lambda}\frac{(R_g^2)_{\lambda=1}}{3} \end{aligned} \qquad \text{(full affine)} \tag{10.49}$$

$$\begin{aligned} (R_g^2)_{\parallel} &= \left[\frac{1+\lambda^2}{2}\right]\frac{(R_g^2)_{\lambda=1}}{3} \\ (R_g^2)_{\perp} &= \left[\frac{1+\lambda}{2\lambda}\right]\frac{(R_g^2)_{\lambda=1}}{3} \end{aligned} \qquad \text{(end affine)} \tag{10.50}$$

$$\begin{aligned} \left(R_g^2\right)_\parallel &= \left[\frac{f+2+(f-2)\lambda^2}{2f}\right]\frac{\left(R_g^2\right)_{\lambda=1}}{3} \\ \left(R_g^2\right)_\perp &= \left[\frac{f-2+(f+2)\lambda^2}{2f\lambda}\right]\frac{\left(R_g^2\right)_{\lambda=1}}{3} \end{aligned} \qquad \text{(phantom)} \qquad (10.51)$$

Results expressed as $\alpha_\parallel = [3(R_g^2)_\parallel/(R_g^2)_{\lambda=1}]^{1/2}$ and $\alpha_\perp = [3(R_g^2)_\perp/(R_g^2)_{\lambda=1}]^{1/2}$ as functions of the stretch ratio λ are compared with the model predictions in Figure 10.17a for endlinked PDMS ($M_s = 10,500$, $f = 4$, made and tested without diluent). The results evidently agree nicely with the phantom model, the changes predicted by the other two being much too large. Results for this and two other PDMS networks[106] are compared with the phantom predictions in Figure 10.17b, and again the agreement is good. Similar results were reported for endlinked polyisoprene networks by Yu et al. in 1987[110].

Unfortunately, not much general information about network properties can be drawn from these studies. The results for other studies of endlinked tetrafunctional PDMS networks[17,33,111] suggest that the set used in the study above contain many imperfections. For example, the values of G_o for $M_s \sim 10,000$ and $M_s \sim 20,000$ are 30 to 50% lower than in those other studies, indicating the likelihood of incomplete linking and, consequently, a reduced active strand fraction and much less entanglement trapping. Assuming this to be the case, the observations in Figures 10.17a and 10.17b become quite reasonable: The entanglement contribution is small, the features of molecular structure causing affine displacements are thereby reduced, and only phantom response is seen. The polyisoprene study reports 6 to 8% solubles from their linking chemistry[110], so active fractions and trapped entanglement contributions are reduced and phantom behavior is found. In both cases, what can be concluded is that the microscopic deformations are in rough accord with expectations.

No subsequent SANS studies utilizing more carefully prepared and characterized endlinked networks have appeared, that is, definitive endlinked network results that can distinguish between models have yet to be made. An interesting new approach, random crosslinking of labeled–nonlabeled block copolymers, has been taken up recently by Richter, et al.[112–116]. Even here, however, the linking chemistry used is rather imperfect, with sol fraction $s \sim 0.03$[114] or even more[116], implying significant dangling fractions, $f_d \gtrsim 0.20$, and low trapped entanglement fractions, $T_e \lesssim 0.50$.

The subject of scattering from deformed systems for all types of networks has in fact been clouded for some time by a peculiar phenomenon. As seen with the two-dimensional detectors used to study deformation effects, the SANS patterns that result go by the visually descriptive names of "lozenges" and "butterflies." These patterns have been the subject of extensive speculation[117–119]. A good summary of the salient facts is given by Boué et al.[120]. Read and McLeish[121,122] (see also

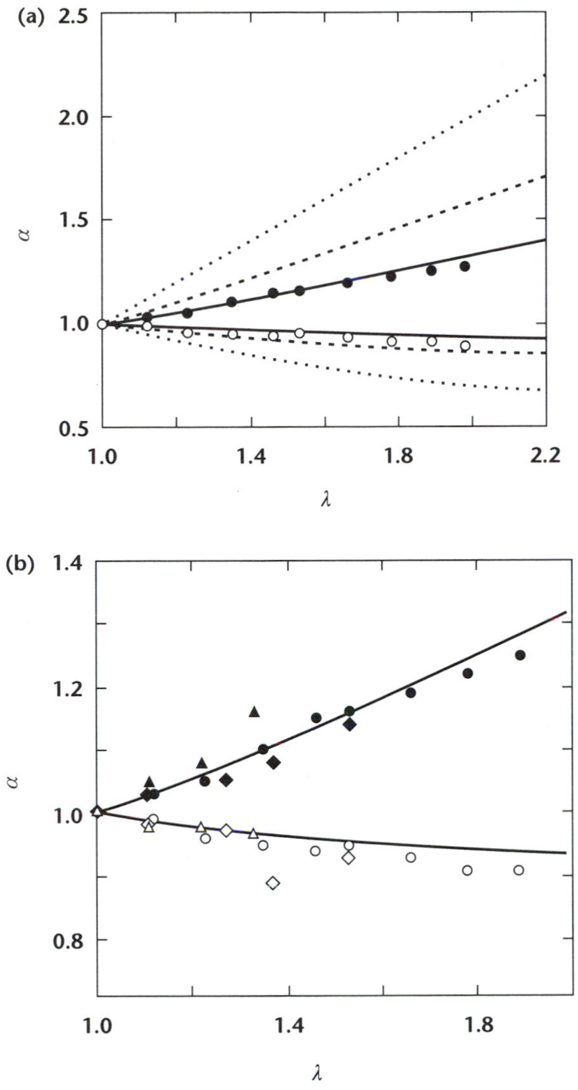

FIGURE 10.17 Comparison of SANS-based strand size ratios, perpendicular (filled symbols) and parallel (open symbols) to the stretch direction as functions of macroscopic stretch ratio for endlinked polydimethylsiloxane networks with the predictions of various network models. Data from Belzung et al.[106]. (a) Comparison of data for an undiluted network $M_s = 10{,}500$ with the fully affine (···), junction affine (— — —), and phantom (—) predictions, Eqs. 10.49–10.51. (b) Comparison of data for three networks—undiluted $M_s = 10{,}500$ (○, ●), undiluted $M_s = 6{,}100$ (△▲), and $M_s = 10{,}500$ swollen to give $\phi = 0.6$ (◊, ♦)—with the phantom network prediction Eq. 10.51.

discussion[123–125]) have proposed that the patterns are related to the presence of elastically inactive dangling structures, the inevitable result of imperfectly formed networks. If the Read-McLeish explanation is indeed the correct one, the need for shifting SANS studies to more perfect networks to avoid extraneous clutter becomes imperative.

10.4.2 Macroscopic Features

As described earlier, a considerable amount of experimental information is available on the equilibrium properties of elastomeric networks.

- The initial shear modulus G_o depends quite generally on network structure and trapped entanglement concentration through the Langley equation, Eq. 9.90.
- As exemplified by the data for a natural rubber network in Figure 10.4, the stress-strain relationship for modest uniaxial compressions, $0.5 \lesssim \lambda \lesssim 1$, is reasonably well described by the neo-Hookean form, $\sigma \propto \lambda^2 - 1/\lambda$. As shown below, however, the values of $\sigma/(\lambda^2 - 1/\lambda)$ for most other systems decrease somewhat with increasing compression.
- The behavior in uniaxial extension commonly departs from the phantom form and, for modest strains, $1 \lesssim \lambda \lesssim 2$, is described reasonably well by the Mooney-Rivlin formula [Eq. 10.5].
- The Ferry-Kan equation, Eq. 10.17, based on network properties for many network types, species, and crosslink densities, strongly suggests that the ratio C_2/C_1 is related to the contribution of trapped entanglements to the initial modulus.
- The addition of diluent alters the Mooney-Rivlin coefficients in apparently universal ways, C_1 obeying the phantom–affine network formulas and C_2 decreasing toward zero much more rapidly. Both swollen networks and supercoiled networks are more neo-Hookean in stress-strain behavior than those formed and tested without diluents.
- The quantitative rules governing equilibrium swelling behavior remain unclear.
- Unlike the other aspects of network behavior, the temperature dependence of the modulus, predicted with phantom-affine network models alone, appears to be universally correct for real networks.

These are the macroscopic facts that molecular models of entangled networks must somehow comprehend. The following section, dealing with various entanglement models and their properties, refers extensively to the items in this list.

10.5 Modeling Uncrossability

Entanglement interactions, the effects arising from the mutual uncrossability of covalent chain backbones, have been considered briefly in earlier parts of this chapter.

Entanglement trapping through macrocycle formation in the ordinary course of crosslinking and the contribution of such topological species to the initial modulus were discussed in Sections 9.3.2 and 9.7. The Ferry-Kan formulation (Section 10.1.3) demonstrated a correlation between trapped entanglements and Mooney-Rivlin coefficients. None required a specific set of entanglement attributes beyond assumptions such as pairwise interaction. The remainder of this chapter deals with the attempts made to represent entanglement microscopically, in some simple yet physically sensible way that includes enough relevant detail to work out testable predictions.

Busse introduced the idea of chain entanglement in 1932[126]. Noting the elastic character over short time intervals of even uncrosslinked natural rubber, he proposed the existence of a temporary network structure with junctions consisting of regions in which chains are interlocked but not chemically bonded, thereby permitting *chain slippage* around or along one another to occur. In 1940, Treloar pointed out that some sort of physical coupling would follow simply and naturally from an overlapping random arrangement of long flexible chains in the melt state[127]. Flory appears to have been the first to consider permanently trapped entanglements in networks, envisioning loose loops or encirclements that increase the elastic modulus by behaving as additional constraints on the network strands[128].

In 1967, Edwards suggested that the fundamental basis for entanglement effects on network elasticity is *topological classification,* the permanence of spatial relationships among the network macrocycles[129]. (This aspect was discussed briefly under entanglement trapping factor in Section 9.3.2.) He pointed out that both looped and nonlooped classes of otherwise identical structures contribute to response, loops inducing an attraction between macrocycles and nonloops introducing a repulsion. He went on to prove an important theorem: that the full effect of permanent partitioning into topological classes conferred by uncrossability always increases the modulus relative to that for a phantom network of the same structure. In 1976, Deam and Edwards considered the effects of classification in greater detail[130].

Edwards pointed out another feature of molecular uncrossability in 1967[131], namely the confining effect of chain-filled surroundings on the conformations available to a network strand. His *confinement* view of entanglement eventually grew into the *tube model,* the representation that mutual uncrossability constrains all segments of each network strand to lie within some distance—the *tube diameter*—of the path of that strand through its surroundings when the network was formed.

These various ideas have inspired three rather distinct types of molecular representation for the entanglement interaction—classification models, slip-link models, and the confinement or tube models. The three are illustrated in Figures 10.18 through 10.20. Of them, slip-links are the most explicit and easiest to analyze. Slip-links also capture the intuitive notion that entanglements have more freedom to adjust to deformation and hence are mechanically more compliant than chemical

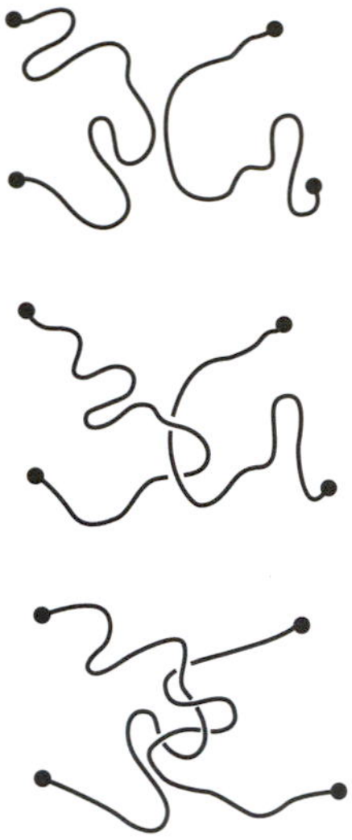

FIGURE 10.18 Topological classification of network strands.

junctions. Slip-links, however, have the disadvantage of being wrong in principle, because they unrealistically require the loops always to touch at some point. Topological classification, on the other hand, has the advantage of being almost certainly correct as a general principle, but it is elusive and very difficult to implement in a rigorous way. Tube models occupy an intermediate position, being more tractable than classification models, yet containing essential features such as softness relative to chemical links and the capability of imposing constraints upon certain types of spatial fluctuations. Tube models also became one of the founding principles for

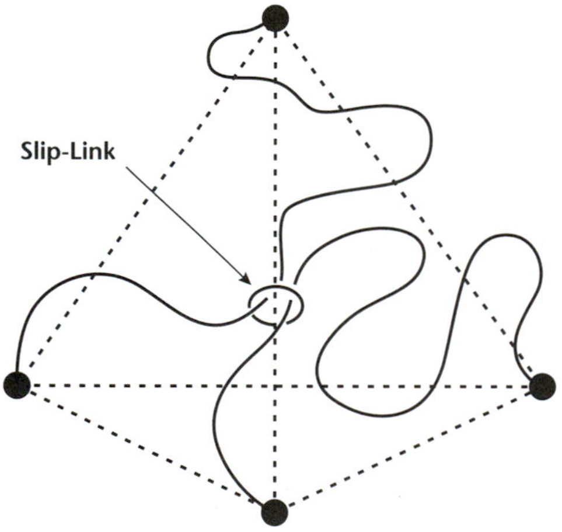

FIGURE 10.19 The slip-link tetrahedron.

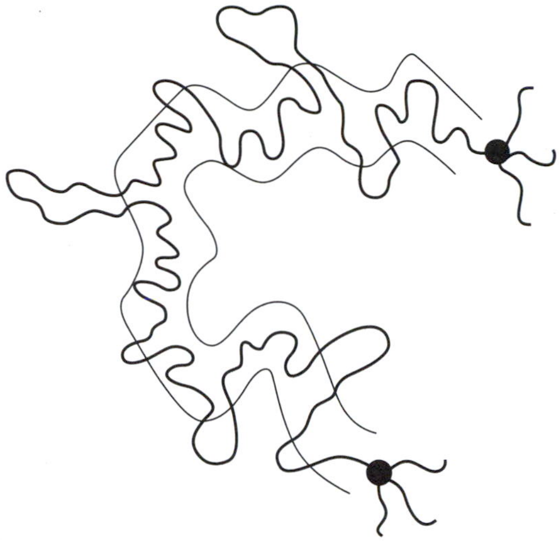

FIGURE 10.20 The tube model of topological constraints for the case of network strands.

dynamic theories of entangled polymer liquids—the *Doi-Edwards theory*[132] and its many extensions—and thus offer at least the possibility of connections between network and liquid-state parameters and properties. Examples of all three types are considered in the following sections, along with a fourth, the constrained junction model.

Finally, a selection of data for testing the entangled network theories is included here. Higgs and Gaylord[133] performed the valuable service of gathering and sorting the literature data on stress in uniaxial extension and compression[8–11]. Added to their compilation are some results obtained by Rennar and Oppermann[17,18]. Data from other extension–compression studies[12,14,134], although not included here, are consistent with this body of data. The selected data are expressed as dimensionless stress $S(\lambda) = \sigma / G_o(\lambda^2 - 1/\lambda)$ and corresponding extension ratio in Table 10.9. The following formula, a generalized version of the Langley formula Eq. 9.90, was used as a framework for theory testing:

$$\frac{\sigma}{(\lambda^2 - 1\lambda)G_o} = \frac{(\nu - h\mu)RT}{G_o} + \frac{T_e G_N^o Y(\lambda)}{G_o} \tag{10.52}$$

or in terms of dimensionless stress, Eq. 10.3,

$$S(\lambda) = (1 - \varphi) + \varphi Y(\lambda) \tag{10.53}$$

$$\varphi = T_e G_N^o / G_o \tag{10.54}$$

TABLE 10.9 Dimensionless stress[(a)] and approximate entanglement contribution[(b)] from uniaxial extension and compression data for polymeric networks[8–11,17,18]

λ	$S_{exp}(\lambda)$	$Y_{exp}(\lambda)$	λ	$S_{exp}(\lambda)$	$Y_{exp}(\lambda)$	λ	$S_{exp}(\lambda)$	$Y_{exp}(\lambda)$
0.143	0.923	0.846	0.666	1.059	1.118	1.434	0.887	0.774
0.143	0.927	0.854	0.697	1.024	1.048	1.496	0.945	0.890
0.161	1.000	1.000	0.779	1.068	1.136	1.522	0.849	0.698
0.165	0.920	0.840	0.811	1.084	1.168	1.541	0.920	0.840
0.229	0.956	0.912	0.876	1.025	1.050	1.623	0.869	0.738
0.230	0.949	0.898	0.904	1.018	1.036	1.626	0.821	0.642
0.232	0.960	0.920	0.904	1.059	1.118	1.764	0.815	0.630
0.270	1.024	1.048	0.947	1.013	1.026	1.849	0.785	0.570
0.326	0.967	0.934	1.006	1.002	1.004	2.210	0.743	0.486
0.332	0.945	0.890	1.046	0.996	0.992	2.216	0.769	0.538
0.333	0.948	0.896	1.084	0.968	0.936	2.228	0.740	0.480
0.354	1.006	1.012	1.106	0.987	0.974	2.496	0.743	0.486
0.459	1.031	1.062	1.106	0.981	0.962	2.738	0.681	0.362
0.478	1.001	1.002	1.131	0.948	0.896	2.831	0.686	0.372
0.478	1.003	1.006	1.209	0.920	0.840	3.236	0.668	0.336
0.496	1.017	1.034	1.223	0.947	0.894	3.621	0.680	0.360
0.570	1.042	1.084	1.236	0.945	0.890	3.643	0.751	0.502
0.571	1.039	1.078	1.345	0.890	0.780	3.817	0.683	0.366
0.587	1.049	1.098	1.350	0.882	0.764	3.957	0.780	0.560
0.664	1.057	1.114	1.397	0.928	0.856	4.102	0.640	0.280

(a) $S_{exp}(\lambda) = \sigma / G_o(\lambda^2 - 1/\lambda)$, dimensionless stress data from selected studies.
(b) $Y_{exp}(\lambda)$ is the fraction of the stress supported by chain entanglement, estimated from Eq. 10.53 with $\varphi = T_e G_N^o / G_o = 0.5$: $Y_{exp}(\lambda) = z\, S_{exp}(\lambda) - 1$.

in which φ is the fraction of the initial shear modulus contributed by topological interactions, and $Y(\lambda)$ is the dimensionless stress contributed by topological interactions, expressed as a function of stretch ratio with $Y(\lambda) = 1$ at $\lambda = 1$, and thereby available for testing theory-based predictions.

This form for expressing the stress, introduced by Heinrich and Straube[135], has the advantage of providing a natural bridge between experiment, where typically both chemical structure and network topology contribute significantly to G_o, and theory, from which only a purely topological contribution is usually calculated. Unfortunately, values of φ are not available for most of the networks represented in Table 10.9. A visual inspection of those data suggest, however, that $S(\lambda) \equiv \sigma / G_o(\lambda^2 - 1/\lambda)$ levels off near 0.5 at large extensions, whereas most theories, focused on the entanglement contribution alone, go to zero for large λ. Thus, perhaps $\varphi = 1/2$ is an acceptable approximation for the many networks represented in the compilation. With that stipulation, (admittedly tenuous), experimental values of the entanglement contribution

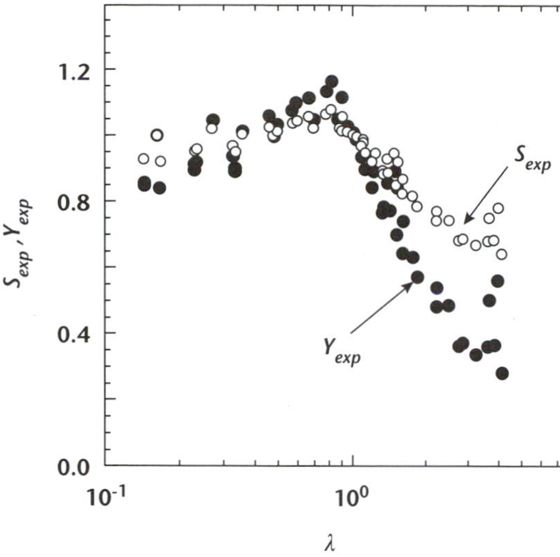

FIGURE 10.21 Literature data from Table 10.9 on dimensionless stress $S(\lambda) = \sigma/G_o(\lambda^2 - 1/\lambda)$ for uniaxial extension and compression and the normalized entanglement contribution $Y(\lambda)$ derived from them using Eq. 10.52 and the $\varphi = 0.5$ approximation.

$Y(\lambda) = 2S(\lambda) - 1$ can be extracted from $S(\lambda)$, the experimental dimensionless stress. The experimental values of $Y(\lambda)$ so obtained are listed along with those of $S(\lambda)$ in Table 10.9. Both functions are displayed in Figure 10.21.

10.5.1 Topological Classification

The first analytically solved example of a classification problem, looped and not-looped fractions for a random coil macrocycle and an infinite rod, was published in 1967 by Prager and Frisch[136]. These researchers found that looped probability density versus separation distance behaves similarly to the complimentary error function: $p(r) \sim \mathrm{erfc}(ar/R_g)$, where R_g is the radius of gyration of the macrocycle and a is some constant of order unity. They calculated the force–displacement relationship for a rod–macrocycle ensemble and found a force constant K_{ent} for small displacements about 1.66 times larger than that for a macrocycle attached permanently to the rod at one point, then softening with increasing displacements. For the simplest case of topological capture—rigid rings of radius R—the enlooped probability is essentially linear: $p(r) = 1 - 2r/R$ for $r < 2R$[137]. Frank-Kamenetskii et al. evaluated $p(r)$ for pairs of random-coil macrocycles by Monte-Carlo simulation[138], finding for that case a rough fit to the cubic exponential form, $p(r) \propto \exp[-a(r/R_g)^3]$. Everaers and

Kremer confirmed that observation with molecular dynamics simulations[139]. They also demonstrated a very important new result: comparable magnitudes of force constant for isolated entangled pairs, $K_{ent} \sim 1.3kT$, and total topological modulus contribution per entanglement at melt densities, $\sim 0.85kT$. This suggests that topological theories of network elasticity can capture most of the entanglement effect by including nothing more elaborate than the pair-wise contributions.

The pair-enloopment probability can in general be written $p(r/R_o)$, where R_o is some pair-separation range within which couple formation is likely, a distance on the order of the macrocycle size $(R_g)_r$. Graessley and Pearson[143] examined classification effects with some trial expressions for the loop probability. Thus, for the free energy of deformation per pair for ensembles of looped and nonlooped macrocyle pairs[137],

$$\Delta A = -kT\left[p(r)\ln\frac{p(r')}{p(r)} + [1-p(r)]\ln\frac{1-p(r')}{1-p(r)}\right] \tag{10.55}$$

where $p(r)$ and $p(r')$ represent looped probabilities for the separation distances before and after deformation. Affine uniaxial deformation was treated, and the various trial forms produced a diversity of behavior. The exponential form, $p(r) = \exp(-ar/R_o)/R_o$, led to stress-strain behavior resembling that found experimentally—a Mooney-Rivlin softening in extension and approximate neo-Hookean behavior in compression. For all trial expressions, the contribution to initial modulus had the form:

$$G_o = CkT\nu_r^2 R_o^3 \tag{10.56}$$

where ν_r is the number density of macrocycles in the network, and C is some dimensionless number of order unity.

The form of Eq. 10.56 is easy to understand physically. Implicit in the discussion so far is the simplifying assumption that only pairwise interactions of macrocycles need be considered. The product $\nu_r R_o^3$ is the *macrocycle overlap parameter*, a quantity proportional to the average number of macrocycles lying within the pervaded volume of any macrocycle. Hence, the quantity $\nu_r^2 R_o^3$, and therefore the modulus, is proportional to the number density of macrocycle–macrocycle interactions.

In 1985, Iwata used the principle of topological classification to express the stress in uniaxial extension for both as-formed and swollen networks[141]. His predictions led to $G_o \propto \phi^{2/3}$, with only an insignificant C_2 contribution. His result for the initial modulus of as-formed networks is interesting. He expressed G_o as the sum of two universal contributions: the phantom modulus G_{ph} and a macrocycle interaction term. In the terminology used here,

$$G_o = (\nu - \mu)kT + a(\nu - \mu)^2 R_o^3 kT \tag{10.57}$$

where $(\nu - \mu)$ is ν_r, the number density of macrocycles (the cycle rank of a network divided by its volume), and a is a parameter of order unity that depends on the pattern of network connectivity. Iwata undertook to calculate from first principles both a and the relationship between macrocycle–macrocycle range R_o and network strand size $(R_g)_s$ for one particular pattern. He compared his result, expressed as reduced modulus $G_o/\nu kT$ versus strand overlap concentration $\phi_s = [4\pi(R_g^3)_s \nu/3]$, with data for well-characterized networks of three species: polydimethylsiloxane PDMS[34], 1,4-polybutadiene PBD[45], and ethylene–propylene copolymer EP[46], all with average functionality $\bar{f} = 2\nu/\mu$ of about 4. A single curve was found to accommodate the data, although with considerable scatter. Remarkably, the predicted and observed slopes differ by only about a factor of 2.

This is an alternative to the traditional interpretation of entanglement contributions described in Section 9.7, involving the Langley equation [Eq. 9.90] and trapping factor ideas. It has the attractive feature that flexible chain networks should obey the same equation regardless of the plateau modulus of the species. If correct, it should also apply without reference to network perfection and the entanglement trapping factor. Using Eq. 10.57 does not avoid the necessity of careful structural characterization, however, because applying it requires estimates of the elastically active strand and junction concentrations. Also, as shown in Figure 10.22, a simplified version

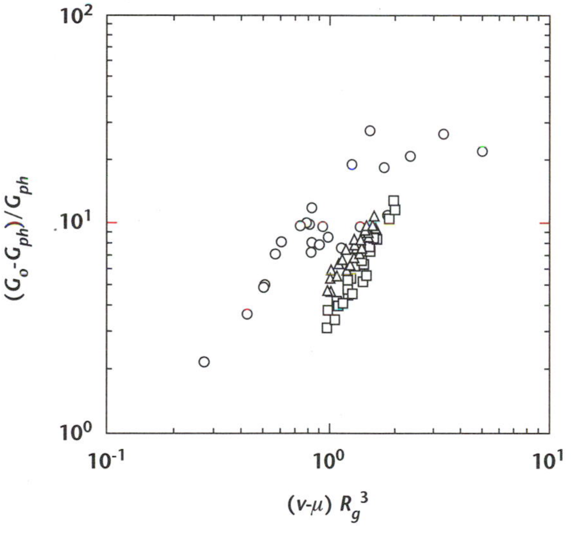

FIGURE 10.22 Examination of predictions based on the Iwata topological classification model. Data are shown for polydimethylsiloxane networks (○) from Patel et al.[33], 1,4-polybutadiene networks (□) from Dossin and Graessley[45], and ethylene–propylene copolymer networks (△) from Pearson and Graessley[46].

of this approach, which does not require detailed topological calculations, is only marginally successful in unifying the network data. The macrocycle interaction term was isolated by plotting $(G_o - G_{ph})/G_{ph}$ as a function of $(\nu - \mu)(R_g^3)_s$, the assumption being that the ratio $R_o/(R_g)_s$ is the same for all networks. The data sources in this case are the Dossin and Pearson dissertations for PBD[73] and EP[74] and Patel et al. for PDMS[33], including results for both the model and imperfect networks. Leaving aside the scatter, two kinds of differences remain. The data for different species are displaced from one another, and even within each species $(G_o - G_{ph})/G_{ph}$ appears not to be directly proportional to $(\nu - \mu)(R_g^3)_s$. Omitting the less certain data decreases the scatter but does not change the conclusions. Thus, this simplified approach to the classification problem appears to be fruitless.

Network simulations have now advanced to the point of permitting the evaluation of both stress-strain properties and topological classification probabilities. The initial results of Kremer and Everaers are consistent with the macrocycle interaction framework and offer the possibility of significant progress in the future[139,142].

10.5.2 Slip-Links

The slip-link model and some of its properties were described briefly by Graessley in 1974[137]. In 1977, Graessley and Pearson[143] demonstrated numerically that the model yields uniaxial extension–compression behavior that closely resembles the experimental observations—strain softening in extension, nearly neo-Hookean in compression. They also explored the effects of swelling and supercoiling and showed that the extensional behavior becomes more neo-Hookean with swelling (C_2/C_1 diminishes as Q increases), again as found experimentally. The ratio C_2/C_1 increases with supercoiling, contrary to experiment, but perhaps the experiments merely reflects reduced trapped entanglement (slip-link) concentrations in solution-formed networks, as was speculated in Section 10.1.2.

In 1981, Ball et al. applied the replica formalism[130] to an ensemble of independent slip-link tetrahedra and obtained an analytical expression for the average free energy of deformation per slip-link[144]:

$$\Delta A = \frac{kT}{2} \sum_{x,y,z} \left[\frac{(1+\eta)\lambda_i^2}{1+\eta\lambda_i^2} + \ln\left(1+\eta\lambda_i^2\right) \right] \tag{10.58}$$

The parameter η defines the permitted slip-link range along their strands; $\eta = 0$ corresponds to a crosslink and $\eta = \infty$ to an object whose deformation free energy depends only on volume changes: $\Delta A = kT \ln \lambda_x \lambda_y \lambda_z$. For uniaxial extension–compression,

$$\sigma/(\lambda^2 - 1/\lambda) = \nu kT \Lambda[(1+\eta)(1-\lambda\eta^2)\Lambda + \eta] \tag{10.59}$$

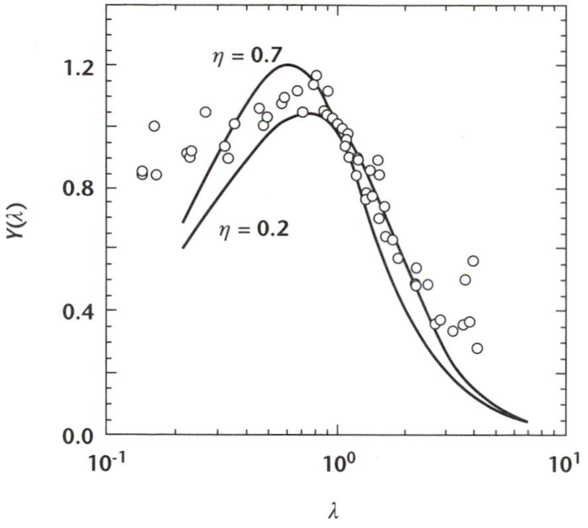

FIGURE 10.23 Comparison of experimental tension–compression data with predictions of the Ball et al. slip-link model[144].

where $\Lambda = \lambda/(1+\eta\lambda^2)(\eta+\lambda)$, and υ is the number density of slip-links in the network. The initial modulus is $G_o = \nu kT/(1+\eta)^2$, so using Eq. 10.53, the dimensionless stress contributed by entanglement for the Ball et al. model is:

$$Y(\lambda) = \frac{(1+\eta)^2\lambda}{(\eta+\lambda)(1+\eta\lambda^2)}\left[\frac{(1+\eta)(1-\lambda\eta^2)\lambda}{(\eta+\lambda)(1+\eta\lambda^2)} + \eta\right] \tag{10.60}$$

The behavior of $Y(\lambda)$ for some values of the slip range parameter η is compared with experimental data (Table 10.9) in Figure 10.23. The calculated results are insensitive to η in the extension region, and they describe the experimental behavior in that region rather well. Indeed, Thirion and Weil[145] found that the prediction for $\eta = 0.4$ described the stress–extension data of cis-polyisoprene networks rather well. Evidently, however, agreement is poor in the compression region for all values of η. Edwards and Vilgis extended the model by including finite chain length effects[146]. The effect of swelling on stress-strain behavior was not explored in these later slip-link calculations.

A significant virtue of the Ball et al. slip-link analysis is its demonstration by example that even networks of Gaussian strands can exhibit stress-strain behavior that is not neo-Hookean. That result provides at least some justification for finding that thermoelastic predictions based on Gaussian strand models are in excellent agreement with data from networks whose stress-strain properties are distinctly not neo-Hookean (Section 10.3).

10.5.3 Constrained Junction Models

The tube models discussed in the following section consider mainly the uncrossability constraints on the lateral fluctuations of network strands. The fluctuations of network junctions are also subject to uncrossability constraints. Ronca and Allegra were the first to consider junction constraints and particularly what affect deformation might have upon them[147]. Their ideas became the basis for the Flory constrained junction model[148,149], also as developed later by Flory and Erman[150]. This model, in effect, permits the fluctuation-span coefficient h (Section 9.2.4) to vary with deformation. It is proposed that the span of junction fluctuations increases with deformation, thereby reducing the modulus and hence providing another molecular interpretation of departures from the neo-Hookean form. The stored energy density is thus expressed formally as:

$$W = [\nu - h(\lambda_1, \lambda_2, \lambda_3)\mu]kT\left(\frac{\lambda_1^2 + \lambda_2^2 + \lambda_3^2 - 3}{2}\right) \tag{10.61}$$

where $0 \leq h(\lambda_1, \lambda_2, \lambda_3) \leq 1$.

This interpretation of stress-strain data confines the initial modulus roughly to the range $(\nu - \mu)kT < G_o < \nu kT$. As is evident from Table 9.6, but especially from Table 9.4 the ratio $G_o/\nu kT$ exceeds unity for many networks, sometimes by a considerable margin. Accordingly, the constrained junction model is of little conceptual value in understanding the properties of networks with large trapped-entanglement contributions. Erman and Monnerie[151,152] have developed a version of the model that includes strand–strand interactions as well, in effect bringing the $T_e G_e$ term in Eq. 9.90 back into consideration. Horkay and McKenna[153] have shown that reasonable fits to stress-strain data can be achieved with the Erman-Monnerie version. That approach may be useful as a correlation scheme, but like many other models, it lacks predictive ability.

10.5.4 Tube Models

Many versions of the tube model exist, all treating the interaction in roughly the same way: Uncrossability acts as a constraining field of force distributed all along the length of the network strands. The field penalizes excursions of a strand from its *primitive path*—a random-walk trajectory running from one network junction to the other through the mesh of surrounding strands, a path established permanently for each strand at the time of network formation. The free energy penalty increases with excursion amplitude, and, as suggested in Figure 10.20, the field is commonly approximated as an uncrossable tube of constraints, the tube centerline being the primitive path of the strand. Both path step length and tube diameter are of the order of the mesh size a, a parameter related to the plateau modulus[132]. Comparisons of the various versions with the experimental data on Table 10.9 are shown in Figures 10.24a, b, and c. Discussion is deferred to the end of the section.

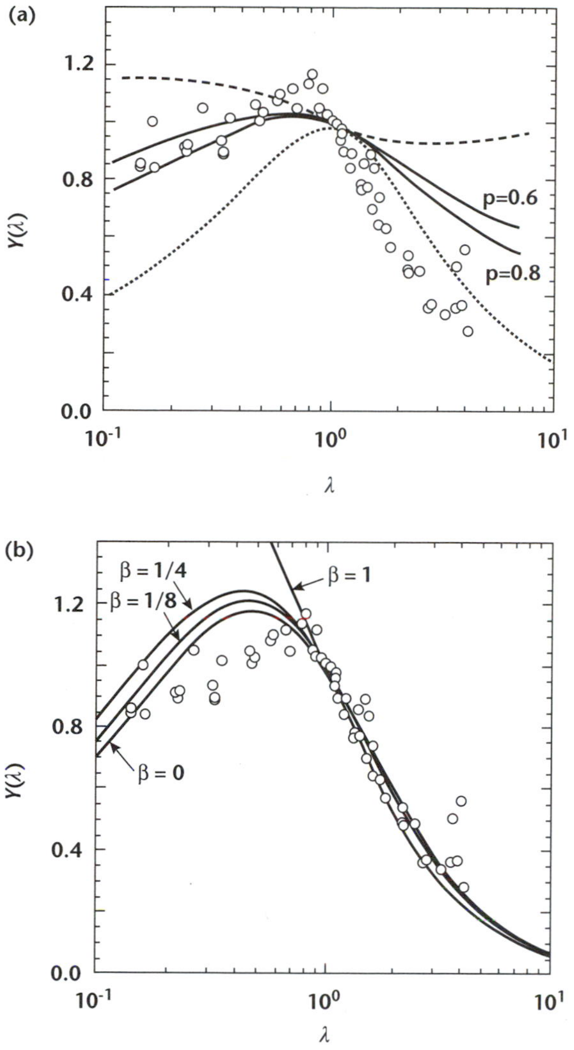

FIGURE 10.24 Comparison of experimental tension–compression data with the predictions of various tube models. The open circles indicate experimental data from Table 10.9. (a) The modified Marrucci affine model[154], shown by the dotted curve; the Marrucci constant diameter model[155], shown by the dashed curve; and the strand fluctuation model[140,156], shown by the solid curves. (b) The Heinrich-Straube model[161].

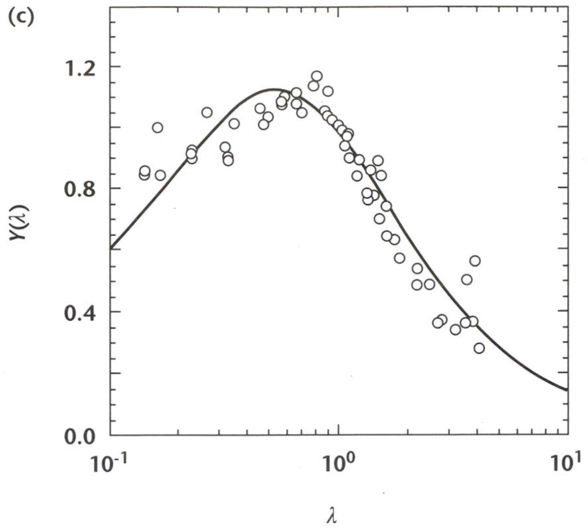

FIGURE 10.24 (*Continued*) (c) The Rubinstein-Panyukov model[165].

Affine models. Marrucci was the first to apply the tube model to network elasticity[154]. He pointed out an ambiguity that is still not fully resolved. Any affine deformation changes the length of the tube because its centerline, the primitive path, is always stretched[132], but does the tube diameter (the span of lateral fluctuations) also change with deformation? It could be argued that the tube volume should be conserved for any volume-conserving deformation, and hence, on average, the tube diameter should become smaller. However, it could just as well be argued that the tube diameter does not change with deformation. The tube diameter defines the lateral span of topological confinement, a property that could reasonably be expected to depend only on the density of strands—the *mesh size*—and hence to be unchanged by any volume-conserving deformation.

The free energy of deformation is different for each of the two possibilities, and Marrucci considered both. For the affine version[154] he obtained:

$$Y(\lambda) = \frac{1}{(2+r)}\left[2 + r\left(\frac{3}{\lambda^2 + 2\lambda^{-1}}\right)^{1/2}\right] \tag{10.62}$$

in which r is the ratio of tube diameter for dynamic properties in the melt to its value in the network. Some ambiguity exists, however, and one interpretation is that r is equivalent to the entanglement trapping factor T_e, discussed in Chapter 9, and as such should lie in the range $0 < r < 1$. In that range, however, the dependence on λ is too weak to explain the data in Table 10.9. In a perfect network, $r = 1$, and for that case $Y(\lambda)$ goes from unity at $\lambda = 1$ to $Y(\lambda) = 2/3 +$ const.$/\lambda$ at high extensions,

$\lambda \gg 1$, and $\Upsilon(\lambda) = 2/3 +$ const. $\lambda^{1/2}$ at high compressions $\lambda \ll 1$. The constant term inside the brackets of Eq. 10.62 is simply too large. However, the discussion also allows the interpretation that only the second term in the brackets of Eq. 10.62 is the entanglement contribution. In this case, the dimensionless entanglement contribution is what will henceforth be called the *modified affine model:*

$$\Upsilon(\lambda) = \left(\frac{3}{\lambda^2 + 2\lambda^{-1}}\right)^{1/2} \tag{10.63}$$

This formula is compared with the data in Table 10.9 in Figure 24a, for discussion with others below.

Constant mesh density models. Marrucci developed the constant mesh-size tube model first[155]. It is essentially the Doi-Edwards entangled liquid model[132] applied to networks. For uniaxial extension–compression, the elastic energy density from topological interactions is:

$$W = 3T_e G_N^{\circ}(I^2 - 1)/2 \tag{10.64}$$

where:

$$I(\lambda) = \frac{1}{\lambda^{1/2}} \int_0^1 \left[1 + (\lambda^3 - 1)u^2\right]^{1/2} du \tag{10.65}$$

The topological stress contribution can be obtained using Eq. 9.53. The algebraically complex expression that results[140] can then be used to evaluate $\Upsilon(\lambda)$. Comparison with the extension–compression data is shown in Equation 10.24a.

The strand in a tube picture can be viewed in another way[140]. Think of the strand and its primitive path as being expressed in units of some small size, less than the tube diameter but otherwise left unspecified. The strand has n_s such units, many more than n_p, the primitive path number, so $n_s - n_p$ of the strand units are surplus and available to form looplike fluctuations projecting out all along the path. Deformation increases the path length, so $n_s - n_p$ becomes smaller, fewer and smaller projections occur, and the entropy associated with the fluctuations is decreased. That tube model variant has the following elastic energy density for uniaxial extension–compression:

$$W = \frac{3T_e G_N^{\circ}\left[I(\lambda) - 1\right]}{2}\left[1 + \frac{I(\lambda) - 1}{6p}\right] \tag{10.66}$$

where p is $(n_s - n_p)_{\circ}/n_s$, the surplus fraction of strand units in the undeformed network, and $I(\lambda)$ is as defined in Eq. 10.65. Obtaining $\Upsilon(\lambda)$ as described previously for the constant mesh density model[156], leads to the comparison with data shown for a few values of p in Figure 10.24a.

Empirical models. Other tube versions include the Gaylord-Douglas model[157,158], the Heinrich-Straube model[135,159–161], and the Wagner model[162]. All appear to be expansions upon the Marrucci idea of affine tube diameter changes with deformation[154]. The molecular motivation for the forms they use is unfortunately unclear. Like the models described earlier, these models contain adjustable parameters and some can describe the uniaxial extension–compression data in Table 10.10 fairly well.

Consider as an example the Heinrich-Straube model, for which the dimensionless stress contribution is:

$$Y(\lambda) = \frac{2(\lambda^{\beta/2} - \lambda^{-\beta})}{\beta(\lambda^2 - \lambda^{-1})} \tag{10.67}$$

where the parameter β is suggested to lie in the range $0 < \beta < 1$. The two limits of the Heinrich-Straube formulation are:

$$Y(\lambda) = \frac{3 \ln \lambda}{\lambda^2 - \lambda^{-1}} \qquad \beta = 0 \tag{10.68}$$

$$Y(\lambda) = \frac{2}{1 + \lambda^{3/2}} \qquad \beta = 1 \tag{10.69}$$

Those values of β and two intermediate ones are used in the comparison with extension–compression data in Figure 10.23b. The calculated behavior in extension is insensitive to β and describes the data in that region rather well. Klüppel has reported good agreement with the $\beta = 1$ form for natural rubber networks[163], but only extensional data were available in that study. Presumably other values would have worked nearly as well. The compression region is more discriminating. Values of β greater than about 0.3 are clearly unsatisfactory there.

Nonaffine models. In 1997, Rubinstein and Panyukov introduced a nonaffine tube model[164]. According to their analysis of the deformation microscopics, the span of the fluctuations that define the tube diameter in a densely entangled network is related to the macroscopic stretch ratios λ_x, λ_y, λ_z in what seems a counterintuitive manner: $a_i = a_o \lambda_i^{1/2}$, where $i = x, y, z$, and a_o is the rest state diameter. Thus, the Rubinstein-Panyukov tube expands in those portions of the tube that are stretched, contrary to the contraction in the Marrucci affine model, and for reasons that are not obvious. In any case, this leads them to an expression for the stored energy density,

$$W = T_e G_N^o \sum_{x,y,z} \left(\lambda_i + \frac{1}{\lambda_i} \right) \tag{10.70}$$

The application of Eq. 9.53 for uniaxial extension–compression yields:

$$Y(\lambda) = \frac{1}{1 - \lambda^{1/2} + \lambda} \tag{10.71}$$

Unlike the other entanglement models, this one contains no adjustable constants.

For analytical models, the Mooney-Rivlin coefficients in the small strain region can be calculated using:

$$\begin{aligned} 2C_2 &= (\partial[\sigma/(\lambda^2 - 1/\lambda)]/\partial(1/\lambda))_{\lambda=1} \\ 2C_1 &= G_o - 2C_2 \end{aligned} \tag{10.72}$$

Equation 10.52 arises naturally in the Rubinstein-Panyukov model because they assume the additivity of chemical network and entanglement contributions. With that relationship and Eq. 10.72, the following expressions are obtained:

$$\begin{aligned} 2C_1 &= (\nu - h\mu)kT + \frac{1}{2}T_e G_N^o \\ 2C_2 &= \frac{1}{2}T_e G_N^o \end{aligned} \tag{10.73}$$

These relationships are consistent with the Ferry-Kan equation and essentially the same as those found in the Dossin and Pearson studies[45,46]. Indeed, Eq. 10.71 fits the extensional portion of the extension–compression data very well, but its variation with compression is too weak.

Rubinstein and Panyukov have recently proposed an extended version of their model, the *slip-tube network*[165]. They note that the microscopics of that earlier version had made no provision for equilibration of coil stretches along the tube. Equilibration along the tube is an important process in the tube theory for entangled liquids[132], and there is no reason to omit it for entangled networks. They devised a method to include that process, which results in an algebraically complex expression for the stress, but again one that for highly entangled networks contains no adjustable parameters. They developed an exact solution, showing that it led to the limiting forms, $Y(\lambda) \propto \lambda^{-1}$ for large extension and $Y(\lambda) \propto \lambda^{1/2}$ for large compressions. They went on to show that the exact solution is adequately described by the following approximate formula:

$$Y(\lambda) = \frac{1}{0.74\lambda + 0.61\lambda^{-1/2} - 0.35} \tag{10.74}$$

Rubinstein and Panyukov developed a body of literature data for testing Eq. 10.74, estimating as well as possible the values of $(\nu - h\mu)kT$ and $T_e G_N^o$ for each data set, and found excellent agreement in both extension and compression.

The predictions of Eq. 10.74 are compared in Figure 10.24c with the extension–compression data from Table 10.8. The agreement is remarkable. Expressions for the Mooney-Rivlin coefficients from Eqs. 10.72 and 10.74 are essentially the same as those in Eq. 10.73, except now the factors of 1/2 are 0.565 and 0.435 for C_1 and C_2.

Tube model comparisons. Like the original Marrucci tube model, the constant mesh density version, compared with data in Figure 10.24a, is essentially indistinguishable from the neo-Hookean form and hence inconsistent with experiment. Even subtracting off a strain-independent portion—the device used to obtain the modified affine model—doesn't improve matters much. The modified Marrucci model, Equation 10.63, gives a reasonable description of extensional behavior, but its prediction for compression shows too much softening. The fluctuating strand model, on the other hand, is reasonably consistent in compression but, for physically reasonable values of the parameter p, not soft enough in extension. As shown in Figure 10.24b, the Heinrich-Straube model with parameter value $\beta = 0$, Eq. 10.68, describes both compression and extension behavior rather well. The meaning of β, and how it might change with other modes of deformation is a worry, however.

The 2002 Rubinstein-Panyukov model[165] is attractive for several reasons. As shown in Figure 10.24c, it fits the uniaxial extension–compression data nicely, and it does so without parameter adjustment. Although sometimes difficult to follow in detail, the 2002 R-P development is motivated by a molecular picture of the deformation mechanism, so generalization to deal with response to such deformations as swelling and supercoiling without new parameters is a real possibility. At this point, the 2002 R-P model seems to offer the best hope for a molecularly based network model that encompasses the experimental observations on entangled network elasticity.

10.5.5 Commentary

Some comparisons have been made of predictions by various entangled network models with uniaxial extension–compression data. Several capture at least some aspects of the behavior one might plausibly attribute to entanglements. Most have an adjustable parameter and, by a proper choice, one can sometimes achieve reasonable fits to experimental results. How to confirm such parameters independently, by obtaining them from other types of experiments or by demonstrating their universality, is the difficult point. Thus, it is unclear what is accomplished by data fitting beyond demonstrating that the model in question is not grossly inconsistent with uniaxial extension–compression behavior. So far, the models have been virtually useless in providing a molecular understanding of even the most elementary effects, such as those associated with swelling. Despite the hope offered by the recent Rubinstein-Panyukov model,[165] the subject of uncrossability modeling in networks seems a long way from settlement.

REFERENCES

1. Treloar L.R.G. 1975. *The Physics of Rubber Elasticity*, 3rd ed., Oxford: Oxford University Press.
2. Ferry J.D. 1980. *Viscoelastic Properties of Polymers*, 3rd ed., New York: J. D. Wiley & Sons.
3. Mullins L. 1950. *J. Phys. Colloid Chem.* 54:209.
4. Mark J.E. 1994. *Acc. Chem. Res.* 27:271.
5. Erman B. and Mark J.E. 1997. *Structures and Properties of Rubberlike Networks*, New York: Oxford University Press.
6. Rivlin R.S. 1948. *Phil. Trans. Roy. Soc.* A241:379.
7. Rivlin R.S. 1956. In *Rheology, Theory and Applications*, Eirich F.R. (ed.). London: Academic Books, p. 351.
8. Treloar L.R.G. 1944. *Trans. Faraday Soc.* 40:59.
9. Rivlin R.S. and D.W. Saunders. 1951. *Proc. Roy. Soc.(London)* A243:251.
10. Erman B. and P.J. Flory. 1978. *J. Polym. Sci., Polym. Phys. Ed.* 16:1115.
11. Pak H. and P.J. Flory. 1979. *J. Polym. Sci., Polym. Phys. Ed.* 17:1845.
12. Gottlieb M., C.W. Macosko, and T.C. Lepsch. 1981. *J. Polym. Sci., Polym. Phys. Ed.* 19:1603.
13. Kawabata S., K. Matsuda, K. Tei, and H. Kawai. 1981. *Macromolecules* 14:154.
14. Xu P. and J.E. Mark. 1991. *Makromol. Chem.* 192:567.
15. van der Hoff B.M.E. and E.J. Buckler. 1967. *J. Macromol. Sci. (Chem.)* A1:747.
16. Mooney M. 1940. *J. Appl. Phys.* 11:582.
17. Oppermann W. and N. Rennar. 1987. *Prog. Polym. Coll. Sci.* 75:49.
18. Rennar N. 1988. Doctoral thesis, Universität Clausthal.
19. Grest G.S., M. Putz, R. Everaers, and K. Kremer. 2000. *J. Non-Cryst. Solids* 274:139.
20. Krigbaum W.R. and R.-J. Roe. 1965. *Rub. Chem. Tech.* 38:1039.
21. Dušek K. and W. Prins. 1969. *Adv. Polym. Sci.* 6:1.
22. Mark J.E. 1975. *Rub. Chem. Tech.* 48:495.
23. Price C. 1976. *Proc. Royal Soc. (London)* A351:331.
24. Allen G. 1976. *Proc. Royal Soc. (London)* A351:381.
25. Allen G., M.J. Kirkham, J. Padget, and C. Price. 1971. *Trans. Faraday Soc.* 67:1278.
26. Llorente M.A. and J.E. Mark. 1979. *J. Chem. Phys.* 71:682.
27. Gumbrell S.M., L. Mullins, and R.S. Rivlin. 1953. *Trans. Faraday Soc.* 49:1495.
28. Price C., G. Allen, F. De Candia, M.C. Kirkham, and A. Subramaniam. 1970. *Polymer* 11:486.
29. De Candia F. 1972. *Macromolecules* 5:102.
30. De Candia F., L. Amelino, and C. Price. 1972. *J. Polym. Sci., Part A-2* 10:975.
31. Johnson R.M. and J.E. Mark. 1972. *Macromolecules* 5:41.
32. Sivasailam K. and C. Cohen. 2000. *J. Rheology* 44:897.
33. Patel K.P., S. Malone, C. Cohen, J.R. Gillmor, and R.H. Colby. 1992. *Macromolecules* 25:5241.
34. Gottlieb M., C.W. Macosko, G.S. Benjamin, K.O. Meyers, and E.W. Merrill. 1981. *Macromolecules* 14:1039.
35. Chen Z., C. Cohen, and F.A. Escobedo. 2002. *Macromolecules* 35:3296.
36. Urayama K. and S. Kohjiya. 1996. *J. Chem. Phys.* 104:3352.
37. Urayama K., T. Kawamura, and S. Kohjiya. 1996. *J. Chem. Phys.* 105:4833.
38. Ferry J.D. and H.-C. Kan. 1978. *Rubber Chem. Tech.* 51:731.

39. Taylor C.R., R. Greco, O. Kramer, and J.D. Ferry. 1976. *Trans. Soc. Rheol.* 20:141.
40. Langley N.R. 1968. *Macromolecules* 1:348.
41. Mullins L. 1959. *J. Appl. Polym. Sci.* 2:1.
42. Bristow G.M. 1965. *J. Appl. Polym. Sci.* 9:1571.
43. Kraus G. and G.A. Moczygemba. 1964. *J. Polym. Sci.* A2:277.
44. Mark J.E. 1970. *J. Polym. Sci.* C31:97.
45. Dossin L.M. and W.W. Graessley. 1979. *Macromolecules* 12:123.
46. Pearson D.S. and W.W. Graessley. 1980. *Macromolecules* 13:1001.
47. Meyers K.O., M.L. Bye, and E.W. Merrill. 1980. *Macromolecules* 13:1045.
48. Kan H.-C. and J.D. Ferry. 1979. *Macromolecules* 12:494.
49. Kan H.-C. and J.D. Ferry. 1980. *Macromolecules* 13:1313.
50. Hvidt S., O. Kramer, W. Batsberg, and J.D. Ferry. 1980. *Macromolecules* 13:933.
51. Frenkel J. 1938. *Acta Physico-Chimica, U.R.S.S.* 9:235.
52. Haller W. 1931. *Kolloid Z.* 56:257.
53. Flory P.J. and J. Rehner. 1943. *J. Chem. Phys.* 11:521.
54. Flory P.J. 1949. *J. Chem. Phys.* 17:303.
55. Flory P.J. 1950. *J. Chem. Phys.* 18:108.
56. Flory P.J. 1953. *Principles of Polymer Chemistry.* Ithaca, NY: Cornell University Press.
57. Mark J.E. 1970. *J. Amer. Chem. Soc.* 92:7252.
58. Froelich D., D. Crawford, T. Rozek, and W. Prins. 1972. *Macromolecules* 5:100.
59. McKenna G.B., K.M. Flynn, and Y. Chen. 1990. *Polymer* 31:1937.
60. Gent A.H. and R.H. Tobias. 1982. *J. Polym. Sci.: Polym.Phys. Ed.* 20:2317.
61. Tobias R.H. 1982. Doctoral thesis, Akron University.
62. Liu X., B. Bauer, and R. Briber. 1997. *Macromolecules* 30:4704.
63. Gilra N., C. Cohen, R.M. Briber, B.J. Bauer, R.C. Hedden, and A.Z. Panagiotopoulos. 2001. *Macromolecules* 34:7773.
64. Brotzman R.W. and B.E. Eichinger. 1982. *Macromolecules* 15:531.
65. Neuburger N.A. and B.E. Eichinger. 1988. *Macromolecules* 21:3060.
66. Zhao Y. and B.E. Eichinger. 1992. *Macromolecules* 25:6988.
67. Zhao Y. and B.E. Eichinger. 1992. *Macromolecules* 25:6996.
68. Moore C.G. and W.F. Watson. 1956. *J. Appl. Polym. Sci.* 19:237.
69. Kraus G. 1956. *Rubber World, October,* 67.
70. Kraus G. 1956. *Rubber World, November,* 254.
71. Bristow G.M. and W.F. Watson. 1558. *Trans. Faraday Soc.* 54:1731.
72. Sheehan C.J. and A.L. Bisio. 1966. *Rub. Chem. Tech.* 39:149.
73. Dossin L.M. 1978. Doctoral thesis, Northwestern University.
74. Pearson D.S. 1978. Doctoral thesis, Northwestern University.
75. Malone S.P., C. Vosburgh, and C. Cohen. 1993. *Polymer* 34:5149.
76. Obukhov S.P., M. Rubinstein, and R.H. Colby. 1994. *Macromolecules* 27:3191.
77. Kuwahara N., T. Okazawa, and M. Kaneko. 1968. *J. Polym. Sci.: Pt. C* 23:543.
78. Pütz M., K. Kremer, and R. Everaers. 2000. *Phys. Rev. Lett.* 84:298.
79. Joule J.P. 1859. *Phil. Trans. Roy. Soc.* 149:91.
80. Anthony R.L., R.H. Caston, and E. Guth. 1942. *J. Phys. Chem.* 46:826.
81. James H.M. and E. Guth. 1943. *J. Chem. Phys.* 11:455.
82. Boggs F.W. 1952. *J. Chem. Phys.* 20:1952.
83. Flory P.J. 1961. *Trans. Faraday Soc.* 57:829.
84. Allen G., U. Bianchi, and C. Price. 1963. *Trans. Faraday Soc.* 59:2493.
85. Christensen R.G. and C.A.J. Hoeve. 1970. *J. Polym. Sci.: Pt. A-1* 8:1503.

86. Penn R.W. 1970. *Trans. Soc. Rheol.* 14:509.
87. Gee G. 1946. *Trans. Faraday Soc.* 42:585.
88. Graessley W.W. and L.J. Fetters. 2001. *Macromolecules* 34:7147.
89. Boyce P.H. and L.R.G. Treloar. 1970. *Polymer* 11:21.
90. Mark J.E. 1976. *J. Polym. Sci.: Macromol. Rev.* 11:135.
91. Volkenstein M.V. and O.B. Ptitsyn. 1955. *Zh. Tekh. Fiz.* 25:662.
92. Volkenstein M.V. 1963. *Configurational Statistics of Polymer Chains.* New York: J. Wiley & Sons.
93. Flory P.J., C.A. Hoeve, and A. Ciferri. 1959. *J. Polym. Sci.* 34:337.
94. Flory P.J., A. Ciferri, and R. Chiang. 1961. *J. Am. Chem. Soc.* 83:1023.
95. Fischer E.W., G.R. Strobl, M. Dettenmaier, M. Stamm, and N. Steidle. 1979. *Discuss. Faraday Soc.* 68:26.
96. Delouche B. and E.T. Samulski. 1981. *Macromolecules* 14:575.
97. Sotta P., B. Deloche, J. Herz, A. Lapp, D. Durand, and J.-C. Rabadeux. 1987. *Macromolecules* 20:2769.
98. McLoughlin K., C. Szeto, T.M. Duncan, and C. Cohen. 1996. *Macromolecules* 29:5475.
99. Hedden R.C., E. McCaskey, C. Cohen, and T. M. Duncan. 2001. *Macromolecules* 34:3285.
100. Doi M., D.S. Pearson, J. Kornfield, and G. Fuller. 1989. *Macromolecules* 22:1488.
101. Doi M. and H. Watanabe. 1991. *Macromolecules* 24:740.
102. Ylitalo C.M., J.A. Kornfield, G. Fuller, and D.S. Pearson. 1991. *Macromolecules* 24:749.
103. Gent A.H. 1969. *Macromolecules* 2:262.
104. Beltzung M., C. Picot, P. Rempp, and J. Herz. 1982. *Macromolecules* 15:1594.
105. Beltzung M., J. Herz, and C. Picot. 1983. *Macromolecules* 16:580.
106. Beltzung M., C. Picot, and J. Herz. *Macromolecules* 17:663.
107. Gilra N., A.Z. Panagiotopoulos, and C. Cohen. 2001. *J. Chem. Phys.* 115:1100.
108. Bastide J., Boué, F. and Buzier M. 1989. In *Molecular Basis of Polymer Networks,* Baumgartner A. and Picot C.E. (eds.). Berlin: Springer-Verlag, pp. 48–64.
108a. Benoit H., R. Duplessix, R. Ober, M. Daoud, J.C. Cotton, B. Farnoux, and G. Jannink. 1975. *Macromolecules* 8:451.
109. Pearson D.S. 1977. *Macromolecules* 10:696.
110. Yu H., Kitano T., Kim C.Y., Amis E.J., Chang T., Landry M.R., Wesson J.A., Han C.C., Lodge T.P., and Glinka C.J. 1987. In *Advances in Elastomers and Rubber Elasticity,* Lai J. and Mark J.E. (eds.). New York: Plenum Publ. Co., pp. 407–420.
111. Macosko C.W. and G.S. Benjamin. 1981. *Pure Appl. Chem.* 53:1505.
112. Straube E., V. Urban, W. Pyckhout-Hintzen, and D. Richter. 1994. *Macromolecules* 27:7681.
113. Straube E., V. Urban, W. Pyckhout-Hintzen, D. Richter, and C.J. Glinka. 1995. *Phys. Rev. Lett.* 74:4464.
114. Westermann S., V. Urban, W. Pyckhout-Hintzen, D. Richter, and E. Straube. 1996. *Macromolecules* 29:6165.
115. Pyckhout-Hintzen W., S. Westermann, V. Urban, D. Richter, and E. Straube. 1997. *Physica B* 234–236:236.
116. Westermann S., W. Pyckhout-Hintzen, D. Richter, E. Straube, S. Egelhauf, and R. May. 2001. *Macromolecules* 34:2186.
117. Bastide J., L. Leibler, and J. Prost. 1990. *Macromolecules* 23:1821.

118. Ramzi A., F. Zielinski, J. Bastide, and F. Boué. 1995. *Macromolecules* 28:3570.
119. Mendes E., R. Oeser, C. Hayes, F. Boué, and J. Bastide. 1996. *Macromolecules* 29:5574.
120. Boué F., Bastide J., and Buzier M. 1989. In *Molecular Basis of Polymer Networks*, Baumgartner A. and Picot C.E. (eds.). Berlin: Springer-Verlag, pp. 65–81.
121. Read D.J. and T.C.B. McLeish. 1997. *Phys. Rev. Lett.* 79:87.
122. Read D.J. and T.C.B. McLeish. 1997. *Macromolecules* 30:6376.
123. Westermann S., V. Urban, W. Pyckhout-Hintzen, D. Richter, and E. Straube. 1998. *Phys. Rev. Lett.* 80:5449.
124. Read D.J. and T.C.B. McLeish. 1998. *Phys. Rev. Lett.* 80:5450.
125. Mergell B. and R. Everaers. 2001. *Macromolecules* 34:5675.
126. Busse W.F. 1932. *J. Phys. Chem.* 36:2862.
127. L.R.G. Treloar. 1940. *Trans. Faraday Soc.* 36:538.
128. Flory P.J. 1944. *Chem. Rev.* 35:51.
129. Edwards S.F. 1967. *Proc. Phys. Soc., London* 91:513.
130. Deam R.T. and S.F. Edwards. 1976. *Philos. Trans. R. Soc. London, Ser. A* 280:317.
131. Edwards S.F. 1967. *Proc. Phys. Soc., London* 92:9.
132. Doi M. and Edwards S.F. 1986. *The Theory of Polymer Dynamics.* Oxford: Oxford University Press.
133. Higgs P.G. and R.J. Gaylord. 1990. *Polymer* 31:70.
134. Mott P.H. and C.M. Roland. 1996. *Macromolecules* 29:6941.
135. Heinrich G., E. Straube, and G. Helmis. 1988. *Adv. Polym. Sci.* 85:33.
136. Prager S. and H.L. Frisch. 1967. *J. Chem. Phys.* 46:1475.
137. Graessley W.W. 1974. *Adv. Polym. Sci.* 16:1.
138. Frank-Kamenetskii M.D., A.V. Lukashin, and A.V. Vologodskii. 1975. *Nature* 258:398.
139. Everaers R. and K. Kremer. 1996. *Phys. Rev. E* 53:R37.
140. Graessley W.W. 1982. *Adv. Polym. Sci.* 47:67.
141. Iwata K. 1985. *J. Chem. Phys.* 83:1969.
142. Everaers R. 1999. *New J. Phys.* 1:1.
143. Graessley W.W. and D.S. Pearson. 1977. *J. Chem. Phys.* 66:3363.
144. Ball R.C., M. Doi, S.F. Edwards, and M. Warner. 1981. *Polymer* 22:1010.
145. Thirion P. and T. Weil. 1984. *Polymer* 25:609.
146. Edwards S.F. and T. Vilgis. 1986. *Polymer* 27:483.
147. Ronca G. and G. Allegra. 1975. *J. Chem. Phys.* 63:4990.
148. Flory P.J. 1977. *J. Chem. Phys.* 66:5720.
149. Flory P.J. 1979. *Polymer* 20:1317.
150. Flory P.J. and B. Erman. 1982. *Macromolecules* 15:800.
151. Erman B. and L. Monnerie. 1989. *Macromolecules* 22:3342.
152. Erman B. and L. Monnerie. 1992. *Macromolecules* 25:4456.
153. Horkay F. and McKenna G.B. 1996. In *Physical Properties of Polymers Handbook*, Mark J.E. (ed.). Woodbury, NY: AIP Press, p. 379.
154. Marrucci G. 1981. *Macromolecules* 14:434.
155. Marrucci G. 1979. *Rheologica Acta* 18:193.
156. Barkofsky P. and Sebastian J. 1998. Private communication.
157. Gaylord R.J. and J.F. Douglas. 1987. *Polym. Bull.* 18:347.
158. Gaylord R.J. and J.F. Douglas. 1990. *Polym. Bull.* 23:529.
159. Heinrich G. and E. Straube. 1983. *Acta Polymerica* 34:589.

160. Heinrich G. and E. Straube. 1984. *Acta Polymerica* 35:115.
161. Heinrich G. and E. Straube. 1987. *Polym. Bull.* 34:247.
162. Wagner M.H. 1994. *J. Rheol.* 38:655.
163. Klüppel M. 1992. *Prog. Polym. Coll. Sci.* 90:137.
164. Rubinstein M. and S. Panyukov. 1997. *Macromolecules* 30:8036.
165. Rubinstein M. and S. Panyukov. 2002. *Macromolecules* 35:6670.

Appendix A—Symbols

Standard international units are used throughout the book. Units such as Poise for viscosity, Angstroms for length and calories for energy have been avoided. Certain symbols have been reserved for a single entity throughout; specifically, k always denotes the Boltzmann constant, N_a the Avogadro number, and R the universal gas constant. Molecular weight appears constantly, and in a harmless reversion to previous practice, molecular weight M is treated as a dimensionless quantity, thereby avoiding a needless repetition of units such as *mol* and the Dalton. Energy density and pressure, entities with the same fundamental dimensions, also appear frequently. A convenient range for the topics treated here is 10^6 Pascal, or *MPa*; occasionally the equivalent units of 10^6 *Joules/meter*^{3}, or $J\,cm^{-3}$, seem more appropriate to the topic at hand and used instead.

The following is a list of the major symbols used in the book. In most cases a reference is included as a guide to the location to its initial use.

a	Virial coefficient size ratio (Eq. 6.63)
a_2,a_3	Microsolution virial coefficients (Eq. 6.20)
A	Helmholtz free energy (Sec. 2.1.3)
A	Cross-section area (Sec. 9.4)
A_2, A_3	Second and third virial coefficients (Eq. 5.5)
B_n, B_w	Number- and weight-average branches per molecule (Eq. 1.11)
c	Prigogine parameter (Sec. 3.6)
c,c_i	Solute weight concentration (Eq. 3.15, Sec. 5.1.1)
c^*	Overlap concentration (Sec. 5.1.1)
C	Number of mixture components (Sec. 3.1)
C_1,C_2	Mooney-Rivlin coefficients (Eq. 10.5)
C_{sc}	Coherent scattering contrast factor (Eq. 4.143)

$C_{LS}, C_{SANS}, C_{SAXS}$	Light, neutron, x-ray contrast factors (Eqs. 5.23, 5.45, 5.46)
C_∞	Characteristic Ratio, (Eq. 4.30)
D_o	Diffusion coefficient in the dilute limit (Eq. 5.55)
f	Branch point functionality (Sec. 4.3.1)
f_a	Elastically active network fraction (Eq. 9.40)
f_d	Dangling network fraction (Eq. 9.40)
f_e	Energetic fraction of the restoring force (Eq. 10.35)
f_v	Free volume fraction (Eq. 3.108)
Δf_v	Free volume difference for components (Eq. 3.109)
F	Force (Sec. 9.4)
$F_i, F(M)$	Number distribution function (Sec. 1.4.1)
$F_2(T), F_3(T)$	Temperature dependence of virial coefficients (Eq. 6.15)
$F(r)$	Random coil spring force (Eq. 4.85)
g, g_θ, g_{rw}	Coil size ratio for good, theta and random walk (Eq. 6.63)
g', g'_θ, g'_{rw}	Viscometric size ratio for good, theta and random walk (Eq. 6.63)
g_m	Free energy of mixing density (Eq. 3.27)
$g(r)$	Pair distribution function (Eq. 2.30)
G	Gibbs free energy (Eq. 3.1)
G_o	Shear modulus for small deformations (Eq. 9.55)
G_N^o	Plateau modulus (Sec. 9.7.1)
ΔG_m	Gibbs free energy of mixing (Eq. 3.1)
h, h_θ, h_{rw}	Hydrodynamic size ratio for good, theta and random walk (Eq. 6.63)
ΔH_m	Enthalpy of mixing (Eq. 3.2)
I_D, I_o	Detector and incident intensity (Eq. 2.45)
$I_r(q)$	Reduced scattering intensity (Eq. 2.53)
k	Boltzmann constant, $1.381 \times 10^{-23}\, J\, K^{-1}$
k_H	Huggins coefficient (Eq. 5.64)
K	Degrees Kelvin
K	Random coil spring constant (Sec. 4.6.1)
l	Backbone bond length (Chs. 1,4)
l_o	Persistence length (Eq. 4.36)
l_K	Kuhn step length (Eq. 4.32)
l_p	Helfand packing length (Eq. 4.41)
l_s	Statistical segment length (Eq. 4.33)
L	Backbone contour length (Eq. 4.11)
L	Test specimen length (Sec. 9.4)
L_o	Initial (rest) length (Sec. 9.4)

m_o	Molecular weight per mer (Eq. 1.1) or backbone bond (Table 4.2)
M	Polymer molecular weight (Ch. 1)
M_n, M_w, M_z	Number-, weight- and z-average molecular weights (Eqs. 1.5,1.7,1.8)
$M^{\ddagger}$	Crossover molecular weight to coil expansion (Eq. 6.33)
n	Number of monomeric units per chain (Ch. 1)
n	Refractive index (Sec. 5.1.2)
$n_K^{\ddagger}$	Number of Kuhn steps to coil expansion onset (Eq. 6.62)
N	Number of monomeric units per chain (Ch. 8)
N_a	The Avogadro number, 6.022×10^{23}
$p_i(r)$	Pair distribution for random walks (Eq. 4.111)
P	Pressure (Ch. 2)
P_a	Ambient pressure, $\sim 0.102\ MPa$
P_c	Critical pressure (Ch. 2)
P_n, P_w	Number- and weight-average polymerization index (Sec. 1.4.1)
$P(q)$	Component form factor (Eq. 2.61)
P^*	Characteristic pressure (Eq. 2.99)
$\tilde{P}$	Reduced pressure (Eq. 2.98)
P_{hs}	Hard sphere liquid pressure (Sec. 2.2.3)
q	Scattering vector magnitude (Eq. 2.43)
Q	Swell ratio (Eq. 10.26)
$Q(q)$	Intermolecular scattering factor (Eq. 2.62)
$Q(n,V,T)$	Partition function (Eq. 2.14)
r	Distance
$\mathbf{r}$	Position vector (Sec. 2.1.2)
R	Universal gas constant, $8.3145\ J\ K^{-1}$
R_o	Network macrocycle size (Sec. 10.5.1)
R_g	Radius of gyration (Sec. 4.1.2)
$(R_g)_o$	R_g unperturbed by excluded volume (Sec. 6.1.1)
$(R_g)_\theta$	R_g at the theta condition (Sec. 2.2.2)
$R_g^{\ddagger}$	R_g at the onset of coil expansion (Eq. 7.24)
$\langle R^2 \rangle^{1/2}$	Root-mean-square end-to-end distance (Sec. 4.1.1)
R_h	Hydrodynamic radius (Eq. 6.48)
R_t	Thermodynamic radius (Eq. 6.36)
R_v	Viscometric radius (Eq. 6.49)
s	Sol fraction (Sec. 9.2.2)
S	Entropy (Ch. 2)
$S(q)$	Static structure factor (Sec. 4.7.2)
$S(\lambda)$	Dimensionless stress (Eq. 10.3)

ΔS_m	Entropy of mixing (Eq. 3.3)
ΔS_{com}	Combinatorial entropy of mixing (Sec. 3.3)
T	Temperature
T_{mp}	Normal melting temperature (Table 2.1)
T_{bp}	Normal boiling temperature (Table 2.1)
T_c	Critical temperature (Ch. 2)
T_e	Entanglement trapping factor (Sec. 9.3.2)
T_m	Midrange temperature (Eq. 3.75)
T_{sp}	Spinodal temperature (Eq. 7.100)
T_θ	Temperature at the theta condition (Eq. 6.32)
T^*	Characteristic temperature (Eq. 2.99)
T_{ginz}	Ginzburg temperature (Eq. 7.116)
$\tilde{T}$	Reduced temperature (Eq. 2.98)
$\boldsymbol{u}_0, \boldsymbol{u}_D$	Incident beam and detector direction vectors (Sec. 2.1.5)
$\boldsymbol{u}_i$	Velocity of particle i (Sec. 6.1.2)
U	Internal energy (Ch. 2)
$\boldsymbol{U}_i$	Local liquid velocity at particle i (Sec. 6.1.2)
υ	Volume per sphere (Eq. 2.88)
υ_f	Free volume per sphere (Eq. 2.88)
υ_p, υ_s	Molecular volumes of polymeric mers and solvent (Eq. 6.21)
υ_{occ}	Occupied volume (Sec. 6.1.1)
υ_{per}	Pervaded volume (Eq. 6.2.1)
υ_{rcp}	Volume per sphere at random close packing (Eq. 2.88)
υ^*	Characteristic volume (Eq. 2.99)
$\tilde{\upsilon}$	Reduced volume (Eq. 2.98)
$\Delta \upsilon_e$	Fractional change in volume with mixing (Eq. 3.10)
V	Volume (Ch. 2)
V_0, V_f	Initial and final mixture volume (Eq. 3.105)
V_{sc}	Scattering volume (Sec. 2.1.5)
V_i	Molar volume of species i (Eq. 3.8)
V_d	Molar volume of diluent (Eq. 10.22)
V_p, V_s	Molar volume of polymer and solvent (Sec. 7.2.1)
$\overline{V}_i$	Partial molar volume of component i (Eq. 3.9)
V_{rcp}	Random close packed volume (Sec. 2.3)
V_{ref}	Reference volume for interaction parameters (Eq. 3.64) and networks (Eq. 9.82)
ΔV_m	Volume change on mixing (Eq. 3.8)
$V_n(r)$	Probability density for random walk end-to-end distance (Sec. 4.5.1)
w	Gel fraction (Sec. 9.2.2)
w_i	Weight fraction of mixture component i (Eq. 3.14)
W	Elastic energy density (Sec. 9.4)

$W_i, W(M)$	Weight distribution function (Sec. 1.4.1)
$W'(M)$	Logarithmic weight distribution (Eq. 1.12)
x_i	Mole fraction of component i (Eq. 3.14)
X, X_{ij}	Interaction density coefficient (Sec. 3.4)
X_E	Excess interaction density coefficient (Eq. 8.56)
$Y(\lambda)$	Topological contribution to $S(\lambda)$ (Eq. 10.53)
z	Number of nearest neighbors (Sec. 3.3)
Z_o	Molecular friction coefficient (Eq. 5.53)
$Z(n,V,T)$	Configuration integral (Eq. 2.17)
α	Thermal expansion coefficient, $-(d\ln\rho/dT)_P$
α_1, α_s	Thermal expansion coefficient for pure solvent (Sec. 3.6.2, Eq. 7.80)
α_h	Hydrodynamic size expansion factor (Sec. 6.2.2)
α_R	End-to-end expansion factor (Sec. 6.2.2)
α_s	Coil size expansion factor (Sec. 6.2.2)
α_t	Thermodynamic size expansion factor (Sec. 6.2.3)
α_υ	Viscometric size expansion factor (Sec. 6.2.2)
β	Isothermal compressibility, $(d\ln\rho/dP)_T$
$\hat{\beta}$	Binary cluster integral (Eq. 6.13)
$\tilde{\beta}$	Scaling coefficient of order unity (Eq. 7.48)
γ	Size-length power law exponent (Eq. 6.3)
γ	Crosslink index (Eq. 9.18)
$\delta, \delta_i, \delta_{CED}$	Component solubility parameter (Eq. 3.94)
δ_{PVT}	Solubility parameter from PVT data (Eq. 8.1)
ε	Potential well depth (Sec. 2.1.1)
$\Delta\varepsilon_{ij}$	Exchange energy per pair (Eq. 3.91)
ζ_o	Monomeric friction coefficient (Sec. 6.1.2)
$\eta_s, \eta(c)$	Solvent and solution viscosity (Sec. 5.2.2)
$[\eta]$	Intrinsic viscosity (Sec. 5.2.2)
$[\eta]^{\ddagger}$	Onset intrinsic viscosity for coil expansion (Eq. 7.25)
θ	Spherical coordinates angle (Sec. 2.1.2)
θ	Scattering angle (Sec. 2.1.5)
κ	Temperature coefficient of coil dimensions (Eq. 4.31)
λ	Wavelength of incident beam (Sec. 2.1.5)
λ	Branch points per unit molecular weight (Eq. 6.69)
λ	Stretch ratio (Sec. 9.4)
μ	Number density of elastically active network junctions (Sec. 9.3.1)
μ_i	Chemical potential of mixture component i, (Eq. 3.5)
μ_i^o	Chemical potential of pure component i, (Eq. 3.5)
ν	Number density of scattering sites (Sec. 2.1.5) and molecules (Sec. 2.2.2)

ν	Number density of elastically active network strands (Sec. 9.3.1)
ξ	Excluded volume screening length (Sec. 7.1.2)
ξ_{oz}	Scattering correlation length (Sec. 7.3.1)
$(\xi_{sm})_o$	Critical correlation length for simple mixtures (Sec. 7.5.3)
π, π_{op}	Osmotic pressure (Sec. 3.1)
π_{el}	Elastic pressure (Eq. 6.19)
Π_{CED}	Cohesive energy density (Sec. 2.3.5)
Π_{IP}	Internal pressure (Sec. 2.3.5)
ρ	Mass density (Ch. 2)
ρ_c, ρ_L, ρ_S	Pure component critical, liquid and solid mass density (Ch. 2)
σ	Molecular size scale (Sec. 2.1.1)
σ	Stress (Sec. 9.4)
σ_s, σ_t	Shear and tensile stress (Sec. 9.4)
$d\Sigma/d\Omega$	Differential scattering cross section (Eq. 2.52)
τ_E, τ_E^{-1}	Experimental time and rate scales (Sec. 1.3.2)
τ_S, τ_S^{-1}	Structural time and rate scales (Sec. 1.3.2)
ϕ	Spherical coordinates angle (Sec. 2.1.2)
ϕ	Packing fraction (Sec. 2.2.3)
ϕ, ϕ_i	Solute volume fraction and volume fraction of component i (Sec. 3.1)
ϕ_{self}	Self volume fraction (Eq. 7.4)
ϕ_{ref}	Reference volume fraction (Sec. 10.1.2)
ϕ^*	Overlap volume fraction (Sec. 7.1.2)
$\phi^{\ddagger}$	Crossover volume fraction to fully screened (Sec. 7.1.2)
Φ	Fox-Flory parameter (Eq. 6.52)
χ, χ_{ij}	Interaction parameter for simple mixtures (Sec. 3.4)
χ_c	Critical point interaction parameter (Sec. 3.4.2)
χ_θ	Theta condition interaction parameter (Sec. 7.5.1)
χ_h, χ_s	Enthalpic and entropic parts of the interaction parameter (Eq. 3.72)
χ_{scat}	Scattering-derived interaction parameter (Eq. 7.78)
χ_{FV}	Interaction parameter from free volume mismatch (Eq. 7.80)
χ_m	Interaction parameter at T_m (Eq. 3.75)
ψ	Interpenetration function (Eq. 6.52)
$\psi(r)$	Intermolecular potential energy function (Sec. 2.1.1)
Ψ	Mandelkern-Flory parameter (Eq. 6.52)
φ_i	Volume fraction comonomer i in a copolymer (Sec. 8.5)
φ	Fraction of network modulus from entanglements (Eq. 10.54)
ω	Wave frequency of incident beam (Eq. 2.40)
Ω	Number of distinguishable configurations (Eq. 2.23)

Subject Index

Author Index

C

L